Health Monitoring of Structural Materials and Components

Health Monitoring of Structural Materials and Components

Methods with Applications

Douglas E. Adams
Purdue University, USA

John Wiley & Sons, Ltd

Other Wiley Editorial Offices

John Wiley & Sons Inc., 111 River Street, Hoboken, NJ 07030, USA

Jossey-Bass, 989 Market Street, San Francisco, CA 94103-1741, USA

Wiley-VCH Verlag GmbH, Boschstr. 12, D-69469 Weinheim, Germany

John Wiley & Sons Australia Ltd, 42 McDougall Street, Milton, Queensland 4064, Australia

John Wiley & Sons (Asia) Pte Ltd, 2 Clementi Loop #02-01, Jin Xing Distripark, Singapore 129809

John Wiley & Sons Canada Ltd, 6045 Freemont Blvd, Mississauga, ONT, Canada L5R 4J3

Wiley also publishes its books in a variety of electronic formats. Some content that appears in print may not be
available in electronic books.

Anniversary Logo Design: Richard J. Pacifico

British Library Cataloguing in Publication Data

A catalogue record for this book is available from the British Library
ISBN 978-0-470-03313-5 (HB)

Typeset in 10/12 pt Times by Thomson Digital Noida, India
Printed and bound in Great Britain by Antony Rowe Ltd, Chippenham, Wiltshire
This book is printed on acid-free paper responsibly manufactured from sustainable forestry
in which at least two trees are planted for each one used for paper production.

To Jean, Caroline, and Madeleine for their inspiration.

And to Mom and Dad for their patience.

Contents

Preface   xiii

Acknowledgments   xv

1 Introduction   1
   1.1   Basics of Health Monitoring   1
   1.2   Commercial Needs for Health Monitoring Technology   3
   1.3   Defense Needs for Health Monitoring Technology   4
   1.4   Technical Approach to Health Monitoring   7
   1.5   Definitions of Common Terminology   11
   1.6   Comparison of Nondestructive Testing (NDT)
         and Health Monitoring Techniques   16
   1.7   Potential Impact of Health Monitoring Technologies   18
   1.8   Overview of Technical Areas in Health Monitoring   21
   1.9   Summary   23
   References   24
   Problems   25

2 Modeling Components   29
   2.1   Modeling Needs   29
   2.2   First-Principle Models   30
         2.2.1   Component Vibration Models   30
         2.2.2   Vibration Natural Frequencies and Modal Deflection Shapes   33
         2.2.3   Free Vibration Response   35
         2.2.4   Forced Vibration Response (Frequency Response Models)   37
         2.2.5   Impedance and Compliance Models   40
         2.2.6   Transmissibility Forced Response Models   44
         2.2.7   Nonlinear Dynamic Models   48
         2.2.8   Wave Propagation Models (One Dimensional)   52
                 2.2.8.1   Analytical Solution for Longitudinal Waves   52
                 2.2.8.2   Longitudinal Wave Propagation Finite Element Model   54
                 2.2.8.3   Analytical Solution for Rod with Transverse Waves   58
                 2.2.8.4   Transverse Wave Propagation Finite Element Model   59
         2.2.9   Wave Propagation Models (Two Dimensional)   62

2.3	Data-Driven Models		65
	2.3.1	Experimental Time Domain Models	66
		2.3.1.1 Direct Parameter Models	66
		2.3.1.2 Restoring Force and Phase-Plane Models	68
		2.3.1.3 Discrete Time Models	71
	2.3.2	Experimental Frequency Response Models	74
		2.3.2.1 Frequency Response Sensitivity Functions	76
		2.3.2.2 Virtual Force Models	81
	2.3.3	Experimental Modal Vibration Models	84
	2.3.4	Other Data-Driven Models	88
2.4	Load Models		88
	2.4.1	External Mechanical Excitations	88
		2.4.1.1 Impulsive Excitations	89
		2.4.1.2 Narrowband Excitations	91
		2.4.1.3 Broadband Random Excitations	92
	2.4.2	Acoustic Pressure, Temperature and Other Environmental Loads	94
2.5	Summary		99
References			101
Problems			103

3   Modeling Damage   **107**
3.1	Static Damage Models		107
	3.1.1	Fasteners and Joints	107
	3.1.2	Cracking	109
	3.1.3	Plastic Deformation, Penetration and Erosion	112
	3.1.4	Delamination, Debonding and Separation	113
	3.1.5	Creep and Buckling	115
	3.1.6	Corrosion and Oxidation	116
		3.1.6.1 Fiber Pull Out and Fiber Breakage	117
	3.1.7	Matrix Cracking	119
	3.1.8	Microstructural Changes	119
3.2	Dynamic Models for Damage		120
	3.2.1	Phenomenological Models	120
	3.2.2	Generalized Damage Growth Models	121
3.3	Failure Models		122
3.4	Performance Models		124
3.5	Summary		124
References			125
Problems			125

4   Measurements   **127**
4.1	Measurement Needs		127
4.2	Data Environment		129
	4.2.1	Amplitude and Frequency Ranges	129
	4.2.2	Nature of Data	133
	4.2.3	Environmental Factors	136

4.3   Transducer Attachment Methods   137
      4.3.1   Durability   137
      4.3.2   Stability   139
      4.3.3   Directionality   140
      4.3.4   Frequency Range (Wavelength)   141
4.4   Transducers   145
      4.4.1   Overview of Sensors and Actuators   146
      4.4.2   Passive Sensors   150
              4.4.2.1   Resistance Strain Gauge Model   150
              4.4.2.2   Piezoelectric Accelerometer Model   152
              4.4.2.3   Transmission Models (Cable, Amplifier and Power Supply) 155
      4.4.3   Active Piezoelectric Transducers (Actuators)   159
      4.4.4   Other Types of Sensors   162
      4.4.5   Transducer Placement and Orientation   163
              4.4.5.1   Observability Criterion   167
              4.4.5.2   Controllability Criterion   169
4.5   Data Acquisition   170
      4.5.1   Common Errors   171
      4.5.2   Aliasing   174
      4.5.3   Leakage   178
      4.5.4   Channel Limitations in Data Acquisition   179
4.6   Summary   180
References   181
Problems   183

5   Data Analysis   187
5.1   Data Analysis Needs and Framework   187
5.2   Filter Data   187
      5.2.1   Time Domain Filters   188
      5.2.2   Frequency Domain Filters   193
      5.2.3   Spatial Filters   194
5.3   Estimation of Unmeasured Variables (State Inference)   201
5.4   Temporal Analysis   205
      5.4.1   Statistical (Nondeterministic) Analysis   205
      5.4.2   Deterministic Analysis   210
5.5   Transformation of Data   216
      5.5.1   Spectral (Frequency) Analysis   217
      5.5.2   Higher-Order Spectral Analysis   223
      5.5.3   Analysis Using Other Spectral Transformations   226
      5.5.4   Time-Frequency Analysis   226
5.6   Averaging of Data   234
      5.6.1   Cyclic Averaging   234
      5.6.2   Frequency Response Function Estimation   236
      5.6.3   Averaging of Data in Rotating Systems   240
5.7   Spatial Data Analysis   242
      5.7.1   Modal and Operational Deflection Patterns   243

		5.7.2	Transfer Path Analysis	246
		5.7.3	Multidirectional Data	250
		5.7.4	Triangulation	254
	5.8	Feature Extraction		257
		5.8.1	Model-Based Feature Extraction (Damage)	258
		5.8.2	Model-Based Feature Extraction (Loading)	260
		5.8.3	Dimensionality of Feature Sets	262
		5.8.4	Statistical Models for Features	265
	5.9	Variability Analysis		270
	5.10	Loads Identification		280
		5.10.1	Overview	280
		5.10.2	Estimation Errors	281
		5.10.3	Conditioning of Loads Identification Algorithms	282
	5.11	Damage Identification		287
		5.11.1	Damage Detection	287
		5.11.2	Damage Localization	290
		5.11.3	Damage Quantification	292
	5.12	Regression Analysis for Prognosis		295
		5.12.1	Physics-Based Methods	296
		5.12.2	Data-Driven Methods	297
	5.13	Combining Measurement and Data Analysis		297
	5.14	Summary		298
	References			299
	Problems			301
6	**Case Studies: Loads Identification**			**305**
	6.1	Metallic Thermal Protection System Panel		305
		6.1.1	Data-Driven	305
		6.1.2	Physics-Based	308
	6.2	Gas Turbine Engine Wire Harness and Connector		311
	6.3	Fuselage Rivet Process Monitoring		313
	6.4	Large Engine Valve Assembly		316
	6.5	Suspension with Loosening Bolt		319
	6.6	Sandwich Panel Undergoing Combined Thermo-Acoustic Loading		322
	6.7	Summary		327
	References			327
	Problems			328
7	**Case Studies: Damage Identification**			**329**
	7.1	Vibration-Based Methods		329
		7.1.1	Metallic Thermal Protection System Panel	329
			7.1.1.1 Passive Method	329
			7.1.1.2 Active Methods	333
		7.1.2	Gas Turbine Engine Wire Harness and Connector	344
		7.1.3	Suspension System	349

	7.2	Wave Propagation-Based Methods	350
		7.2.1 Wheel Spindle	351
		7.2.2 Ceramic Tile	356
		7.2.3 Gamma Titanium Aluminide Sheet	359
		7.2.4 Aluminum Plate	364
	7.3	Damage Identification Under Load	367
		7.3.1 Metallic Panel Under Thermo-Acoustic Loading	369
		7.3.2 Aluminum Plate Under Vibration Loading	371
	7.4	Summary	375
		References	376
		Problems	377
8	**Case Studies: Damage and Performance Prediction (Prognosis)**		**379**
	8.1	S2 Glass Cylinder (Performance Prediction)	379
	8.2	Stability Bar Linkage (Damage Growth Modeling)	383
	8.3	Summary	393
		References	393
		Problems	394
Appendix A			**395**
Appendix B			**421**
Index			**455**

Preface

This textbook presents a set of methods and applications in three areas of health monitoring for structural materials and components: (1) loads identification, (2) *in situ* damage identification (diagnostics), and (3) damage and performance prediction (prognostics). The applications focus on using vibration and wave propagation measurements for health monitoring as opposed to electromagnetic, thermal, or other measurement variables that are also indicative of component health.

The book aims to provide readers with a summary of the technical skills and practical understanding required to solve new problems encountered in this emerging field. It is written for newcomers who would like a review of the basics in modeling, measurement, signal processing, and data analysis. The book should also appeal to more experienced readers who are looking for a reference text to address current and future challenges. The book is a manual for conceptualizing, designing, and operating health-monitoring systems for mechanical and structural systems.

Concepts in modeling, measurements, and data analysis are conveyed through examples because students tend to respond better to materials when they are linked to applications. The book focuses on component-level health monitoring; however, issues in health monitoring at the system level are also incorporated. The book includes detailed case studies of certain approaches and also provides a state-of-the-art review of other methods from the literature. However, the focus is on an integrated approach to health monitoring, using techniques with which the author is most familiar, to avoid misapplication of the techniques developed by others. The primary challenge in health monitoring is the integration of many different methods in modeling, measurement, and data analysis. Because one book cannot possibly cover all viable methods for health monitoring, a foundation is presented here on which readers can build. Suitable references to the literature and other textbooks on core health-monitoring skills are provided to assist the readers in increasing the depth of their understanding.

In the introductory chapter, terminology is defined, the state-of-the-art is reviewed, and an approach to health monitoring of structural materials and components is presented. In the next four technical chapters on modeling and prediction and measurement and data analysis, the core technical approaches for health monitoring are summarized. Then the final three chapters discuss case studies in loads/damage identification and performance prediction, where the principle advantages of online health monitoring are demonstrated using real-world examples.

As mentioned above, the book focuses on using vibration and wave propagation response data for identifying loads and damage.

Mechanical and structural system applications incorporated in the textbook examples include engine components (valves, wire harnesses), automotive accessories (wheels, spindles, suspensions, exhaust), aircraft parts (fuselage, bonded repairs), spacecraft components (metallic and ceramic thermal protection systems, fuel tanks), and defense system components (body armor, composite vehicle armor, rocket motor casings). This wide variety of systems should appeal to aerospace, civil, mechanical, agricultural, defense, and other communities of engineers.

The companion website for the book is http://www.wiley.com/go/adams_health. The MATLAB® computer programs, which are used throughout the book, can be downloaded from this website. The reference list from the Appendix is also listed on the website.

Acknowledgments

They say no one writes a book by oneself. This is an understatement in my case.

First, I gratefully acknowledge the students with whom I have worked through the years. These students have taught me a great deal about the methods presented in this textbook and have also helped me to explain how the methods work. Specifically, I thank Timothy Johnson, Shankar Sundararaman, Muhammad Haroon, Harold Kess, Jonathan White, Jason Hundhausen, Chulho Yang, Hao Jiang, Nathaniel Yoder, Emily Prewett, Madhura Nataraju, Gavin McGee, Rebecca Brown, Carlos Escobar, and Claudia Ellmer for their hard work and dedication to research and learning.

Second, I thank all of the researchers whose work has been summarized primarily in the Appendix in an attempt to present a review of the state-of-the art.

Third, I thank the researchers and praticioners with whom I have worked and from whom I have learned so much because of their collaboration and friendship. At the risk of leaving someone out, I want to make special mention of Dr. Charles Farrar from Los Alamos National Laboratory, Prof. Keith Worden from University of Sheffield, Prof. Daniel Inman from Virginia Tech, Prof. Udo Peil from Technical University of Braunschweig, Prof. Michael Todd from University of California San Diego, Prof. Fu-Kuo Chang from Stanford University, Prof. Daryll Pines from University of Maryland, Dr. Kumar Jata from Air Force Research Laboratory Materials and Manufacturing Directorate, Mr. Mark Derriso from Air Force Research Laboratory Air Vehicles Directorate, Dr. Shawn Walsh from Army Research Laboratory, Mr. Elias Rigas from Army Research Laboratory, Mr. Matt Triplett from U. S. Aviation and Missile Command, Mrs. Pam Brown and Mrs. Dorothy Foley from the Stryker Program Management Office, Dr. Lane Miller from Lord Corporation, Mr. Michael Lally and Mr. Richard Bono from The Modal Shop, Mr. Jim Lally from PCB Piezotronics, Dr. Grant Gordon from Honeywell, Mr. Matthew Bedwell from Caterpillar, Ms. Christine Lutz from ArvinMeritor, and Mr. Thomas Ryan from Rolls-Royce Corporation. I also thank sponsors who have funded proprietary research and who cannot be recognized by name.

Fourth, I want to acknowledge the industry and government sponsors who have supported the research my students and I have done in health monitoring. Without these sponsors, I would have never been exposed to the interesting and practical case studies presented in this book. Health-monitoring technologies for structural components and materials will advance only if visionaries in industry and government agencies encourage the use of such technologies.

Finally, I thank my family for indulging me to write this book.

1
Introduction

1.1 BASICS OF HEALTH MONITORING

Health monitoring is the scientific process of nondestructively identifying four characteristics related to the fitness of an engineered component (or system) as it operates:

(a) the operational and environmental loads that act on the component (or system),
(b) the mechanical damage that is caused by that loading,
(c) the growth of damage as the component (or system) operates, and
(d) the future performance of the component (or system) as damage accumulates.

Knowledge of characteristics (a)–(c) is combined with component design criteria and system operational specifications to identify characteristic (d), which determines whether or not a component will perform satisfactorily in the future. Health monitoring technologies must be nondestructive (e.g. should not involve cutting a part open to inspect it) and are ideally implemented online with embedded hardware/software in an automated manner as a system operates. Humans constantly apply health monitoring methods in their everyday lives. For example, a 'purr' or 'humm' is a sign that an automobile engine is working properly; on the contrary, a 'clank' or a 'clunk' is a sign that the engine should be fixed. This type of comparison between healthy and unhealthy signatures is the foundation of many health monitoring techniques. Figure 1.1 illustrates the basic aspects of health monitoring applied to an unmanned aerial vehicle (UAV).

Loads due to landing, aerodynamic forces, engine thrust and foreign object impact are shown acting on the aircraft. These loads cause the aircraft to respond by deforming and vibrating during operation. For example, compressive loads in the landing gear during touch down transmit through the aircraft to the stabilizer struts. Sensors inside the aircraft measure signals, which could indicate the vibration amplitude and frequency content along with the strain and so on. There are also environmental factors including corrosive and temperature effects, for instance, in the propeller bearings. These environmental factors can accelerate how quickly damage such as spalling (flaking of material) in the bearings initiates and grows. Measurements of environmental variables such as temperature might be made in the engine housing, whereas measurements of ultraviolet radiation or humidity might be made on the wing spar because composites can degrade when these levels are high. When a crack forms in the metal stabilizer strut as shown, changes in the vibration response indicate the presence of a crack. Damage in components such as the wing spar can consist, for example, of

Health Monitoring of Structural Materials and Components: Methods with Applications   D. Adams
© 2007 John Wiley & Sons, Ltd

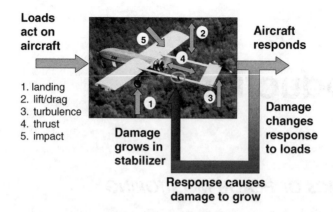

Loads
act on
aircraft

1. landing
2. lift/drag
3. turbulence
4. thrust
5. impact

Aircraft
responds

Damage
grows in
stabilizer

Damage
changes
response
to loads

**Response causes
damage to grow**

Figure 1.1   Aspects of health monitoring in an unmanned aerial vehicle

cracks or corrosion if the wing is metallic or matrix cracks, delaminations, or broken fibers if the wing is composite. Algorithms are needed to process the vibration data in order to assess the damage.

One key aspect to note in this example application is that when the aerodynamic loading on the stabilizer at location 3 is high, the crack in the strut affects the vibration response of the aircraft more than when the loads are low. *In other words, health monitoring technologies sometimes detect damage more readily when loads are acting on a component to accentuate the damage.* This aspect of health monitoring is one reason why monitoring may be preferred over inspections that are carried out offline (Hundhausen *et al.*, 2004). In this case, offline inspections are performed on the ground when the aircraft is not operating. As the aircraft continues to respond, damage continues to grow in the strut, resulting in an ever changing load to the right side strut of the stabilizer. Damage grows in the future according to the current severity of damage and the nature of the loading applied to the component in the future. This aspect of damage accumulation in health monitoring is important because it results in a redistribution of the loads to or away from damaged areas (Haroon and Adams, 2005). *In other words, damage accumulation can result in changes to the way damaged components respond, suggesting that both the loading and damage should be monitored continuously in a system in order to predict the future operational capability of its various components.*

Models play a crucial role when loads are estimated, damage is identified and the future performance of a component is predicted. For example, without first-principle physics-based models, it would be necessary to acquire response data for each type of damage at various damage levels in the UAV in order to quantify the damage level (i.e. length of crack in the strut). If models are used, damage can be quantified with relatively few datasets (Johnson *et al.*, 2004). On the contrary, if a damage mechanism such as corrosion that was not modeled appears in a component such as the stabilizer strut, then the model developed for cracks would incorrectly quantify the damage. In these instances, the data that is acquired should be relied on more heavily than first-principle models. *In other words, there are cases in which health monitoring should be based more on first-principle models and other cases when health monitoring should be based more on data.* Usually, a combination of data-driven and physics-based approaches is most effective when developing health monitoring algorithms.

1.2   COMMERCIAL NEEDS FOR HEALTH MONITORING TECHNOLOGY

Commercial applications are driving the need for health monitoring technologies. For example, Figure 1.2(a) shows a large truck used at excavation sites for hauling debris. Manufacturers of products such as this truck are now leasing their products instead of selling them. This new business approach, which is similar to the one employed by rental car companies and is referred to as 'power by the hour', requires that manufacturers maintain the performance of their customer's products in order to make a profit. For example, it is estimated that a 2 % increase in downtime per year for 20 of the trucks in Figure 1.2(a) would cost the manufacturer $13 million.

Condition-based maintenance (CBM) is the application of health monitoring technologies for the purpose of scheduling service and maintenance for products according to the condition of those products as opposed to a fixed time table. For example, the operator's manual for an automobile usually specifies that the car be serviced every 2000–3000 miles in the first few years of operation. However, the oil condition at 2000 miles might not warrant replacement, or in some cases, the suspension or battery might indicate that service is needed prior to 2000 miles. If service were scheduled when the oil begins to lose too much viscosity or the battery begins to lose charge, the car would operate more efficiently and would cost less to maintain in the long run. CBM would make it possible for manufacturers to service their products more efficiently in a timely manner to avoid catastrophic failures and delays. Manufacturers could also order parts just-in-time through autonomic logistics to avoid failure and downtime if health monitoring information were readily available.

CBM would also apply to gas turbine engines in commercial aircraft containing discs, rotors, bearings, wire harnesses (Figure 1.2(b)) and fuselage components. Health monitoring information could help to increase safety and reduce delays in air traffic. For example, in December 2005, an Alaska Airlines plane flight number 536 traveling from Seattle, WA, to Burbank, CA, was struck by a baggage cart prior to take off, but the operator failed to notify pilots. The plane lost pressure in the fuselage at 26 000 ft 20 min after taking off and was subsequently forced to make an

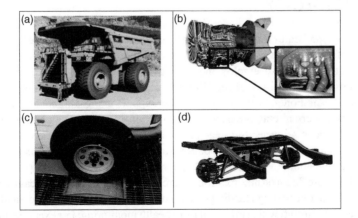

Figure 1.2   (a) Mining truck used for excavation (courtesy 2003 Pan American Damage Prognosis Workshop), (b) gas turbine engine with wire harnesses and connector panel, (c) radial tire subject to bead and tread damage, and (d) lightweight integrated truck suspension system with many components

emergency landing triggering major delays in regional air traffic. If effective health monitoring technologies had been installed on this aircraft, those technologies would have automatically notified crew about the impact, thereby helping to avoid the risk to passengers and travel delays.

In many other commercial applications, health monitoring technologies are affecting the development of products such as radial rubber tires (Figure 1.2(c)) and integrated suspension systems (Figure 1.2(d)). For example, tire recalls in 2000 focused national attention on tire regulations because the old standards from 1967 were written when 99 % of tires had bias plies. The US Congress responded to public concerns with the Transportation Recall Enhancement, Accountability and Documentation (TREAD) Act. Then the National Highway Traffic Safety Administration (NHTSA) issued new stringent requirements for tire durability with an emphasis on radial tires and bead area damage where the tire meets the rim. In tire fatigue endurance testing, health monitoring technologies for detecting bead area damage are needed to reduce the time and variation of manual inspections (Johnson and Adams, 2006). Centrifugal forces and heating from the rotation of the tire accentuate the effects that defects have on the tire response along the bead. These load interaction effects suggest that health monitoring techniques may be preferable to offline inspection methods for which these operational loads are not applied. Such technologies for reliability testing could also eventually be implemented in passenger cars and trucks to monitor tire health continuously.

Similarly, suspension systems in cars and trucks are being designed in new ways to reduce weight/cost and increase handling performance and reliability. Integrated suspension systems such as the one shown in Figure 1.2(d) are lighter, resulting in better fuel economy vehicles. The durability of integrated suspensions must be ensured through rigorous testing. Because these suspensions have many components and are geometrically complex, dynamic loads and mechanical damage in components are difficult to identify with traditional multibody modeling and manual inspection techniques (Haroon, 2007). Health monitoring technologies are needed to process the data acquired during these lengthy durability tests. Online data processing can quicken durability tests, making them less expensive for manufacturers.

1.3 DEFENSE NEEDS FOR HEALTH MONITORING TECHNOLOGY

This book focuses on health monitoring of structural components in commercial applications involving aircraft, ground vehicles, machinery and other types of mechanical systems. However, defense applications can also employ monitoring technology. For example, $59 billion in maintenance costs made up 14 % of the US Department of Defense budget in 2004 (Navarra, 2004). In the US Air Force, costs to operate and support aging aircraft are rising and new procurements of modern aircraft are declining. If aircraft were serviced when needed based on the vehicle health and mission needs instead of at fixed intervals, the readiness of aircraft could be ensured and sustainment costs would drop. For example, the F-16 requires as many as 25 h of maintenance per flight hour (Malley, 2001). As aircraft such as the C-17 cargo plane shown in the top of Figure 1.3(a) are flown on new missions, there is a heightened need for health monitoring to ensure that new loads due to rough landing strips and environmental factors (e.g. sand and smoke) are considered. New aircraft will also benefit from health monitoring. For example, the morphing air vehicle shown in Figure 1.3(a) will be articulated in nature with numerous components all of which could become damaged. Health data concerning these aircraft could be used to determine if missions with a particular loading profile could be successfully performed.

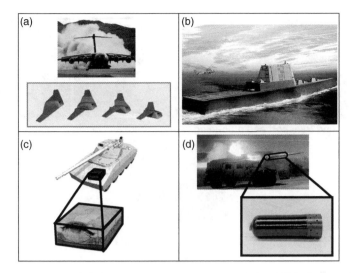

Figure 1.3   (a) Morphing aircraft design showing change in shape and C-17 aircraft landing on desert air strip with high loading to landing gear, (b) illustration of US Navy DDX destroyer, (c) lightweight composite armor and (d) composite missile casings

The US Navy also has an increased focus on mission readiness of their DDG class ships as well as their new DDX platforms shown in Figure 1.3(b). Ships contain the hull in addition to numerous hydraulic, mechanical and electrical subsystems, which enable vital functions from firefighting to propulsion. Knowledge of the hull and subsystem health prior to a mission could be used to determine if the ship's readiness is high enough to justify deployment. If health monitoring systems aboard the ship indicated that replacement parts are needed, then the ship could call ahead to the next available port to ensure that parts are delivered on time to enable repairs. This mode of operation where health monitoring systems are used to efficiently order parts is sometimes referred to as *autonomic logistics*. As the new DDX naval destroyer is designed and tested, health monitoring algorithms for processing thousands and perhaps millions of channels of sensor data must be written so that commanders can ascertain the capability of these ships prior to deployment and after deployment as the mission is updated to anticipate threats.

There are also needs for health monitoring technologies to ensure the readiness of defense systems used by the US Army. For example, new vehicle and body armors such as the materials shown in Figure 1.3(c) as well as new precision attack missiles composed of composite materials as shown in Figure 1.3(d) will be used in the future. In these new lighter weight defense systems, the Army Research Laboratory estimates that 85 % of all field damage is caused by some kind of impact event (Walsh *et al.*, 2005). Figure 1.4(a) illustrates a variety of transportation and storage-type incidents that can introduce impacts to missile containers. When impacts occur, there is a drop in burst strength of the composite missile casing as shown in Figure 1.4(b). By identifying the magnitudes of loads acting on these missiles, their ability to sustain a given firing pressure could be determined using this type of design data.

Ground vehicles present another interesting need for health monitoring because they are continuously being modified to combat new threats on the battlefield. For example, ground vehicles such as the one shown in Figure 1.5(a) experience large dynamic loads to the suspension. Additional weight due to armor and extra payload causes higher dynamic loading

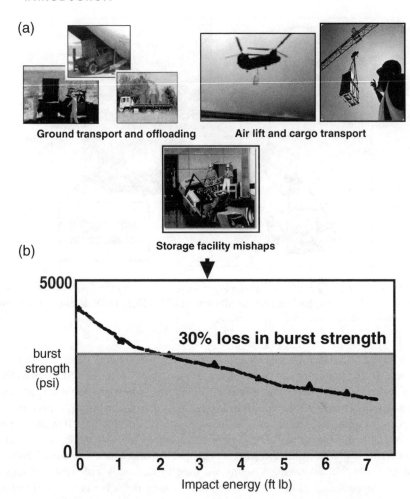

Figure 1.4   (a) Types of events that cause impacts of composite components and (b) typical loss in burst strength due to impact of missile composed of carbon filament (courtesy Walsh *et al.*, 2005)

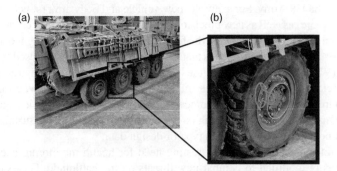

Figure 1.5   (a) Military ground vehicle and (b) spindle that experiences higher dynamic loading when slat armor is added to vehicle (Ackers *et al.*, 2006, SPIE)

when the vehicle encounters rough terrain as drivers steer to avoid obstacles. These higher loads can cause damage in the drive train components such as the spindle shown in Figure 1.5(b). Changes in loading and any resulting damage in areas such as this spindle could be monitored to ensure the readiness of these vehicles. In many applications, the ability to monitor the performance of hard-to-reach areas such as drive train spindles, which cannot be accessed without tearing down the wheel, is the primary motivation for implementing health monitoring technology (Ackers *et al.*, 2006).

1.4   TECHNICAL APPROACH TO HEALTH MONITORING

Health monitoring technologies for systems such as the gas turbine engine wire harness and connector panel shown in Figure 1.2(b) involve four key elements as defined in Section 1.1: (a) loads identification, (b) damage identification, (c) damage growth prediction and (d) prediction of the effects of damage growth on component (or system) performance. Figure 1.6

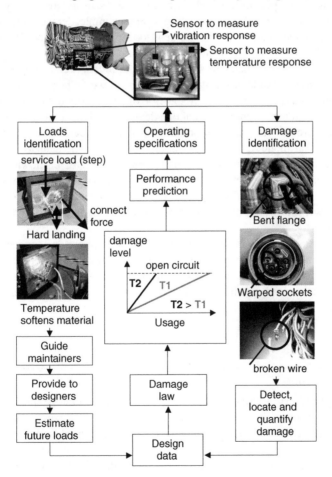

Figure 1.6   Health monitoring architecture consisting of loads identification, damage identification (diagnostics), damage and performance prediction (prognostics) for gas turbine engine wire harness and connectors

presents a health monitoring architecture comprised of these four elements as they apply to harnesses and connectors. Wire harnesses and connectors are an ideal application for health monitoring because temperatures are low enough to enable sensing (150 °C) and the wire harness affects the performance of many other components in the engine (Stites *et al.*, 2006). Various aspects of the architecture for health monitoring of wire harnesses and connectors are described next.

Sensing

The response of the wire harness and connector subsystem to operational and environmental loading inside the engine must be monitored either continuously or frequently enough to provide meaningful health data. For example, it might be sufficient to sample the temperature at a slow rate like once per minute; however, it would be necessary to sample the vibration much faster like once per millisecond. In many cases, sensors are already conveniently integrated into components by manufacturers. For example, temperature and vibration sensors (accelerometers) are found in different locations of the engine. However, if there are no models that relate measurements at these locations to the conditions on the connector panel or harness of interest, then new sensors could be added. Pre-existing sensors can also pose limitations in amplitude or sensitivity; new sensors would be required in this case.

Sensors can be *passive* in nature, as in the case of a temperature thermocouple. Sensors can also be *active* in nature, meaning that one device called an *actuator* could transmit an acoustic signal through the connector or harness, whereas a passive sensor could measure the response to that input signal. For example, a piezo-ceramic actuator could be driven with a voltage signal that produces an acoustic 'ping', which travels down a wire harness, and a second piezo-ceramic sensor could measure the response on the opposite end of the harness. The main advantage of active sensing is that the input amplitude, frequency and so on, to the sensor can be controlled leading to higher signal-to-noise ratios and more repeatable data.

Sensors for measuring one variable such as the shock response of a connector that occurs during a hard landing are also sensitive to other variables such as temperature, electromagnetic disturbances and so on. Sensors must be selected to minimize sensitivity to sources of variation and maximize sensitivity to the loading parameters and damage mechanisms of interest.

Loads identification

The integrated sensors described above are used to monitor the operational and environmental loads that act on the wire harnesses, connectors, and the wires and pins contained within. Loads illustrated in Figure 1.6 include the following: hard landing shock loads that pull down on the harness, causing stresses in the connector; impact loads due to engine technicians who inadvertently step on the connector in order to gain access to different areas of the engine for service and maintenance purposes; and outward pulling loads on the harness that stress the connector and potentially wires inside the harness during servicing. In this application, it is important to monitor continuously for these types of loads. Some of the loads occur when the engine is operating, whereas others occur during pre-flight and post-flight system checks.

In general, loads could be *cyclic* or *transient* in nature. For example, hard landings cause transient vertical loads on the harness, resulting in stresses to the connector threads and pins. On the contrary, steady operation of the engine causes cyclic stresses in the connector and its pins. Environmental loading can also cause substantial changes in the operational mechanical loads that are measured. For example, when temperatures inside the engine compartment increase, the connector stress relief apparatus and the connector panel soften. This softening of the material (decrease in elastic modulus) produces larger response amplitudes when loads act on the harness, resulting in larger stresses to the connector.

After identifying the magnitudes, frequency ranges and so on of these various types of loads using algorithms that process raw sensor data, the information obtained can be used for a variety of purposes. First, an unusually high force level on a connector could be reported to maintainers, who could then inspect that connector to ensure it is connected properly and that there are no broken pins causing open circuits. Second, data collected on loads during operation could be provided to designers, who could use this information to revise design specifications. For example, the shock spectrum on connectors in different locations could be revised in specifications so that suppliers could improve future designs of the connector stress relief. Third, load data from the past can be used to predict what the loads are likely to be in the future. This future loading information is then used to make damage and performance predictions.

Damage identification

The integrated sensors described above are also used to identify potential damage in the connector and harness. Damage identification is often referred to as *diagnostics* or *state awareness*. In this application, it is important to at least isolate damage (or *faults* as they are sometimes called) in the harness/connector, because otherwise components attached to the harness may be replaced unnecessarily. For example, if the bent flange in Figure 1.6 that is introduced when a technician inadvertently steps on it results in an open or intermittent circuit to a starting motor for the engine, then that motor might be removed and replaced mistakenly. In addition, there are many possible damage locations in the wire harness and connector subsystem, so an automated fault detection system would be desirable.

Damage identification consists of several steps including *signal processing* and *feature extraction*. Signal processing methods are applied to raw data that is acquired from the integrated sensors in order to produce valid information for diagnostics. For example, it might be convenient to transform time data acquired from vibration sensors on the connector panel into frequency data using Fourier analysis techniques (discussed later). Then this frequency data can be used to extract features, which relate to a model of the damage mechanism of interest. For example, a model of the connector and panel could indicate that a loose connector causes a shift in a vibration frequency of the panel. Then the feature of interest would be this frequency value.

One of the main challenges in damage detection is the variability caused by changes in operational and environmental loading and the resulting variations in features used for diagnostics. These variations can cause false positive or negative indications of damage (called *false alarms*), which should be minimized according to the end user's health monitoring specifications. For example, if it is essential to identify every instance of a potential failure in the harness/connector no matter how improbable, then diagnostic algorithms should be designed in that manner. When selecting signal processing and feature extraction methods in health monitoring algorithms, the extent to which these methods amplify or suppress sources of variability must be considered. For example, if changes in engine RPM and temperature cause unwanted changes in frequency data, then those areas of the dataset should be discarded when attempting to identify damage.

In addition to detecting the presence of damage, damage identification algorithms should ideally also provide an indication of the damage location and level. For example, a damage location algorithm for the connector panel such as the one at the top of Figure 1.6 should indicate which of the seven connectors shown is damaged. The task of damage location usually requires a model of the component and data from multiple sensors. Similarly, the task of damage quantification nearly always requires a physics-based model of the component. Also, damage is more quantifiable using active sensing where the sensor signals can be compared to a measured input signal as described above.

Damage identification information can then be provided to maintainers to trigger additional inspections, maintenance or orders for replacement parts. This information is also used in the subsequent damage and performance predictions.

Damage prediction

After the loading and damage information is obtained, this information can be combined using some sort of damage law to predict how quickly damage will accumulate in the future. Design data that indicates the types of materials involved, geometry of the parts and so on is always needed to carry out the damage prediction task. For example, design data might indicate that the flange shown in Figure 1.6 can sustain a certain amount of deformation before the wires snap against the sockets. A solid mechanics model of the flange could be combined with information about the current amount of deformation from previous loads. This updated model of the flange could then be used to predict the amount of time the connector could continue to operate with steady (cyclic) and transient loads anticipated in the future.

There are numerous types of damage laws for metallic, composite, ceramic and other materials. Each damage law (equation) is comprised of parameters that account for geometrical properties of the component (length, width, thickness) and loading characteristics (stress level, frequency, direction). In many cases, changes in damage can be predicted using trending models, which are based almost exclusively on previous loading data and trends in damage identification information. This kind of data-driven prediction works well if loads and the response behavior of the component in the future are very similar to those observed in the past. In other cases, changes in damage must be predicted using physics-based models that apply more broadly to different operational scenarios than those experienced in the past. For example, if the connector responds differently to small and large operating loads, then a physics-based model would be needed to describe changes in the bent flange for these different loading levels.

Performance prediction

The damage prediction law shown in Figure 1.6 (middle) illustrates that environmental factors also affect predictions. Two different curves are shown corresponding to a low-temperature case (T1) and a higher temperature case (T2). The usage of a component indicates the amount of time it has been operating or the number of loading cycles a component undergoes. The damage level is the prediction being made. When the damage level reaches a certain threshold, design data would be used to declare that the component has failed. In this case, failure would mean an open circuit might occur when the wires inside the connector bend and break. When it is predicted that the connector will fail if it continues to operate, this prediction is referred to as a performance prediction or a *prognosis*.

This performance prediction could then be utilized along with the operating specifications for the engine to determine if the connector should be replaced or can perform for a certain number of future flights. In addition, this information might be used to guide the crew who operate the aircraft and engine. For example, if the flange were in imminent danger of failing, then the pilots could be instructed to avoid hard landings, which might cause the flange to bend to the point where the wires break, and an open circuit condition is reached.

1.5 DEFINITIONS OF COMMON TERMINOLOGY

Health monitoring is an emerging field of engineering, but there is already a vast terminology associated with it. The first term that must be defined is *damage*. In this book, damage is defined as *a permanent change in the mechanical state of a structural material or component that could potentially affect its performance*. Common sources of damage in materials and structural components include the following:

Material examples:
Micro-structural defects (dislocations, voids, inclusions)
Oxidation/corrosion (loss of material)
Nitridation (build up of material)
Exfoliation (pealing)
Erosion (pitting, spalling, abrasive wear)
Yielding/creep (plasticity, loss in modulus)
Residual stress
Cracking (fatigue, matrix, ply)

Structural examples:
Fastening fault (weld crack, bolt preload, broken rivet)
Adhesive fault (de-bonding, delamination, separation)
Clearance change (gap)
Instability (thermo-mechanical buckling)

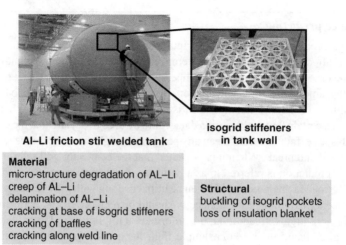

Al–Li friction stir welded tank

**isogrid stiffeners
in tank wall**

Material
micro-structure degradation of AL–Li
creep of AL–Li
delamination of AL–Li
cracking at base of isogrid stiffeners
cracking of baffles
cracking along weld line

Structural
buckling of isogrid pockets
loss of insulation blanket

Figure 1.7 Types of damage in Al–Li friction stir welded tank (Ackers *et al.*, 2006, DEStech Publications, Inc.)

For illustration purposes, this definition of damage is applied below to an Al–Li cryo-tank, a thermal protection system panel and a large engine valve. First consider the large Al–Li tank shown in Figure 1.7. It is apparent from the picture of this tank that material and structural damage of the types listed could occur anywhere along the wall or dome. In fact, components such as this tank are called *unitized structures* because they do not use fasteners around which damage often localizes. Damage mechanisms such as these that can occur anywhere within a component are referred to as *global*. Global damage is difficult to detect and usually requires that sensors be distributed throughout the component. For example, acceleration sensors could be placed along and around the tank to measure different shapes in which the tank vibrates. In some cases, elastic waves can be propagated through the tank to detect damage with fewer sensor sets. This particular component is also said to be *damage intolerant* because even the smallest crack in the wall could be catastrophic due to pressurization of liquid fuel contained within the tank. The tank would be less sensitive to lost insulation, for example, over a short period of time during a launch if it became cracked in the location where insulation is lost.

Also note that large static acceleration forces (*g* forces) acting on the tank during launch cause inward or outward buckling of the tank. This buckling deformation causes relatively small defects in the wall to appear larger (open up) when health monitoring damage detection methods are applied (Sundararaman *et al.*, 2005). Recall that this type of change in how damage is perceived in sensor response data when components operate under load is one advantage of online damage detection techniques.

Next consider the metallic thermal protection system panel shown in Figure 1.8. This panel is a next-generation thermal barrier component with mechanical attachments, which are easier to install and repair, leading to faster turnaround times relative to the bonded ceramic tiles that are used in the Space Shuttle Orbiter (Hundhausen, 2004). Because this panel is constructed from multiple layers, it has damage mechanisms at the interfaces of all its subcomponents. For example, the face sheets (top and bottom) can debond from the honeycomb core. Components such as this panel are sometimes referred to as *parasitic* because they are attached to other structures.

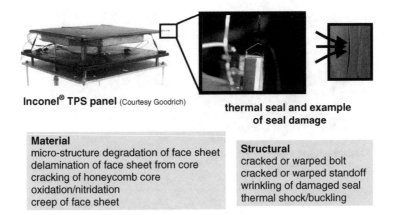

Inconel® TPS panel (Courtesy Goodrich)

thermal seal and example of seal damage

Material	**Structural**
micro-structure degradation of face sheet	cracked or warped bolt
delamination of face sheet from core	cracked or warped standoff
cracking of honeycomb core	wrinkling of damaged seal
oxidation/nitridation	thermal shock/buckling
creep of face sheet	

Figure 1.8 Types of damage in Inconel® thermal protection system panel

Damage mechanisms are often more localized in predetermined locations in parasitic components. For example, fasteners (bolts) can crack as can the mechanical struts in this panel. As in the Al–Li tank structure, the thermal protection system panel exhibits a sensitivity to load during damage detection. When the panel is heated during operation to as high as 1500 °F surface temperatures, the face sheet softens and the temperature gradient across the panel causes bending moments to act on the attachment bolts. These bending moments can be quite high, leading to a greater sensitivity of vibration measurements to small losses in preload through the bolt due to cracking. On the contrary, these high temperatures and fluctuations in temperature can also cause false positive indications of damage. To avoid these false readings, a thermo-mechanical model of the panel must be used or baseline data must be acquired at various measured temperatures. This baseline information can then be used to distinguish between changes in the panel response due to damage and changes due to temperature fluctuations.

Finally consider the engine valve shown in Figure 1.9. This valve is positioned on the intake airline of a large marine diesel engine. The purpose of the valve is to close off the air supply to the engine in the event of a fire or other emergency situation. Loads that act on the valve include low-frequency vibrations on the order of several gs ($1g = 9.81$ m/s^2) due to the engine crankshaft rotation and temperatures as high as 120 °F. The valve is also adjacent to the turbocharger, which introduces higher frequency vibrations.

Each of these loads and their combined effects on the dynamic response of the valve will be discussed in more detail later in the book. Damage mechanisms that result from these loads include those listed in Figure 1.9. Note that the bushing shown in the figure is clearly damaged due to impacts that erode the material. Other not so obvious damage includes the solenoid, which loses its pull force capability as temperature increases, and the internal valve mechanisms, which fuse together in a cold welding (material transfer) process due to the vibrations introduced by the turbocharger. This latter example of damage demonstrates that damage and failure in a given component are often caused by subtle interactions within the system (McGee and Adams, 2002).

In some cases, this definition of damage may not fit the situation. For instance, manufacturing flaws, such as assembly errors in the thermal protection system panel or the heat-affected zone in a friction stir weld, could be considered damage. However, these flaws could also be

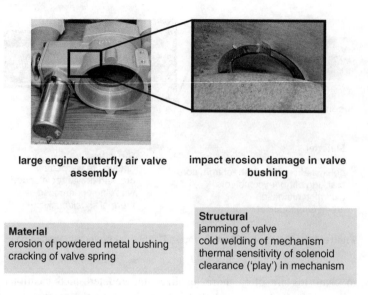

Figure 1.9 Types of damage in intake valve for large engine (Hundhausen *et al.*, 2005, John Wiley & Sons)

treated as part of the baseline (healthy) component. Squeaks and rattles in automobiles are less obvious examples of manufactured flaws. These sources of noise result in warranty claims to car companies and could be considered damage. However, it is customary to refer to manufacturing flaws as *functional degradation* if they affect performance but do not result in component failures.

In other cases, damage may be present but would not likely result in failure under ordinary types of loading. For example, a crack at the tip of a wing spar would not ordinarily be stressed to the point where it could cause structural failure of the spar. Similarly, the reinforced concrete support column shown in Figure 1.10 at the rear has experienced damage to the concrete due to high shear loading during the 1994 Northridge earthquake; however, the reinforcement bars in the column have not been damaged. This column might sustain traffic loads, which push down on the column. However, the damage the column has sustained might be sufficient to result in failure in the event of another earthquake so the column would be flagged as damaged in an inspection.

Figure 1.11 illustrates a generic example of this type of scenario where a component is damaged but does not lose its ability to perform under dynamic loading. In Figure 1.11(a), damage in the center mechanical element is not stressed during operation of the component, because the damaged area is not being strained. In this case, the damage could possibly be ignored for health monitoring purposes because the damage does not affect the performance of the component. On the contrary, this same damaged area is severely strained in Figure 1.11(b), resulting in a potential loss in performance under load. When defining damage in a given application, the nature of the operational and environmental loads must be considered.

Other commonly used terms in health monitoring are defined below. More terms will be defined as they are needed throughout the book.

Damage diagnosis: The process of identifying damage in structural materials and systems.

Figure 1.10   Damage to highway overpass reinforced support column during 1994 Northridge, CA, earthquake (courtesy Prof. F. Seible, University of California, San Diego) showing cracking in concrete of column in rear with less significant rebar damage

Damage prognosis: The process of predicting the future probable capability of a structural material or system in an online manner, taking into account the effects of damage accumulation and estimated future loading.

Failure: Instant at which a structural material or component has no remaining useful life.

Functional degradation: A reduction in performance of a component during operation that does not lead to failure.

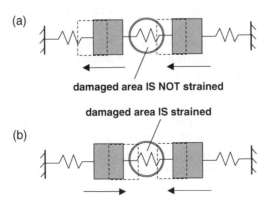

Figure 1.11   Damage in two different operating conditions: (a) damaged area is not strained by the motion and (b) damaged area is strained by the motion

Structural health monitoring (SHM): Health monitoring applied to structural systems in a variety of applications including mechanical, aerospace, civil, marine and agricultural systems.

Health and usage monitoring systems (HUMS): Same as health monitoring with application to rotorcraft and other types of machinery.

Integrated systems health management (ISHM): The process through which the operation of complete systems including mechanical, electrical, firmware and other subsystems are managed online using diagnostics and prognostics.

Integrated systems health monitoring: Health monitoring applied to any complete dynamic system such as an aircraft or ground vehicle system including all of its subsystems.

Intelligent maintenance: Same as condition-based maintenance where the condition of a structural material or system is used to schedule service and maintenance in an automated manner.

Nondestructive evaluation (NDE): The process of assessing the current damage state of a structural material or component without accelerating the damage.

Prognostic health management (PHM): Logistic processes through which the operations of structural components are managed using diagnostics and prognostics.

Product life-cycle management (PLM): The use of diagnostics, prognostics and software tools to manage the development and operational phases of a component.

Reliability forecasting: The process of predicting the future probable capability of a structural material or component in an offline manner using damage and future loading information.

1.6 COMPARISON OF NONDESTRUCTIVE TESTING (NDT) AND HEALTH MONITORING TECHNIQUES

NDT is the offline implementation of NDE methodologies for assessing the damage state of a structural material or component without accelerating the damage. There are many methods for NDT including dye penetrant testing, modal testing, radiography, ultrasonics, infrared thermography, eddy current and X-ray tomography (Hellier, 2001; Grandt, 2004). Health monitoring and NDT methodologies are similar in many ways. Both methods seek to identify the damage state of a structural material or component; however, health monitoring does this online as a component operates and also identifies loading and applies prognosis to predict future performance.

Health monitoring should be viewed as a complementary method to available NDT methods, which can be used to corroborate health monitoring data and perform more precise inspections of local areas of a component. For example, if a health monitoring system identifies that the UAV in Figure 1.1 has cracking in the stabilizer strut when the plane is flying, then an NDT method such as ultrasonics could be used to identify an accurate measure of the crack depth when the plane is on the ground. NDT would be applied locally to the strut producing a clearer image of the crack length, orientation and so on. Health monitoring of the entire UAV would be applied globally, leading to better awareness of the plane's operational health but less precise damage information. A complete comparison of these two NDE technologies is illustrated in Figure 1.12.

As shown in Figure 1.12, both methods provide information to operators for structural control (real-time changes) and structural design (future modifications) to improve the performance of components even if they should become damaged. In fact, one of the

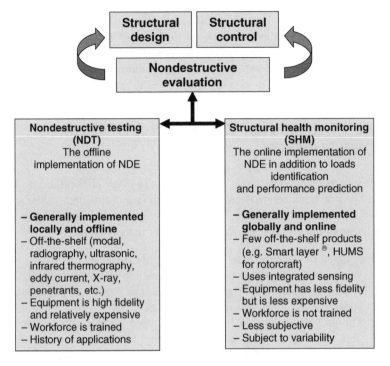

Figure 1.12 Comparison of NDT and health monitoring technologies

primary benefits of health monitoring technologies is that they can be used to optimize the weight, durability and other aspects of a component without sacrificing reliability.

For example, health monitoring data from the engine valve discussed in Figure 1.9 indicated that the valve could become damaged, resulting in a failure to close. This information was used to redesign the valve as shown in Figure 1.13. The new design was selected because it used a guillotine-type closure mechanism, which overcame the reduction in solenoid pulling

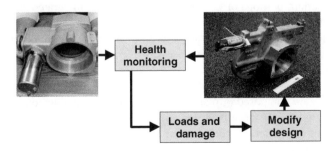

Figure 1.13 Previous generation engine valve and modified design with guillotine-type valve action to assist in valve closure

force with increases in temperature. The new design also avoided impacts that might hinder valve closure. Information about the vibration loading due to the valve position near the turbocharger also influenced future engine designs. This feedback of information about operational and environmental loading, damage and failure modes is an important role of health monitoring.

1.7 POTENTIAL IMPACT OF HEALTH MONITORING TECHNOLOGIES

Health monitoring technologies should not be applied in every application. Instead, six key factors should be evaluated before deciding that health monitoring approaches such as those presented in this book are appropriate. These factors are listed below along with examples taken from the structural components discussed earlier in the chapter.

Factor 1: Operational or environmental loading is highly variable.
  (a) When loads vary significantly, load interaction effects complicate the detection of damage using NDT (e.g. heating of thermal protection panel causes apparent damage level in cracked bolt to vary, Figure 1.8).
  (b) The growth of damage is also complicated by load interaction effects (e.g. transient overloads act on a cracked wheel axle in a mining truck, causing the crack tip to blunt followed by cyclic loads, Figure 1.2(a)).
  (c) The response of a structural material and component varies significantly, resulting in nonlinear, time-varying effects that complicate the health management of a component.

Factor 2: Loads, material, structural, damage or failure models are uncertain.
  (a) Complete usage databases do not exist for some new materials (e.g. filament wound carbon composites, Figure 1.3(d)) or structural concepts (e.g. morphing aircraft, Figure 1.3(a)).
  (b) Uncertainties in these areas can result in large health management uncertainties requiring some sort of diagnostic and prognostic information.

Factor 3: Designs are either significantly over or under conservative.
  (a) Health monitoring technologies can be used to assist in scaling back the weight in certain portions of the structural system (e.g. integrated suspension system, Figure 1.2(d)).
  (b) These technologies can also be used to manage risk in systems where lower factors of safety are used by necessity to enable operation (e.g. aircraft systems, spacecraft).

Factor 4: Systems are composed of many interconnected, interacting components.
  (a) Knowledge of more than one structural material or component is required in order to predict the performance of one particular component (e.g. integrated suspension system, Figure 1.2(d)).
  (b) Uncertainties in the interactions among these components lead to changes in system performance.

Factor 5: Unanticipated changes can occur over the system life cycle.
   (a) Health monitoring helps to manage risk due to structural modifications, inadvertent loads (e.g. service procedures as in the wire harness, Figure 1.6) and design or manufacturing flaws (e.g. rivet patterns that are skewed, leading to stress concentrations and cracking).

Factor 6: Damage can occur in hard-to-reach areas.
   (a) NDT methods cannot be applied to these areas unless the system is dismantled (e.g. wheel assembly in Figure 1.5 where spindle is buried).
   (b) Dismantling of a structural system can compromise its performance.

If these six factors are considered at a minimum and a health monitoring technology is implemented successfully, the end user of a structural system can realize some or all of the benefits illustrated in Figure 1.14. These potential benefits are described below with some examples.

(1) Reduced risk to operators and structural systems because of more accurate assessment of system health prior to operation (e.g. detecting cracks in Space Shuttle solid rocket boosters prior to launch).
(2) Performance optimization through prognostics-driven control (e.g. wear in satellite gyro bearings is opposed by adjusting the attitude of the satellite to redistribute lubricant around the bearings in order to reduce wear).
(3) Reduced risk through life-extending operation and control based on the use of health monitoring information (e.g. cracking in spindle of ground vehicle is identified and vehicle is removed from heavy service to avoid failure, Figure 1.5).
(4) Reduced cost for servicing structural systems and ordering parts (logistics) based on condition-based maintenance scheduling (e.g. damage in drive train or engine of mining truck is detected prior to failure and parts are pre-ordered to avoid costly

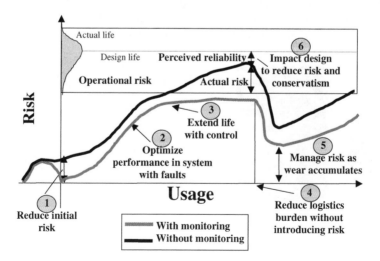

Figure 1.14 Benefits in system development, operation and renewal that can be realized using health monitoring technologies

downtime, Figure 1.2(a)). Asset availability may also be increased due to accurate estimation of remaining life and provisioning of just-in-time spares for field repair.

(5) Lower cumulative risk for sustaining the capability, or readiness, of a structural system after maintenance actions are taken (e.g. when wire harness and connector are serviced, there may be residual damage due to connection forces that bend pins over time or inadvertent impacts on connectors that cause damage to accumulate, Figure 1.6).

(6) Reduced design conservatism through online determination of future performance across a fleet of components or systems (e.g. missile casings are monitored throughout a military deployment and health monitoring data indicates that the weight could be reduced by removing layers of filament, Figure 1.3(d)). Traditional design approaches make conservative assumptions about the life across a fleet of components.

There are many health monitoring systems available for diagnosing faults in *rotating* systems. In some cases, these systems can be retrofitted to structural components to acquire and analyze data for health monitoring purposes. For example, consider the LANSHARC™ hardware (black box) shown in Figure 1.15(a) on the left. This system utilizes one vibration acceleration sensor placed on the nonrotating spindle housing of a lathe (shown in the right of Figure 1.15(a)). The acceleration data from this sensor is processed to calculate the value shown in Figure 1.15(b) as a function of part number. When the vibration level increases, this value increases. This particular machine would be taken out of service at 949 and 1600 cycles to replace the cutting tool. This kind of industrial machinery health monitoring system can save time and money by avoiding scrap parts and downtime in a production facility (Schiefer *et al.*, 2001). The system in Figure 1.15(a) is used in an automotive manufacturing facility.

There are fewer technologies commercially available for health monitoring of structural components. Consider the health monitoring technology illustrated in Figure 1.16. Figure 1.16(a) shows a planar component with an embedded SMART Layer® consisting of piezoelectric transducers that can transmit and receive elastic waves. This data is acquired and analyzed using a laptop computer and data acquisition system. Strips of the SMART Layer® can be installed in many different types of components. For example, Figure 1.16(b) shows a strip installed along a row of riveted joints around which cracks can develop and 'link up' causing fatigue failure of the aluminum lap joint. Figure 1.16(c) shows a small strip cut to fit an

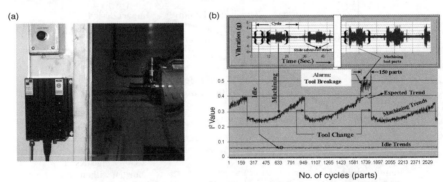

Figure 1.15   (a) LANSHARC™ condition-based maintenance system for monitoring spindle cutting tool and (b) unusual vibrations signatures indicate that tool needs to be replaced (Hundhausen *et al.*, 2005, Society for Experimental Mechanics, Inc.) (courtesy R. Bono, The Modal Shop Inc. of PCB Group, Cincinnati, OH)

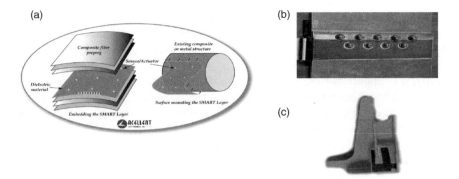

Figure 1.16   (a) SMART Layer® technology for embedding passive and active piezoelectric transducers into component, (b) panel with riveted joints around which cracks can develop and (c) landing gear with strips of sensors (courtesy Dr A. Kumar, Accelent Technologies Inc., Sunnyvale, CA)

aircraft landing gear part for health monitoring of loads and potential fatigue cracking in that component.

1.8 OVERVIEW OF TECHNICAL AREAS IN HEALTH MONITORING

The core technical areas of health monitoring discussed in this book include *modeling*, *measurements*, *data analysis* and *prediction*. These areas are broken down with subtasks in the flow chart shown in Figure 1.17. The first task in the flow chart is to define the health monitoring problem. The subtasks in the flow chart will be described in more detail in later sections of the book. A general point to note concerning these tasks is that they are interrelated. For example, models are developed for use in interrogating data and predicting performance. If models are not properly chosen to suppress sources of variability in measurement data, then damage will not be identified correctly.

To illustrate the interrelationships between the health monitoring technical areas, consider the thermal protection system example shown in Figure 1.18. In this system, models must be selected to describe panel damage such as a cracked bolt in the attachment strut. Models of the mechanical (vibration, acoustic) and environmental (temperature) loading applied to the panel must also be selected. These models are important because they are used in turn to select sensors, which can survive the temperatures behind the panel where sensors are installed. A typical launch vehicle would be covered with panels such as this one; therefore, it would be necessary to distribute sensors across the vehicle. With this distributed measurement network, models describing the panel loads would be even more critical for use in intelligently processing the measured data.

These models would be equally important for quantifying damage in the bolt (e.g. loss in preload) and accounting for sources of variability due to the thermo-mechanical loads. Variability affects both the sensor measurement and the computational algorithm used to interrogate data; therefore, steps to reduce this variability must be applied simultaneously in both the measurement and data analysis subtasks. Finally, the uncertainty in a prediction depends on the uncertainty in all preceding steps. By considering each of these technical elements in parallel, the propagation of uncertainty from element to element can be minimized in order to forecast crack growth in the bolt using the health monitoring models, data and features describing the bolt damage.

Finally, it is important to refer to the literature in health monitoring because there are continually new advancements in this field as technologies for sensing, computational mechanics and other areas emerge. There are numerous references available including those from the annual *Structural Health Monitoring International and European Workshops* (references given for 2004 and 2005 proceedings). The annual *SPIE Conference on Smart Structures and Nondestructive Evaluation* also publishes proceedings, which

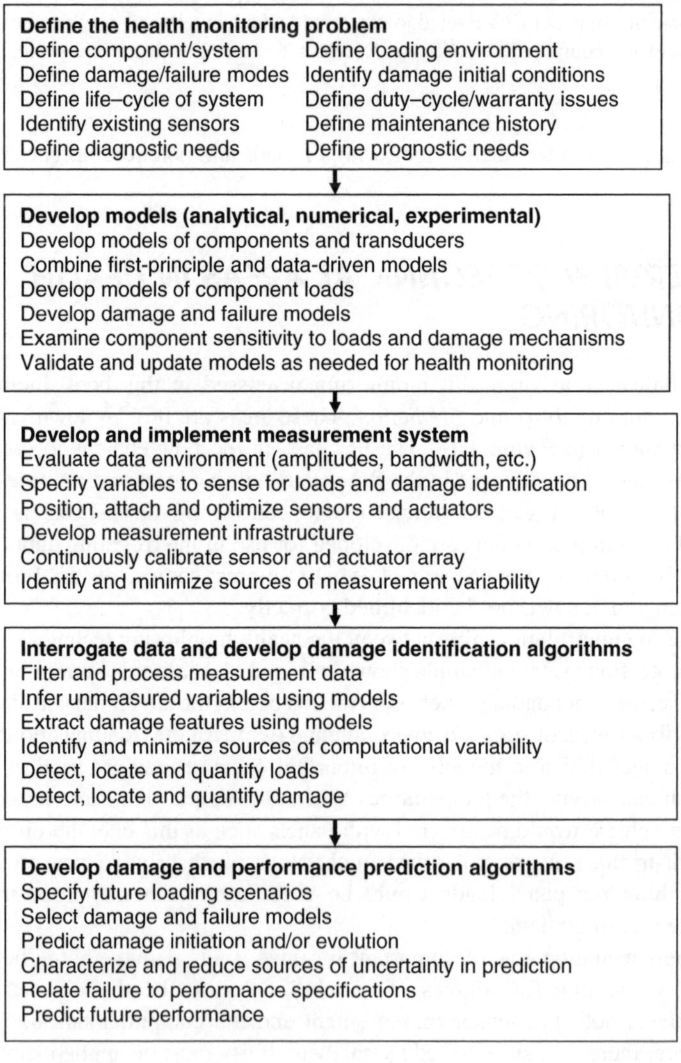

Define the health monitoring problem

Define component/system	Define loading environment
Define damage/failure modes	Identify damage initial conditions
Define life–cycle of system	Define duty–cycle/warranty issues
Identify existing sensors	Define maintenance history
Define diagnostic needs	Define prognostic needs

↓

Develop models (analytical, numerical, experimental)

Develop models of components and transducers
Combine first-principle and data-driven models
Develop models of component loads
Develop damage and failure models
Examine component sensitivity to loads and damage mechanisms
Validate and update models as needed for health monitoring

↓

Develop and implement measurement system

Evaluate data environment (amplitudes, bandwidth, etc.)
Specify variables to sense for loads and damage identification
Position, attach and optimize sensors and actuators
Develop measurement infrastructure
Continuously calibrate sensor and actuator array
Identify and minimize sources of measurement variability

↓

Interrogate data and develop damage identification algorithms

Filter and process measurement data
Infer unmeasured variables using models
Extract damage features using models
Identify and minimize sources of computational variability
Detect, locate and quantify loads
Detect, locate and quantify damage

↓

Develop damage and performance prediction algorithms

Specify future loading scenarios
Select damage and failure models
Predict damage initiation and/or evolution
Characterize and reduce sources of uncertainty in prediction
Relate failure to performance specifications
Predict future performance

Figure 1.17 Flow chart of technical areas in health monitoring

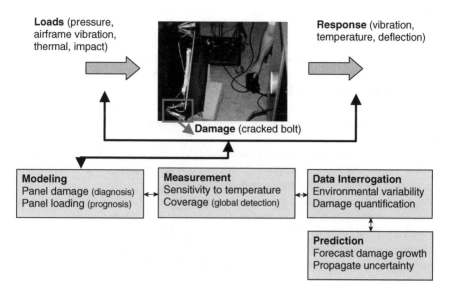

Figure 1.18 Interrelationships between technical areas of health monitoring for a metallic thermal protection system panel

contain a review of recent advancements in the field. Appendix B provides an overview of the technical journals and conferences that highlight developments in health monitoring. This book does not make an attempt to cover all of the methods available. Instead, examples are given of how to apply a basic methodology for health monitoring using vibration and wave propagation techniques involving the steps in Figure 1.17. Appendix B also provides a broad sampling of literature that focuses on many other health monitoring approaches.

1.9 SUMMARY

This chapter has defined the health monitoring process and technical elements involved in that process for a number of different applications. Key benefits of health monitoring were also discussed. A summary of key points in the chapter is given below.

- Health monitoring involves the *identification of operational and environmental loads, mechanical damage, the growth of damage, and the future performance of the component as damage accumulates.*
- Health monitoring technologies sometimes *detect damage more readily when loads are acting on a component to accentuate the damage.*
- *Damage accumulation can result in changes to the way damaged components respond, suggesting that both the loading and the damage should be monitored continuously* in the system in order to predict the future operational capability of its various components.

- There are cases in which health monitoring should be based more on *a priori models* and other cases when health monitoring should be based more on *data*.
- Health monitoring technologies can be used to implement *condition-based maintenance, which reduces the ownership costs for structural materials and systems*, and other business practices including autonomic logistics.
- Health monitoring diagnostic and prognostic approaches can be instrumental in product development for *identifying the durability of components and minimizing weight*.
- Sensing can be accomplished using *passive and active approaches*; passive approaches are relatively simple but more susceptible to measurement variability, and active approaches use actuation signals to better locate and quantify damage.
- The loads identification task can provide maintainers with information needed to service components or conduct additional *targeted inspections* in localized areas.
- Damage identification, also called *diagnostics*, aims to *detect, locate and quantify damage*; *signal processing and feature extraction* are two subtasks that require the use of models to distinguish damage from measurement variability, locate damage and quantify damage levels.
- Damage can be *local or global* in nature; local damage is restricted to a certain small region of a component, whereas global damage could occur anywhere in the component.
- Damage prediction, or *prognosis*, requires that damage laws be used; model-based prediction uses physics laws to forecast the growth of damage, whereas data-driven prediction uses trends to forecast the growth of damage.
- The term 'damage' is defined according to the *extent to which mechanical changes degrade the performance* of a component.
- Health monitoring and NDT are complementary technologies; NDT is implemented *offline and locally*, whereas health monitoring is implemented *online and globally*.
- Health monitoring has significant impacts on the *design and control* of structural systems in addition to the maintenance of those systems.
- Health monitoring should only be applied in cases where the investment in technology is justified for the reasons discussed in Section 1.7.
- Health monitoring involves *modeling, measurement, data analysis and prediction* tasks summarized in Figure 1.17.

The next two chapters discuss modeling and prediction techniques. Then issues in measurement methodologies and sensing technologies for health monitoring of structural materials and components are discussed in subsequent chapters. Data analysis issues are addressed for loads and damage identification. Then many of these techniques are applied to health monitoring case studies covering loads, damage and performance prediction in the final three chapters.

REFERENCES

Ackers, S., Kess, H., White, J., Johnson, T., Evans, R., Adams, D.E. and Brown, P. (2006) 'Crack detection in a wheel spindle using modal impacts and wave propagation', *Proceedings of the SPIE Conference on Nondestructive Evaluation and Smart Structures and Materials*, Chulha Vista, CA, Vol. 6177, paper no. 61770B.

Adams, D.E. (2005) 'Prognosis applications and examples', Chapter 18 in *Damage Prognosis*, Inman, D. and Farrar, C. (Eds), John Wiley & Sons Ltd., Chichester, West Sussex, England.

Boller, C. and Staszewski, W. (Ed.) (2004) *Structural Health Monitoring*, DEStech Publications Inc., Lancaster, Pennsylvania, USA.

Chang, F.K. (Ed.) (2005) *Structural Health Monitoring: Advancements and Challenges for Implementation*, DEStech Publications Inc., Lancaster, Pennsylvania, USA.

Grandt, A. (2004) *Fundamentals of Structural Integrity*, John Wiley & Sons, New York, NY.

Haroon, M. (2007) *A Methodology for Mechanical Diagnostics and Prognostics to Assess Durability of Ground Vehicle Suspension Systems*, Doctoral Thesis, School of Mechanical Engineering, Purdue University, West Lafayette, Indiana.

Haroon, M. and Adams, D.E. (2005) 'Active and event driven passive mechanical fault identification in ground vehicle systems', *Proceedings of the American Society of Mechanical Engineers International Mechanical Engineering Congress and Exposition*, Orlando, FL, paper no. 80582.

Hellier, C. (2001) *Handbook of Nondestructive Evaluation*, McGraw-Hill, New York.

Hundhausen, R. (2004) *Mechanical Loads Identification and Diagnostics for a Thermal Protection System Panel in a Semi-Realistic Thermo-Acoustic Operating Environment*, Masters Thesis, School of Mechanical Engineering, Purdue University, West Lafayette, Indiana.

Hundhausen, R.J., Adams, D.E., Derriso, M., Kukuchek, P. and Alloway, R. (2004) 'Loads, damage identification and NDE/SHM data fusion in standoff thermal protection systems using passive vibration based methods', *Proceedings of the 2004 European Workshop on Structural Health Monitoring*, Munich, pp. 959–966.

Johnson, T. and Adams, D.E. (2006) 'Rolling tire diagnostic experiments for identifying incipient bead damage using time, frequency, and phase-plane analysis', *Society of Automotive Engineering World Congress on Experiments in Automotive Engineering*, Detroit, paper no. 2006-01-1621.

Johnson, T., Yang, C., Adams, D.E. and Ciray, S. (2004) 'Embedded sensitivity functions for identifying damage in structural systems', *J. Smart Mater. Struct.*, **14**, 155–169.

Malley, M. (2001) *Methodology for Simulating the Joint Strike Fighter's (JSF) Prognostics and Health Management System*, Masters Thesis, Air Force Institute of Technology.

McGee, C. and Adams, D.E. (2002) 'Multiple equilibria and their effects on impact damage in an air-handling assembly', *Nonlinear Dyn.*, **27**(1), 55–68.

Navarra, K. (2004) 'FLTC #11: anticipatory support for air & space fleet readiness', Materials and Manufacturing Directorate Nondestructive Evaluation Branch (program review), Wright Patterson Air Force Base, Ohio, June.

Schiefer, M., Lally, M. and Edie, P. (2001) 'A smart sensor signal conditioner', *Proceedings of the International Modal Analysis Conference*, Society of Experimental Mechanics, pp. 1284–1290.

Stites, N., Adams, D.E., Sterkenburg, R., Ryan, T. (2006) 'Loads and damage identification in gas turbine engine wire harnesses and connectors', *Proceedings of the European Workshop on Structural Health Monitoring*, Granada, Spain, pp. 996–1003.

Sundararaman, S., Adams, D.E. and Jata, K. (2005) 'Structural health monitoring studies of a friction stir welded Al–Li plate for cyrotank', *Proceedings of TMS (The Minerals, Metals & Materials Society) Symposium on Materials Damage Prognosis*, New Orleans, LA, pp. 269–278.

Sundararaman, S., Haroon, M., Adams, D.E., Jata, K. (2004) 'Incipient damage identification using elastic wave propagation through a friction stir welded Al–Li interface for cryogenic tank applications', *Proceedings of the European Workshop on Structural Health Monitoring*, Munich, Germany, pp. 525–532.

Walsh, S., Pergantis, C. and Triplett, M. (2005) 'Assessment of sensor technology and structural health monitoring strategies for a composite missile motor case prognostic and diagnostic system', *ATO Prognostics and Diagnostics for the Future Force*, IV.LG.2004.01.

PROBLEMS

(1) Consider the police ceramic body armor vest shown in Figure 1.19. List the potential operational and environmental loads that can produce damage in this armor. Also, list the likely damage mechanisms. What could loads identification information be used for in this

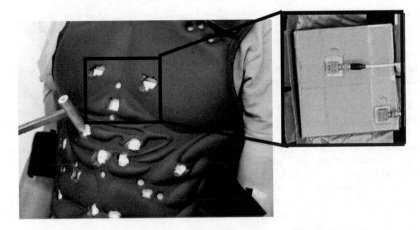

Figure 1.19   Ceramic body armor instrumented with response sensors

application? What about damage identification information? What would be the challenges in predicting the life of this armor?

(2) Consider the passenger truck shown in Figure 1.20. List the potential reasons from the list of factors in Section 1.7 that would justify the use of health monitoring in this application. Consider both consumer trucks and trucks used in commercial shipping businesses.

Figure 1.20   Passenger truck

(3) Read about one of the following NDT technologies in a reference and summarize the key advantages and disadvantages of those technologies: dye penetrant, ultrasonics, infrared thermography and radiography.

(4) Consider the rolling tire shown in Figure 1.21. Explain how a crack in the side of the tire changes as the tire is rolling. Include the effects of centrifugal forces, effects of tire tread deflection and temperature in your explanation.

(5) Consider the wrapped ceramic thermal protection system tile shown in Figure 1.22. It consists of a powdered ceramic core wrapped in a composite vest and is placed on a strain isolation pad, which is affixed to an aluminum airframe. List the potential operational and environmental loads that can produce damage in this tile. Also, list the likely damage mechanisms.

Figure 1.21   Rolling tire with bead damage

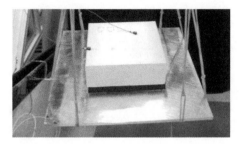

Figure 1.22   Wrapped ceramic tile bonded to aluminum plate

(6) Consider the windmill power generation system shown in Figure 1.23(a). List the potential operational and environmental loads that can produce damage in the rotor, transmission, electrical systems and tower (shown in Figure 1.23(b)). Also, list the likely damage mechanisms and examine the potential benefits of a health monitoring system for this application.

(a)   (b)

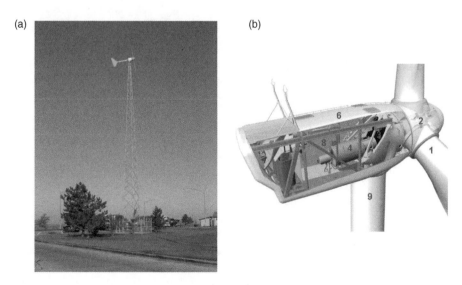

Figure 1.23   (a) Windmill station and (b) schematic of nacelle showing various electro-mechanical components

2
Modeling Components

2.1 MODELING NEEDS

Modeling affects every aspect of health monitoring. *Models enable the conversion of raw data into information about the health of a structural material or component*. For example, consider the metallic thermal protection system panel shown in Figure 2.1. This panel is constructed from Inconel®, a high-temperature metal that would protect aluminum airframes during space flight. The panel is shown instrumented with accelerometers, which measure the motion of the panel over a certain amplitude and frequency range. An impact is shown acting on the panel in addition to aerodynamic pressure loads. Damage in the form of cracking in one of the panel standoff attachments is also indicated in the figure. Modeling in this application is needed for all of the following reasons:

- In order to properly choose the appropriate accelerometers (or other type of sensor), *load models* of the panel are needed to estimate the response amplitudes and frequencies of the panel. For example, an accelerometer with a 1g limit[1] should not be chosen if the load model indicates responses greater than 1g. These estimates could be based on data acquired during tests or on first-principle models of the panel.
- *Sensor models* are needed to determine if a sensor will provide reliable measurements over the entire range of operating conditions. For example, a model describing the change in sensitivity of the sensor with variations in temperature is needed; otherwise, changes in response data due to temperature swings may be misinterpreted as damage.
- A model of the *sensor hardware* is needed that is used to install the accelerometer. Different attachment methods (glue, epoxy and screw) have different characteristics that vary with signal frequency. These differences are important in health monitoring.
- An *input–output model* of the panel is needed to describe how the panel responds to dynamic forces. This model can be used to estimate the location and magnitude of impacts due to foreign objects such as micrometeorites.
- Similarly, a model of the *panel's vibration response* to aerodynamic pressure loading is needed for diagnosing (detecting, locating and quantifying) damage to the panel. To *detect* damage, a statistical model would be needed to distinguish healthy and damaged panels. To *locate* damage, measured data must be interpreted with a distributed model of the different

---

[1]One g of acceleration is equal to $9.81\,\text{m/s}^2$; acceleration is usually expressed in terms of g's.

Health Monitoring of Structural Materials and Components: Methods with Applications   D. Adams
© 2007 John Wiley & Sons, Ltd

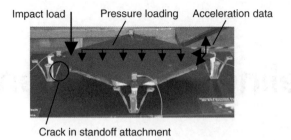

Impact load   Pressure loading   Acceleration data

Crack in standoff attachment

Figure 2.1   Needs in modeling for health monitoring of thermal protection panel

panel components. To *quantify* damage, a first-principle model would be needed to relate changes in response to the damage level. A *baseline model* of the panel is usually developed before damage occurs. Then, the damage is identified by comparing new data to the behavior of this baseline model.

- In addition to the panel model, the *standoff attachment mechanics* must also be modeled to ensure that if the fasteners (bolts) and/or standoffs vary from panel to panel, then changes in the observed response data are interpreted correctly.
- After the loading and panel models are identified, a *progressive model of the damage* is needed to predict the rate at which damage is accumulating as a function of loading. This model could be based on the observed trends in the damage or on first-principle models of specific damage mechanisms (e.g. crack, loosening bolt, etc.).
- Finally, a *predictive model* of future panel performance must be developed by combining load models with damaged panel models and performance models.

In this chapter, different types of dynamic models will be introduced to describe the vibrations, input–output dynamics, elastic wave propagation, and combined load responses of structural components and measurement hardware. Some models will be derived using first principles (called physics-based models), whereas other models will be derived from test data (called data-driven models). The next chapter discusses damage.

2.2   FIRST-PRINCIPLE MODELS

2.2.1   Component Vibration Models

Dynamic models are useful for health monitoring because they describe how structural components with and without damage continuously respond to the operating environment. Dynamic models are often used to analyze how structural components *vibrate*, or oscillate. For example, vibration data can be analyzed using these models to identify impact loads and mechanical damage in the form of oxidation (change in mass) or loss of preload in a bolted joint (change in stiffness). Vibration models are derived using either the Newton–Euler equations or Lagrange's equations (Tse *et al.*, 1978; Thomson and Dahleh, 1998). This book uses the Newton–Euler equations because internal forces are revealed in these equations. Also, undergraduate courses in dynamics discuss these methods. In addition to providing differential equation models of structural components, the Newton–Euler equations will also be used to analyze measurement data.

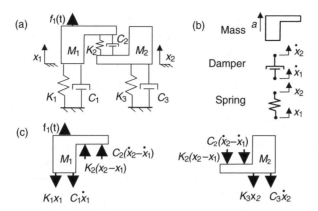

Figure 2.2 (a) Two lumped masses undergoing dynamic forcing, (b) types of elements in structural systems, and (c) free body diagrams for structural components

Consider the structural component shown in Figure 2.2(a). This model could represent a simplified automotive suspension component, a thermal protection panel, or another component that vibrates. The displacements $x_1(t)$ and $x_2(t)$ describe the translational movement of the two masses M_1 and M_2 up and down from their resting, or *equilibrium*, positions. $x_1(t)$ and $x_2(t)$ are called the *outputs*, or *responses*, of the system. $f_1(t)$ is the applied force and is called the *input*, or *excitation*. This model is called a *two degree of freedom* model because there are two independent coordinates, $x_1(t)$ and $x_2(t)$, that describe all of the component's possible motions. In general, the number of *degrees of freedom* (DOFs) is defined as the minimum number of variables needed in a model to describe the motions of interest in a component. When modeling, the objective should not be to choose complicated models with many DOFs. Instead, simple models that describe the behaviors of primary interest should be chosen because these models are most useful. In this example, the model contains *lumped parameters* at certain locations; however, *continuous parameter* models can also be developed as in Sections 2.2.8 and 2.2.9 for modeling vibrations as well as wave propagation.

This system contains three types of elements: inertia elements (masses), dissipative elements (dampers) and elastic elements (springs). These elements are illustrated in Figure 2.2(b). The mathematical laws that relate the force, f, applied to or across these elements and the resulting motions are as follows:

$$\text{Inertia}: f = M \times a \tag{2.1a}$$

$$\text{Damping}: f = C \times \Delta v \tag{2.1b}$$

$$\text{Stiffness}: f = K \times \Delta x \tag{2.1c}$$

where M is the mass in kg, C is the viscous damping coefficient in N·s/m, K is the spring constant in N/m, a is the acceleration, $\ddot{x}$, of mass M, and Δx and Δv are the relative displacement $(x_2 - x_1)$ and velocity $(\dot{x}_2 - \dot{x}_1)$, respectively, across the spring and damper[2]. The damping in Equation (2.1b) is called *viscous* damping because it relates the damping force

---

[2]A dot above a variable as in $\dot{x}$ is used to denote the derivative with respect to time, dx/dt.

to the velocity across the damper. Also note that both the spring and damper forces are *linear*, or proportional, functions of the displacement and velocity. In many structures, there are *nonlinear*, or nonproportional, elements due to joints, materials or damage. Damage is nearly always nonproportional in nature. Gravity has been ignored in this model because it is a static force that will not play a significant role in health monitoring[3].

The mass and stiffness parameters in Equations (2.1a)–(2.1c) can usually be calculated knowing the material properties and geometric characteristics of a component. For example, each of the mass coefficients in the model in Figure 2.2(a) could be estimated by weighing the mass $\times$ gravity of the thermal protection system panel shown in Figure 2.1 and dividing by two. The stiffness coefficients, K_1 and K_3, could be estimated by calculating the bending stiffness of the struts that support the panel. The remaining stiffness K_2 could be estimated by calculating the vertical shear stiffness of the panel.

The free body diagrams of the system are shown in Figure 2.2(b). Recall that free body diagrams illustrate all of the forces both external and internal that act on the DOFs in a system. Note that spring and damper elements oppose the motions of inertias. Also note that the forces acting between the two masses are equal and opposite because of Newton's Third Law. When a component becomes damaged, the internal forces in the free body diagrams of Figure 2.2(b) change; therefore, a dynamic model that describes how these forces change will be useful for health monitoring.

Newton's Second Law of motion states that the sum of the forces acting on a body of constant mass is equal to the mass times its acceleration, $\sum Forces = ma$. This law is applied to both the masses in Figure 2.2(b) in order to derive the system *equations of motion*. For the first mass,

$$\sum F = M_1 a_1$$
$$= f_1(t) - K_1 x_1 - C_1 \dot{x}_1 + K_2(x_2 - x_1) + C_2(\dot{x}_2 - \dot{x}_1) \tag{2.2}$$

and for the second mass,

$$\sum F = M_2 a_2$$
$$= -K_3 x_2 - C_3 \dot{x}_2 - K_2(x_2 - x_1) - C_2(\dot{x}_2 - \dot{x}_1) \tag{2.3}$$

It is more convenient to write these two equations in vector–matrix form as follows. This set of equations is referred to as a *time domain* model of the component.

$$\begin{bmatrix} M_1 & 0 \\ 0 & M_2 \end{bmatrix} \begin{Bmatrix} \ddot{x}_1 \\ \ddot{x}_2 \end{Bmatrix} + \begin{bmatrix} C_1 + C_2 & -C_2 \\ -C_2 & C_2 + C_3 \end{bmatrix} \begin{Bmatrix} \dot{x}_1 \\ \dot{x}_2 \end{Bmatrix}$$
$$+ \begin{bmatrix} K_1 + K_2 & -K_2 \\ -K_2 & K_2 + K_3 \end{bmatrix} \begin{Bmatrix} x_1 \\ x_2 \end{Bmatrix} = \begin{Bmatrix} f_1(t) \\ 0 \end{Bmatrix} \tag{2.4a}$$

$$[M]\{\ddot{x}\} + [C]\{\dot{x}\} + [K]\{x\} = \{f\} \tag{2.4b}$$

---

[3]In some cases, static gravitational forces do affect the growth of damage as in a building foundation.

The acceleration and velocity variables are both expressed in terms of derivatives of the displacement variable in these equations. Equation (2.4b) is a shorthand notation for the equations of motion including the mass matrix $[M]$, damping matrix $[C]$, stiffness matrix $[K]$, displacement vector $\{x\}$ and forcing vector $\{f\}$. Note the mass matrix is diagonal because the displacement coordinates being used are absolute coordinates; in other words, the coordinates are defined with respect to an inertial reference frame. When relative coordinates are used, the mass matrix will not be diagonal. Also, note the diagonal entries in the damping and stiffness matrices are determined by considering which dampers and springs are connected to a given DOF. The off-diagonal entries of these matrices are determined by considering which dampers and springs connect the DOFs.

2.2.2 Vibration Natural Frequencies and Modal Deflection Shapes

Equations (2.4a) and (2.4b) relate the external forces to the response motions of the two-DOF component in Figure 2.2(a). This model can be used in many ways. For example, it can be used to determine the *natural frequencies* and *modal deflection shapes* of the model. These characteristics are useful because when components become damaged, the natural frequencies and deflection shapes change as the damage grows. In addition, if the damage is in a region of high strain where one of the component's natural deflection shapes is excited during operation, then the damage may be more evident in measured response data. If a prediction of the existence of useful life in the component is made, then shifts in natural frequencies must be considered when making the prediction; otherwise, the operating loads will not properly excite damage in the component. If shifts in natural frequencies are not included, the prediction may overestimate or underestimate useful life.

To compute the modal frequencies and modal deflection shapes, the equations of motion are rewritten in *state–space form* as follows:

$$\frac{d}{dt}\left\{ \begin{matrix} \{x\} \\ \{\dot{x}\} \end{matrix} \right\} = \left[\begin{matrix} [0] & [I] \\ -[M]^{-1}[K] & -[M]^{-1}[C] \end{matrix} \right] \left\{ \begin{matrix} \{x\} \\ \{\dot{x}\} \end{matrix} \right\} + \left[\begin{matrix} [0] \\ [I] \end{matrix} \right] \left\{ \begin{matrix} f_1(t) \\ 0 \end{matrix} \right\} \tag{2.5a}$$

$$= [A]\left\{ \begin{matrix} \{x\} \\ \{\dot{x}\} \end{matrix} \right\} + [B]\left\{ \begin{matrix} f_1(t) \\ 0 \end{matrix} \right\} \tag{2.5b}$$

where $[A]$ is called the *system matrix*, and $[B]$ is the *input matrix*. Then the natural frequencies and modal deflection shapes are found by calculating the eigenvalues (λ) and eigenvectors ($\{\{x\}^{T}\{\dot{x}\}^{T}\}$) of the system matrix using the following equation:

$$[A]\left\{ \begin{matrix} \{x\} \\ \{\dot{x}\} \end{matrix} \right\} = \lambda\left\{ \begin{matrix} \{x\} \\ \{\dot{x}\} \end{matrix} \right\} \tag{2.6}$$

The eigenvalue problem associated with solving Equation (2.6) is discussed in many books on numerical methods (Strang, 1988). The MATLAB® file 'twodofeig.m' provided on the CD accompanying this book demonstrates how to calculate eigenvalues and eigenvectors for the two-DOF component for a given set of mass, damping and stiffness parameters. Appendix A describes the command structure in this simulation code. For this component model, there are four eigenvalues, $\lambda_{1,2}$ and $\lambda_{3,4}$ and four eigenvectors, $\{\{x_{1,2}\}^{\mathrm{T}}\{\dot{x}_{1,2}\}^{\mathrm{T}}\}$ and $\{\{x_{3,4}\}^{\mathrm{T}}\{\dot{x}_{3,4}\}^{\mathrm{T}}\}$, which come in pairs of two[4]. In general, the number of eigenvalues is equal to $2n$, where n is the number of DOFs. There are two distinct natural frequencies, one for each DOF in the system. The natural frequencies for this component are the *imaginary parts* of the first and third eigenvalues, $\mathrm{imag}(\lambda_1)$ and $\mathrm{imag}(\lambda_3)$. These frequencies are called the *damped* natural frequencies if the component has damping (*undamped* natural frequencies for no damping). The *real part* of an eigenvalue is usually negative and is called the *damping factor*, $-\sigma$. It determines the rate at which the response at a natural frequency decays and also prevents the response from becoming large if the component is excited at a natural frequency.

The modal deflection shapes are found in the first two rows of the first and third eigenvectors, $\{x_1\}$ and $\{x_3\}$. It is customary to normalize each shape so that the largest element of each is unity. It is also customary to label the two modal deflection shapes as $\{\psi_1\}$ and $\{\psi_2\}$. For example, 'twodofeig.m' provides the following four eigenvalues, two natural frequencies, two damping factors and two deflection shapes for the specific set of mass, damping and stiffness parameters given in the simulation file:

$$\lambda_{1,2} = -0.06 \pm 3.2\mathrm{rad/sec} \rightarrow \omega_{d1} = 3.2\mathrm{rad/sec}, \sigma_1 = 0.06\mathrm{rad/sec}, \{\psi_1\} = \begin{Bmatrix} 1 \\ 1 \end{Bmatrix} \quad (2.7a)$$

$$\lambda_{3,4} = -0.26 \pm 7.1\mathrm{rad/sec} \rightarrow \omega_{d2} = 7.1\mathrm{rad/sec}, \sigma_2 = 0.26\mathrm{rad/sec}, \{\psi_2\} = \begin{Bmatrix} 1 \\ -1 \end{Bmatrix} \quad (2.7b)$$

The first mode shape[5], $\{1\ 1\}^{\mathrm{T}}$, corresponds to the *in-phase motion* of the two DOFs, whereas the second shape, $\{1-1\}^{\mathrm{T}}$, corresponds to the *out-of-phase motion*. These two mode shapes are illustrated in Figure 2.3. Note that a proportional damping model has been assumed in this example: $[C] = \alpha[M] + \beta[K]$, where α and β are real constants. This type of damping leads to mode shapes that are entirely real, called *real normal modes*.

The significant differences between these two deflection shapes emphasize why this information is useful for health monitoring. Figure 2.3(a) illustrates that measurements of response data that follow the first deflection shape would not make it possible to detect damage in the spring, K_2, connecting the two DOFs because that spring does not deform during that

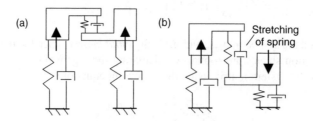

Figure 2.3 (a) First modal deflection shape and (b) second modal deflection shape showing stretching of spring connecting two DOFs

---

[4]Each pair of eigenvalues is a complex conjugate pair meaning that if, $\lambda_1 = a + bj$, then $\lambda_2 = a - bj$.
[5]The superscript symbol $\{1\ 1\}^{\mathrm{T}}$ denotes the transpose of the row vector producing a column vector.

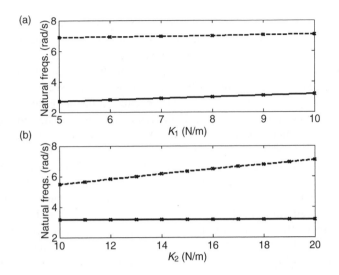

Figure 2.4 (a) Change in the two natural frequencies (ω_{d1} (—), ω_{d2} (- - -)) with changes in K_1 and (b) change in two natural frequencies with changes in K_2

motion. On the contrary, the second deflection shape in Figure 2.3(b) would cause this spring to compress and stretch; therefore, measurements that follow the second deflection shape would make it possible to detect damage in K_2. Figure 2.4 examines this aspect of damage detection by showing how the two natural frequencies of this system change as K_1 changes (Figure 2.4(a)) and as K_2 changes (Figure 2.4(b)). Note that K_1 causes small changes in both the natural frequencies; however, K_2 only causes changes in the second natural frequency. K_2 does not cause the first natural frequency to change because K_2 is not stretched in the first mode shape.

2.2.3 Free Vibration Response

The natural frequencies and modal deflection shapes define the *free response* of a component. The free response of the two-DOF model in Figure 2.2(a) occurs when initial conditions are applied to each DOF. There are four initial conditions, one on each displacement ($x_1(0), x_2(0)$) and one on each velocity ($\dot{x}_1(0), \dot{x}_2(0)$). Regardless of the specific values of the initial conditions, the free response of this two-DOF component can always be written in the following way using the two natural frequencies, damping factors, and modal deflections extracted in Equations (2.7a) and (2.7b):

$$\begin{Bmatrix} x_1(t) \\ x_2(t) \end{Bmatrix} = B_1\{\psi_1\}e^{-\sigma_1 t}\sin(\omega_{d1}t + \phi_1) + B_2\{\psi_2\}e^{-\sigma_2 t}\sin(\omega_{d2}t + \phi_2) \qquad (2.8)$$

for any linear two-DOF component with constant parameters, and

$$\begin{Bmatrix} x_1(t) \\ x_2(t) \end{Bmatrix} = B_1\begin{Bmatrix} 1 \\ 1 \end{Bmatrix}e^{-0.06t}\sin(3.2t + \phi_1) + B_2\begin{Bmatrix} 1 \\ -1 \end{Bmatrix}e^{-0.26t}\sin(7.1t + \phi_2) \qquad (2.9)$$

for the two-DOF component parameters used in the 'twodofeig.m' simulation code. Each group of terms in these equations is called a *mode of vibration*. Each mode of vibration has its own natural frequency, damping factor, and modal deflection shape. Also, each mode of vibration has two undetermined constants, B_1, ϕ_1 and B_2, ϕ_2. These constants must be calculated by applying initial conditions from a given operational scenario.

The free response of a component is usually excited in operation by an excitation that changes suddenly. For example, if the model in Figure 2.2(a) represents a thermal protection system panel, then a meteorite that strikes the panel and bounces off will excite the free response. This type of *impulsive* excitation causes a sudden change in the velocity of the point on the panel that is struck. Therefore, an impulse can be modeled as an initial condition on the velocity. Consider a scenario when a meteorite strikes M_1 in the two-DOF model, which is initially at rest. Suppose the impact causes M_1 to suddenly reach 1 m/s velocity. The initial conditions in this scenario would be $x_1(0) = 0\,\text{m} = x_2(0)$, $\dot{x}_1(0) = 1$ and $\dot{x}_2(0) = 0\,\text{m/s}$. The simulation file 'twodoffree.m' uses MATLAB$^{®}$ to compute the free response of the two-DOF model for this set of initial conditions. The two displacement responses in this impact scenario are shown in Figure 2.5. x_1 is shown with a solid line (–) and x_2 is shown with a dashed line (- - -). There are several important characteristics to observe in these responses:

- The responses both oscillate and decay rather slowly due to small amounts of damping in the two modes of vibration; recall $\sigma_1 = 0.06$ and $\sigma_2 = 0.26\,\text{rad/s}$, which are small.
- The responses both exhibit two frequencies of oscillation between 0 and 4s but then the second mode of vibration dies out leaving only the first mode in the responses. Based on the results depicted in Figure 2.4(b), which shows that the natural frequency of the first mode does not vary with K_2, the response data beyond 4s cannot be used to detect damage connecting the two DOFs. This result illustrates why models are helpful.
- The oscillation frequencies are the damped natural frequencies, $\omega_{d1} = 3.2\,\text{rad/s}$ and $\omega_{d2} = 7.1\,\text{rad/s}$; without damping they become the undamped natural frequency, ω_{nr}. These two frequencies are related through the *damping ratio*, ζ, for the rth mode of vibration: $\omega_{dr} = \sqrt{1 - \zeta^2}\,\omega_{nr}$. Also, the damping factor is given by $\sigma_r = \zeta\omega_{nr}$.

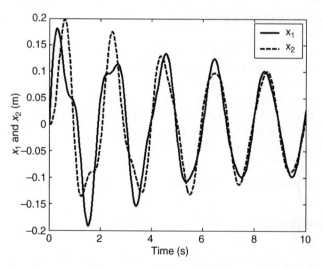

Figure 2.5 Plots of the two displacement variables for a simulated impact on M_1

- The impact on M_1 causes x_1 to change suddenly followed soon after by x_2, and then the two DOFs move somewhat synchronously as time progresses. This delay in time between the impact and the response at x_2 could be used in practice to *locate* the impact.
- At certain points in time near $t = 1.7\,\text{s}$ and $t = 2.2\,\text{s}$, there is a significant difference between the motions of the first and second DOFs. These differences would lead to potential damage caused by straining in the material connecting the two DOFs.

2.2.4 Forced Vibration Response (Frequency Response Models)

The free vibration response analyzed in Section 2.2.3 is determined almost entirely by the characteristics of the system. In nearly all health-monitoring applications, however, operational loads also affect component responses. *Frequency response function (FRF)* models are useful for understanding and interpreting the forced response of a component to a wide range of excitations. To construct an FRF model, the vector–matrix equations of motion of the two-DOF model from Equation (2.4) are used. Although operational excitations usually contain more than one sinusoidal component, FRF models are derived at a single frequency. It is also assumed that the steady-state response is at the same frequency as the excitation. To simplify the algebra, the force and response are expressed in terms of complex numbers as follows:

$$
\begin{Bmatrix} f_1(t) \\ f_2(t) \end{Bmatrix} = \begin{Bmatrix} F_1 \\ F_2 \end{Bmatrix} e^{j\omega t}
$$
$$
\begin{Bmatrix} x_1(t) \\ x_2(t) \end{Bmatrix} = \begin{Bmatrix} X_1 \\ X_2 \end{Bmatrix} e^{j\omega t}
$$

$$(2.10\text{a}, \text{b})$$

where F_1 and F_2 and X_1 and X_2 are phasors (complex numbers) of the excitation forces and response displacements, respectively, and ω is the excitation frequency in rad/s.

Each of the phasors describes the amplitude and phase of the corresponding force (or response). For example, the force applied on DOF x_1 can be written in either of the following ways:

$$f_1(t) = |F_1| \cos(\omega t + \sphericalangle F_1) \leftrightarrow f_1(t) = \text{Real}(F_1 e^{j\omega t}) \tag{2.11}$$

If the phasor is $F_1 = a + bj$, then the amplitude of the force is $|F_1| = \sqrt{a^2 + b^2}$ and the phase of the force is $\sphericalangle F_1 = a \tan b/a$. Similar expressions apply to the other phasor. If Equations (2.10a) and (2.10b) are substituted into the equations of motion, Equation (2.4), the following sequence of operations can be carried out:

$$[M]\left(\frac{\text{d}^2}{\text{d}t^2}\begin{Bmatrix} X_1 \\ X_2 \end{Bmatrix} e^{j\omega t}\right) + [C]\left(\frac{\text{d}}{\text{d}t}\begin{Bmatrix} X_1 \\ X_2 \end{Bmatrix} e^{j\omega t}\right) + [K]\begin{Bmatrix} X_1 \\ X_2 \end{Bmatrix} e^{j\omega t} = \begin{Bmatrix} F_1 \\ F_2 \end{Bmatrix} e^{j\omega t} \tag{2.12a}$$

$$\left[-\omega^2[M] + j\omega[C] + [K]\right]\begin{Bmatrix} X_1 \\ X_2 \end{Bmatrix} = \begin{Bmatrix} F_1 \\ F_2 \end{Bmatrix} \tag{2.12b}$$

The common exponential factor on both sides of Equation (2.12b) was removed to arrive at this result. The matrix on the left hand side of Equation (2.12b) is called the *impedance matrix*. In fact, the model in Equation (2.12b) is called the *impedance model* of the system. If the impedance matrix is inverted and multiplied on the left of both sides of the equation, the *frequency response function matrix*, $[H(\omega)]$, for the two DOF model is obtained:

$$\begin{Bmatrix} X_1 \\ X_2 \end{Bmatrix} = [-\omega^2[M] + j\omega[C] + [K]]^{-1} \begin{Bmatrix} F_1 \\ F_2 \end{Bmatrix}$$

$$\begin{Bmatrix} X_1(\omega) \\ X_2(\omega) \end{Bmatrix} = [H(\omega)] \begin{Bmatrix} F_1(\omega) \\ F_2(\omega) \end{Bmatrix} \qquad (2.13\text{a, b, c})$$

$$\begin{Bmatrix} X_1(\omega) \\ X_2(\omega) \end{Bmatrix} = \begin{bmatrix} H_{11}(\omega) & H_{12}(\omega) \\ H_{21}(\omega) & H_{22}(\omega) \end{bmatrix} \begin{Bmatrix} F_1(\omega) \\ F_2(\omega) \end{Bmatrix}$$

In moving from the first line of this equation to the second line, the frequency dependence of each phasor was included in the response and excitation vectors. In moving from the second line to the third line of this equation, the FRF matrix was expanded using a subscript notation. There are four FRFs in the FRF matrix. Subscripts are attached to the individual FRFs to indicate which excitation contributes to which response. For example, the FRF $H_{pq}(\omega)$ in this matrix relates the excitation force at DOF q to the response displacement of DOF p. Equation (2.13c) relates the amplitudes and phases of the input external forces to the amplitudes and phases of the output displacements as a function of frequency in the steady state. Equation (2.13c) is called a *frequency domain* model.

   If the model parameters are known, the FRF matrix can be computed by, first, constructing the impedance matrix and, second, inverting that matrix as a function of frequency. The simulation code 'twodoffrf.m' performs these operations. The two-DOF model in Figure 2.2(a) has one excitation force $f_1(t)$ applied to M_1; therefore, there are two FRFs relating this excitation to the two responses: $H_{11}(\omega)$ and $H_{21}(\omega)$. $H_{11}(\omega)$ relates $f_1(t)$ to $x_1(t)$, whereas $H_{21}(\omega)$ relates $f_1(t)$ to $x_2(t)$. 'twodoffrf.m' was used to generate the plots shown in Figure 2.6(a)–(d). The top two plots are the relative magnitude and phase between $f_1(t)$ and $x_1(t)$. The bottom two plots are the relative magnitude and phase between $f_1(t)$ and $x_2(t)$. There are several important characteristics to observe in these plots:

- The peaks occur near the damped natural frequencies calculated in Equations (2.7a) and (2.7b). These are the *resonant frequencies* where the component responds with large *amplitudes*. Note that the frequency axis f is in units of Hertz (cycles/s), which can be converted to frequency ω in rad/s with the expression $\omega = 2\pi f$. For example, the first peak is at 3.2 rad/s, or 0.5 Hz, as shown in Figure 2.6(a).
- Changes in the phase plots occur at these same resonant frequencies. In general, responses lag behind the excitation forces more as they pass through more resonances.
- The nature of the forced response at the two DOFs $x_1(t)$ and $x_2(t)$ depends on the excitation force. If the excitation force contains one frequency at 1 Hz, then the responses will contain one frequency at 1 Hz; if the force contains frequencies between 0.2 and 2 Hz, then the responses will also contain those frequencies.
- The FRF magnitude at the second resonant frequency is smaller than at the first resonant frequency; therefore, if the excitation spans a broad frequency range, the first mode of vibration would usually dominate in the response of this component.

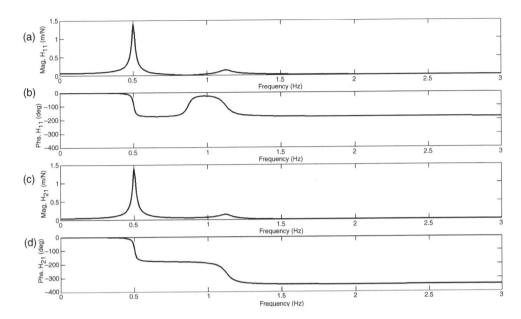

Figure 2.6 (a) Magnitude and (b) phase of $H_{11}(\omega)$ relating f_1 to x_1, and (c) magnitude and (d) phase of $H_{21}(\omega)$ relating f_1 to x_2

- The FRF magnitude at low frequencies below 0.25 Hz is dominated by the static stiffness of the component; the magnitude at resonant frequencies is dominated by the damping; and the magnitude at high frequencies beyond 1.5 Hz is dominated by the mass.

   As in the natural frequency plots in Figure 2.4(a) and (b), which show that the first mode of vibration was insensitive to changes in stiffness K_2, FRFs can also be used to determine which frequencies in the response are sensitive to which types of damage. For example, consider the result in Figure 2.7, which shows how the magnitude and phase of $H_{11}(\omega)$ changes when K_2 changes from 20 to 15 N/m (25% decrease). These plots were generated using the simulation code 'twodoffrf.m'. This type of damage might occur because of a delamination in a sandwich material causing a reduction in elasticity around the delaminated area. The magnitude plot shows that the FRF does not vary below 0.75 Hz when K_2 changes. This result suggests that FRF measurements below 0.75 Hz would not be useful in detecting damage in the region of the component connecting the two DOFs. In contrast, the magnitude and phase plots show that the frequency range between 0.75 and 1.5 Hz is sensitive to this type of damage. When the stiffness decreases, the second resonant frequency decreases as well. This simple type of model is useful for selecting the frequency range of an excitation for damage detection or for developing a damage detection algorithm that analyzes frequency response data.

   Other types of damage can also be investigated using FRF models. For example, consider oxidation damage, which causes a change in mass due to the loss in density of a material undergoing high-temperature heating for long periods of time. Figure 2.8 shows the $H_{11}(\omega)$ magnitude and phase plots for a 25% loss in mass of M_2 from 1 to 0.75 kg. For this type of damage, both of the resonant peaks shift upward due to the decrease in mass. This result suggests that FRF measurement data throughout most of the frequency range shown in the plot could be used to detect oxidation damage.

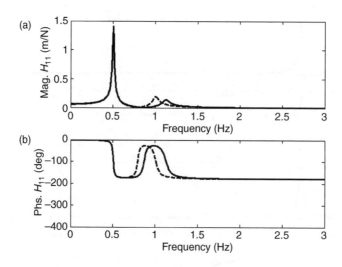

Figure 2.7   (a) Magnitude and (b) phase of $H_{11}(\omega)$ with $K_2 = 20\,\text{N/m}$ (_) and $K_2 = 15\,\text{N/m}$ (- - -) showing change with stiffness due to loss in elasticity (delamination)

2.2.5 Impedance and Compliance Models

In many applications, FRFs can be used to model the effects of coupling among different components in a system. For example, coupling can occur between sensors and the specimens on which sensors are mounted. Consider the two-DOF system in Figure 2.2(a). Suppose that a sensor is mounted on mass M_2 to measure its response as shown in Figure 2.9(a). The sensor has mass (M_s), damping (C_s) and stiffness (K_s). Figure 2.9(b) shows two different types of acceleration sensors mounted to a wheel using a superglue adhesive. In some cases, sensors do

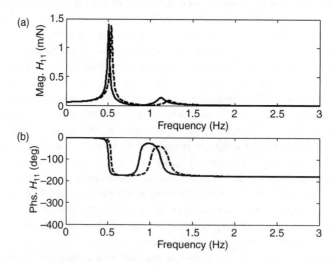

Figure 2.8   (a) Magnitude and (b) phase of $H_{11}(\omega)$ with $M_2 = 1\,\text{kg}$ (_) and $M_2 = 0.75\,\text{kg}$ (- - -) showing change with mass due to loss in density (oxidation)

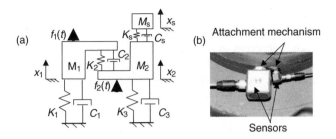

Figure 2.9 (a) Two lumped masses undergoing dynamic forcing with sensor installed on DOF 1 and (b) typical acceleration sensor with attachment method

not affect the component response, but in other cases, sensors do influence the component response leading to measurement errors. For instance, the small sensor shown on the right in Figure 2.9(b) would probably not affect the component response at low frequencies because the mass of the sensor is very small relative to the component. In contrast, this same sensor would affect the response at higher frequencies where even the small mass of the sensor influences the component response. The stiffness of the attachment mechanism can also influence the component response. The effects that one component has on another component's response can also be modeled in suspension systems and attached panels like the one shown in Figure 2.1.

If a model (first principle or data driven) is already available for the component to which a sensor will be attached (as was done in Equation (2.4) for the two-DOF system), then sensor models can be incorporated using the methods described in this section. Two approaches can be used. In the first approach called the *impedance method*, the effects of the added component are incorporated into the impedance model of the existing component. For the two-DOF system, Equation (2.12b) contains the impedance model. In order to add in the effects of the sensor, the sensor and attachment mechanism must be modeled. Figure 2.10 shows the

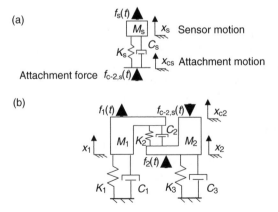

Figure 2.10 (a) Simplified model of sensor and attachment mechanism, and (b) a two-DOF system model showing forces and motions associated with the attached sensor

schematic for a simplified model of the sensor being considered here. There is an attachment motion, x_{cs}, for the sensor and a sensor motion, x_s. A force, $f_{c\text{-}s,2}$, is also shown acting on the sensor attachment. The subscripts in $_{c\text{-}s,2}$ denote the connection force by mass M_2 acting on the connection to mass M_s. A second force, f_s, is included in the schematic because in some applications, actuators attached to the components produce actuation forces. The impedance model for the sensor is given by

$$(-\omega^2 M_s + j\omega C_s + K_s)X_s = F_s + (j\omega C_s + K_s)X_{cs} \tag{2.14a}$$

$$\text{where } F_{c\text{-}s,2} = (j\omega C_s + K_s)(X_{cs} - X_s)^{\cdot} \tag{2.14b}$$

Having modeled the sensor and attachment, the model in Equation (2.14) can now be combined with the two-DOF system model by enforcing the following two constraints:

$$F_{c\text{-}s,2} = F_{c\text{-}2,s} \tag{2.15a}$$

$$X_{cs} = \overset{\cdot}{X}_{c2} \tag{2.15b}$$

The phasor $F_{c\text{-}2,s}$ denotes the connection force acting on mass M_2 by the sensor. The phasor X_{c2} denotes the motion of the connection on mass M_2. These two constraints ensure that Newton's Third Law (action–reaction) is satisfied and that the sensor attachment does not come loose. When these two constraints are enforced in the impedance model for the two-DOF system, the following impedance model is obtained for the resulting three-DOF system with the sensor attached:

$$\begin{bmatrix} -\omega^2 M_1 + j\omega(C_1 + C_2) + K_1 + K_2 & -j\omega C_2 - K_2 & 0 \\ -j\omega C_2 - K_2 & -\omega^2 M_2 + j\omega(C_2 + C_3 + C_s) + K_1 + K_2 + K_s & -j\omega C_s - K_s \\ 0 & -j\omega C_s - K_s & -\omega^2 M_s + j\omega C_s + K_s \end{bmatrix} \begin{Bmatrix} X_1 \\ X_2 \\ X_s \end{Bmatrix} = \begin{Bmatrix} F_1 \\ F_2 \\ F_s \end{Bmatrix} \tag{2.16}$$

The impedance matrix for this set of equations can then be inverted to obtain the FRFs for the two-DOF system with a sensor attached. Figure 2.11 shows the magnitudes of the two FRFs between the excitation force f_1 and the responses x_1 and x_2 with (- - -) and without (_) a sensor attached to M_2. The simulation code 'twodofwsensor.m' was used to generate these plots. The sensor has a mass that is 5% of M_2 and the attachment stiffness is five times the stiffness of the spring K_2. Note the shift in resonant frequencies downward due to the mass of the sensor. This type of shift is not uncommon especially at higher frequencies of excitation where even small masses produce significant inertia forces. This result suggests that health-monitoring methods must take into account the effects that sensors have on components.

In the second approach called the *compliance method*, the effects of the added component are incorporated directly into the frequency response model of the existing component. The same steps are taken as in the impedance approach. The sensor FRF model is developed first. All of the FRFs between both forces applied to the sensor model and the corresponding displacements must be found. Equations (2.14a) and (2.14b) provide the means to calculate

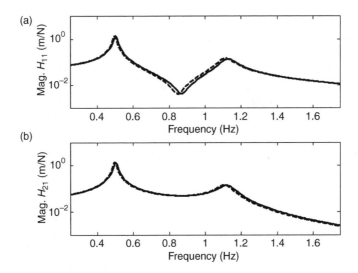

Figure 2.11 Magnitude of (a) $H_{11}(\omega)$ and (b) $H_{21}(\omega)$ for sensor with 5% mass of M_2 and attachment with five times the stiffness of K_2 showing effects on FRFs with (- - -) and without (—) sensor

these FRFs as demonstrated in the following equations:

$$\begin{bmatrix} -\omega^2 M_s + j\omega C_s + K_s & -j\omega C_s - K_s \\ -j\omega C_s - K_s & j\omega C_s + K_s \end{bmatrix} \left\{ \begin{array}{c} X_s(\omega) \\ X_{cs}(\omega) \end{array} \right\} = \left\{ \begin{array}{c} F_s(\omega) \\ F_{c-s,2}(\omega) \end{array} \right\} \tag{2.17a}$$

$$\left\{ \begin{array}{c} X_s(\omega) \\ X_{cs}(\omega) \end{array} \right\} = \begin{bmatrix} H_{ss}(\omega) & H(\omega) \\ H_{cs,s}(\omega) & H_{cs,c}(\omega) \end{bmatrix} \left\{ \begin{array}{c} F_s(\omega) \\ F_{c-s,2}(\omega) \end{array} \right\} \tag{2.17b}$$

The FRFs in Equation (2.17b) are found by inverting the matrix on the left hand side of Equation (2.17b). Then the corresponding FRF model for the two-DOF system is written as in the previous section; however, the force and motion at the point where the sensor connects to M_2 must now be included in the equations:

$$\left\{ \begin{array}{c} X_1(\omega) \\ X_2(\omega) \\ X_{c2}(\omega) \end{array} \right\} = \begin{bmatrix} H_{11}(\omega) & H_{12}(\omega) & H_{1c}(\omega) \\ H_{21}(\omega) & H_{22}(\omega) & H_{2c}(\omega) \\ H_{21}(\omega) & H_{22}(\omega) & H_{2c}(\omega) \end{bmatrix} \left\{ \begin{array}{c} F_1(\omega) \\ F_2(\omega) \\ F_{c-2,s}(\omega) \end{array} \right\} \tag{2.18}$$

In this set of equations, the third row is identical to the second row because the motion X_{c2} is the same as X_2. Finally, Equations (2.17b) and (2.18) are combined by applying the constraints in Equations (2.15a) and (2.15b). The objective in combining these equations is to eliminate the connection forces, $F_{c-2,s} = F_{c-s,2}$, and motions, $X_{c2} = X_{cs}$. This process leads to a solution for the connection force:

$$F_{c-2,s}(\omega) = \frac{H_{21}(\omega)F_1(\omega) + H_{22}(\omega)F_2(\omega) - H_{cs,s}(\omega)F_s(\omega)}{H_{cs,c}(\omega) - H_{2c}(\omega)}, \tag{2.19}$$

which can be substituted into the first line of Equation (2.17a) and the first two lines of Equation (2.18) to obtain the following FRF compliance model:

$$\left\{ \begin{array}{c} X_1(\omega) \\ X_2(\omega) \\ X_s(\omega) \end{array} \right\} = \left[\begin{array}{ccc} H_{11} + \dfrac{H_{1c}H_{21}}{H_{cs,c} - H_{2c}} & H_{12} + \dfrac{H_{1c}H_{22}}{H_{cs,c} - H_{2c}} & -\dfrac{H_{1c}H_{cs,s}}{H_{cs,c} - H_{2c}} \\[3mm] H_{21} + \dfrac{H_{2c}H_{21}}{H_{cs,c} - H_{2c}} & H_{22} + \dfrac{H_{2c}H_{22}}{H_{cs,c} - H_{2c}} & -\dfrac{H_{2c}H_{cs,s}}{H_{cs,c} - H_{2c}} \\[3mm] \dfrac{H_{sc}H_{21}}{H_{cs,c} - H_{2c}} & \dfrac{H_{sc}H_{22}}{H_{cs,c} - H_{2c}} & H_{ss} - \dfrac{H_{sc}H_{cs}}{H_{cs,c} - H_{2c}} \end{array} \right] \left\{ \begin{array}{c} F_1(\omega) \\ F_2(\omega) \\ F_s(\omega) \end{array} \right\}.$$

$$(2.20)$$

This FRF model can be used in ways similar to the impedance model to predict how sensors will affect the FRF of a component. This type of approach can also be used to determine how changes in the boundary conditions of a component (how a panel is mounted, how a link is attached, etc.) will affect the frequency response of that component. As described in subsequent chapters, many applications involve changes in the boundary conditions of the component being monitored. Impedance and compliance models developed in this section can be utilized to determine the effects of those changes. In addition, the compliance approach *can be used experimentally when only the FRF measurements are available*. On the contrary, impedance modeling can only be applied when all mass, damping and stiffness parameters are known.

2.2.6 Transmissibility Forced Response Models

The frequency response function model in Section 2.2.4 was used to relate external forces to component steady-state vibration responses. In many applications, external forces are not measured; therefore, a model that relates the vibration response at one DOF to the response at another DOF is useful (Johnson, 2002). This type of model is called a *transmissibility* model and will be derived in this section.

Consider the three-DOF model shown in Figure 2.12. This model is an extension of the two-DOF model already examined. Suppose that the spring K_2 decreases due to damage in that particular region of the component. Although an FRF model could detect this type of damage, FRFs are not generally useful for locating damage except in the simplest components. Transmissibility models are useful for locating damage because they are *only functions of the elements in the transmission path connecting two DOFs* (Johnson and Adams, 2002).

To examine this characteristic of transmissibility functions, the equations of motion for the three-DOF model in Figure 2.12 are derived. Then the FRF model is calculated and the ratios

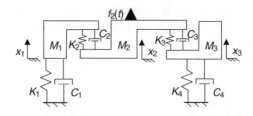

Figure 2.12   Model of three lumped masses undergoing dynamic forcing

of these FRFs are formed to calculate transmissibility relationships. The equations of motion for the three-DOF model are given by

$$
\begin{bmatrix} M_1 & 0 & 0 \\ 0 & M_2 & 0 \\ 0 & 0 & M_3 \end{bmatrix} \begin{Bmatrix} \ddot{x}_1 \\ \ddot{x}_2 \\ \ddot{x}_3 \end{Bmatrix} + \begin{bmatrix} C_1 + C_2 & -C_2 & 0 \\ -C_2 & C_2 + C_3 & -C_3 \\ 0 & -C_3 & C_3 + C_4 \end{bmatrix} \begin{Bmatrix} \dot{x}_1 \\ \dot{x}_2 \\ \dot{x}_3 \end{Bmatrix}
$$

$$
+ \begin{bmatrix} K_1 + K_2 & -K_2 & 0 \\ -K_2 & K_2 + K_3 & -K_3 \\ 0 & -K_3 & K_3 + K_4 \end{bmatrix} \begin{Bmatrix} x_1 \\ x_2 \\ x_3 \end{Bmatrix} = \begin{Bmatrix} 0 \\ f_2(t) \\ 0 \end{Bmatrix} \quad \text{(2.21a, b)}
$$

$$
[M]\{\ddot{x}\} + [C]\{\dot{x}\} + [K]\{x\} = \{f\}
$$

There are three mass parameters and four damping and stiffness parameters, and three DOFs in these equations of motion. The steps in Section 2.2.4 can be followed to calculate the FRF matrix for this set of equations. The result of carrying out these steps is given in the following equation:

$$
\begin{Bmatrix} X_1(\omega) \\ X_2(\omega) \\ X_3(\omega) \end{Bmatrix} = \begin{bmatrix} H_{11}(\omega) & H_{12}(\omega) & H_{13}(\omega) \\ H_{21}(\omega) & H_{22}(\omega) & H_{23}(\omega) \\ H_{31}(\omega) & H_{32}(\omega) & H_{33}(\omega) \end{bmatrix} \begin{Bmatrix} 0 \\ F_2(\omega) \\ 0 \end{Bmatrix}. \quad \text{(2.22)}
$$

The FRF matrix is a 3 by 3 matrix. The excitation force vector only has one nonzero entry because the other two forces are zero in the schematic diagram. The individual entries, $H_{pq}(\omega)$, of this matrix can be found by inverting the impedance matrix, $-\omega^2[M] + j\omega[C] + [K]$, for the three-DOF model. However, the entries in the first and third columns of the FRF matrix do not need to be calculated because the excitation force is applied at DOF 2 (see Figure 2.12). The individual entries of the second column of the FRF matrix are found by computing the adjoint matrix of $-\omega^2[M] + j\omega[C] + [K]$, taking the transpose and then dividing by the determinant of the impedance matrix. The entries in the second column of the FRF matrix are found from this procedure to be

$$
H_{12}(\omega) = \frac{-(j\omega C_2 + K_2)(-\omega^2 M_3 + j\omega(C_3 + C_4) + K_3 + K_4)}{\Delta(\omega)} \quad \text{(2.23a)}
$$

$$
H_{22}(\omega) = \frac{(-\omega^2 M_1 + j\omega(C_1 + C_2) + K_1 + K_2)(-\omega^2 M_3 + j\omega(C_3 + C_4) + K_3 + K_4)}{\Delta(\omega)} \quad \text{(2.23b)}
$$

$$
H_{31}(\omega) = \frac{(j\omega C_3 + K_3)(-\omega^2 M_1 + j\omega(C_1 + C_2) + K_1 + K_2)}{\Delta(\omega)} \quad \text{(2.23c)}
$$

where the expression $\Delta(\omega)$ is the determinant of the impedance matrix. There are two important characteristics to note in these FRF expressions:

(1) Each FRF has $\Delta(\omega)$ in its denominator. $\Delta(\omega)$ is a function of every mass, damping and stiffness parameter in the three-DOF model; therefore, every FRF will change when the system is damaged, no matter where the damage is located.

(2) The ratios of FRFs do not contain $\Delta(\omega)$; $\Delta(\omega)$ cancels out when the ratio is taken between two FRFs. This result suggests that ratios of FRFs, called transmissibility functions, will not be a function of all of the system parameters; therefore, transmissibility functions can be used to isolate (locate) the effects of damage.

To demonstrate characteristic (2), the ratios of the FRFs in Equations (2.23a)–(2.23c) can be computed as follows:

$$T_{21}(\omega) = \frac{H_{22}(\omega)}{H_{12}(\omega)} = \frac{-\omega^2 M_1 + j\omega(C_1 + C_2) + K_1 + K_2}{j\omega C_2 + K_2} \qquad (2.24a)$$

$$T_{32}(\omega) = \frac{H_{32}(\omega)}{H_{22}(\omega)} = \frac{j\omega C_3 + K_3}{-\omega^2 M_3 + j\omega(C_3 + C_4) + K_3 + K_4} \qquad (2.24b)$$

$$T_{31}(\omega) = \frac{H_{32}(\omega)}{H_{12}(\omega)} = \frac{(j\omega C_3 + K_3)(-\omega^2 M_1 + j\omega(C_1 + C_2) + K_1 + K_2)}{(-\omega^2 M_3 + j\omega(C_3 + C_4) + K_3 + K_4)(j\omega C_2 + K_2)} \qquad (2.24c)$$

$T_{pq}(\omega)$ is the transmissibility function between DOFs q and p. This function describes the relative response amplitude and phase between these two DOFs. The simulation code 'threedoftrans.m' was used to generate plots of the magnitudes of these three functions. Figure 2.13 shows the transmissibility magnitudes for $T_{21}(\omega)$, $T_{32}(\omega)$ and $T_{31}(\omega)$. Unlike the FRF magnitude plots discussed previously, the transmissibility plots do not exhibit one peak

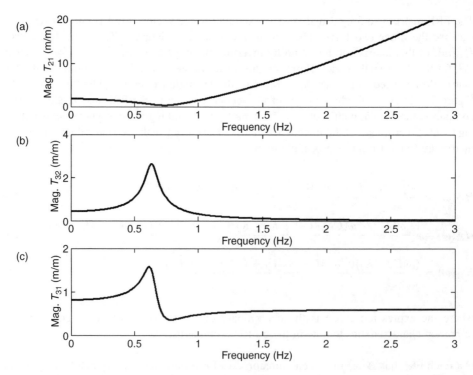

Figure 2.13 Magnitudes of transmissibility functions (a) $T_{21}(\omega)$, (b) $T_{32}(\omega)$ and (c) $T_{31}(\omega)$ with units of m/m for excitation applied at DOF 2

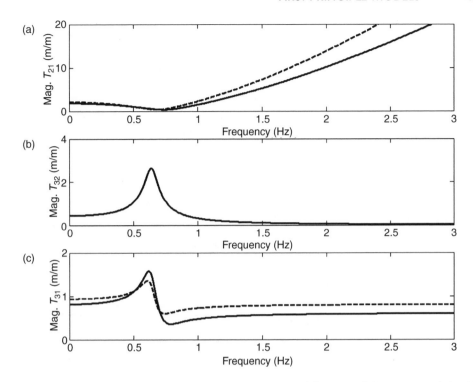

Figure 2.14 Magnitudes of transmissibility functions (a) $T_{21}(\omega)$, (b) $T_{32}(\omega)$ and (c) $T_{31}(\omega)$ with $K_2 = 12\,\text{N/m}$ (—) and $K_2 = 9\,\text{N/m}$ (- - -) showing change with stiffness due to loss in elasticity (delamination) for excitation applied at DOF 2

for each DOF in the model. The peaks in transmissibility plots are not actually resonances of the system, but are *frequencies of high transmission* from one DOF to another.

Equations (2.24a)–(2.24c) suggest that transmissibility functions can be useful for locating damage using health-monitoring data. In general, $T_{pq}(\omega)$ is only a function of the mass, damping and stiffness parameters that are located in the path connecting DOFs q and p. For example, $T_{21}(\omega)$ is not a function of C_3 and K_3 because those elements are not between DOFs 1 and 2. To illustrate this characteristic of transmissibility functions, consider a damage mechanism such as a delamination where K_2 changes from 12 to 9N/m. Figure 2.14 shows the three transmissibility functions before (–) and after (- - -) this change occurs due to damage. Note that the first and third transmissibility plots change when K_2 changes because the paths connecting DOFs 1 and 2, and DOFs 1 and 3 contain K_2. On the contrary, the second plot corresponding to the $T_{32}(\omega)$ transmissibility function from Equation (2.24b) does not vary when K_2 changes because K_2 is not between DOFs 2 and 3. This result illustrates that the changes observed in transmissibility functions can be used to locate damage in components.

Transmissibility functions can also be used to locate damage associated with changes in mass properties (density). For example, Figure 2.15 shows the same three transmissibility function magnitudes before (–) and after (- - -) M_3 changes from 1.1 to 0.8kg. As in the previous case for stiffness, the transmissibility function $T_{21}(\omega)$ does not change when M_3 changes because M_3 is not in the path connecting DOFs 1 and 2.

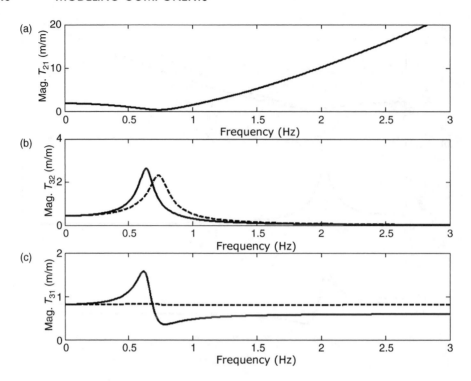

Figure 2.15 Magnitudes of transmissibility functions (a) $T_{21}(\omega)$, (b) $T_{32}(\omega)$ and (c) $T_{31}(\omega)$ with $M_3 = 1.1\,\mathrm{kg}$ (—) and $M_3 = 0.8\,\mathrm{kg}$ (- - -) showing change with mass due to loss in density (oxidation)

The results in this section for transmissibility were based on an excitation force applied at DOF 2. Depending on where the excitation force is applied, these results vary somewhat. In general, if the excitation is not applied at the boundary of a system, the conclusions drawn in this section will apply.

2.2.7 Nonlinear Dynamic Models

The models described in the previous sections contained linear damping and stiffness elements. In linear models, the damping and stiffness forces vary proportionally with relative velocity and displacement coordinates. Many components exhibit damping and stiffness forces that vary nonproportionally with the response motions. In some cases, nonlinearities are present before damage occurs; in other cases, nonlinearity appears after the component becomes damaged (Adams and Farrar, 2002).

For example, if the model in Figure 2.2(a) describes a panel that is mechanically attached to a substructure, the fastener used to attach the panel might exhibit Coulomb friction type damping nonlinearities due to rubbing of the interface between the fastener and the panel. As the fastener loosens, the Coulomb friction effects would escalate due to increased rubbing as the two surfaces slide even more. In order to model and simulate these nonlinearities, numerical methods are needed. Consider the model in Figure 2.2(a) with Coulomb friction nonlinearity at the base of the attachment spring, K_3. The equations of motion of the panel with this nonlinearity are given below:

$$[M]\{\ddot{x}\} + [C]\{\dot{x}\} + [K]\{x\} = \{f\} + \{f_n\} \tag{2.25}$$

where the nonlinear force, $\{f_n\}$, due to friction is given by

$$\{f_n\} = \left\{ \begin{array}{c} 0 \\ -\mu \mathrm{sgn}(\dot{x}_2) \end{array} \right\} \tag{2.26}$$

The matrices in Equation (2.25) were provided in Equation (2.4a). The nonlinear force has one zero entry and one nonlinear entry, which is a function of the velocity of M_2. The sgn($\cdot$) function is the *signum* function, which is defined as follows:

$$\mathrm{sgn}(\dot{x}_2) = \left\{ \begin{array}{l} +1 \text{ for } \dot{x}_2 > 0 \\ -1 \text{ for } \dot{x}_2 < 0 \end{array} \right. . \tag{2.27}$$

The coefficient μ in Equation (2.26) determines the magnitude of the Coulomb friction force. When this nonlinear friction force acts in addition to the linear viscous damping force due to C_3 in the two-DOF model, the force characteristic in Figure 2.16 is obtained. The contributions from linear and nonlinear damping forces are both evident in this plot.

In order to calculate the FRFs for this nonlinear two-DOF model, a numerical simulation must be carried out. FRFs in nonlinear systems change when the level of the excitation force and response amplitude change. Figure 2.17 is a block diagram of a Simulink® model 'two_dof_model_nl.mdl' provided in the Appendix. This model was run using the MATLAB® simulation code 'twodofnlfrf.m'. The FRFs between the excitation force at M_1 and the responses of M_1 and M_2 are shown in Figure 2.18(a) and (b). Note that these FRF magnitudes are plotted on a log axis causing a change in their appearance from the plots in Figure 2.6. The solid curves (–) are FRF magnitude plots where no Coulomb friction force is included in the simulation ($\mu = 0\,\mathrm{N}$). The dashed curves (- - -) correspond to a case when

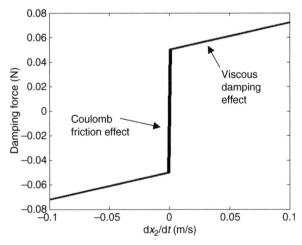

Figure 2.16 Total damping force versus velocity of M_2 in two-DOF model with Coulomb friction in the joint

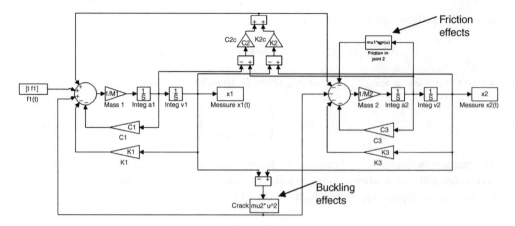

Figure 2.17 Simulink® model[6] for numerical simulation of nonlinear two-DOF model

Coulomb friction is included in the simulation ($\mu = 0.05$ N). Note the significant differences between the two sets of FRFs near the resonant peaks. These differences could be utilized to detect loosening of the bolted fastener that attaches M_2 to the substructure.

The loose fastener example introduced a nonlinear damping force; however, damage due to delamination, cracking and other mechanisms introduce nonlinear stiffness forces as well. Consider the scenario illustrated in Figure 2.19(a)–(c). A panel is shown in Figure 2.19(a) after it has been heated and suddenly cooled causing buckling because of residual stresses in the panel. This scenario could also occur if a composite component were impacted causing plastic deformation.

The two DOFs M_1 and M_2 used to model the panel vibration are shown undergoing relative motions in Figure 2.19(b) and (c). In Figure 2.19(b) the relative motions cause the buckling to

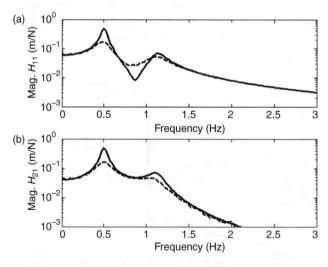

Figure 2.18 Magnitudes of FRFs between f_1 and (a) x_1, and (b) x_2 with (- - -) and without (__) the nonlinear Coulomb friction force show drops at resonant peaks

---

[6]This Simulink® model is described in detail in the Appendix; the left hand side of the model describes the equation of motion for DOF 1 and the right hand side describes the equation of motion for DOF 2.

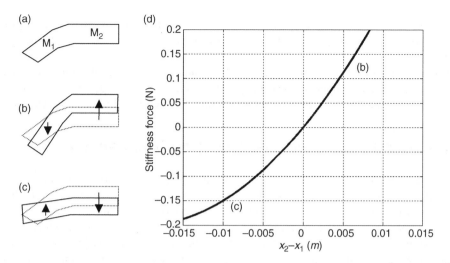

Figure 2.19 (a) Buckled panel due to residual stresses, (b) panel vibration causes increased buckling of material, (c) vibration causes decreased buckling and (d) plot of stiffness force during these motions as a function of relative displacement

increase leading to stiff resistance forces. In Figure 2.19(c) the motions cause the buckling to decrease leading to less stiffening and lower forces. This asymmetry in the stiffness during vibration is captured by the stiffness force shown in Figure 2.19(d). For positive x_2-x_1, the stiffness grows; for negative x_2-x_1, the stiffness drops. A quadratic stiffness force, $\mu \cdot (x_2 - x_1)^2$ N, is used to model this behavior. The 'twodofnlfrf.m' simulation code was used to calculate the FRFs for the two-DOF model subject to this type of nonlinearity. The results of this simulation are shown in Figure 2.20. As in the previous nonlinear FRF plots, the linear FRFs are shown with solid curves (–), and the nonlinear FRFs are shown with dashed

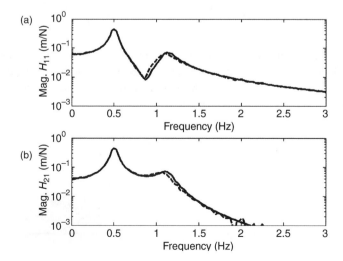

Figure 2.20 Magnitudes of FRFs between f_1 and (a) x_1, and (b) x_2 with (- - -) and without (_) nonlinear buckling stiffness force show changes near second resonant peak

curves (- - -). The shift at the second resonance occurs because buckling causes a reduction in the effective stiffness that resists the component response in the second modal deflection shape.

There are many other methods for modeling and simulating the dynamic behavior of nonlinear structural components (Worden and Tomlinson, 2001). Several other methods appropriate for nonlinear modeling will be described in Section 2.3 on data-driven models.

2.2.8 Wave Propagation Models (One Dimensional)

The models developed in the previous sections described the vibrations of components to external forces. Vibrations can be thought of as a sum of natural frequencies and mode shapes, which are sensitive to changes in components due to damage. Vibrations can also be thought of as *interference patterns* set up by propagating waves as they travel through components. These propagating waves are useful for health monitoring because they encounter damage along the propagation path helping to detect and locate damage. In this section, one-dimensional elastic waves are discussed for homogeneous material components.

2.2.8.1 Analytical Solution for Longitudinal Waves

The one-dimensional component of length L shown in Figure 2.21(a) has volumetric density $\rho(x)$ and modulus $E(x)$, which can both vary with position x. This component could be a stabilizer bar link in an automotive suspension system (Figure 2.21(b)), a crank arm in an engine, or a cable stay on a bridge. The longitudinal displacement of the component cross section with area $A(x)$ at position x along the component is $u(x,t)$. A force is not included in the diagram because usually forces are equated to boundary conditions using the relationship, $f(t) = -EA\varepsilon(x_o, t)$, where $\varepsilon(x_o, t) = \partial u(x,t)/\partial x|_{x=x_o}$ is the strain at the location x_0 where the force is applied. When Newton's Second Law is applied to a small section of the component at position x, the equation of motion is

$$\frac{\partial}{\partial x}\left(EA(x)\frac{\partial u(x,t)}{\partial x}\right) = \rho(x)A(x)\frac{\partial^2 u(x,t)}{\partial t^2} \tag{2.28}$$

This partial differential equation of motion is called the *wave equation*. E, ρ and A usually vary along the length due to component geometry or damage; however, the fundamental aspects of wave propagation can be understood by assuming these parameters are constant. The simplified wave equation can then be written as follows:

$$\frac{\partial^2 u(x,t)}{\partial x^2} = \frac{1}{E/\rho}\frac{\partial^2 u(x,t)}{\partial t^2} = \frac{1}{c^2}\frac{\partial^2 u(x,t)}{\partial t^2} \tag{2.29}$$

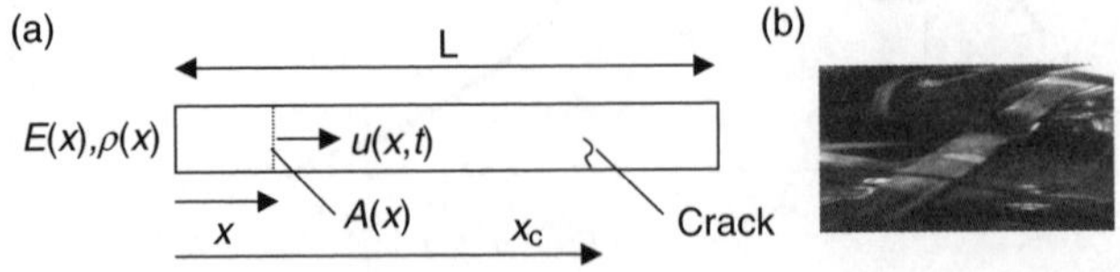

Figure 2.21 (a) One-dimensional model of (b) suspension link with longitudinal motion

The ratio of modulus to density is defined to be c^2, where c is the longitudinal *speed of sound* in the component:

$$c = \sqrt{E/\rho} \text{ units m/s} \tag{2.30}$$

Equation (2.29) is solved using the *separation of variables* method by substituting a solution for $u(x, t)$ that contains two parts: one part $(H(x))$ that varies with x along the component and a second part $(G(t))$ that varies with time t. This solution is given below:

$$u(x, t) = H(x)G(t) \tag{2.31}$$

When this expression is substituted into the wave equation, the two parts of the solution separate out as follows into two ordinary differential equations:

$$\frac{\partial^2 (H(x)G(t))}{\partial x^2} = \frac{1}{c^2} \frac{\partial^2 (H(x)G(t))}{\partial t^2} \tag{2.32a}$$

$$\frac{c^2}{H(x)} \frac{d^2 H(x)}{dx^2} = \frac{1}{G(t)} \frac{d^2 G(t)}{dt^2} \tag{2.32b}$$

The left hand side of Equation (2.32b) is only a function of x, and the right hand side is only a function of t; therefore, both sides of the equation must be equal to the same constant. This constant is chosen to be $-\omega^2$, the circular frequency in rad/s:

$$\frac{c^2}{H(x)} \frac{d^2 H(x)}{dx^2} = -\omega^2 = \frac{1}{G(t)} \frac{d^2 G(t)}{dt^2} \tag{2.33}$$

Then these two equations are solved for $H(x)$ and $G(t)$:

$$\frac{d^2 H(x)}{dx^2} + \frac{\omega^2}{c^2} H(x) = 0 \Rightarrow H(x) = B_1 e^{j\frac{\omega}{c}x} + B_2 e^{-j\frac{\omega}{c}x} \tag{2.34a}$$

$$\frac{d^2 G(t)}{dt^2} + \omega^2 G(t) = 0 \Rightarrow G(t) = B_3 e^{j\omega t} + B_4 e^{-j\omega t} \tag{2.34b}$$

Finally, these solutions are substituted back into Equation (2.31):

$$u(x, t) = B_5 e^{j(\omega t - \frac{\omega}{c}x)} + B_6 e^{j(\omega t + \frac{\omega}{c}x)} \tag{2.35a}$$

$$= B_5 e^{j(\omega t - kx)} + B_6 e^{j(\omega t + kx)} \tag{2.35b}$$

where B_5 and B_6 are combinations of the B_k coefficients in Equations (2.34a) and (2.34b), and k is called the *wave number* and is equal to $\omega/c.k = \omega/c$ is called the *dispersion relation*. c is called the *phase velocity* of a wave at frequency ω. If there are groups of frequencies in a wave front, then the *group velocity* c_g is found from $c_g = d\omega/dk$. Note that the longitudinal group velocity is constant for the rod. With units of rad/m, the wave number can be thought of as a *spatial frequency*; that is k indicates the number of periods in a wave that span a certain distance. Also note that $c = f\lambda$, where f is the frequency in Hz and λ is the wavelength with $\lambda = 2\pi/k$. B_k are found by applying initial and boundary conditions.

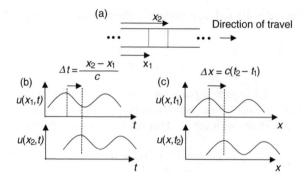

Figure 2.22 (a) Component extending to infinity in both directions and (b,c) example wave traveling to the right with speed c

Equation (2.35b) shows that every solution to the wave equation is the sum of two traveling waves. The wave from the $e^{j(\omega t - kx)}$ term travels to the right; the wave from the $e^{j(\omega t + kx)}$ term travels to the left. Waves are established by excitations applied to the component. Consider the case where waves travel in a component that is infinitely long in both directions. This scenario is considered first because it avoids reflections that complicate the waveforms. Figure 2.22 shows an example wave at one frequency traveling to the right as in the first term of Equation (2.35b). In Figure 2.22(a), two different positions along the component are indicated. Then in (b), the longitudinal displacement is shown as a function of time. It is seen that as the wave pushes material to the right at position x_1, the material to the right at position x_2 responds later in time by an amount $(x_2 - x_1)/c$. The opposite is true of the wave traveling to the left. In (c), the wave is observed to travel from time t_1 to t_2 over the distance $c(t_2 - t_1)$. Two important things to note are that (1) *the wave in* Figure 2.22(b)and(c) *travels without changing shape and* (2) *every frequency wave travels with the same speed (equal to the phase velocity in this case).* A wave like this one that travels the same speed regardless of frequency is called *nondispersive*. In addition to being nondispersive, the other reason the wave in Figure 2.22 maintains its shape is that damping was not considered in the wave equation. If the component is finite with boundaries, then waves traveling to the left add to waves traveling to the right producing *standing waves*. These standing waves produce vibrations, which were examined in previous sections. Also note that when λ is small, smaller defects in the component can be detected because waves interact more with defects in the component.

2.2.8.2 Longitudinal Wave Propagation Finite Element Model

Although the wave equation can be solved analytically for traveling and standing waves when E, ρ and A are constant, in health-monitoring applications, it is necessary to model components where these parameters vary due to either the geometry of the component or damage in the component. For example, a crack at location x_c in Figure 2.21 changes the cross-sectional area of the component at that location. In order to model nonuniform potentially damaged one-dimensional components like this one, a discretized model of the component can be used instead of the continuous model.

To develop this discrete model of the component, rod elements like the one shown in Figure 2.23(a) are considered. This rod element is required to have constant cross-sectional

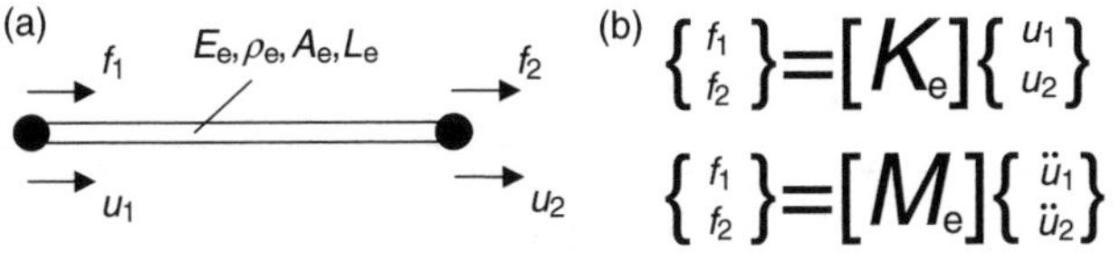

(a)

f_1 E_e, ρ_e, A_e, L_e f_2

u_1 u_2

(b) $\left\{ \begin{matrix} f_1 \\ f_2 \end{matrix} \right\} = [K_e] \left\{ \begin{matrix} u_1 \\ u_2 \end{matrix} \right\}$

$\left\{ \begin{matrix} f_1 \\ f_2 \end{matrix} \right\} = [M_e] \left\{ \begin{matrix} \ddot{u}_1 \\ \ddot{u}_2 \end{matrix} \right\}$

Figure 2.23 (a) One-dimensional rod element undergoing longitudinal motion and (b) stiffness and mass relationships that define the rod element

area, modulus and density. The density ρ for this model is in units of mass per unit length. Nonuniform components like the one in Figure 2.21 can be modeled by interconnected rods with different properties. It can be shown using strength of materials that between its two *nodes* (dark circles on either end), this rod behaves like a spring with spring constant $K_e = A_e E_e / L_e$. The subscript 'e' is used to denote element properties. Thomson and Dahleh (1998) also show that the equations relating the forces, f_1 and f_2, applied at the nodal DOFs of the rod to the displacements, u_1 and u_2, at those DOFs are

$$\left\{ \begin{matrix} f_1 \\ f_2 \end{matrix} \right\} = \begin{bmatrix} K_e & -K_e \\ -K_e & K_e \end{bmatrix} \left\{ \begin{matrix} u_1 \\ u_2 \end{matrix} \right\} \text{where } K_e = \frac{A_e E_e}{L_e} \tag{2.36}$$

Figure 2.23(b) illustrates this stiffness matrix relationship for the rod element. If it is assumed that the velocity along the element varies linearly with distance along the rod, it can also be shown that the forces applied on either end of the rod are related to the accelerations at those locations as follows:

$$\left\{ \begin{matrix} f_1 \\ f_2 \end{matrix} \right\} = \begin{bmatrix} M_{11} & M_{12} \\ M_{21} & M_{22} \end{bmatrix} \left\{ \begin{matrix} \ddot{u}_1 \\ \ddot{u}_2 \end{matrix} \right\} \text{where } M_{11} = M_{22} = \frac{\rho_e L_e}{3} \text{ and } M_{12} = M_{21} = \frac{\rho_e L_e}{6} \tag{2.37}$$

There are many different ways to model linear damping in the rod. To model viscous damping, a *proportional* viscous damping matrix can be computed as a linear combination of the mass and stiffness matrices:

$$[C] = \alpha[M] + \beta[K] \tag{2.38}$$

where α and β are real constants. In many cases involving damage, a *nonproportional* viscous damping model can be used where the equality in Equation (2.38) does not hold. *Structural damping* is also routinely used to model dissipation in solid media.

Consider the component of total length L shown in Figure 2.24(a) with a crack in the location indicated. Seventeen elements each with length $L/17$ are used to model the component as shown in Figure 2.24(b). The crack is modeled as a reduction in cross-sectional area, $A_c = 0.8A$, at element 8 where A is the nominal area and A_c is the cracked area. The density per unit length and modulus are constant. The component is supported on either end by rubber bushings that provide a stiffness K_s that resists longitudinal motions.

The finite element model of this component is assembled by combining the full mass matrix and full stiffness matrix from each of the 17 elemental models in sequence. These elemental matrices are combined into *global* matrices by enforcing the constraints given previously in

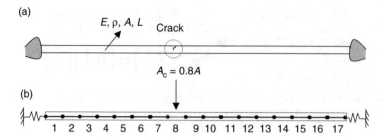

Figure 2.24   (a) Cracked component suspended by bushings on either end and (b) finite element model of component with 17 elements (crack in element 8)

Equations (2.15a) and (2.15b). Recall that the constraints require that motions at connection DOFs be equal, and forces be equal and opposite to satisfy Newton's Third Law. The procedure for assembling the finite element model of the rod is provided in the simulation code 'rod1dwave.m.' Conceptually, the matrices for all elements overlap and add along the diagonal of the total component mass and stiffness matrices. Using this methodology, the resulting mass matrix is then calculated to be

$$[M] = \begin{bmatrix} M_{11} & M_{12} & 0 & \cdots & 0 \\ M_{12} & 2M_{11} & M_{12} & \cdots & 0 \\ 0 & M_{12} & 2M_{11} & \cdots & \vdots \\ \vdots & \vdots & \vdots & \ddots & M_{12} \\ 0 & 0 & \cdots & M_{12} & M_{11} \end{bmatrix}_{17\times17} \tag{2.39}$$

and the finite element model stiffness matrix for the component is given by

$$[K] = \begin{bmatrix} K_e + K_s & -K_e & 0 & \cdots & 0 \\ -K_e & 2K_e & -K_e & \cdots & 0 \\ 0 & -K_e & 2K_e & \cdots & \vdots \\ \vdots & \vdots & \vdots & \ddots & -K_e \\ 0 & 0 & \cdots & -K_e & K_e + K_s \end{bmatrix}_{17\times17} \tag{2.40}$$

The model of the crack is buried inside of this stiffness matrix in the four entries that fill the block of rows (8,9) and columns (8,9). Those four entries are equal to

$$[K(8,9)] = \begin{bmatrix} 1.8K & -0.8K \\ -0.8K & 1.8K \end{bmatrix} \tag{2.41}$$

Figure 2.25 shows the results of running the simulation code 'rod1dwave.m'. The material used in this simulation is aluminum with modulus 70GPa and density 2700kg/m^3. The cross-sectional area used is 6.5cm^2 (1in. × 1in.). The plot in (a) is the longitudinal excitation force in Newton's applied at the left end (node 1). This force is commonly used in health

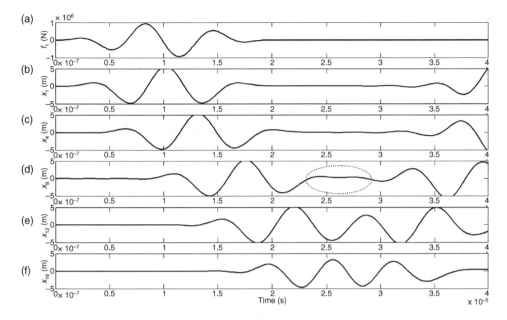

Figure 2.25 Longitudinal excitation force (a) and wave propagation response in aluminum rod (5100m/s) without damping at nodal points (b) 1, (c) 4, (d) 8, (e) 12 and (f) 16

monitoring based on wave propagation because it contains a short band of frequencies (see Section 2.2.8.4). In this example, a sinusoidal excitation pulse with frequency content starting at 40kHz and ending at 240kHz is used. The longitudinal displacement responses at nodes 1, 4, 8, 12 and 16 of the finite element model for a pulse excitation are shown in Figure 2.25(b)–(f). The entire length of time shown is only 40μs. The first thing to note is that the response at node 1 closely follows the excitation force applied because the force is applied at node 1. Then the initial wave excitation propagates from node 1 to nodes 4, 8 and 12, and finally on to node 16. The time taken by the wave to travel from one end to the other is determined by c. c can be calculated as approximately $L/20\,\mu s = 5000\,\text{m/s}$. In Figure 2.25(b)–(d), note that the waveforms are nearly identical except for shifts in time. The waveforms are the same because this wave is nondispersive and retains its shape. However, at (d) in the circled region, reflections from the boundary on the right begin to influence the waveform leading to differences in (e) and (f) in the wave shape due to reflections.

When the crack is included in the model using a reduction in cross-sectional area, the results in Figure 2.26 are obtained after subtracting the propagating waves from the component with and without the crack included. It is expected that the crack will cause a reflection in the waveform that will cause changes in the propagating waves. In Figure 2.26, the earliest difference in waveforms is seen at node 8 around 10μs (circled) because the crack is located at that node. All other nodal responses on both sides of the crack also exhibit differences; however, the differences at other nodal positions occur later in time than the difference at node 8 where the crack is located.

Finite element models of this type can also be used equally well to model vibrations, which arise as the propagating waves reflect and interfere with one another in the rod.

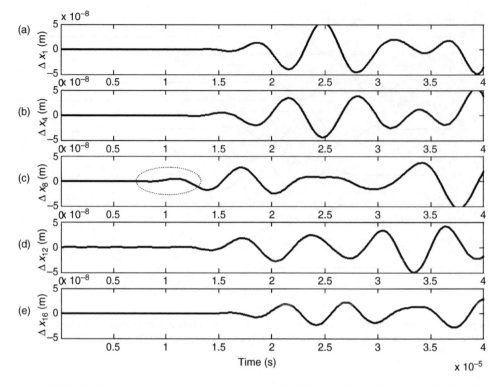

Figure 2.26 Difference in wave propagation response at nodal points (a) 1, (b) 4, (c) 8, (d) 12 and (e) 16 before and after crack is placed in element 8

2.2.8.3 Analytical Solution for Rod with Transverse Waves

The one-dimensional homogeneous material component of length L shown in Figure 2.27 is the same as the rod in Section 2.2.8.1; however, transverse (bending) motions, $u(x, t)$, are considered in this section instead of longitudinal motions. The cross-sectional area, $A(x)$, and area moment of inertia, $I(x)$, are indicated in the figure. $f(x, t)$ is a force per unit length applied to the rod in the positive displacement direction. The wave equation of motion for this type of displacement of the rod is given by

$$-\frac{\partial^2}{\partial x^2}\left(E(x)I(x)\frac{\partial^2 u(x, t)}{\partial x^2}\right) + f(x, t) = \rho(x)A(x)\frac{\partial^2 u(x, t)}{\partial t^2} \tag{2.42}$$

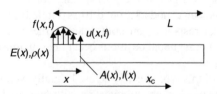

Figure 2.27 One-dimensional component undergoing transverse motion

or for constant modulus, inertia, density and area

$$-\frac{\partial^4 u(x,t)}{\partial x^4} + \frac{1}{EI}f(x,t) = \frac{\rho A}{EI}\frac{\partial^2 u(x,t)}{\partial t^2} = \frac{1}{c^2 I_A}\frac{\partial^2 u(x,t)}{\partial t^2} \qquad (2.43)$$

c is the speed of sound given by Equation (2.30). I_A is equal to I/A. Note that this equation is fourth order in $u(x,t)$. The additional second-order effect is due to bending in a direction perpendicular to the direction of wave motion. In order to determine the effects of this transverse bending on propagating waves, consider the case when $f(x,t) = 0$. The solution in Equation (2.31) leads to the following two equations for $H(x)$ and $G(t)$:

$$\frac{d^4 H(x)}{dx^4} + \frac{\omega^4}{v^4}H(x) = 0 \Rightarrow H(x) = B_1 e^{j\frac{\omega}{v}x} + B_2 e^{-j\frac{\omega}{v}x} + B_3 e^{\frac{\omega}{v}x} + B_4 e^{-\frac{\omega}{v}x} \qquad (2.44a)$$

$$\frac{d^2 G(t)}{dt^2} + \omega^2 G(t) = 0 \Rightarrow G(t) = B_5 e^{j\omega t} + B_6 e^{-j\omega t} \qquad (2.44b)$$

where $v = \sqrt{\omega c I_A}$ is the phase velocity of the wave at frequency ω.
    When these two expressions are substituted into Equation (2.31), the solution is

$$\begin{aligned} u(x,t) &= B_7 e^{j\left(\omega t - \frac{\omega}{v}x\right)} + B_8 e^{j\left(\omega t + \frac{\omega}{v}x\right)} + \left(B_9 e^{\frac{\omega}{v}x} + B_{10} e^{-\frac{\omega}{v}x}\right)e^{j\omega t} \\ &= B_7 e^{j(\omega t - k(\omega)x)} + B_8 e^{j(\omega t + k(\omega)x)} + \left(B_9 e^{k(\omega)x} + B_{10} e^{-k(\omega)x}\right)e^{j\omega t} \end{aligned} \qquad (2.45)$$

where B_7, B_8, B_9 and B_{10} are combinations of the B_k coefficients in Equation (2.44), and $k(\omega)$ is the wave number. As in the previous case, B_k are found by applying initial and boundary conditions. This solution indicates that a transverse wave with frequency ω_1 travels with a different speed than a wave at ω_2 unlike in longitudinal waves. Waves that propagate in one direction but displace perpendicular to the direction of travel therefore exhibit *dispersion*. *Dispersion results in a spreading of waveforms as lower frequencies pull waves back and higher frequencies push waves forward.* Dispersion makes it difficult to interpret measurements when identifying damage. Transverse bending stiffness (fourth-order term in the wave equation) is the physical cause of dispersion in this case resulting in more stiffness and higher speeds for higher frequency waveforms.

2.2.8.4 Transverse Wave Propagation Finite Element Model

As in the longitudinal wave propagation model of the rod, the transverse model can be discretized using finite element methods. Figure 2.28(a) illustrates the transverse finite

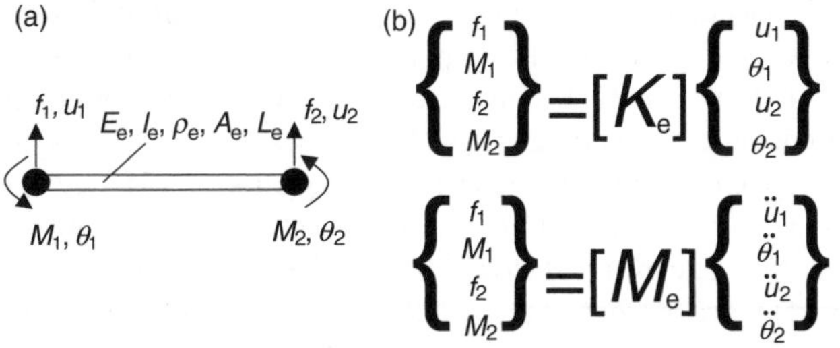

Figure 2.28 (a) One-dimensional rod element undergoing transverse motions and (b) stiffness and mass relationships that define the bending rod element

element subject to forces and moments applied at its nodes. An assumption is made in this model that the element undergoes pure bending, which implies that the length of the element must be larger than its width and height to avoid errors due to large shear forces through the cross section. The forces and moments in Figure 2.28(a) produce transverse displacements and angular rotations. The relationships between these forces and motions are defined by the mass and stiffness matrices below, which are derived in many textbooks on vibrations and finite element modeling:

$$
\left\{ \begin{array}{c} f_1 \\ M_1 \\ f_2 \\ M_2 \end{array} \right\} = \frac{E_e I_e}{L_e^3} \left[\begin{array}{cccc} 12 & 6L_e & -12 & 6L_e \\ 6L_e & 4L_e^2 & -6L_e & 2L_e^2 \\ -12 & -6L_e & 12 & -6L_e \\ 6L_e & 2L_e^2 & -6L_e & 4L_e^2 \end{array} \right] \left\{ \begin{array}{c} u_1 \\ \theta_1 \\ u_2 \\ \theta_2 \end{array} \right\} \quad \text{and} \qquad (2.46)
$$

$$
\left\{ \begin{array}{c} f_1 \\ M_1 \\ f_2 \\ M_2 \end{array} \right\} = \frac{\rho_e L_e}{420} \left[\begin{array}{cccc} 156 & 22L_e & 54 & -13L_e \\ 22L_e & 4L_e^2 & 13L_e & -3L_e^2 \\ 54 & 13L_e & 156 & -22L_e \\ -13L_e & -3L_e^2 & -22L_e & 4L_e^2 \end{array} \right] \left\{ \begin{array}{c} \ddot{u}_1 \\ \ddot{\theta}_1 \\ \ddot{u}_2 \\ \ddot{\theta}_2 \end{array} \right\} \qquad (2.47)
$$

To demonstrate the differences between transverse propagating waves and longitudinal waves, the code 'rod1dtranswave.m' was used to simulate the response of an aluminum component with $\rho = 2700\,\text{kg/m}^3$ and $E = 70\,\text{GPa}$. This component is 1m long, which prevents wave reflections from the right end in the simulation. The rod has a 1 cm × 0.1 cm rectangular cross section and I_e is given by $bh^3/12$. Thirty-four transverse rod elements each with length $L_e = 3\,\text{cm}$ were used in the finite element model. The mass and stiffness matrices for this model were assembled as in the previous case along the diagonal by enforcing two constraints at each node. An excitation force with 1e6N amplitude, 4kHz center frequency, and 3kHz bandwidth was applied at node 1 on the left end of the component. The frequency spectrum describing the amplitude of this excitation force as a function of frequency is shown in Figure 2.29. This type of excitation is often used to identify damage based on wave propagation because it is *narrowband* (i.e. a small group of frequencies make up the waveform). This narrowband excitation produces traveling wave packets like those shown previously for the rod (Figure 2.25).

Figure 2.30 shows the corresponding excitation force time history (a) and transverse displacements at nodes 1, 4, 8, 12 and 16 (b)–(f). These plots show that the waveform becomes distorted as it travels. The two circled areas illustrate the decrease in amplitude of one peak in the propagating wave and increase in another peak. Equation (2.45) can be used to explain these transverse displacement waves. For example, the large initial drop in amplitude from node 1 to 4 is due to the fast decay with respect to distance along the rod caused by the last exponential term in Equation (2.45) (an *evanescent wave*). Likewise, the decrease and increase in amplitude at the beginning and end of the wave packet is due to the dependence of wave speed on frequency in the first two terms of Equation (2.45). Recall the speed of each frequency wave is $v = \sqrt{\omega c I_A}$; therefore, the distortions in Figure 2.30(b)–(f) occur due to dispersion as lower frequencies travel slower than higher frequencies. The distortions toward the end of each nodal wave response time window are also due to dispersion. In this problem, damping was not included. If damping is included, amplitudes decay with distance to an even greater extent.

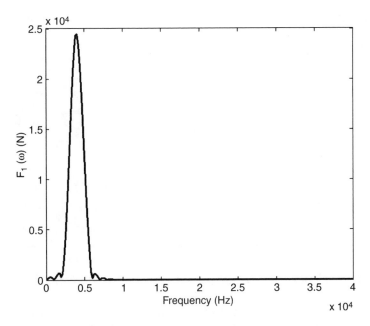

Figure 2.29   Amplitude of excitation force as a function of frequency

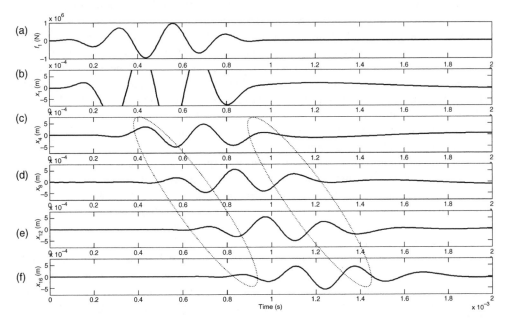

Figure 2.30   Transverse excitation force (a) and wave propagation response in aluminum rod without damping at nodal points (b) 1, (c) 4, (d) 8, (e) 12 and (f) 16

Table 2.1 Bulk propagating waves in infinite solid media (Reproduced by permission of Shankar Sundararaman, Purdue University)

Wave type	Particle displacement
Longitudinal wave (bulk wave) Displacement profile: $w_x(x,t) = A\sin(\omega t - kx)$ Propagation velocity: $V_P = \sqrt{(2\mu + \lambda)/\rho}$	
Shear wave (bulk wave) Displacement profile: $w_z(x,t) = A\sin(\omega t - kx)$ Propagation velocity: $V_S = \sqrt{\mu/\rho}$	

Source: Sundararaman (2003), with permission from Mr. Shankar Sundararaman

2.2.9 *Wave Propagation Models (Two Dimensional)*

In two dimensions, the same modeling techniques from the one-dimensional cases described above are used to obtain analytical and finite element models. Propagating waves are a rich subject with many textbooks available for study (Graff, 1991); however, the basic aspects of two-dimensional wave propagation are often sufficient for health monitoring. Sundararaman (2003) provides a detailed summary of elastic waves in solid media. There are two different types of elastic waves: *bulk* waves and *guided* waves.

Bulk waves occur in infinite solid media without boundaries. Table 2.1 summarizes the two types of bulk waves that can propagate in solids: *longitudinal* waves and *shear* waves. The displacement patterns as well as the propagation velocities are provided. $\mu = E/[2(1 + v)]$ is called the *shear modulus* and $\lambda = Ev/[(1 + v)(1 - 2v)]$ is the *bulk modulus*. These propagating waves are relevant if the component being inspected has very large dimensions relative to a wavelength. Table 2.2 lists some of the terminologies associated with propagating waves in solid media in general. These definitions apply to bulk and guided waves. The displacement and velocity quantities are expressed as vectors following convention. Concepts in phase velocity and group velocity are natural extensions of the definitions introduced earlier for one-dimensional propagating waves in rods and beams.

Guided waves occur in finite media with boundaries (plates, shells, etc.). There are two basic types of two-dimensional finite material components. One type of component is called a *membrane* and the other type is called a *plate*. Simple membranes do not possess any bending stiffness; therefore, waves in these membranes do not exhibit dispersion (i.e. wave group velocities are the same regardless of frequency). More complicated membranes, however, can exhibit dispersion for certain boundary conditions and material properties. Examples of membranes include thin diaphragms and inflatable structures. Let ρ be the density per unit area and T be the tension force per unit length. Then the wave equation of motion for a rectangular membrane with constant material and geometric properties and no forcing function (Kinsler *et al.*, 1982) is found to be similar to the equation of motion for a longitudinal rod (Equation (2.28)):

$$\frac{\partial^2 u(x,y,t)}{\partial x^2} + \frac{\partial^2 u(x,y,t)}{\partial y^2} = \frac{1}{c^2}\frac{\partial^2 u(x,y,t)}{\partial t^2} \text{ with } c = \sqrt{\frac{T}{\rho}} \qquad (2.48)$$

Table 2.2 Terminology of propagating waves in solid media summarized from Nayfeh (Reproduced by permission of Shankar Sundararaman, Purdue University, 2003)

Description	Equation(s)/Illustration
• *Particle velocity* (v): The simple harmonic velocity of a particle about its equilibrium point. It is equal to the rate of change of particle displacement, w. • *Phase velocity* (v_{ph}): The propagation velocity of a single frequency or the velocity of energy transfer across particles along the carrier wave. • *Group velocity* (v_{gr}): The propagation velocity of a group of waves or the velocity of energy transfer across particles along the modulation envelope. • If the phase velocity is not equal to the group velocity, the waveform tends to be *dispersive*. • *Skew angle*: Angle between the direction of wave propagation and the group velocity.	$\mathbf{v} = \partial \mathbf{w}/\partial t$ $v_{ph} = \omega/\mathbf{k}$ $v_{gr} = \dfrac{\partial \omega}{\partial \mathbf{k}} \quad \dfrac{v_{ph}}{1 - \dfrac{\omega}{v_{ph}} \cdot \dfrac{\partial v_{ph}}{\partial \omega}}$ $\omega =$ Frequency $\mathbf{k} =$ Wavenumber

In isotropic media, the particle motion is polarized either parallel or normal to the direction of propagation. In anisotropic media, on the contrary, waves tend to 'bend' as they propagate through the medium and this bending has directional dependence. The directional dependence of the propagation velocities is provided by *slowness surfaces*, which portray the variation of the inverse of the wave speeds with direction.

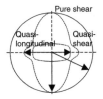

• *Reflection*: When a wave hits a boundary, a portion of the energy is turned back into the medium with a phase change dependent on the material properties.
• *Refraction*: When a wave hits a boundary, a portion of the energy is transmitted into the new medium with a phase change dependent on the material properties of the medium at the interface.
• *Snell's law* can be used to calculate the extent of reflection and refraction given the angle of incidence and the phase speed.
• *Diffraction*: When a wave encounters a discontinuity, it tends to bend and get scattered along a path dependent on the orientation, shape and size of the discontinuity.

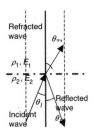

The main difference between this equation and the equation of longitudinal wave propagation in a rod is the appearance of a second position variable, y, due to the rectangular nature of the membrane. The separation of variables method can be applied to the membrane leading to the following form for the transverse displacement solution:

$$u(x, y, t) = B_1 e^{j(\omega t - k_x x - k_y y)} + B_2 e^{j(\omega t + k_x x + k_y y)} + B_3 e^{j(\omega t - k_x x + k_y y)} + B_4 e^{j(\omega t - k_x x + k_y y)} \qquad (2.49)$$

where $k^2 = k_x^2 + k_y^2 = \omega^2/c^2$. The two wave number constants define the propagation spatial frequencies in the two rectangular directions.

Plates possess bending stiffness; therefore, plate waves of transverse displacement exhibit dispersion as they travel. Dispersion can be generally thought of as a *loss in elastic wave energy in the propagation direction leading to changes in group speed with frequency.* The wave equation for a rectangular plate of thickness h, modulus E, and Poisson's ratio v, is given by (Meirovitch, 1967)

$$\frac{\partial^4 u(x, y, t)}{\partial x^4} + \frac{\partial^4 u(x, y, t)}{\partial y^4} = \frac{1}{c^2} \frac{\partial^2 u(x, y, t)}{\partial t^2} \text{ with } c = \sqrt{\frac{D}{\rho}} \qquad (2.50)$$

where D is the flexural rigidity of the plate, $D = Eh^3/12\sqrt{(1 - v^2)}$. The solution is then of the following form:

$$u(x, y, t) = B_1 e^{j(\omega t - k_x(\omega)x - k_y(\omega)y)} + \ldots + B_2 e^{j\omega t} e^{-k_x(\omega)x - k_y(\omega)y} + \ldots \qquad (2.51)$$

There are many types of plate waves depending on which of these terms contributes when the boundary conditions are applied. For example, if the plate is semi-infinite (extends to $\pm\infty$ in the x direction but from 0 to ∞ in the y direction) with air as the boundary at $y = 0$, then *Rayleigh (surface)* wave solutions are obtained. These waves exponentially decay into the depth of the plate eventually causing dissipation of waves traveling along the surface of the plate as well. Table 2.3 provides the relevant information for Rayleigh wave solutions including the phase velocity and displacement patterns. *Lamb* waves, on the contrary, occur when plates have upper and lower boundaries (finite plates). These are dispersive waves. Lamb waves can be symmetric or antisymmetric depending on the nature of the strains inside the plate that propagate the wave. Table 2.3 also provides information about these two types of Lamb waves. Two-dimensional waves are simulated in plates later in Chapter 4 (see Section 5.7.4).

Baranek (1988) provides a rigorous overview of each of these kinds of waves. In general, wave propagation is quite complicated. Internal boundaries within a component or damage that causes waves to scatter can only be modeled using sophisticated methods. The discussion here has also been limited to flat rectangular membranes and plates; however, the same principles apply to curvilinear components (Soedel, 1994). For circular components, for instance, the wave equation is solved using *Bessel functions* instead of exponentials as in rectangular plates. When components are curved, they can often be treated as flat as long as the wavelength is much less than the radius of curvature. For composite components, the mass and stiffness matrices required to construct finite element models for wave propagation can be found in Jones (1975) and elsewhere.

Table 2.3 Guided propagating waves in solid media summarized from Schmerr (Reproduced by permission of Shankar Sundararaman, Purdue University, 2003)

Wave type	Particle displacement
Rayleigh wave (semi-infinite surface wave) Displacement: $w_x = \left(jkAe^{-az} - bBe^{-ba}\right)e^{j(\omega t - kx)}$ $\qquad w_z = \left(-aAe^{-az} - jkBe^{-bz}\right)e^{j(\omega t - kx)}$ where $a = \sqrt{(k^2 - \omega^2/V_P^2)}$, $b = \sqrt{(k^2 - \omega^2/V_S^2)}$ Wave equation: $(2 - V_R^2/V_S^2)^2 = 4(1 - V_R^2/V_P^2)^{1/2}(1 - V_R^2/V_S^2)^{1/2}$ Phase velocity (V_R): $V_R = \dfrac{0.862 + 1.14v}{1 + v}V_S$	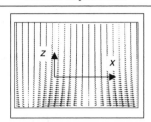
Symmetric Lamb wave (plate wave) displacement profile: $w_x = Ak_{Ls}\left(\dfrac{\cos h(az)}{\sin h(ad)} - \dfrac{2ab}{k_{Ls}^2 + b^2}\dfrac{\cos h(bz)}{\sin h(bd)}\right)e^{j(\omega t - k_{Ls}x + \frac{\pi}{2})}$ $w_z = -Aa\left(\dfrac{\sin h(az)}{\sin h(ad)} - \dfrac{2k_{Ls}^2}{k_{Ls}^2 + b^2}\dfrac{\sin h(bz)}{\sin h(bd)}\right)e^{j(\omega t - k_{Ls}x)}$ Here, $\omega d = 2\pi(37)$ kHz-mm k_{Ls} = Symmeteric wavenumber	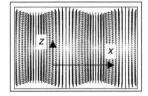
Antisymmetric Lamb wave (plate wave) displacement profile: $w_x = Bk_{La}\left(\dfrac{\sin h(az)}{\cos h(ad)} - \dfrac{2ab}{k_{La}^2 + b^2}\dfrac{\sin h(bz)}{\cos h(bd)}\right)e^{j(\omega t - k_{La}x + \frac{\pi}{2})}$ $w_z = -Ba\left(\dfrac{\cos h(az)}{\cos h(ad)} - \dfrac{2k_{La}^2}{k_{La}^2 + b^2}\dfrac{\cos h(bz)}{\cos h(bd)}\right)e^{j(\omega t - k_{La}x)}$ Here, $\omega d = 2\pi(37)$ kHz-mm k_{La} = Antisymmeteric wavenumber	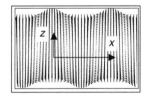

2.3 DATA-DRIVEN MODELS

The models discussed in Section 2.2 were derived using physical laws based on first principles. In many applications, *data-driven* models must be developed to describe the dynamic behavior of structural components. Data-driven models are often based on first principles, but the health-monitoring information is extracted directly from data. This section discusses these models and ways in which they can be applied to measurement data. Much of the discussion below is based on the time and frequency domain models described in the sections above. Details on measurements and data analysis to obtain these models will be discussed in the next chapter.

2.3.1 Experimental Time Domain Models

The time domain model introduced in Equation (2.4) assumed that all of the mass, damping and stiffness parameters were known beforehand. Only then can the model be used to study the free and forced response to enable health monitoring of the component. It is often the case that these parameters are unknown; therefore, the parameters must be estimated in order to develop a component model. The three methods discussed below are only a subset of the methods available for estimating the parameters in time domain models. The models can then be used for health monitoring to identify loading or damage and track changes in damage with time.

2.3.1.1 Direct Parameter Models

When monitoring the health of structural materials and components, it is necessary to determine changes in model parameters due to damage. Consider the two-DOF model with the equations of motion given in Equation (2.4b). This model consists of eight unique parameters: two masses, three damping coefficients, and three stiffness coefficients. By rewriting the equations of motion as follows, the model parameters can be grouped into one vector:

$$\begin{bmatrix} \ddot{x}_1 & 0 & \dot{x}_1 & \dot{x}_1 - \dot{x}_2 & 0 & x_1 & x_1 - x_2 & 0 \\ 0 & \ddot{x}_2 & 0 & \dot{x}_2 - \dot{x}_1 & \dot{x}_2 & 0 & x_2 - x_1 & x_2 \end{bmatrix}_{2\times 8} \begin{Bmatrix} M_1 \\ M_2 \\ C_1 \\ C_2 \\ C_3 \\ K_1 \\ K_2 \\ K_3 \end{Bmatrix}_{8\times 1} = \begin{Bmatrix} f_1(t) \\ 0 \end{Bmatrix}_{2\times 1} \qquad (2.52)$$

The two equations in vector–matrix form in Equation (2.52) are written with subscripts on the vectors and matrices to emphasize the number of equations and number of parameters involved. These equations can then be used to estimate the model parameters.

   In order to estimate the eight parameters in the two-DOF model, the input and output time domain data must be measured at an adequate number of time points. For instance, one time point is not enough to estimate eight parameters using Equation (2.52) because there are only two equations. There must be *at least as many equations as unknown parameters*. To generate additional equations, Equation (2.52) is written at various points in time $k = 1, 2, \ldots, N_t$ where N_t is the total number of time points used. The set of equations below shows how these equations are expressed one after another in matrix form:

$$\begin{bmatrix} \ddot{x}_1(t_1) & 0 & \dot{x}_1(t_1) & \dot{x}_1(t_1) - \dot{x}_2(t_1) & 0 & x_1(t_1) & x_1(t_1) - x_2(t_1) & 0 \\ 0 & \ddot{x}_2(t_1) & 0 & \dot{x}_2(t_1) - \dot{x}_1(t_1) & \dot{x}_2(t_1) & 0 & x_2(t_1) - x_1(t_1) & x_2(t_1) \\ \ddot{x}_1(t_2) & 0 & \dot{x}_1(t_2) & \dot{x}_1(t_2) - \dot{x}_2(t_2) & 0 & x_1(t_2) & x_1(t_2) - x_2(t_2) & 0 \\ 0 & \ddot{x}_2(t_2) & 0 & \dot{x}_2(t_2) - \dot{x}_1(t_2) & \dot{x}_2(t_2) & 0 & x_2(t_2) - x_1(t_2) & x_2(t_2) \\ \vdots & \vdots & \vdots & \vdots & \vdots & \vdots & \vdots & \vdots \\ \vdots & \vdots & \vdots & \vdots & \vdots & \vdots & \vdots & \vdots \\ \ddot{x}_1(t_{N_t}) & 0 & \dot{x}_1(t_{N_t}) & \dot{x}_1(t_{N_t}) - \dot{x}_2(t_{N_t}) & 0 & x_1(t_{N_t}) & x_1(t_{N_t}) - x_2(t_{N_t}) & 0 \\ 0 & \ddot{x}_2(t_{N_t}) & 0 & \dot{x}_2(t_{N_t}) - \dot{x}_1(t_{N_t}) & \dot{x}_2(t_{N_t}) & 0 & x_2(t_{N_t}) - x_1(t_{N_t}) & x_2(t_{N_t}) \end{bmatrix}_{2N_t \times 8}$$

$$\times \left\{ \begin{array}{c} M_1 \\ M_2 \\ C_1 \\ C_2 \\ C_3 \\ K_1 \\ K_2 \\ K_3 \end{array} \right\}_{8 \times 1} = \left\{ \begin{array}{c} f_1(t_1) \\ 0 \\ f_1(t_2) \\ 0 \\ \vdots \\ \vdots \\ f_1(t_{N_t}) \\ 0 \end{array} \right\}_{2N_t \times 1} , \quad D_1 P = D_2 \qquad (2.53)$$

where D_1 and D_2 are the matrix and vector on the left and right hand sides of the equation and P is the vector of model parameters to be estimated. When D_1 is a square matrix, it can be inverted to estimate the parameters; however, better parameter estimates are usually obtained by using more equations in an overdetermined solution for P to reduce the errors in each equation due to measurement error and modeling errors. If $2N_t > 8$, then the *pseudo-inverse* of D_1 can be used to calculate the parameter estimates that minimize the sum of the squared error across all of the equations. This calculation is written as follows:

$$P = D_1^{+} D_2 \qquad (2.54a)$$

$$D_1^{+} = (D_1^{T} D_1)^{-1} D_1^{T} \qquad (2.54b)$$

where D_1 is called the pseudo-inverse of D_1, and D_1^{T} is the transpose of the D_1 matrix, that is each entry $D_{1,ij}^{T}$ of D_1^{T} is equal to $D_{1,ji}$. *Direct Parameter Estimation* (DPE) is the formal name of the procedure used to estimate the model parameters directly from the time domain equations of motion in this way (Mohammad *et al.*, 1992).

For the two-DOF model, the simulation code 'twodofdpe.m' was used to simulate the response of the two-DOF model to a random[7] excitation force, $f_1(t)$. In the simulation, the velocity and displacement variables were calculated; then the acceleration variables were estimated by differentiating the velocity variables. In practice, differentiation is not recommended and is rarely needed because acceleration data is most often acquired and can be integrated as described in Chapter 3 to estimate the velocity and displacement variables. The x_1 and x_2 displacement responses to the random excitation force are shown in Figure 2.31. One thousand time points were used by 'twodofdpe.m' in Equation (2.53) to estimate the parameter vector, P. The comparison between the actual parameter values and estimated values are given in Table 2.4. There are only small errors in the damping coefficient estimates because the acceleration time histories were obtained by differentiating and because of the numerical errors introduced in the simulation. This method can be used to estimate parameters as damage progresses. The method can also be used to estimate the forces applied to a component. The method can also be extended when the equations of motion are nonlinear (Haroon *et al.*, 2005).

---

[7]A uniform distribution was used in MATLAB® to generate this excitation force; however, other types of random excitations can also be generated, for example normal (Gaussian) distribution.

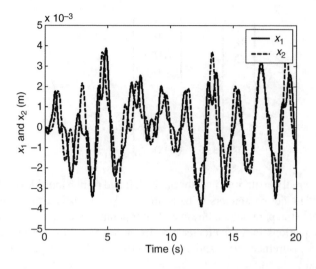

Figure 2.31   Time responses x_1 (__) and x_2 (- - -) to a random excitation $f_1(t)$

2.3.1.2   Restoring Force and Phase-Plane Models

The time domain data used for direct parameter models can also be used to model how damping or stiffness forces within a component change due to damage. These models are called *restoring force models* or *force–state maps*. Consider the two-DOF model discussed in Section 2.2.7. This model was used earlier to analyze how the vibration response of the panel changes if there is damage due to friction at a mechanical attachment fastener or damage due to buckling of the panel. The two equations of motion of that model can be rewritten into the following form:

$$M_1\ddot{x}_1 = \sum_{M_1} F(x_1, x_2, \dot{x}_1, \dot{x}_2) \tag{2.55a}$$

$$M_2\ddot{x}_2 = \sum_{M_2} F(x_1, x_2, \dot{x}_1, \dot{x}_2) \tag{2.55b}$$

where the two terms on the right hand sides of these equations describe the forces that act on the two masses as a function of the displacements and velocities. These forces are called the *restoring forces* because they restore the two DOFs to their static equilibrium positions. The nature of these forces will change if damage occurs. In some cases, the restoring forces grow or shrink in amplitude; however, in other cases, the restoring forces exhibit nonlinear

Table 2.4   Comparison of actual parameters and direct parameter model parameter estimates for two-DOF system using Equation (2.54a)

	M_1	M_2	C_1	C_2	C_3	K_1	K_2	K_3
Actual	1	1	0.11	0.20	0.11	10	20	10
Estimated	1	1	0.12	0.21	0.12	10	20	10

characteristics when damage occurs. Note that Equations (2.55a) and (2.55b) are simply Newton's Second Law expressed in a slightly different form.

The significance of Equations (2.55a) and (2.55b) is that the left hand sides of the equations are functions of the acceleration variables only. Because acceleration measurements are relatively easy to make in most applications, these equations can be used to construct restoring force models because the left hand sides of the equations can be directly calculated. One option to construct these models is to utilize the same approach as in Section 2.3.1.1 with direct parameter models. If the mass values are known, the restoring forces on the right hand sides can be decomposed into damping and stiffness force terms, and then the parameters within these terms can be estimated (Masri and Caughey, 1979).

A second option is to utilize *projection models* of the restoring forces (Haroon *et al.*, 2004). These projection models are highly descriptive and can be used to identify loads and damage in components. Restoring force projections are generated by plotting the acceleration variable $(\ddot{x}_1, \ddot{x}_2)$ versus the velocity and displacement variables $(x_1, x_2, \dot{x}_1, \dot{x}_2)$. These plots describe the damping and stiffness forces acting on a DOF in graphical form. In these plots, the forces are scaled by the mass of each DOF. Restoring force projections are most clear when the excitation and response time histories are sinusoidal.

The simulation code 'twodofrf.m' was used to generate the restoring force projection models shown in Figure 2.32. Figure 2.32(a) was generated at 2rad/s excitation frequency whereas (b) was generated at 8rad/s. Consider the two restoring force curves in (a). One of the curves looks like an ellipse; this curve is for the two-DOF model with no Coulomb friction acting on M_2. This restoring force projection model is valid at 2rad/s. It describes how the damping force acting on M_2 changes as a function of its velocity. The ellipse in (b) describes the same damping force at 8rad/s. As discussed in Section 2.2.7, a bolted interface like the one shown in Figure 2.1 for the thermal protection system panel can loosen causing slipping and Coulomb friction. The 'twodofrf.m' code was used to simulate Coulomb friction in the 'two_dof_model_nl.mdl' Simulink® model. Recall that Figure 2.16 showed how friction

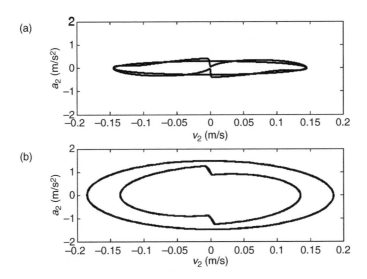

Figure 2.32   Restoring force projections for damping force on M_2 as a function of v_2 at (a) 2rad/s and (b) 8rad/s excitation frequency with and without Coulomb friction

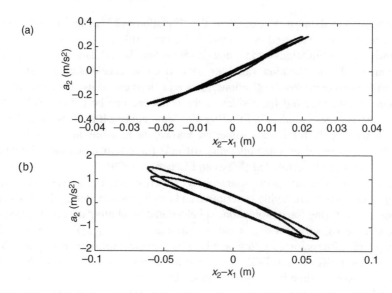

Figure 2.33 Restoring force projections for stiffness force on M_2 as a function of $x_2 - x_1$ at (a) 2rad/s and (b) 8rad/s excitation frequency with and without buckling

produces large forces for small velocities. The distorted restoring force projections shown in Figure 2.32(a) and (b) exhibit a stick-slip characteristic for small velocities because of this Coulomb friction characteristic. Changes in restoring force projections can be used to characterize damage in components as in this example.

A similar restoring force projection model can be constructed for the stiffness restoring force that acts between the two DOFs. Figure 2.33 shows this model for a 2rad/s (Figure 2.33(a)) and 8rad/s (Figure 2.33(b)) excitation frequency. In this model, the acceleration of M_2 versus the relative displacement $x_2 - x_1$ is plotted. At the top and bottom of this figure there are elliptical restoring force projections, which describe the baseline stiffness restoring force acting between the two DOFs. When the quadratic stiffness nonlinearity is added to the simulation as described earlier (see Figure 2.19), the distorted restoring force characteristics are obtained. Although these distortions are more difficult to see in (a) at 2rad/s, (b) at 8rad/s clearly shows the stiffening and softening that occur due to the simulated effects of buckling.

Instead of plotting acceleration variables versus velocity and displacement variables to generate restoring force projection models, the velocity of one variable versus the displacement variable can be plotted. In fact, many more combinations are possible where one variable versus another is plotted. These types of plots can also be used as models to describe dynamic data and are referred to as *phase–plane trajectories*. As with the restoring force projection models, phase–plane trajectories experience a change in shape and size as the damping and stiffness restoring forces change due to damage. For example, Figure 2.34 shows the phase–plane trajectory for M_2. These results correspond directly to the results already presented in Figure 2.33 for the buckled panel simulation.

Note that at 2rad/s, there are only subtle changes in the phase–plane trajectory; however, an interesting shift in the trajectory is seen at 8rad/s (Figure 2.34(b)) due to the quadratic stiffness force introduced by the buckling of the panel. This quadratic stiffness is asymmetric as shown in Figure 2.19(d) leading to 'softer' and 'harder' sides to the spring connecting the two DOFs.

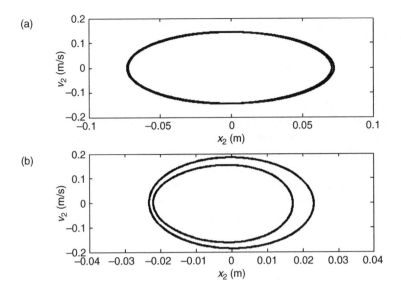

Figure 2.34 Phase–plane trajectories on M_2 at (a) 2rad/s and (b) 8rad/s excitation frequency with and without buckling showing asymmetric stiffness in x_2

This asymmetry in stiffness introduces an asymmetry in the phase plane-trajectory causing the ellipse to squeeze toward negative displacement.

2.3.1.3 Discrete Time Models

Measured data for health monitoring is usually sampled at discrete time points. The time domain models described in the previous sections were continuous-time models. Although those models can also be applied to data that are sampled at discrete time points, discrete time models are often more convenient because they can relate input forces to output responses in structural components *even when all of the response variables are not measured*. To motivate the development of discrete time models, consider the two equations of motion in Equation (2.4). If the Laplace transform[8] is applied to these two equations, the following set of equations is obtained in vector–matrix form:

$$\begin{bmatrix} M_1 s^2 + (C_1 + C_2)s + K_1 + K_2 & -sC_2 - K_2 \\ -sC_2 - K_2 & M_2 s^2 + (C_2 + C_3)s + K_2 + K_3 \end{bmatrix} \begin{Bmatrix} X_1(s) \\ X_2(s) \end{Bmatrix} = \begin{Bmatrix} F_1(s) \\ 0 \end{Bmatrix}$$
(2.56)

Using the second of these equations, $X_2(s)$ can be eliminated in terms of $X_1(s)$ in the first equation as follows:

$$(M_1 s^2 + (C_1 + C_2)s + K_1 + K_2)X_1(s) + (-sC_2 - K_2)\frac{sC_2 + K_2}{M_2 s^2 + (C_2 + C_3)s + K_2 + K_3}X_1(s) = F_1(s)$$
(2.57)

---

[8]The Laplace transform is related to other transforms described in Chapter 5. If $y(t)$ is the time domain signal, then the Laplace transform of $y(t)$ is $\int_0^\infty y(t)e^{-st}dt$. Zero initial conditions are assumed.

which can be transformed back into the time domain to produce the following fourth order differential equation:

$$c_1 \ddddot{x}_1 + c_2 \dddot{x}_1 + c_3 \ddot{x}_1 + c_4 \dot{x}_1 + c_5 x_1 = d_1 \ddot{f}_1 + d_2 \dot{f}_1 + d_3 f_1 \qquad (2.58)$$

The coefficients c_k and d_k are functions of the various model parameters. Note that the response of DOF 1 appears with four derivatives and the excitation force appears with two derivatives.

Next, it is assumed that the displacement variable $x_1(t)$ is sampled at discrete time points, $t = n\Delta T$, where ΔT is the separation between time samples and n is an integer that determines the sample number. In order to represent the dynamical relationship between the input force and output motion given in Equation (2.58), there are many ways in which to calculate the time derivatives using discrete time data. One possibility is to use the backward difference formulas given below for the velocity and acceleration in terms of the displacement at discrete time points:

$$\dot{x}(n\Delta T) \cong \frac{x(n\Delta T) - x((n-1)\Delta T)}{\Delta T}$$

$$\ddot{x}(n\Delta T) \cong \frac{x(n\Delta T) - 2x((n-1)\Delta T) + x((n-2)\Delta T)}{(\Delta T)^2} \qquad (2.59a, b)$$

When these expressions are substituted into Equation (2.58), the displacement x_1 at the present time is found to be related to the displacements x_1 and input force f_1 at the present time as well as previous points in time. The resulting equation is called a *difference equation of motion*:

$$\begin{aligned} x_1(n) = &a_1 x_1(n-1) + a_2 x_1(n-2) + a_3 x_1(n-3) + a_4 x_1(n-4) \\ &+ b_1 f_1(n) + b_2 f_1(n-1) + b_3 f_1(n-2) \end{aligned} \qquad (2.60)$$

where the coefficients a_i and b_i are functions of the mass parameters, damping parameters, stiffness parameters, and sample time step. a_i are called *autoregressive* coefficients, and b_i are called *moving average* coefficients. The ΔT time step was dropped for simplicity in the notation. A similar equation can be written for $x_2(n)$. This expression is significant because it suggests that to develop a discrete time model, one needs measurements of the input, $f_1(n)$, and only the output variable of interest, $x_1(n)$.

As in the direct parameter estimation process, Equation (2.60) is written for a certain number of time samples, N_t, in order to estimate the discrete time model parameter vector, $\{a_1 a_2 a_3 a_4 b_1 b_2 b_3\}^T$, that minimizes the sum of the squared errors across all of the equations. The simulation code 'twodofdtm.m' was used to simulate the response of the two-DOF system model to a sum of sinusoids: $f_1(k) = \sin(2\pi k\Delta t) + \sin(4\pi k\Delta t) + \sin(6\pi k\Delta t)$. Figure 2.35(a) shows the responses at $x_1(n)$ and $x_2(n)$ at the time instants $k\Delta T$ for this excitation force. The discrete time points of data are indicated with x and o symbols. The code was then used to estimate the coefficients in Equation (2.60) and then predict the response of the model at DOF $x_1(n)$. Figure 2.35(b) shows the agreement between predicted values of the displacement $(+)$ and actual values (x).

The general rules for selecting the spacing between discrete time data and the number of time points to consider for parameter estimation are provided by Ljung (1987). In general, smaller time steps are not necessarily better because they lead to dependent equations in the parameter estimation process and poor parameter estimates. The frequency of the time step should be chosen approximately two to three times the highest frequency in the time domain

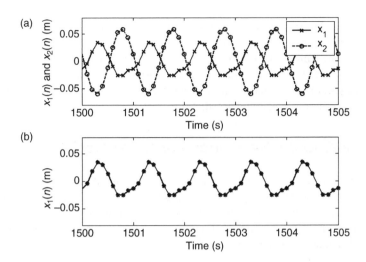

Figure 2.35 (a) Discrete time response of two DOFs to sinusoidal excitation and (b) actual (x) and predicted $(+)$ response $x_1(n)$ at time instants $n\Delta T$

signal being modeled. The number of time shifts considered is a function of noise in the data and the highest delay in the discrete time model. For the two-DOF example just considered, there were four time delays in the output variable and three time delays in the input variable producing seven model coefficients. In this case, a minimum of seven averages are required. However, the simulation code 'twodofdtm.m' uses $3 \times 7 = 21$ averages (time shifts).

Regarding noise on the data, Figure 2.36 shows the effects of random noise added to the sinusoidal excitation force to increase the variance in the response data from sample to sample. The 2% added noise has little effect, but the 5% and 10% noise do cause the prediction to

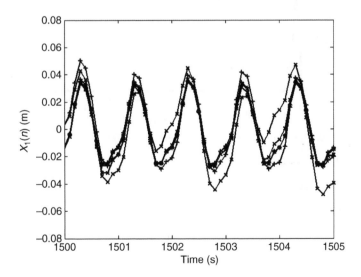

Figure 2.36 Predicted $(+)$ and actual (x) discrete time response of $x_1(n)$ for 2%, 5% and 10% random noise added to excitation force

experience errors. By increasing the number of time points considered in the parameter estimation algorithm, these errors can be reduced if the noise is *stationary* (statistical properties do not vary with time). The parameter estimation algorithm in 'twodofdtm.m' normalizes the data matrices by the standard deviation of each variable in Equation (2.60) to improve the accuracy of the parameter estimates and reduce the sensitivity to small errors in the variables. This practice is common in parameter estimation.

2.3.2 Experimental Frequency Response Models

The FRFs in Section 2.2.4 were calculated from a first-principle model; however, FRFs are often acquired experimentally using input and output data. A few methods by which these FRF measurements can be utilized to construct models for loads and damage identification in health monitoring are discussed in the following sections. Measurement issues associated with obtaining FRFs and ensuring that they are of sufficient quality in the first place are discussed in Chapter 5.

Two requirements of health monitoring in structural components are to identify component loads and component damage. FRFs are useful for fulfilling both of these requirements. Consider the three-DOF model given in Figure 2.12. Suppose it was desirable to estimate the excitation force applied to this model. The excitation could be due to either buffeting in an aircraft component or an impact acting on a thermal protection system tile. If the FRFs for this system are measured before the excitation forces are applied, then Equation (2.22) can be used to estimate the responses obtained from such an excitation force. On the contrary, Equation (2.22) can also be used to estimate the applied forces that produce the measured responses.

For example, consider the impact force, $f_2(t)$, shown in Figure 2.37(a) that occurs at DOF 2. The simulation code 'threedoffrfload.m' was used to generate and apply this excitation force

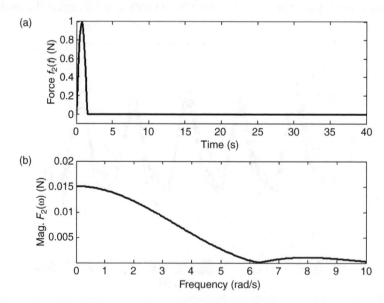

Figure 2.37   (a) Impact force time history acting on three-DOF model and (b) frequency domain representation of force showing broad frequency content

to the three-DOF system model in Figure 2.12. The impact is relatively short occurring over approximately 1 s. Figure 2.37(b) shows the corresponding *frequency spectrum* (magnitude) of this excitation force. The frequency spectrum of a time domain signal such as $f_2(t)$ describes the energy in a signal as a function of frequency. Another way of interpreting a frequency spectrum is that the time signal can be calculated by summing up the Fourier series components at all frequencies in the spectrum. In general, shorter more crisp time domain forces lead to frequency spectra that extend across a broader frequency range. The way in which spectra are computed will be discussed in Chapter 5.

If this spectrum is substituted into Equation (2.22), then the responses in Figure 2.38 are obtained. Note that the time response at DOF 2 (- - -) in Figure 2.38(a) has the largest response because the impact occurs at that DOF. In the frequency spectra (Figure 2.38(b)), the resonances of the three-DOF system model near 2.5 and 4.5 rad/s are responding to the impact force. Again note that the spectrum at DOF 2 has the largest amplitude because the impact occurs at that DOF. This FRF model can also be applied in the reverse direction. For example, if the responses in Figure 2.38 are available from measured data, then the relationship in Equation (2.22) can be inverted to yield

$$\left\{ \begin{array}{c} F_1(\omega) \\ F_2(\omega) \\ F_3(\omega) \end{array} \right\} = \left[\begin{array}{ccc} H_{11}(\omega) & H_{12}(\omega) & H_{23}(\omega) \\ H_{21}(\omega) & H_{22}(\omega) & H_{23}(\omega) \\ H_{31}(\omega) & H_{32}(\omega) & H_{33}(\omega) \end{array} \right]^{-1} \left\{ \begin{array}{c} X_1(\omega) \\ X_2(\omega) \\ X_3(\omega) \end{array} \right\} \tag{2.61}$$

In this example, this equation would provide estimates of forces acting on all three DOFs; however, the forces acting on DOFs 1 and 3 would be much smaller (and zero in theory) than the force acting on DOF 2. This type of *inverse problem* where excitation forces are found

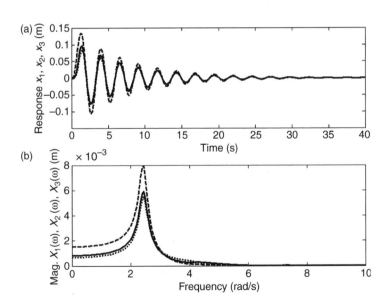

Figure 2.38   (a) Impact response time histories of three DOFs (–) x_1, (- - -) x_2 and (...) x_3, and (b) frequency domain representation of responses showing that resonances have been excited by impact force

using measured responses instead of the other way around can provide good results but is also sensitive to the measurements used (Cardi *et al.*, 2006). In later chapters, methods for solving inverse problems in the proper way by conditioning the measurements to obtain accurate results will be described.

2.3.2.1 Frequency Response Sensitivity Functions

In the previous section, an FRF model of a structural component was utilized to estimate the responses of a three-DOF system to an impact type load. That same model could also be used to estimate the impact force that produces a set of measured responses. In this section, the same FRF model is utilized to predict how individual FRFs change when damage occurs (Yang, 2004). Conversely, the model can also be used to estimate the level of damage that causes a change in measured FRFs (Johnson *et al.*, 2004). To demonstrate how FRFs can be used in this way, the single-DOF model shown in Figure 2.39(a) will be considered. The model has three parameters, M, C and K, that change in different ways depending on the damage mechanism. For example, the stiffness K may increase if this model describes a panel that buckles due to warping. Warping introduces curvature into panels causing them to stiffen. On the contrary, cracking in a component causes a reduction in its cross-sectional area and load-carrying capability by reducing the stiffness (refer to Chapter 3). If an FRF model of the component is measured or calculated analytically, the change in this FRF model as a consequence of these stiffness changes can be determined.

The FRF model for the single-DOF system shown in Figure 2.39(a) is given by

$$X(\omega) = H(\omega)F(\omega)$$

$$\text{where } H(\omega) = \frac{1}{K - M\omega^2 + j\omega C} \qquad (2.62a, b)$$

Figure 2.40 shows the magnitude (a) and phase (b) of this FRF as a function of frequency for three different values of K: 10 N/m (–), 9 N/m (- - -) and 11 N/m (...). These plots were

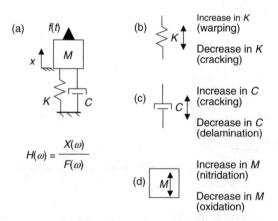

Figure 2.39   (a) Single-DOF system model, hypothetical changes in (b) stiffness, (c) damping and (d) mass due to damage

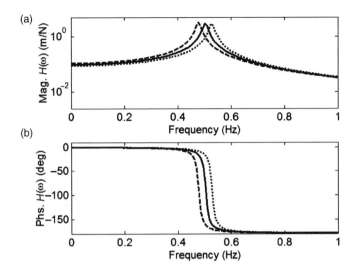

Figure 2.40   FRF magnitude (a) and phase (b) for single-DOF system model

generated using the simulation code 'onedofsens.m'. Note that the amplitude at the original resonant frequency drops when the stiffness is increased or decreased. This decrease in amplitude occurs because no matter how the stiffness changes, the change causes the resonant frequency to shift. For frequencies lower than the resonant frequency, decreases in stiffness cause increases in the magnitude. This result is expected because at low frequencies, the FRF approaches $1/K$ and the component behaves like a static spring with stiffness K. For frequencies higher than the resonant frequency, the stiffness has little effect on the magnitude because the mass dominates at higher excitation frequencies. Also note that the phase of the FRF increases for all frequencies when the stiffness increases but decreases for all frequencies when the stiffness decreases. The amplitude and phase changes are always largest near the resonant frequency.

These observed changes in the FRF magnitude and phase can be identified using sensitivity analysis. To determine the sensitivity of the FRF to changes in stiffness, the partial derivative of $H(\omega)$ in Equation (2.62b) is taken with respect to K as follows:

$$\frac{\partial H(\omega)}{\partial K} = -\frac{1}{(K - M\omega^2 + j\omega C)^2} \qquad (2.63a, b)$$
$$= -H^2(\omega)$$

The second line of this equation indicates that changes in the FRF with stiffness are only a function of the FRF. In other words, *it is not necessary to know the initial stiffness to determine how the FRF changes when the stiffness changes.* As mentioned previously, many model parameters for a component are often unknown; therefore, this sensitivity formula can be useful in health monitoring. For example, if it is assumed that the changes in $H(\omega)$ and K are small percentages of their original values at a given frequency, then Equation (2.63b) can be expressed in terms of *finite differences* instead of partial differentials. The sensitivity to stiffness can then be expressed as follows:

$$\frac{\Delta H(\omega)}{\Delta K} \cong -H^2(\omega) = S_{H,K}(\omega) \tag{2.64a}$$

$$\text{where } \Delta H(\omega) = H_2(\omega) - H_1(\omega) \tag{2.64b}$$

$$\Delta K = K_2 - K_1 \tag{2.64c}$$

where Δ denotes the difference between the FRF and spring constant before and after the damage causes a change in K, $H_2(\omega)$-$H_1(\omega)$ is the difference between the FRFs for the two different values of K, and $K_2 - K_1$ is the difference between the two values of K. The function $S_{H,K}(\omega)$ is shorthand notation for the sensitivity of $H(\omega)$ to the changes in K. This function is called the *embedded sensitivity function* with respect to K because the sensitivity information is embedded inside of $H(\omega)$. This sensitivity function is most accurate for small changes in K (less than approximately 3% of the original stiffness).

Figure 2.41 (generated using 'onedofsens.m') shows the magnitude (a) and phase (b) of this sensitivity function. Note the following characteristics:

- The magnitude and phase experience the largest changes at resonance as expected from the result in Figure 2.40.
- The magnitude experiences a lesser change at low frequencies and very little change at high frequencies as expected because the stiffness dominates for frequencies below resonance and the mass dominates for frequencies above resonance.

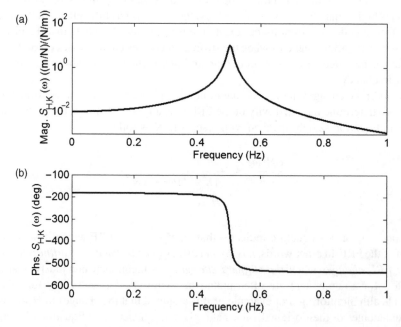

Figure 2.41 Magnitude (a) and phase (b) of embedded sensitivity function $S_{H,K}$ for single-DOF system model

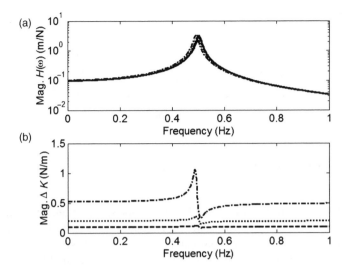

Figure 2.42 (a) Magnitude of single DOF FRF for 1%, 2% and 5% decreases in K, and (b) estimates of ΔK using Equation(2.64a)

- The phase experiences negligible changes for low and high frequencies because the sensitivity phase is near $-180°$ for low frequency and $-540°$ for high frequency, which result in no effective change in phase.

The embedded sensitivity function has other uses in modeling for health monitoring. For example, by using $S_{H,K}(\omega)$ and the change in $H(\omega)$ (i.e. ΔH), the change in K (i.e. ΔK) can be found. In other words, damage can be quantified. For 1% (-0.1N/m), 2% (-0.2N/m) and 5% (-5N/m) decreases in K, the simulation code 'onedofesens.m' was used to estimate the change in stiffness ΔK by applying Equation (2.64a). Figure 2.42(a) shows the magnitudes of the FRF for each of these stiffness values including the original value of 10N/m. As expected, the resonant frequency shifts downward in frequency as the stiffness decreases. Figure 2.42(b) shows the estimates of ΔK for each percentage decrease. The absolute value of ΔK is shown. For a 1% decrease in stiffness, the estimate is nearly constant with frequency at the correct value of 0.1N/m. However, when ΔK reaches -0.2N/m, the estimate is constant only below and above the resonant frequency. Near the resonant frequency, the estimate is in error because a 2% change in stiffness violates the assumption of small ΔH somewhat. Again, when the stiffness change grows to -5% (-5N/m), the accuracy of the estimate decreases near the resonance. However, the estimate of ΔK for low frequencies and high frequencies is nearly perfect. This example shows that embedded sensitivity functions should only be applied away from resonances to estimate changes in stiffness that are relatively small compared to the original stiffness.

The same methodology described above for stiffness can be applied to identify changes in damping and mass. However, the sensitivity function models for damping and mass are given by the following partial derivatives:

$$\frac{\partial H(\omega)}{\partial C} = -(j\omega) \times H^2(\omega) \tag{2.65a}$$

$$\frac{\partial H(\omega)}{\partial M} = -(j\omega)^2 \times H^2(\omega) \tag{2.65b}$$

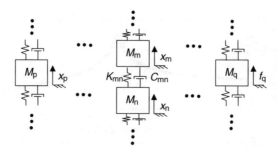

Figure 2.43   General form of lumped parameter multiple-DOF system model

Note that the damping sensitivity function is equal to the stiffness sensitivity function multiplied by $j\omega$, and the mass sensitivity function is equal to the stiffness sensitivity function multiplied by $(j\omega)^2$. Sensitivity functions can also be used to determine which among these parameters (mass, damping or stiffness) has caused a change in the FRFs. Each sensitivity function has a different shape as a function of frequency, and these differences can be used to select the proper sensitivity function to estimate the parameter change. The problems at the end of the chapter provide an example of this technique.

Most structural components have more than one DOF; consequently, formulas for calculating embedded sensitivity functions for models with two or more DOFs are also needed in practice. Figure 2.43 shows a general lumped parameter multiple-DOF model with various masses, dampers and springs. Connections among the DOFs can be arbitrary with springs and dampers in various directions to simulate restoring forces in shear, bending and so forth. The sensitivity of FRF $H_{pq}(\omega)$ with respect to changes in stiffness K_{mn} between DOFs m and n is given by the following formula (Yang $et\ al.$, 2004):

$$\frac{\partial H_{pq}(\omega)}{\partial K_{mn}} = -(H_{pm} - H_{pn}) \times (H_{qm} - H_{qn}) \qquad (2.66a)$$

$$\frac{\partial H_{pq}(\omega)}{\partial C_{mn}} = (j\omega) \times \frac{\partial H_{pq}(\omega)}{\partial K_{mn}} \qquad (2.66b)$$

$$\frac{\partial H_{pq}(\omega)}{\partial M_{m0}} = (j\omega)^2 \times \frac{\partial H_{pq}(\omega)}{\partial K_{m0}} \qquad (2.66c)$$

where $H_{j0}(\omega) = 0$. This formula can be used to estimate changes in stiffness, damping or mass as in the single-DOF example. For instance, consider the three-DOF system model in Figure 2.12. Assume the excitation force is applied at DOF M_2. Suppose that the stiffness K_3 increases by 8% due to buckling. Because the excitation is applied at DOF 2 as shown in Figure 2.12, one convenient form of Equation (2.66a) can be obtained by choosing $p = q = 2$. The two subscripts m and n must be chosen so as to bracket the damaged element that is of interest. Note that m cannot be equal to n because that does not bracket an element. In this example, K_3 is sandwiched between DOFs 2 and 3; therefore, $m = 2$ and $n = 3$. With these choices for the subscripts in Equation (2.66a), the following sensitivity formula is obtained:

$$\begin{aligned}\frac{\partial H_{22}(\omega)}{\partial K_{23}} &= -(H_{22} - H_{23}) \times (H_{22} - H_{23}) \\ &= -(H_{22} - H_{32})^2\end{aligned} \qquad (2.67a, b, c)$$

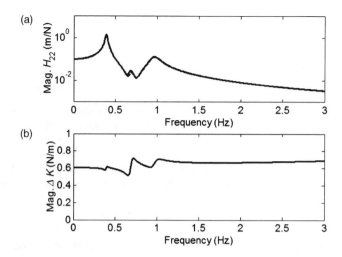

Figure 2.44 (a) Magnitude of $H_{22}(\omega)$ for 8% increase (0.64N/m) in K_3, and (b) estimate of ΔK using Equation (2.67a)

The second line of this equation was obtained by ensuring that reciprocity holds for the FRFs when the input and output DOFs are interchanged, that is $H_{23} = H_{32}$. The simulation code 'threedofesense.m' was used to calculate and plot the $H_{22}(\omega)$ FRF for the original stiffness (–) and modified stiffness (- - -) due to damage in Figure 2.44(a). Note the small shift upward in frequency in the range from 0.6 to 1Hz. When Equation (2.67a) was used to estimate the change in stiffness that caused this shift in the FRF, the estimate shown in Figure 2.44(b) was obtained. The exact value of the stiffness change was 0.64N/m; therefore, the estimate low in frequency and high in frequency is relatively accurate. The estimate near the peaks in the FRF has error because the 8% shift in stiffness violates the assumption made in converting the partial differentials to finite differences.

This sensitivity method could also be useful for updating component models to account for the effects of damage on the dynamic response of components to loads. For example, if a reduction in stiffness is identified, then the stiffness of the component in the damaged area could be reduced to make more accurate predictions of future load-carrying capability. More will be said about prediction later in the chapter.

2.3.2.2 Virtual Force Models

The sensitivity function model described in the previous section was used to estimate the level of damage due to changes in mechanical properties such as mass or stiffness. The method in this section describes damage as a force introduced in the vicinity of the damage. Figure 2.45 illustrates this concept. The figure shows a damaged component on the left with cracking in one location causing a change in stiffness and damping in addition to corrosion in a different location causing a reduction in mass. This component is then modeled as the sum of (1) a healthy component with no damage and (2) a set of mechanical forces applied in the vicinity of the damage to account for the localized effects of the damage. These forces are called *virtual forces* because they are used to model the change in local mechanical properties due to damage (White, 2006).

Figure 2.45   Damaged component on the left is equivalent to an undamaged component plus a set of virtual mechanical forces that modify the local internal force distribution (Hundhausen *et al.*, 2005, DEStech Publications, Inc.)

For a general multiple-DOF system model with damage, the FRF relationship between the response displacement vector, $\{X(\omega)\}$, and external force vector, $\{F(\omega)\}$, is given by

$$\{X(\omega)\} = [H_{dam}(\omega)]\{F(\omega)\} \tag{2.68}$$

where $[H_{dam}(\omega)]$ is the FRF matrix (assumed to be square) of the damaged system model. Using the concept in Figure 2.45, Equation (2.68) can be rewritten in terms of the undamaged FRF matrix, $[H(\omega)]$, and the virtual force vector, $\{F_v(\omega)\}$, as follows:

$$\begin{aligned} \{X(\omega)\} &= [H(\omega)]\{F(\omega)\} + [H(\omega)]\{F_v(\omega)\} \\ &= [H(\omega)](\{F(\omega)\} + \{F_v(\omega)\}) \end{aligned} \tag{2.69a, b}$$

By decomposing the response in this manner, it has been assumed that the damage can be modeled using a localized change in force on the component. This type of virtual force model can describe both linear and nonlinear forms of damage (see Section 2.2.7).

Although Equation (2.69a) can be used to compute the virtual force vector to identify damage, it is often more convenient to calculate the virtual force in terms of the undamaged and damaged FRF matrices. In order to calculate the virtual force using FRF matrices, $\{F_v(\omega)\}$ in Equation (2.69a) is expressed in terms of the external forcing vector, $\{F(\omega)\}$, using a virtual force projection matrix, $[H_v(\omega)]$, as follows:

$$\{F_v(\omega)\} = [H_v(\omega)]\{F(\omega)\} \tag{2.70}$$

Then this projection matrix can be computed with the following sequence of manipulations:

$$\begin{aligned} \{X(\omega)\} &= [H(\omega)]\{F(\omega)\} + [H(\omega)][H_v(\omega)]\{F(\omega)\} \\ &= [H(\omega)]([I] + [H_v(\omega)])\{F(\omega)\} \\ &= [H_{dam}(\omega)]\{F(\omega)\} \end{aligned} \tag{2.71}$$

Therefore, $[H_v(\omega)] = [H(\omega)]^{-1}[H_{dam}(\omega)] - [I]$

Note that in order to utilize this type of virtual force model, it is necessary to measure the entire healthy FRF matrix (White *et al.*, 2006). Measurement techniques for acquiring this matrix of data will be discussed in later chapters.

Consider a damage scenario for the three-DOF system model in Figure 2.12 where mass M_1 decreases by 5% simulating oxidation of the material in that region of the component. The simulation code 'threedofvforce.m' was used to calculate and plot the FRFs shown in Figure 2.46. These FRFs correspond to an excitation applied at DOF 2. Note that all of the

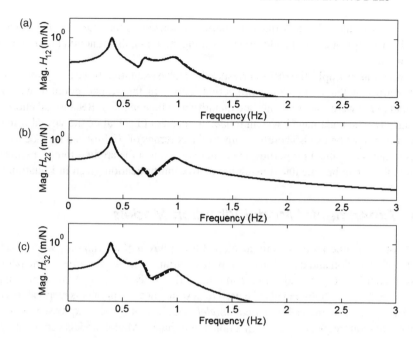

Figure 2.46 FRF magnitudes for three-DOF system model (a) $H_{12}(\omega)$, (b) $H_{22}(\omega)$ and (c) $H_{32}(\omega)$ with excitation at DOF 2 with (- - -) and without (_) change in K_1

FRFs change when the mass decreases. These FRFs must be interpreted using a model such as the virtual force model in order to locate the source of the change. Figure 2.47 shows the magnitudes of the entries in the second column of the virtual force projection matrix, $[H_v(\omega)]$, from Equation (2.71). The line types in the plot correspond to (–) DOF 1, (- - -) DOF 2 and

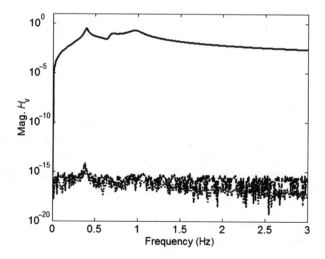

Figure 2.47 Second column of virtual force projection matrix, $[H_v(\omega)]$, for three-DOF system model showing force at DOFs (_) 1, (- - -) 2, and (. . .) 3

(...) DOF 3. Note the only entry that is numerically nonzero is for DOF 1, which is where the damaged mass M_1 is located. In addition to locating the mass, this function also quantifies the change in mass.

In the previous example, the FRFs corresponding to an excitation force at F_2 were examined and the entries of the virtual force projection matrix in the second column, which again correspond to F_2, were used to locate and quantify a change in mass. Because all entries of the FRF matrix for the undamaged system model are needed to calculate the virtual force, more than one force must be used. Therefore, any of the columns of $[H_v(\omega)]$ can be used to identify changes in the mechanical properties due to damage. When the applied forces are measured, the virtual force can be calculated using the matrix multiplication given in Equation (2.70).

2.3.3 Experimental Modal Vibration Models

The previous dynamic models were developed using *physical coordinates* where each DOF consisted of a location and direction of a physical variable like displacement. The parameters in those models were physical mass, damping and stiffness values. A modeling approach based on *nonphysical* DOFs can also be used to describe how components respond to dynamic loads. These models are called *modal models* because they are expressed in terms of component natural frequencies and deflection mode shapes. Modal models can be analytically derived using the methods in Section 2.2.2. In fact, the free vibration response of a component was shown in Section 2.2.3 to be the sum of a component's modes of vibration. In many applications for which an analytical (first-principle) model is unavailable, experimental (data-driven) modal models can be identified more directly from measured data.

The three-DOF system model in Figure 2.12 is considered as an example. The analytical equations of motion for this model were given in Equation (2.21). In order to motivate the development of data-driven modal models, the physical coordinates (x_1, x_2 and x_3) are transformed into modal coordinates, called *principal coordinates* (p_1, p_2 and p_3). The mode shapes are used to make this transformation:

$$\left\{ \begin{array}{c} x_1(t) \\ x_2(t) \\ x_3(t) \end{array} \right\} = [\{\psi_1\} \ \{\psi_2\} \ \{\psi_3\}] \left\{ \begin{array}{c} p_1(t) \\ p_2(t) \\ p_3(t) \end{array} \right\} \tag{2.72a}$$

$$\{x(t)\} = [\Psi]\{p(t)\} \tag{2.72b}$$

where the vectors $\{\Psi_r\}$ are the deflection mode shapes of the component and the matrix of mode shapes, $[\Psi]$, is called the *modal matrix*. This equation was already applied in a different form in Equation (2.8) to describe the free response of a two-DOF system model. If this coordinate transformation is substituted into the model equations of motion (Equation (2.21b)) and both sides of the equation are multiplied by the transpose of the modal matrix, $[\Psi]^T$, then the following sequence of equations is obtained:

$$[M]\{\ddot{x}(t)\} + [C]\{\dot{x}(t)\} + [K]\{x(t)\} = \{f(t)\}$$
$$[M][\Psi]\{\ddot{p}(t)\} + [C][\Psi]\{\dot{p}(t)\} + [K][\Psi]\{p(t)\} = \{f(t)\}$$
$$[\Psi]^T[M][\Psi]\{\ddot{p}(t)\} + [\Psi]^T[C][\Psi]\{\dot{p}(t)\} + [\Psi]^T[K][\Psi]\{p(t)\} = [\Psi]^T\{f(t)\}$$
$$[M_r]\{\ddot{p}(t)\} + [C_r]\{\dot{p}(t)\} + [K_r]\{p(t)\} = \{f_r(t)\}$$

$$\tag{2.73a--d}$$

The matrices $[M_r]$, $[C_r]$ and $[K_r]$ are called the *modal mass, damping and stiffness matrices*, respectively; they are diagonal symmetric matrices. r is the *mode number*. Each of the entries is dependent on how the modal vectors are scaled. For example, one common method for scaling the modal vectors is to make each entry of $[M_r]$ equal to one (this is called *unity modal mass*). The forcing vector $\{f_r(t)\}$ is called the *modal force*.

Each of the equations in Equation (2.73) is uncoupled from all of the other equations. When the FRF analysis method from Section 2.2.4 is applied to this equation, the following relationship is obtained between the modal coordinate phasor and the modal force phasor:

$$\{P(\omega)\} = [-\omega^2[M_r] + j\omega[C_r] + [K_r]]^{-1}\{F_r(\omega)\}$$
$$= [H_r(\omega)]\{F_r(\omega)\} \qquad (2.74a, b, c)$$
$$\text{where } \mathrm{diag}([H_r(\omega)]) = \frac{1/M_r}{\omega_{nr}^2 - \omega^2 + j2\sigma_r\omega}$$

In order to develop modal models based on experimental data, this set of equations is converted back into physical coordinates by substituting Equation (2.72b) into (2.74b) and then multiplying both sides of the resulting equation by the modal matrix:

$$\{X(\omega)\} = [\Psi][H_r(\omega)][\Psi]^T\{F(\omega)\}$$
$$= [H(\omega)]\{F(\omega)\} \qquad (2.75)$$

Each entry $H_{pq}(\omega)$ of the FRF matrix, $[H(\omega)]$, can then be expressed using the modal vector coefficients for the input (q) and output (p) DOFs along with the eigenvalues λ_r, which are sometimes called the *modal frequencies*:

$$H_{pq}(\omega) = \sum_{r=1}^{N} \frac{Q_r\psi_{pr}\psi_{qr}}{j\omega - \lambda_r} + \frac{Q_r^*\psi_{pr}\psi_{qr}}{j\omega - \lambda_r^*}$$
$$= \sum_{r=1}^{N} \frac{A_{pqr}}{j\omega - \lambda_r} + \frac{A_{pqr}^*}{j\omega - \lambda_r^*} \qquad (2.76a, b, c)$$
$$\text{where } A_{pqr} = Q_r\psi_{pr}\psi_{qr} \text{ and } Q_r = \frac{1/M_r}{j2\omega_{dr}}$$

where ψ_{pr} is the *modal vector coefficient* for mode r at DOF p, Q_r is a *modal scale factor*, which depends on how the modal vectors are scaled, and A_{pqr} is called the *modal residue* for output DOF p and input DOF q at mode r. The complex conjugates of A_{pqr} and modal frequencies λ_r are indicated with an asterisk, that is $(a + bj)^* = a - bj$.

This equation can be used to estimate the modal parameters in a data-driven modal model of a component. There are many methods for estimating these parameters (Allemang, 1999). For example, the *peak-pick* (or *quadrature*) method extracts the modal vectors for one mode at a time by assuming that one pair of terms in Equation (2.76a) dominates near a resonant peak in the FRF data. When this assumption is made, the imaginary parts of the FRFs at each resonant frequency are proportional to the modal vector exhibited by the system at that mode. If one column of the FRF matrix is considered, then the excitation DOF is fixed and the FRF with the

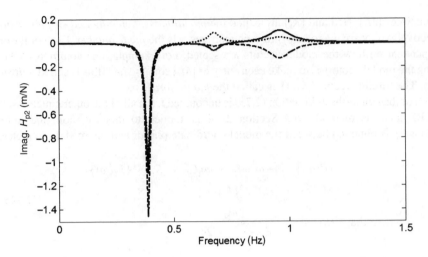

Figure 2.48 Imaginary parts of (–) H_{12}, (- - -) H_{22} and (. . .) H_{32} showing relative deflection pattern at each mode of vibration

single mode assumption is given by

$$H_{pq}(\omega_{dr}) \cong \frac{A_{pqr}}{j\omega_{dr} - \lambda_r} = \frac{Q_r \psi_{pr} \psi_{qr}}{-\sigma_r} = \left(\frac{Q_r \psi_{qr}}{-\sigma_r}\right) \psi_{pr} \tag{2.77}$$

This equation indicates that the FRFs $H_{pq}(\omega_{dr})$ evaluated near the resonant frequency ω_{dr} for fixed q are proportional to the mode shape of the corresponding output DOF p. Furthermore, Q_r is an imaginary number (see Equation (2.76c)). Therefore, the imaginary part of the FRFs can be utilized as the modal vector for mode r.

Consider the three-DOF system model. The simulation code 'threedofmodal.m' was utilized to obtain the FRFs for an excitation applied at DOF 2 of the model (see Figure 2.12). The imaginary parts of these FRFs are plotted in Figure 2.48. The imaginary parts at the damped natural frequencies were also scaled such that the largest amplitude among the DOFs is unity. These values are listed in Table 2.5. Note the following characteristics of these modal deflection shapes:

- The first mode shape consists of all three DOFs moving in phase with DOF 2, which moves more relative to the DOFs 1 and 3. This mode shape would cause all elements in the system model to undergo limited amounts of stretching and compression. Therefore, changes to the dampers and springs connecting DOFs 1, 2 and 3 would not be very visible in the response data.

Table 2.5 Imaginary parts of FRFs used in peak-pick modal parameter estimation for excitation applied at DOF 2 and responses at DOFs 1 to 3

DOF	$\omega_{d1} = 2.45\,\text{rad/s}$	$\omega_{d2} = 4.27\,\text{rad/s}$	$\omega_{d3} = 6.03\,\text{rad/s}$
x_1	−0.75	−0.71	0.82
x_2	−1.00	−0.33	−1.00
x_3	−0.70	1.00	0.36

- The second mode shape consists of larger motions at DOF 3 with less motion at DOF 1 and even less motion at DOF 2. DOF 3 moves out of phase with DOFs 1 and 2 leading to large relative motions across the elements connecting DOFs 3 and 2. At this mode, changes primarily to the damper and spring connecting DOFs 3 and 2 would be most evident in response data.
- The third mode shape consists of larger motions at DOF 2 with large motion across the elements connecting DOFs 1 and 2, and DOFs 2 and 3. At this mode, changes to all damping and stiffness elements would be evident in the response data.

Other methods can be used to extract both the modal frequencies and modal vectors for use in modeling a component. For example, the *local least squares method* uses FRF data in the neighborhood of the damped natural frequencies as in the peak-pick method. Equation (2.77) is rewritten in the following form to develop this method:

$$[H_{pq}(\omega_{dr})\ 1]\left\{\begin{array}{c}\lambda_r\\A_{pqr}\end{array}\right\} = j\omega_{dr}H_{pq}(\omega_{dr}) \tag{2.78}$$

This form is used because it isolates the modal parameters to be estimated (λ_r and A_{pqr}) from the FRF data that can be measured; however, there are two unknowns in Equation (2.78), and these cannot be found using only one equation. Other equations are generated at frequencies just above and below the resonant frequency as in the set of equations below:

$$\begin{bmatrix}H_{pq}(\omega_1) & 1\\H_{pq}(\omega_2) & 1\\\vdots & \vdots\\H_{pq}(\omega_{N_f}) & 1\end{bmatrix}_{N_f\times 2}\left\{\begin{array}{c}\lambda_r\\A_{pqr}\end{array}\right\}_{2\times 1} = \left\{\begin{array}{c}j\omega_1 H_{pq}(\omega_1)\\j\omega_2 H_{pq}(\omega_2)\\\vdots\\j\omega_{N_f} H_{pq}(\omega_{N_f})\end{array}\right\}_{N_f\times 1} \tag{2.79}$$

For fixed q (input DOF), this equation is then generated for all p (output DOFs), and the resulting set of equations is solved for the modal frequency and residue vector, which contains the mode shape as explained previously.

For the three-DOF system model, the simulation code 'threedofmodal.m' was used to extract the modal frequencies and modal vectors by implementing Equation (2.79). The results are given in Table 2.6. The modal vectors were obtained by dividing each residue vector for a given mode by the entry with the maximum magnitude in that vector. The modal frequencies are expressed with units of rad/s including both the real ($-\sigma_r$) and imaginary (ω_{dr}) parts. The estimated damped natural frequencies are very accurate. The mode shapes for modes 1 and 3 are accurate as well; they all are nearly imaginary and have the correct amplitudes. The mode shape for mode 2 is reasonable but has relatively large real parts indicating a *complex mode*.

Table 2.6 Modal frequencies and modal vectors estimated using local least squares parameter estimation for excitation applied at DOF 2 and responses at DOFs 1 to 3

DOF	$\lambda_1 = -0.07 + j2.45$	$\lambda_2 = -0.19 + j4.25$	$\lambda_3 = -0.38 + j6.02$
x_1	$0.02 - 0.74j$	$-0.55 - 0.64j$	$-0.19 + 82j$
x_2	$0.04 - 1.00j$	$-0.10 - 0.29j$	$-0.09 - 1.00j$
x_3	$0.01 - 0.69j$	$-0.45 + 0.89j$	$-0.05 + 0.37j$

A complex mode is one in which there is relative phase between the DOFs as they undergo the modal deflection. Complex modes occur in practice because damping is often nonproportional or nonlinear in nature. In this example, the modes should be real normal modes because the damping is proportional (see Equation (2.38)). The error is due to the coupling that exists between modes. This coupling is evident in Figure 2.48 especially between modes 2 and 3 but was ignored in the local least squares method to extract the modal vectors. One way to improve the results is to use a smaller frequency resolution in the FRF data. This choice leads to more data points close to the resonant peak where the assumption of a single mode is most valid. The results can also be improved by using a multidegree of freedom modal parameter estimation method (Allemang and Brown, 1998).

2.3.4 Other Data-Driven Models

There are many other types of data-driven models. Ljung (1987) describes various time and frequency domain model identification approaches. These models can be used to describe time-invariant and time-varying systems. Worden and Tomlinson (2001) cover a wide range of issues in modeling nonlinear structural components with both time and frequency domain techniques. Volterra series, higher order frequency response functions and other methods are derived and applied to model mechanical systems. The text also reviews fundamentals in parameter estimation using least-squares methods and probabilistic issues dealing with model parameters under uncertainty. Haykin (1994) discusses neural network models for representing a wide range of dynamic systems. Multilayer perceptrons, radial basis function networks and other types of artificial neural network models are developed in that text. All of these methods strive to produce an input–output model that can be used to predict the response of a dynamic system or identify changes in model parameters that have caused changes in the response of that system. Other types of models are implemented in subsequent chapters.

2.4 LOAD MODELS

Before using the models described in the previous sections for health monitoring to identify damage or predict how damage grows, models must also be developed for the loads acting on the component and measurement system. Loads that cause vibration or changes in vibration response with time are of primary interest in this book. Loads cause damage to initiate in the first place and loads can cause sensors and data analysis algorithms to fail; therefore, load models are essential in designing health-monitoring systems. The topic of excitations generated by health-monitoring hardware, that is actuators, for actively interrogating components with damage will be addressed in Chapter 4.

2.4.1 External Mechanical Excitations

The models described above for components are affected by the nature of external forces (and moments) applied to a component. These forces and moments are usually applied to the component at the connections, or boundary conditions, through which the component is attached to the system. In some cases, external forces occur when impacts, pressures and so forth are applied directly to the component. The types of forces also determine the damage mechanisms and failure modes to expect. The following examples illustrate the influence that external forces can have on component models:

- When the preload force level on the ends of the linkage in Figure 2.24 change, the stiffness of the bushings change because bushings are nonlinear in nature leading to changes in traveling wave amplitudes.
- When the vibration frequency of excitation changes in the panel shown in Figure 2.1 due to changes in how the underlying airframe vibrates, different modes of vibration are excited leading to more (or less) sensitivity to certain damage locations in the panel.
- Similarly, the vibration frequency of excitation changes the nature of the restoring force models in Section 2.3.1.2.
- In order to properly estimate parameters in direct parameter and discrete time domain models, all of the frequencies of interest in the models must be sufficiently excited.
- If the location of the applied force changes in a transmissibility function vibration model, so too do the transmissibility functions.
- If the impact force location changes in the example from Section 2.3.2, the measured FRFs must include this new impact location in order to estimate the impact force.
- Different sensor and actuator locations are more (or less) sensitive to certain vibration excitation frequencies due to differences in mode shapes.

2.4.1.1 Impulsive Excitations

In order to classify external excitations, the relationship between excitation time histories and excitation frequency content must be understood. Figure 2.49(a) shows a tire undergoing an excitation force $f(t)$ and corresponding response $x(t)$. Three types of excitation forces are

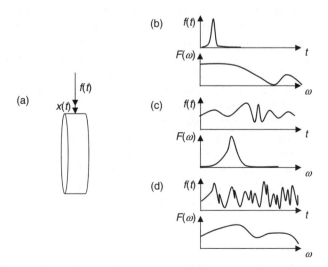

Figure 2.49 Types of mechanical forces acting on tire (a) with force time history and frequency spectra for (b) impacts, (c) narrowband excitations and (d) broadband random excitations

shown in Figure 2.49(b)–(d). The force in (b) is an *impulsive* type force. This force is transient (narrow) in the time domain meaning that the force starts and is over quickly. This type of force might occur if the tire runs over an obstacle (speed bump) or strikes the curb. The amplitude of this impulsive excitation as a function of frequency is shown in (b). Note that the amplitudes are *broadband* in nature extending over a wide frequency range. In the limit as the force becomes very short in time, the Dirac delta function, $\delta(t)$, is used to model this type of load. $\delta(t)$ is defined by

$$\delta(t) = \begin{cases} 0 \text{ for } t < 0- \\ 0 \text{ for } t < 0+ \end{cases} \text{ and } \int_{-\infty}^{+\infty} \delta(t)dt = 1 \qquad (2.80a, b)$$

where $0+$ and $0-$ denote times just to the right and left of $t = 0$, respectively. The Fourier transform, where $\mathbf{F}[f(t)] = \int_{-\infty}^{+\infty} f(t)e^{-j\omega t}dt$ is the Fourier transform of $f(t)$, produces the frequency spectrum of the impulse $\delta(t)$:

$$\mathbf{F}[\delta(t)] = \int_{-\infty}^{+\infty} \delta(t)e^{-j\omega t}dt = 1 \text{ for all } \omega \qquad (2.81)$$

which confirms that impulses span a broad frequency range. This relationship indicates that the impulse possesses energy at all frequencies.

The energy imparted by an impulsive type excitation can be calculated by integrating the product of the force time history and the displacement at the point of application:

$$E = \int_{0}^{x(T)} f(t)dx = \int_{0}^{T} f(t)\dot{x}dt \qquad (2.82)$$

where T denotes the end time and 0 denotes the beginning time of the impulsive excitation. This relationship indicates that a curb impact on a tire, for example, could deliver large energies in one of two ways: (1) through a large force $f(t)$ due to higher speed of the tire upon impact or (2) through a large displacement (or velocity) response $x(t)$ upon impact at an area of the tire such as the sidewall where the stiffness is smaller.

It can be concluded that impulsive excitations excite vibrations and propagating waves across a wide range of frequencies. This type of response data are useful for identifying models of components and identifying damage because impulsive data contain information about the component over a wide range of frequencies. An example of an impulsive excitation for the wire harness connector in Figure 1.6 is a hard landing or impact during service if the connector is stepped on. Impacts on the thermal protection system panel shown in Figure 2.1 are also impulsive in nature. Other impulsive excitations include sudden changes in excitation torque in rotating transmission drivelines. Impacts can cause all of the following types of damage:

- localized damage to the material leading to corrosion and other forms of degradation,
- fatigue cracking over time due to a stress concentration introduced at an impact site,
- denting and penetration leading to reduction in bending stiffness and,
- fiber breakage in composite materials.

2.4.1.2 Narrowband Excitations

Damage due to fatigue is usually caused by excitations in narrow frequency bands that cause components to undergo large forced response amplitudes leading to cracking and other forms of damage. Figure 2.49(c) shows a *narrowband* type of excitation (refer to Section 2.2.8.4). A narrowband excitation for a tire, for example, can occur on a roadway surface with seams and depressions, which are regularly spaced. The regular spacing coupled with the tire rotation produces cyclic forces in certain frequency bands. These forces act on the tread, sidewall and other parts of the tire leading to potential damage. The windmill rotor in Figure 1.23 also experiences cyclic forces due to its rotation. In cyclic excitations that are periodic in nature, it is customary to use Fourier series to model the excitations. The Fourier series for a periodic function $f(t)$ is given by

$$f(t) = \sum_{n=0}^{\infty} [a_n \cos(n\omega_o t) + b_n \sin(n\omega_o t)]$$

$$\text{where } a_n = \frac{2}{T_o} \int_{-T_o/2}^{T_o/2} f(t) \cos(n\omega_o t) dt \tag{2.83}$$

$$b_n = \frac{2}{T_o} \int_{-T_o/2}^{T_o/2} f(t) \sin(n\omega_o t) dt \text{ with } T_o = \frac{2\pi}{\omega_o}$$

where ω_o is the *fundamental frequency* of the force signal, the a_n and b_n coefficients define the contribution of each sinusoid and cosinusoid at frequency $n\omega_o$, and T_o is the *fundamental period* of the force signal. For example, if a tire rotates 8 times per second, then the fundamental frequency is 8Hz or 50rad/s. The tire excitation force would then contain the 50rad/s frequency and its various harmonics as expressed in Equation (2.83). The harmonics appear because as the tire rotates, it experiences forces at various positions in its rotation due to nonuniformity. The plot in Figure 2.49(c) shows a narrowband *spectrum* displaying the amplitude of the force signal as a function of frequency. This spectrum is simply a plot of the coefficients, $\sqrt{a_n^2 + b_n^2}$, as a function of ω. After calculating the coefficients in Equation (2.83), the various modeling techniques discussed above can be applied. In linear models such as the FRF model, each frequency can be analyzed independently. In nonlinear models such as the two-DOF model with Coulomb friction and asymmetric stiffness, simulations must consider all frequencies simultaneously.

In some applications, excitations are not periodic in nature. For example, the excitation shown in Figure 2.30(a) is not periodic although it is comprised of frequency components in a narrow band as shown in Figure 2.29. The Fourier integral transform is used in these cases to model excitations. As mentioned above in Section 2.4.1.1, the Fourier transform of a general forcing function $f(t)$ is given by

$$\mathbf{F}[f(t)] = \int_{-\infty}^{+\infty} f(t) e^{-j\omega t} dt = \int_{-\infty}^{+\infty} f(t) e^{-j2\pi f t} dt \tag{2.84}$$

The Fourier transform is a more general form of the Fourier series in Equation (2.83). In both load models, excitations $f(t)$ are modeled as sums of sines and cosines[9]. However, unlike the

---

[9]Sines and cosines in the Fourier integral are buried inside of the exponential: $e^{-j\omega t} = \cos \omega t - j \sin \omega t$.

Fourier series, the Fourier transform uses an infinitely long fundamental period T to decompose excitations into oscillating components. Also, the Fourier series uses discrete frequencies in its formulation, but the Fourier transform uses a continuous range of frequencies to model $f(t)$ excitations that are transient or do not repeat.

2.4.1.3 Broadband Random Excitations

The passenger truck in Figure 1.20 can experience impulsive excitations due to obstacles and narrowband excitations due to cyclic forces from the roadway. In general, however, the truck experiences a combination of these types of forces. Also, rough roads introduce excitation forces into the truck over a broad range of frequencies. These excitations are random in nature due to fluctuations in specific frequencies and amplitudes that are present at a given instant in time. In order to model excitations like these, probabilistic concepts in random variables are needed. If the Fourier transform, $F(\omega)$, of $f(t)$ is calculated, then the *power* (or *auto*) *spectrum* $G_{ff}(\omega)$ is given by (Bendat and Piersol, 1980)

$$G_{ff}(\omega) = F(\omega) \times F^*(\omega) = |F(\omega)|^2 \tag{2.85}$$

with units of (force)^2. In some applications, it is more common to measure the *auto-spectral* (or *power*) *density* $S_{ff}(\omega)$. $S_{ff}(\omega)$ with units of (force)^2/Hz is given by

$$S_{ff}(\omega) = \frac{2\pi}{\Delta\omega} G_{ff}(\omega) \tag{2.86}$$

The auto-spectral density can be used in conjunction with other methods to generate a probabilistic representation of an excitation force. If an excitation is measured, then its mean value, μ, can be computed using the formula:

$$\mu = \lim_{T\to\infty} \frac{1}{T} \int_0^T f(t)dt \tag{2.87}$$

the variance can be computed as follows:

$$\sigma^2 = \lim_{T\to\infty} \frac{1}{T} \int_0^T (f(t) - \mu)^2 dt \tag{2.88}$$

and the mean square value can also be found:

$$v^2 = \lim_{T\to\infty} \frac{1}{T} \int_0^T f^2(t)dt \tag{2.89}$$

These parameter are related as follows: $\sigma^2 = v^2 - \mu^2$. The auto-spectral density can also be used to calculate v:

$$v^2 = \int_0^\infty S_{ff}(\omega) \frac{1}{2\pi} d\omega \tag{2.90}$$

When the excitation is only measured at discrete points in time for a certain length of time, then *the integrals in these equations are replaced by sampled averages.*

To demonstrate the process of modeling broadband excitations using spectral densities and statistical parameters, consider the excitation imparted to the passenger truck tire in Figure 1.20 by the roadway. This analysis is performed in the simulation code 'bbandpdf.m'. If the excitation time history follows a *Gaussian (or normal) distribution* with mean value μ and standard deviation σ, then the excitation can be expressed as follows using the probability density function $p(f)$:

$$p(f) = \frac{1}{\sigma\sqrt{2\pi}}e^{-(f-\mu)^2/2\sigma^2} \tag{2.91}$$

This formula indicates that the probability $f(t)$ will lie in a certain range is given by

$$P(f \in (f_1, f_2)) = \int_{f_1}^{f_2} p(f)\,\mathrm{d}f \tag{2.92}$$

where the probability density function is integrated over the excitation value range from f_1 to f_2. Consider the forcing time history shown in Figure 2.50(a). This signal has frequency components at 4, 5.5 and 6.5 Hz in addition to Gaussian random noise. It is common in practice to have a few dominant frequencies excited due to rotating machinery or other periodic sources and random components due to more random forces. The frequency spectrum for the excitation is shown in (b). By computing the auto-spectral density function in Equation (2.86), the mean square value can be obtained from Equation (2.90): $V^2 = 9.9\,\mathrm{N}^2$. Then the mean value can be estimated using Equation (2.87), 0.02 N, and the variance follows from the

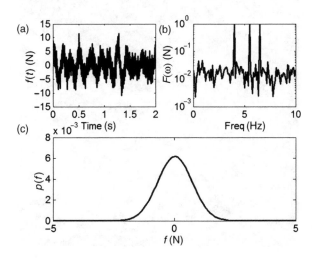

Figure 2.50 (a) Excitation signal with three sinusoidal components at 4, 5.5 and 6.5 Hz in addition to Gaussian random content, (b) Fourier spectrum and (c) probability density

formula $\sigma^2 = v^2 - \mu^2$: $\sigma^2 = 9.9\,\mathrm{N}^2$. These parameters can then be substituted into Equation (2.91) to construct the force probability density function. Conversely, this process can be applied backward using the auto-spectral density and mean value of the excitation from measured data to generate the actual force time history.

Other probability density functions can also be utilized to model excitations including Rayleigh and uniform distributions. Thomson and Dahleh (1998) and Bendat and Piersol (1986) have reviewed these and many other types of probabilistic models in detail.

2.4.2 Acoustic Pressure, Temperature and Other Environmental Loads

Acoustic loads due to dynamic changes in pressure occur on components such as the ceramic tile shown in Figure 1.23 and the metallic panel shown in Figure 2.1. If these loads are severe, they can result in sonic fatigue causing panels to buckle as shown in Figure 2.51. Sonic fatigue is a form of high cycle fatigue that occurs in thin panels exposed to low strain levels ($\sim$100 microstrain) over long periods of time. At the minimum, acoustic loads affect health-monitoring response data and the process of damage identification because they incite vibrations and propagating waves in components across a broad frequency range (Jiang et al., 2006). Acoustic pressure loads are modeled as distributed forces applied across a given area. Force is calculated using the pressure multiplied by the area of application.

Temperature fluctuations affect the health-monitoring process by loading components, making them more susceptible to certain types of damage (e.g. creep and corrosion). If the temperature of a component changes, strains in the component produce stresses that can cause the component to remain deformed even after the temperature returns to normal. In addition,

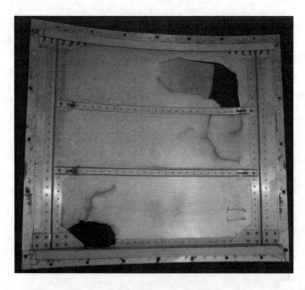

Figure 2.51   Panel that failed after undergoing sonic fatigue due to high acoustic loading (courtesy Dr S. M. Spottswood, Structural Mechanics Branch, AFRL/VASM)

temperature fluctuations cause changes in vibration and wave propagation characteristics, which must be taken into account to avoid false indications of damage in health-monitoring algorithms (Sohn *et al.*, 1999; Kess, 2005). (Young, 2001) provides many results useful for analyzing the strains and stresses produced by temperature loads applied to material components. Temperature fluctuations also cause changes in material modulus and other material properties, which can be ascertained from material data sheets and other material characterization data.

Other environmental factors that load components include the ambient humidity (percent moisture in air), ultraviolet radiation and chemical composition of liquid or gas in which a component is immersed. Changes in humidity primarily affect composite material components. These changes can be modeled by modifying the material density to account for absorption of moisture. Ultraviolet radiation also primarily affects composite materials by causing the matrix in these materials to harden resulting in slight increases in component modulus. Changes in ambient chemical composition primarily cause changes in the laws governing the growth of damage especially corrosive damage.

To demonstrate the need for acoustic and thermal load models, the panel in Figure 2.1 is modeled here as it undergoes typical loading during launch and reentry in aerospace applications. The acoustic loads applied to the panel are selected based on the estimated acoustic loading over the panel surface area (force=pressure×area). The sound pressure levels are selected based on the loading profile illustrated in Figure 2.52. The profile shows the acoustic pressure in dB (with respect to 2e-6Pa) as a function of temperature in degrees Fahrenheit. The amplitude units of pressure are lb/in.^{2} ($\times 10^2$). This approximation is based on actual flight data (Hundhausen, 2004). A stepwise approach is used to model the pressure and temperature. In other words, each variable is sampled and held constant for a given mission segment. An auto-spectral density function is selected for these loads with uniform amplitude over the

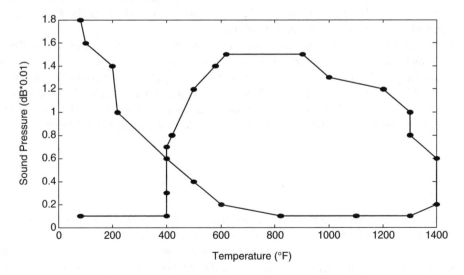

Figure 2.52 Approximate pressure versus temperature launch, exit and reentry profile of aerospace vehicle (Hundhausen *et al.*, 2005, Mr. Jason Hundhausen, LANL)

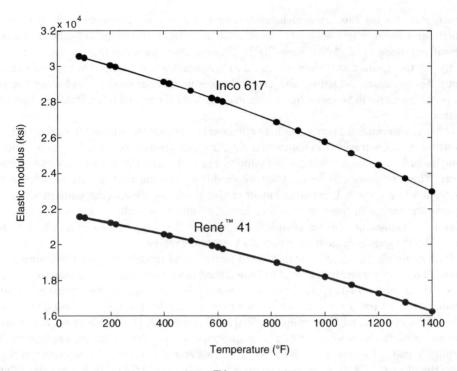

Figure 2.53 Variation in Inco 617 and René^{TM} 41 modulus in ksi with temperature over typical mission profile (Hundhausen *et al.*, 2005, DEStech Publications, Inc.)

simulation frequency range from 0.1 to 1000Hz and a Gaussian distribution. Loads are also applied to the panel centers of mass in the lateral direction to simulate the forces due to vibration of the supporting airframe beneath the panel.

High temperatures cause significant reductions in the modulus of metallic materials. Recall that the modulus of elasticity determines the wave propagation velocity and the natural vibration frequencies of components; therefore, it is important to model this change in modulus. It is assumed that the panel in Figure 2.1 is constructed from Inco 617 and René^{TM} 41 standoffs. Figure 2.53 shows the variation in modulus of both these materials. These variations are listed on material data sheets and can be described using the following empirical equation:

$$\frac{E}{E_{RT}} = 0.9988 - (1.306 \times 10^{-4})\Delta T - (4.298 \times 10^{-8})\Delta T^2 \qquad (2.93)$$

where E is the modulus at an elevated temperature, E_{RT} is the modulus at room temperature (70°F) and ΔT is the difference between these temperatures. Because the temperature varies somewhat from flight to flight, a Gaussian distribution (Equation (2.91)) is used to represent these variations in the temperature model. Also, the standoffs of the panel in Figure 2.1 are beneath the panel whereas the panel surface is exposed to the highest temperatures; therefore,

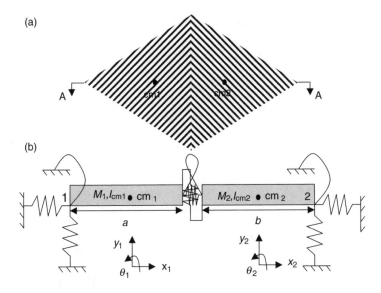

Figure 2.54   (a) Top view of diamond-shaped panel with two lumped inertias spanning cross section AA and (b) side view of six-DOF low-order model at cross section AA (Ackers *et al.*, 2006, Mr. Jason Hundhausen, LANL)

a 30% temperature drop is assumed across the panel due to thermal resistance to heat conduction.

These two load models can now be applied to the panel model. Figure 2.54(a) shows a top view of the panel in Figure 2.1. The two cross-hatched shaded regions have centers of mass cm1 and cm2 that lie on cross section AA. A schematic of the six-DOF model that is used here is illustrated in Figure 2.54(b). It includes two translational (M_1,M_2) and rotational (I_{cm1},I_{cm2}) inertias restrained at their boundaries by out-of-plane, in-plane and rotational springs, which model the bolted standoff panel connections at locations 1 and 2. Coordinates x_1 and x_2 describe the in-plane translational motions of the two inertias; y_1 and y_2 describe the out-of-plane (transverse) motions; and, θ_1 and θ_2 describe the counter-clockwise rotational motions. The matrix equation of motion for the panel is

$$[M]_{6\times6}\{\ddot{q}(t)\}_{6\times1} + [C]_{6\times6}\{\dot{q}(t)\}_{6\times1} + [K]_{6\times6}\{q(t)\}_{6\times1} = \{f(t)\}_{6\times1} \qquad (2.94)$$

where

$$\{q(t)\} = \{x_1(t)\,x_2(t)\,y_1(t)\,y_2(t)\,\theta_1(t)\,\theta_2(t)\}^T \qquad (2.95)$$

and the vector of external forces and moments is

$$\{f(t)\} = \{f_{x1}(t)\,f_{x2}(t)\,f_{y1}(t)\,f_{y2}(t)\,\tau_1(t)\,\tau_2(t)\}^T \qquad (2.96)$$

The mass, stiffness and damping matrices assuming proportional damping are given by

$$[M] = \mathrm{diag}[M_1 \quad M_2 \quad M_1 \quad M_2 \quad I_{cm1} \quad I_{cm2}]$$

$$[K] = \begin{bmatrix} K_2 + K_6 + K_7 & -K_6 & 0 \\ * & K_4 + K_6 + K_7 & 0 \\ * & * & K_1 + K_5 + K_7 \\ * & * & * \\ * & * & * \\ * & * & * \end{bmatrix}$$

$$\begin{bmatrix} 0 & 0 & 0 \\ -K_7 & 0 & 0 \\ -K_5 & -\dfrac{a}{2}K_1 + \dfrac{a}{2}K_5 & \dfrac{b}{2}K_5 \\ K_3 + K_5 + K_7 & -\dfrac{a}{2}K_5 & \dfrac{b}{2}K_3 - \dfrac{b}{2}K_5 \\ * & K_{t1} + K_{t2} + \left(\dfrac{a}{2}\right)^2 K_1 + \left(\dfrac{a}{2}\right)^2 K_5 & -K_{t2} + \left(\dfrac{a}{2}\right)\left(\dfrac{b}{2}\right)K_5 \\ * & * & K_{t3} + K_{t2} + \left(\dfrac{b}{2}\right)^2 K_3 + \left(\dfrac{b}{2}\right)^2 K_5 \end{bmatrix}$$

$$[C] = \alpha[M] + \beta[K]$$

$$(2.97a, b, c)$$

where 'diag' denotes a diagonal matrix and the *entries denote symmetric stiffness coefficients with respect to the diagonal. The parameters in this stiffness (and damping) matrix are attributed to the following mechanical properties:

K_1, K_3 – Transverse (vertical) stiffness of standoffs at corners of panel diamond;
K_2, K_4 – Lateral (axial) stiffness of standoffs at corners of panel;
K_5, K_6 – Transverse and axial stiffnesses of panel at center of panel;
K_7 – Transverse-lateral coupling stiffnesses at corners due to standoffs of panel;
K_{t1}, K_{t3} – Rotational stiffness of standoffs at corners of panel;
K_{t2} – Rotational stiffness of panel at center due to first bending motion;
a, b – Length of two panel segments (inertias) assumed to be equal;
α, β – Proportional damping (mass and stiffness) constants for distributing dissipation.

These parameters were estimated using the direct parameter estimation approach and are provided in the MATLAB file 'panelmatrices.mat'. The MATLAB® simulation code 'panelsimprestemp.m' was used to simulate the response of the panel given the loading profile in Figure 2.52. The results are shown in Figure 2.55. Figure 2.55(a) shows the responses y_1 and y_2 over a certain time period of the dynamic simulation. The response of y_2 is larger due to a loose fastener on the right hand side of the panel at location 2. This response can also be animated using the 'panelsimprestemp.m' code. Figure 2.55(b) shows the change in the six-model natural frequencies with temperature due to the modulus change with temperature during the simulation. These kinds of temperature changes not only disrupt health-monitoring efforts but also lead to changes in panel response, which affect operational failure modes (e.g. thermomechanical fatigue).

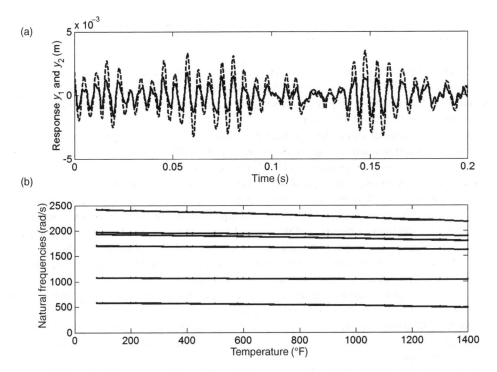

Figure 2.55 (a) y_1 (—) and y_2 (- - -) responses of panel in typical simulation and (b) natural frequencies of model as a function of temperature during simulation

In addition to the effects considered above, temperature changes can also cause components to bend due to boundary condition constraints (such as the standoffs in Figure 2.1). For example, Young (2001) cites Kikukawa (1960), who examined the effects of thermal shock on thermal stresses in components. He showed that if a solid component is suddenly made to change temperature by ΔT, then a stress of level $\Delta TkE/(1 - v)$, where k is the thermal conductivity of the material, forms in the surface of the component. This sudden stress results in damage, transient waves, and vibrations. Kessler (2002) also showed that the phase velocity of a traveling wave at frequency ω for a component with radius of curvature R increases as R and ω decrease:

$$c_{cyl} = c_{ph}\left(1 + (c_{ph}/\omega R)^2\right) \tag{2.98}$$

There are many other implications of temperature changes that are addressed in textbooks on thermal stress analysis.

2.5 SUMMARY

This chapter has examined different types of dynamic models for describing vibrations and wave propagation in structural components. Load models in addition to component models are described. These models are useful in assessing the change in component response when components are damaged and for predicting the response of components to loads. A summary of key points in the chapter is given below.

- Models are essential in health monitoring. *Models enable raw data to be converted into information about the health of a structural material or component.*
- Some models are based on *physical laws* (physics-based) whereas other models are based on measurement data (data-driven).
- *Vibration* models describe the dynamic response of structural components to dynamic loads. *Resonant frequencies and mode shapes* represent the free response, which changes with damage providing a means to identify the degradation.
- *Lumped parameter (discretized) and continuous models* can be used to model vibrations and wave propagation. Vibrations can be thought of as *standing wave patterns setup by interference between propagating waves.* Lumped parameter models produce ordinary differential equations whereas continuous models produce partial differential equations.
- *Inertia, damping and stiffness* are the three elements included in these dynamic models. *Nonlinear* damping and stiffness restoring forces are common in structural components. *Damage nearly always behaves in a nonlinear manner.*
- MATLAB® can be used to analyze the dynamic response of structural components using built-in functions (e.g. eigenvalue analysis, free and forced response, etc.).
- The free vibration response is a *sum of the modes of vibration* where each mode consists of a natural frequency and mode shape. *Impulsive (or transient)* type excitations excite the free vibration response.
- *Frequency response functions (FRFs)* are a ratio between the response and excitation as a function of frequency. FRFs describe the relative amplitude and phase of the dynamic response of a structural component with respect to the excitation forcing amplitude and phase. Resonances in FRFs for linear components occur at the natural frequencies.
- FRF models and impedance models can be combined for different configurations (structural component, sensor, boundary conditions, etc.) to describe the dynamic response of the combined system of components. *Constraints at the connections* are applied to combine these different models.
- Simulink® can be used to simulate the dynamic response of nonlinear vibrating systems to understand the effects of *nonlinear damping and stiffness effects in damage.* Nonlinear friction at bolted interfaces and buckling of panels with asymmetric stiffness are examples of nonlinear damage mechanisms.
- Propagating elastic waves in one, two and three dimensions can be used to identify damage due to changes in speed of propagation in different directions and other factors. The *wave equation* describes the propagation of elastic waves. The *separation of variables* method decomposes the solution of the wave equation into temporal and spatial pieces.
- The *speed of sound* in a material is a function of the modulus and density. The *wave number* is the ratio of frequency to speed of sound; this number represents a spatial frequency. The *group velocity* is found by differentiating the frequency with respect to the wave number; this velocity describes the speed of a group of waves in a certain frequency bandwidth. The speed of sound is equal to the product of frequency in Hz and wavelength.
- *Smaller defects in the component can be detected when the wavelength is small and frequency is high; however, smaller wavelengths also result in more reflections from rough surfaces and boundaries in a component.*

- Dispersion occurs when the propagation energy of a wave front decreases due to motions in other directions. *Dispersion causes the wave front to change shape* because higher frequencies travel faster than lower frequencies making it more difficult to identify damage.
- *Finite element models (FEMs)* can be used to model structural components with discontinuous mass, damping and stiffness properties due to damage or variations in the healthy structural component material and geometrical parameters.
- *Direct parameter models* in the time domain can be used to estimate mass, damping and stiffness parameters from measured excitation and response data taken from structural components. Parameters are estimated using *least squares solution techniques.*
- Restoring force models relate the internal forces within a component due to stiffness and damping to the mass × acceleration of the component. When the acceleration is used to construct the models, *restoring force projections* are obtained. *Phase–plane trajectories* relating the velocity to the displacement response of a component are also useful models for health monitoring.
- *Discrete time models (difference equation models)* can also be estimated from measurement data. The advantage of discrete time models is that all response variables are not needed to estimate the parameters.
- *Experimental frequency response models (also called embedded sensitivity functions)* can be utilized to estimate the change in frequency response function with changes to the mass, damping or stiffness parameters without requiring the starting values of these parameters. Conversely, the changes in parameters can be estimated.
- *Virtual force models* describe damaged components as a sum of the healthy structural component and forces applied locally to represent the damage.
- *Experimental modal vibration models* can be developed from measured data using various modal parameter estimation methods.
- *Load models* for different mechanical excitations (impulses, narrow band cyclic and broad band random) *exercise components and damage differently.* Loads due to other environmental factors including acoustic pressure and temperature are also important to consider in health monitoring.

REFERENCES

Adams, D.E. and Farrar, C.R. (2002) 'Identifying linear and nonlinear damage using frequency domain ARX models', Struct Health Monit, 1(2), 185–201.

Allemang, R. (1999) 'Vibrations: analytical and experimental modal analysis', UC-SDRL.

Allemang, R. and Brown, D. (1998) 'A unified matrix polynomial approach to modal identification', J Sound and Vibration, 211(3), 301–322.

Bendat, J. and Piersol, A. (1980) 'Engineering applications of correlation and spectral analysis', Wiley Interscience, New York.

Bendat, J. and Piersol, A. (1986) 'Random Data: Analysis and Measurement Procedures', Second Edition, John Wiley & Sons, Ltd, New York.

Beranek, L.L., Ed. (1988) 'Noise and Vibration Control', Revised Edition, Iowa: Institute of Noise Control Engineering.

Cardi, A.A., Adams, D.E. and Walsh, S. (2006) 'Locating and quantifying ceramic body armor impact forces on a compliant torso using acceleration mapping', *Proceedings of SPIE Conference on Nondestructive Evaluation and Smart Structures and Materials*, 6177, Paper no. 617714.

Graff, K. (1991) 'Wave motion in elastic solids', Dover Publications, New York.

Haroon, M., Adams, D.E. and Luk, Y.W. (2004) 'A technique for estimating linear parameters of an automotive suspension system using nonlinear restoring force extraction', *Proceedings of the American Society of Mechanical Engineers International Mechanical Engineering Congress and Exposition*, Anaheim, CA, Paper no. 60212.

Haroon, M., Adams, D.E., Luk, Y.W. and Ferri, A. (2005) 'A time and frequency domain approach for identifying nonlinear mechanical system models in the absence of an input measurement', J Sound and Vibration, **283**, 1137–1155.

Haykin, S. (1994) 'Neural networks: a comprehensive foundation', Macmillan College Publishing Company, New York.

Hundhausen, R. (2004) 'Mechanical loads identification and diagnostics for a thermal protection system panel in a semi-realistic thermo-acoustic operating environment', Masters Thesis, School of Mechanical Engineering, Purdue University, West Lafayette, Indiana.

Hundhausen, R.J., Adams, D.E., Derriso, M., Kukuchek, P. and Alloway, R. (2004) Loads, 'Damage identification and NDE/SHM data fusion in standoff thermal protection systems using passive vibration based methods', *Proceedings of the European Workshop on Structural Health Monitoring*, 959–966.

Jiang, H., Adams, D. and Jata, K. (2006) 'Material damage modeling and detection in a homogeneous thin metallic sheet and sandwich panel using passive acoustic transmission', Struct Health Monit, v. 5, pp. 373–387.

Johnson, T. (2002) 'Analysis of dynamic transmissibility as a feature for structural damage detection', Masters Thesis, School of Mechanical Engineering, Purdue University, West Lafayette, Indiana.

Johnson, T.J., Yang, C., Adams, D.E. and Ciray, S. (2004) 'Embedded sensitivity functions for identifying damage in structural systems', *J. Smart Mat. Struct*, **14**, 155–169.

Jones, R. (1999) *Mechanics of Composites* 2nd Edition, Taylor & Francis, Philadelphia, PA.

Kess, H. (2005) 'Investigation of operational and environmental variability effects on damage detection algorithms in heterogeneous (woven composite) plates', *Masters Thesis*, School of Mechanical Engineering, Purdue University, West Lafayette, Indiana.

Kinsler, L., Frey, A., Coppens, A. and Sanders, J. (2000) *Fundamentals of Acoustics*, 4th Edition, John Wiley & Sons, New York, NY.

Kikukawa, M. (1960) 'Factors of stress concentration for notched bars under tension and bending', *Proceedings of the 10th International Congress of Applied Mechanics*, Stresa, Italy, pp. 337–341.

Kessler, S. (2002) 'Piezoelectric-based in-situ damage detection of composite materials for structural health monitoring systems', *Doctoral Thesis*, Department of Aeronautics and Astronautics, Massachusetts Institute of Technology.

Ljung, L. (1998) *System Identification: Theory for the User* 2nd Edition, PTR Prentice-Hall, Upper Saddle River, NJ.

Meirovitch, L. (1967) *Analytical Methods in Vibrations*, MacMillan Publishing Company, New York, NY.

Mohammad, K., Worden, K. and Tomlinson, G. (1992) 'Direct parameter estimation for linear and non-linear structures', *J. Sound Vib*., **152**(3), 471–499.

Masri, S. and Caughey, T. (1979) 'A nonparametric identification technique for nonlinear dynamic problems', *J. Appl. Mech.*, **46**, 433–447.

Nayfeh, A.H. (1995) *Wave Propagation in Layered Anisotropic Media*, Elsevier Science, Amsterdam.

Schmerr Jr., L.W. (1998) *Fundamentals of Ultrasonic Nondestructive Evaluation: A Modeling Approach*, New York: Plenum Press.

Sohn, H., Dzwonczyk, M., Straser, E.G., Kiremidjian, A.S., Law, K.H. and Meng, T. (1999) 'An experimental study of temperature effect on modal parameters of the Alamosa Canyon bridge', *Earthquake Engin. Struct. Dynamics*, **28**, 879–897.

Soedel, W. (1994) *Vibrations of Plates and Shells*, Marcel Decker, New York, NY.

Stark et al., (1997) 'Takens embedding theorems for forced and stochastic systems', *Nonlinear Anal. Theory Method App.*, **30**, 5303–5314.

Strang, G. (1988) *Linear Algebra and Its Applications*, 3rd Edition, Brooks Cole, New York, NY.

Sundararaman, S. (2003) 'Structural diagnostics through beamforming of phase arrays: characterizing damage in steel and composite plates', *Masters Thesis*, School of Mechanical Engineering, Purdue University, West Lafayette, Indiana.

Tse, F.S., Morse, I.E. and Hinkle, R.T. (1978) *Mechanical Vibrations*, 2nd edition, Prentice Hall, Inc. New York, NY.

Thomson, W. and Dahleh, M. (1998) *Theory of Vibration with Applications*, Prentice-Hall, New York, NY.

White, J. (2006) 'Impact and thermal damage identification in metallic honeycomb thermal protection system panels using active distributed sensing with the method of virtual forces', *Masters Thesis*, School of Mechanical Engineering, Purdue University, West Lafayette, Indiana.

White, J. Adams, D.E. and Jata, K. (2006) 'Damage identification in a sandwich plate using the method of virtual forces', *Proceedings of the International Modal Analysis Conference*, St. Louis, MO, Paper #67.

Worden, K. and Tomlinson, G. (2001) *Nonlinearity in Structural Dynamics: Detection, Identification, and Modeling*, IOP
Publishing Ltd., Bristol, UK.

Yang, C. (2004) 'Experimental embedded sensitivity functions for use in mechanical system identification', *Doctoral
Thesis*, School of Mechanical Engineering, Purdue University, West Lafayette, Indiana.

Yang, C., Adams, D. E., Yoo, S. and Kim, H.-J. (2004) 'An embedded sensitivity approach for diagnosing system-level
vibration problems', *J. Sound Vib.*, **269**(22), 1063–1081.

Young, W. (2001) *Roark's Formulas for Stress and Strain*, 7th Edition, McGraw-Hill, Hightstown, NJ.

PROBLEMS

(1) Consider the rigid body panel model shown in Figure 2.56 with mass M, mass moment of inertia I_{cm}, and springs and viscous dampers as indicated. The body translates $(x(t))$ and rotates $(\theta(t))$ when a vertical force is applied at distance a from the center of mass. Derive the equations of motion and put them into matrix form. Assume that the angle of rotation is small such that $\sin\theta \approx 1$. Ignore gravity.

(2) Consider the system in Figure 2.56. Given that the equations of motion are

$$\begin{bmatrix} M & 0 \\ 0 & I_{cm} \end{bmatrix} \begin{Bmatrix} \ddot{x} \\ \ddot{\theta} \end{Bmatrix} + \begin{bmatrix} C_1 + C_2 & -C_1L_1 + C_2L_2 \\ -C_1L_1 + C_2L_2 & C_1L_1^2 + C_2L_2^2 \end{bmatrix} \begin{Bmatrix} \dot{x} \\ \dot{\theta} \end{Bmatrix}$$
$$+ \begin{bmatrix} K_1 + K_2 & -K_1L_1 + K_2L_2 \\ -K_1L_1 + K_2L_2 & K_1L_1^2 + K_2L_2^2 \end{bmatrix} \begin{Bmatrix} x \\ \theta \end{Bmatrix} = \begin{Bmatrix} -f(t) \\ af(t) \end{Bmatrix}$$

find the modal frequencies and modal deflection shapes of the system model for $M = 10\,\text{kg}$, $I_{cm} = 2\,\text{kgm}^2$, $L_1 = L_2 = 9\,\text{cm}$, $a = 3\,\text{cm}$, $K_1 = 10\,\text{kN/m}$ and $K_2 = 14\,\text{kN/m}$. Assume proportional viscous damping (Eq. (2.97c)) with $\alpha = 0$ and $\beta = 001$. Explain the meaning of the modal deflection shapes.

(3) Again consider the system in Figure 2.56. What percentage change in K_2 would produce a 10% shift in the damped natural frequencies? What percentage change in M would produce a 10% shift in the damped natural frequencies? Based on this information, would corrosion of the component be easier or more difficult to detect using vibration measurements than cracking in the support K_2?

(4) Using the equations of motion from Problem (2), calculate and plot (in MATLAB®) the frequency response functions between the force $f(t)$ and the two response variables in the frequency range from 0 to 20Hz.

(5) Consider the two-DOF system model in Figure 2.2(a). Use the MATLAB® simulation code 'twodoffrf.m' in the Appendix to analyze changes in the frequency response functions when C_2 changes by 50%. This type of damping change is called nonproportional when damping is not proportional to stiffness and mass.

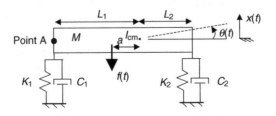

Figure 2.56   Rotational rigid body panel model

(6) Use impedance modeling methods to incorporate a single degree of freedom sensor model positioned at Point A in the system shown in Figure 2.56. Assume the sensor has mass 1 kg and is stud mounted into the body with a stud stiffness 30 kN/m and zero damping. In what frequency range does the sensor most affect the frequency response functions for the system? Where might the sensor be ineffective at measuring the response?

(7) Use the MATLAB® simulation code 'threedoftrans.m' in the Appendix to analyze the transmissibility functions for the three-DOF model in Figure 2.12. Consider a case where K_3 decreases due to damage. Also consider a case where K_1 or K_4 decrease with damage. Over what frequency range should the system be excited in operation to identify damage in all of these elastic elements?

(8) Again use the MATLAB® code 'threedoftrans.m' for the system model shown in Figure 2.12. Move the excitation force from body M_2 to body M_1. Consider decreases in K_2 to simulate damage. How do the transmissibility functions change from those in Figure 2.14? How do the conclusions about the location of damage change? Now consider a case where forces are applied to both M_1 and M_2 (pressure load). How does the conclusion about damage location change?

(9) Use the simulation code 'twodofnlfrf.m' to analyze changes in the frequency response functions for different levels of the excitation f_1. What are the effects on the frequency response for these different excitation levels? How might this sensitivity to excitation level affect the use of models for damage identification?

(10) Apply the boundary conditions for a free-free rod in Equations (2.34a) and (2.34b) and obtain the full longitudinal displacement solution $u(x, t)$ for the two-dimensional wave equation. Free-free boundary conditions correspond to the case where neither end of the rod of length L is attached to anything $(\partial u(x, t)/\partial x = 0$ at $x = 0$ and $L)$. Describe these solutions and what type of response they indicate? (Hint: boundary conditions on the rod result in reflections from both ends)

(11) Consider only the first term in the solution in Equation (2.35b). Sketch the real part of this term for different values of time t as a function of x. What do these plots indicate about the traveling wave? Plot the real part of the first term for different values of position x as a function of t. What do these plots indicate about the traveling wave?

(12) Use the simulation code 'rod1dwave.m' to examine changes in the longitudinal wave response for different positions of the damage. Model the damage using a 20% drop in cross-sectional area in the damaged element. Plot the difference between the wave response for the damaged rod and the wave response for the undamaged rod to determine the location of the damage. How do the results vary when damage is further down the rod versus closer to the source of the excitation pulse?

(13) Again use the simulation code 'rod1dwave.m' to analyze the longitudinal wave propagation response of a damaged rod. Model the damage in different element locations as a 20% reduction in density at those locations. How do the plots showing differences before and after the damage is inserted change from those obtained in the text for changes in cross-sectional area (i.e., stiffness)?

(14) Using the simulation code 'rod1dwave.m,' insert multiple damage locations close together and far apart and plot the differences in wave propagation response before and after the damage is inserted. How well do the plots indicate the presence of multiple versus single damage sites?

(15) Consider the first term in Equation (2.45) indicating the transverse propagating wave in a beam. Sketch the real part of this term for various time values as a function of x. Do the

same for various positions as a function of t. Also examine how these sketches change as the frequency changes. What do these plots indicate about traveling waves along a rod that undergo transverse motions?

(16) Use the simulation code 'rod1dtranswave.m' to examine how the difference between wave propagation response before and after element 10 is modified to have 80% less width than b (due to crack). Conduct the same analysis for 80% less thickness h. Which change has more effect on the results and why?

(17) Use the simulation code 'twodofdpe.m' to estimate the parameters of the two-DOF system model in Figure 2.2 for different numbers of time points, different types of excitation forces at different frequencies and so forth. Explain the errors in the estimated parameters.

(18) Develop a MATLAB® code to estimate the model parameters in the three- DOF system model shown in Figure 2.12. Use a combination of the codes 'twodofdpe.m' and 'threedoftrans.m' to simulate the response of the system and then estimate the parameters. What are the errors in the estimates?

(19) Simulate the restoring forces for the two-DOF system in Figure 2.2 using the 'twodofrf.m' simulation code. Set the nonlinear coefficients to zero. Vary each linear stiffness parameter by 20% and comment on the change in restoring force. Then vary each mass value by 20% and comment on the results. Generate restoring forces for both body M_1 and body M_2. What is the difference between the two sets of restoring forces for the two bodies?

(20) Consider the model shown in Figure 2.57 for a tire with imbalance m at radius r when supported by a suspension with stiffnesses K_1 and K_2. Develop a dynamic model of the response of the spindle in the x and y directions given the frequency of rotation ω. What effect does m have on the vibration levels? In general, at which speeds ω would it be easier to detect an imbalance?

(21) Use the simulation code 'onedofsens.m' to estimate 2, 5, and 15% changes in the mass parameter, M, in the single-DOF model in Figure 2.39(a) based on the frequency response function only. This type of change simulates corrosion damage.

(22) Use the simulation code 'onedofsens.m' to generate a set of frequency response functions for 5% changes in mass, stiffness and damping in the single- DOF system in Figure 2.39(a). Then, compute the difference between the original frequency response function and these functions. Compare these different frequency response functions (magnitudes) to the magnitudes of the sensitivity functions with respect to mass, damping and stiffness. Comment on the comparison. Which sensitivity function matches the difference plots?

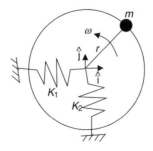

Figure 2.57   Rotational model of suspension and imbalanced tire

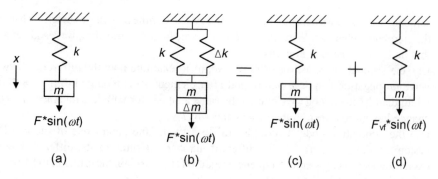

$F^*\sin(\omega t)$

$F^*\sin(\omega t)$

$F^*\sin(\omega t)$ $F_{vf}^*\sin(\omega t)$

(a)   (b)   (c)   (d)

Figure 2.58   (a) Single-DOF system with mass and stiffness, (b) single DOF with damaged values of mass and stiffness, (c) undamaged single-DOF system and (d) system undergoing excitation by virtual damage force

(23) Consider the three-DOF system model in Figure 2.12. Suppose the stiffness K_3 changes by 5%. Use the simulation code 'threedofsense.m' to estimate this change in stiffness based only on the frequency response function before and after the change in stiffness is made. Plot the estimate as a function of frequency.

(24) Again consider the three-DOF system model in Figure 2.12. Suppose the stiffness K_3 changes by 5% and then the stiffness K_1 changes by 5%. Determine how you would use the frequency response functions to estimate these changes in stiffness. The simulation code 'threedofsense.m' can be used to compute the frequency response functions.

(25) Consider the single-DOF system shown in Figure 2.58. In (a), the system is undamaged. Then in (b) the system is shown with modified mass and stiffness parameters simulating the effects of damage. In (c) and (d) the damaged system is decomposed into the undamaged system plus the undamaged system excited by a virtual damage force. Use frequency response methods to show that the virtual force is given by: $F_{vf} = \Delta m\omega^2 H'F - \Delta k H'F$, where H' is the FRF of the damaged system and F is the excitation forcing amplitude at frequency ω.

3

Modeling Damage

There are many different types of damage in metallic and nonmetallic structural material components. This chapter discusses practical modeling techniques for representing many kinds of damage. Both static and dynamic models are discussed. *Static models* are useful for simulating the effects of different types of damage on the component response and interpreting measurement data to identify damage. *Dynamic models* are used to predict the initiation and rate of change (growth) of damage. After describing damage models, brief descriptions of commonly used failure models and criteria are provided. Then a methodology is described for predicting the future performance of a structural component that is damaged.

3.1 STATIC DAMAGE MODELS

Table 3.1 gives a list of damage mechanisms along with static methods for modeling the damage. These examples are described in more detail in the following sections.

3.1.1 Fasteners and Joints

Broken fasteners are the single largest cause of warranty claims in the automobile industry (Bickford, 1990). Fasteners in other applications also become damaged. Self-loosening is the primary source of damage in bolted joints, especially under cyclic transverse loading (Jiang *et al.*, 2003). As a nut is rotated on a bolt's screw thread against a joint, the bolt extends. This extension generates a tension force (or bolt preload). The reaction to this force is a clamp force that causes the joint to become compressed. Equation (3.1) relates the bolt preload, F_p, to the turn angle θ in degrees (Bickford, 1990):

$$F_p = \theta \left(\frac{p}{360}\right) \left(\frac{K_{bolt} K_m}{K_{bolt} + K_m}\right), \tag{3.1}$$

where p is the thread pitch, K_{bolt} is the bolt stiffness and K_m is the joint material stiffness. The equation shows that as the bolt is loosened, the preload decreases, which results in a decrease in the clamp force on the joint. Figure 3.1(a) shows a bolted joint in an automotive suspension. The internal load between the lower ball joint and the top mount decreases as the bolt is

Health Monitoring of Structural Materials and Components: Methods with Applications D. Adams
© 2007 John Wiley & Sons, Ltd

Table 3.1 Examples of material and component damage with modeling approaches

Examples of material and component damage	Damage modeling approach
1. Loss of preload in fastener (suspension in Figure 1.2) Loosening of joint (windmill nacelle fixture in Figure 1.23)	Axial stiffness reduction Interface impacting and sliding
2. Cracking in homogeneous material (spindle in Figure 1.5, clevis Figure 1.2) Ply cracking in layered material (composite armor in Figure 1.3)	Change in cross-sectional area Change in area moment of inertia Local change in damping Impedance change
3. Plastic deformation or penetration (panel in Figure 2.1) Erosion in homogeneous material (valve bushing in Figure 1.9)	Change in stiffness Change in geometry Local change in mass
4. Delamination (Al–Li component in Figure 1.7) Debonding (unwrapping of tile in Figure 1.22) Separation (bead area of tire in Figure 1.21)	Change in bending stiffness Local change in damping Impedance change Eccentric imbalance mass Change in bending stiffness
5. Creep (face sheet on panel in Figure 2.1) Buckling (seal in Figure 1.8)	Stiffness change Change in geometry (area, area moment of inertia, radius of curvature)
6. Corrosion (connector in Figure 1.6) Oxidation (panel in Figure 2.1)	Local change in mass (density) Impedance change
7. Fiber pullout (filament wound canister in Figure 1.3) Fiber breakage (filament wound canister in Figure 1.3)	Local change in damping Stiffness change Impedance change Local change in damping
8. Matrix cracking (composite armor in Figure 1.3)	Modulus change Impedance change
9. Microstructural changes (weld in Figure 1.7) Microcracking (body armor in Figure 1.19)	Local modulus/density change Local change in damping Impedance change

loosened and is the lowest when the bolt is removed completely. If the bolt is cracked, K_{bolt} is decreased to model the crack's effect along the bolt axis:

$$K_{bolt} = \frac{A_c E}{L}, \tag{3.2}$$

where A_c is equal to the nominal cross-sectional area of the bolt, A, minus the area over which the fracture extends due to cracking (see Figure 3.1(a)). If instead, the clamped area on the steering knuckle wears or cracks, K_m is reduced to model the damage.

If the bolt loosens, then the empirical result given by Jiang *et al* (2003) can be used to model the percent reduction in preload, P/P_o, for a given loosening angle θ. When exposed to cyclic loading, a bolted joint undergoes the characteristic change in percent preload shown in Figure 3.1(b). For small numbers of cycles, the loosening effect is subtle but quickly accelerates after a certain number of cycles depending on the bolt. Loosening of the bolt

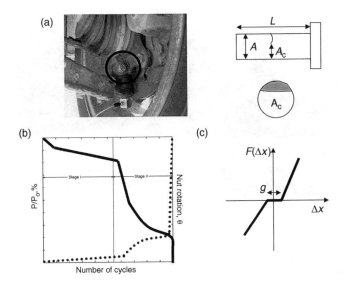

Figure 3.1   (a) Ball joint connecting control arm to steering knuckle with area reduction model of crack (b) self-loosening bolt (Ackers *et al*, 2006, AMSE); percent preload loss (___), nut rotation in deg ($\cdots$) and (c) clearance effect in loose bolt

also introduces sliding and microimpacting across the interface between the bolt and flange. The model for sliding was examined in Section 2.2.7. Figure 2.16 illustrated the type of friction force that appears when sliding occurs across the bolted interface. Figure 3.1(c) illustrates a clearance model for describing the impacts that occur at loose joints along the joint axis. g is the gap formed between the bolt head and seat and Δx is the relative motion between the bolt and seat. This type of discontinuous dead zone stiffness characteristic can be incorporated into Simulink$^{\circledR}$ as in the case of Coulomb friction. Damage in riveted, clamped or other types of fasteners can be modeled using similar methods. This example of fastened joints demonstrates that a variety of damage models are usually needed to describe degradation in a component.

3.1.2 Cracking

Cracks in homogeneous material components can initiate for a variety of reasons including sudden overloads and cyclic fatigue loading. Cracks initiate in areas of a component where there are anomalies (e.g. dislocations) in the material microstructure and/or geometrical stress concentrations (e.g. fillets, holes). The crack length c causes the cross-sectional area moment of inertia and stiffness (axial, bending, torsional) to change. These changes can be modeled using basic concepts from strength of materials. Crack models can then be utilized to determine how the vibrations and wave propagation in material components change when cracked. For example, recall from Section 2.2 that axial and bending stiffness parameters in structural component finite element models are functions of area and area moment of inertia. Also, the phase velocity of transverse bending traveling waves is determined by these parameters ($v = \sqrt{\omega c I_A}$, where $I_A = I/A$).

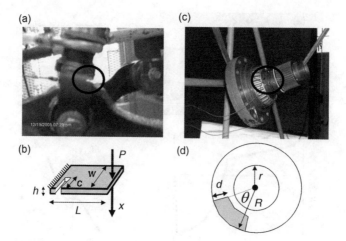

Figure 3.2 (a) Clevis crack at base of suspension strut (b) model of clevis crack (c) cracked area around groove in wheel end spindle and (d) model of crack in spindle

Figure 3.2 shows two examples of cracks in metallic components. Figure 3.2(a) is a clevis of thickness h at the base of an automotive coil over shock module, or strut. A notch of width w is shown highlighted in the clevis. As shown in Figure 3.2(b), the clevis acts as a cantilever beam where its root is at the position of the notch; therefore, the stiffness K_c of the clevis with the notch (and a crack that forms within this notch or in the absence of the notch) can be modeled by changing the cross-sectional area moment of inertia I_c at that position:

$$K_c = \frac{P}{x} = \frac{3EI_c}{L^3} \tag{3.3a}$$

$$\text{with} \quad I_c = \frac{(w-c)h^3}{12}, \tag{3.3b}$$

where P is the axial force transferred from above by the strut and x is the deflection of the clevis. If a component has different boundary conditions than those for the clevis, then different formulas for stiffness can be utilized (Gere and Timoshenko, 1990).

The spindle shown in Figure 3.2(c) is described in Section 1.3. The crack that forms in this spindle spans an angle θ (in deg) with depth d from the outer surface of the annular ring illustrated in Figure 3.2(d). With an inner radius of r and outer radius of R, the cross-sectional area A_d of the spindle at the position of the crack is:

$$A_d = \pi(R^2 - r^2) - \frac{(2rd - d^2)\pi\theta}{360}. \tag{3.4}$$

This area can then be used to construct the stiffness matrix in the axial finite element model described in Section 2.2.8.2. The effect of this reduction in area due to the crack is to shift the lowest longitudinal resonant frequency of the spindle downward. Similarly, the torsional stiffness K_d for the spindle is a function of the polar area moment of inertia at the crack position.

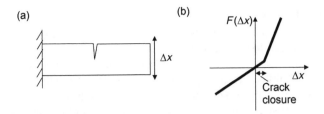

(a)   (b)

$F(\Delta x)$

Δx

Crack
closure

Figure 3.3   (a) Illustration of crack breathing across opening and (b) corresponding nonlinear stiffness restoring force model with bilinear stiffness

The models above describe the effects of a crack as a reduction in linear stiffness due to changes in area or area moment of inertia. However, cracks also introduce other stiffness effects. For example, Figure 3.3(a) shows a cracked component undergoing deflection causing the crack to open and close. Figure 3.3(b) shows the corresponding stiffness restoring force that is experienced by the component as the crack closes leading to higher stiffness for positive deflections. This type of *bilinear* nonlinear force can be incorporated into Simulink® as in the previous cases. Cracks also change the local damping of a material causing the damping to increase due to the heat generated by the opening and closing of the crack. These changes can be described using nonproportional damping models. Structural damping and hysteretic damping are just two of the many damping models available (Tse *et al.*, 1978).

Cracks and many other types of damage can be modeled as changes in *mechanical impedance*. Impedance is strictly speaking the ratio of the applied force on a component (at a certain location in a certain direction) to the velocity response. In some cases, the displacement response is used instead of the velocity. For example, the single DOF system shown in Figure 3.4(a) has the following impedance ($B_X(\omega)$ in terms of displacement and $B_V(\omega)$ in terms of velocity):

$$\frac{F(\omega)}{X(\omega)} = B_X(\omega) = (K - M\omega^2) + j(\omega C)$$

$$\frac{F(\omega)}{V(\omega)} = B_V(\omega) = C + j\left(\omega M - \frac{K}{\omega}\right)$$

(3.5)

If this impedance model (or alternatively an FRF model; see Section 2.2.4) is identified before the clevis becomes cracked, then after it is cracked, as shown in Figure 3.4(b), the impedance can be measured in the vicinity of the crack to identify changes in the local stiffness and damping of the component due to damage. If more detailed information is required, then a discretized model (finite element) can be applied to interpret changes in the damaged impedance measurement.

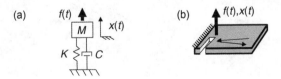

(a) $f(t)$ $x(t)$ (b) $f(t), x(t)$

M

K C

Figure 3.4   (a) Single-DOF system model and (b) force and motion measured in vicinity of crack in clevis when incoming wave is reflected by notch boundary

Impedance and FRF models can be used to characterize changes in the vibration response (standing waves) of a component like the clevis. Changes in impedance due to material and geometric changes introduced by the damage also cause propagating waves to change. Recall from Section 2.2.8 that waves reflect when striking a boundary. Therefore, a wave traveling toward the clevis notch reflects from the notch before it reflects from the clevis support leading to a potential means of characterizing the location and length of the notch. The simulation finite element model 'rod1dtranswave.m' can be used to examine how waves interfere and reflect with notches, cracks, etc. Recall that waves in the one-dimensional longitudinal model reflected (scattered) somewhat due to a change in cross-sectional area representing a crack along the component (see Figure 2.26). In order to fully understand how waves interact with defects, more sophisticated numerical modeling and simulation is required.

3.1.3 Plastic Deformation, Penetration and Erosion

The same methods described above to model cracks using changes in geometry and impedance can be applied to model permanent (plastic) deformations, penetration and erosion in material components. Plastic deformation is usually caused by impacts and other forms of overload (shock, etc.). For example, the panel in Figure 3.5(a) shows dents due to impacts simulating the effects of hail or other foreign object impact (micrometeorites, bird, etc.) in aircraft and aerospace vehicles. When the material is permanently deformed due to these impacts, the local geometry of the component changes leading to changes in local stiffness, damping and possibly even mass distribution (in the case of full penetration). Models similar to those in Equations (3.3) and (3.5) can be used for this type of damage to interpret measurements.

Erosion occurs due to repeated rubbing (abrasion) and impacting between two components. For example, erosion occurs between ball bearings and the inner and outer race of the bearing housing. As the balls and races erode, material flakes enter the lubricant and can be detected to identify this form of damage (often called *spalling*) (Roemer, 2005). Figure 3.5(b) shows a classic case of erosion in a powdered metal bushing through which a butterfly valve rotates. The large divot seen in the bushing is due to impacting during operation of the engine to which this valve is attached. This particular damage can be modeled using a change in the stiffness restoring force that resists the motion of the valve as it impacts the bushing. Figure 3.6 shows how the restoring force upon impact changes as the bushing is eroded increasingly by the impacts. This type of characteristic can be simulated in the manner described previously for friction and asymmetric stiffness nonlinear restoring forces (Section 2.2.7).

(a) (b)

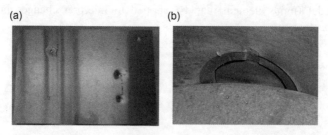

Figure 3.5 (a) Plastic deformation (dents) and penetration due to impact and (b) erosion due to impact between valve and bushing (Ackers *et al.*, 2006, John Wiley & Sons, Inc.)

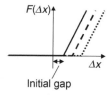

Figure 3.6   Illustration of change in stiffness restoring force with erosion of bushing for increasing levels of erosion depth (___), (- - -) and (. . .)

3.1.4   Delamination, Debonding and Separation

When the layers of a laminated material component separate, the component does not resist bending moments as strongly. Laminated materials derive bending stiffness from the transfer of shear forces from one layer to the adjacent layers; therefore, if the layers separate, the bending stiffness goes down. The damping of the material increases at the location of separation as in the case of a crack because the material undergoes relatively large motions at that point, increasing the local dissipation of energy. However, changes in damping are much more difficult to model and interpret in measurement data because there are so many other factors that influence damping. This section focuses on modeling separation events like *delamination* in laminated composites and *debonding* in sandwich materials.

Figure 3.7(a) shows a radial tire with bead area *separation*, which may or may not involve small cracks in the carcass of the tire. Bead area separations result in delamination between plies of the tire. One practical way to model these separations is by changing the composite elastic modulus of the tire in the radial direction as shown in Figure 3.7(b). The figure shows how the modulus varies as a function of circumferential position around the tire. A step change

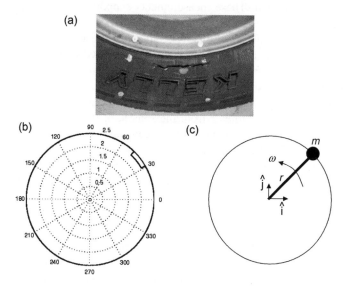

Figure 3.7   (a) Bead area separation in radial tire (b) variation in modulus with circumferential position and (c) imbalance mass to model asymmetric stiffness effects

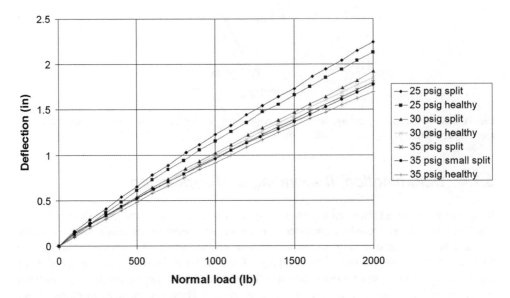

Figure 3.8 Radial deflection versus normal load in tire with separation at bead showing changes in stiffness at various inflation pressures

in one portion of the modulus between 30° and 45° is indicated in the plot. This model is justified because bead area damage changes the stiffness properties of the tire in the radial direction. Figure 3.8 shows a comparison between the spindle load versus radial deflection characteristic of tires at different pressures with and without separation in the bead. At each tire pressure, there is a measurable difference between the stiffness of the healthy tire and the tire with bead area separation. These measurements support the use of the discontinuous modulus model in Figure 3.7(b).

An alternative way to model separation events in tires is to consider the tire action during operation. As a tire rolls, it radially compresses the damaged area once per tire rotation. Because the stiffness of the tire is lower in the circumferential position at which the damage is located, an imbalance force on the tire spindle is produced. This imbalance force can be represented using the schematic in Figure 3.7(c). The schematic shows an eccentric mass m rotating at a frequency ω at a radius r from the spindle of the tire. As this eccentric mass rotates, it creates vertical and horizontal forces causing the spindle to respond in these directions. In fact, with bead area damage in only one sidewall of the tire, the imbalance forces due to separation cause forces that rock the tire from side to side as well. In the plane of the schematic shown, the imbalance forces to model the damage are expressed as

$$f_{\text{imb}}(t) = (m\omega^2 r \cos \omega t)\hat{i} + (m\omega^2 r \sin \omega t)\hat{j}, \tag{3.6}$$

where $\hat{i}$ and $\hat{j}$ are unit vectors. This same approach can be used to model mechanical faults in many types of rotating machines. The expression in Equation (3.6) shows that for higher rotational speeds, bead area damage has a more pronounced effect on the tire response. The equation also shows that when the bead separates at a larger distance r from the spindle on the sidewall, the tire response is more affected by the separation.

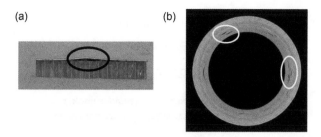

Figure 3.9 (a) Debonding between face sheet and honeycomb core in sandwich panel and (b) delamination of laminated composite cylinder (ultrasonic scan image, courtesy Dr S. Walsh, Army Research Laboratory)

Figure 3.9 shows additional examples of debonding (a) and delamination (b) in a sandwich panel and laminated composite cylinder, respectively. In the sandwich panel on the left, the face sheet has separated from the core because the silicon-based adhesive has melted due to momentary high temperature exposure. This separation lowers the bending stiffness of the panel because the shear forces at the separation site are not transferred between the core and the face sheet. In the cylinder (Figure 3.9), the bending stiffness decreases due to separation in the indicated locations; however, as a pressure vessel, the cylinder relies more on its fibers to provide sufficient radial stiffness.

3.1.5 Creep and Buckling

Spinning blades and shafts undergoing centrifugal loads at elevated temperatures can stretch as the material in the component softens (see Figure 2.53). If the deformation disappears when the temperature and centrifugal load are removed, then no damage has occurred. However, if the deformation remains after loads are removed, then *creep* is said to have occurred and there are often *residual stresses* in the material. In a rotating component, this permanent deformation causes imbalance forces as in the case of the damaged rotating tire. Creep damage can therefore be modeled as in Figure 3.7(c) to determine its effects on the component response during operation. Creep also affects the component cross-sectional area and area moments of inertia leading to changes in both the vibrations and wave propagation within the component (e.g. rotor blade, shaft, etc.). These changes in geometric properties can be modeled as in Section 3.1.2 for cracks. The residual stresses due to creep can also be harmful to the component because cracks can initiate in areas where the residual stresses are the highest.

Components that are not subjected to a static load in operation (as in the case of centrifugal forces) can also creep. If the temperature is high causing the component to expand, then thermal expansion mismatches in the component can cause residual stresses when the high temperature is removed. For example, in the panel shown in Figure 2.1, the face sheet on the top experiences the highest temperature. The core experiences a lesser temperature load due to thermal resistance to conduction through the face sheet. This scenario is illustrated in Figure 3.10(a). When the face sheet expands, it does so against the lower amount of expansion occurring in the core. The core, consequently, experiences a static load at elevated tempera-ture. This high thermal stress at elevated temperature can cause creep (permanent lengthening)

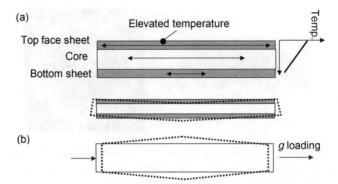

Figure 3.10   (a) Heating, expansion and stresses causing creep in panel and subsequent buckling of the panel and (b) high acceleration static loading causing buckling of cylindrical component resulting in change in radius of curvature

of the core resulting in the type of *buckling* due to residual stress as illustrated in Figure 3.10(a) after temperatures are removed.

Buckling in components can also occur for other reasons. For example, the panel just discussed may also experience very high rates of cooling after being heated to high temperatures. This type of *thermal shock* causes the face sheet to shrink more quickly than the core (over a period of milliseconds). The face sheet can buckle in the process. Also, large dynamic loads (shock) accompanied by high static loading due to either temperature effects or mechanical forces (launch loads) can also cause buckling. For example, Figure 3.10(b) shows a cylindrical component undergoing high acceleration. This scenario occurs when launching rockets to deploy space satellites for communication and other purposes. These rockets can be relatively large with thin walls that can bellow and subsequently buckle during launch. The buckling is sometimes permanent (plastic) resulting in damage. Buckling can be modeled using a reduction in stiffness in the axial direction along the component. The stiffness decreases in this direction because the bending stiffness of a component is smaller than the axial stiffness, and it is the bending stiffness that dominates when buckling occurs as in Figure 3.10(b). Small amounts of buckling can also be modeled as a decrease in the radius of curvature, which causes an increase in the bending stiffness of the component. In addition to changing the wave speed in a component as shown in Equation (2.98), Soedel (1994) has shown that curvature affects natural vibration frequencies. For radii of curvature R_1 and R_2 in the two in-plane directions, the transverse natural frequencies ω_c of a curved component are related to the frequencies ω_f of the flat component by

$$\omega_c^2 = \omega_f^2 + \frac{E}{\rho(1-v^2)}\left(\frac{1}{R_1^2} + \frac{1}{R_2^2} + \frac{2v}{R_1 R_2}\right). \tag{3.7}$$

3.1.6   Corrosion and Oxidation

Corrosion occurs due to environmental chemical loading, which causes pitting, exfoliation and other forms of corrosive damage including hydrogen embrittlement (Grandt, 2004; Jata

(a)    (b)

Figure 3.11 (a) Corrosion of gas turbine wire harness connector and (b) oxidation of titanium plate after being exposed to high temperature thermal soaks (courtesy Dr K. Jata, Air Force Research Laboratory)

and Parthasarathy, 2006). Corrosion at low temperatures is primarily due to solid–liquid interactions and is accelerated by high temperatures. At high temperatures, however, solid–gas interactions are primarily responsible for material loss leading to *oxidation*. Figure 3.11(a) shows a wire harness connector with corrosion inside the connector housing as well as on the pins. Corrosion inside the connector can cause short circuits and, if pins weaken, open circuits. When material components become corroded, material is lost causing a reduction in the cross-sectional area as in the case of a crack. However, in the case of corrosion, the reduction in cross-sectional area can be more wide spread than for localized cracks. Corrosion in the form of pitting also introduces small holes in the surface of a material component. These holes can then serve to initiate cracks. In order to model corrosion damage, methods like those discussed in Section 3.1.2 can be used to modify the cross-sectional area. Alternatively, impedance models like those discussed in that same section can be utilized. The sharp discontinuities at pitting sites scatter propagating waves similar to the way cracks do.

In the case of corrosion damage, the change in mass due to corrosion through the part thickness can also be directly modeled. For example, the sensitivity method in Section 2.3.2.1 can be used to determine the amount of mass loss due to corrosion. The procedure for doing this is the same as for the stiffness example provided in that section. Once this mass loss is known, the density and part geometry can be used to estimate the depth of corrosion for use in health monitoring. The titanium plate shown in Figure 3.11(b) has experienced oxidation on its surface due to several cycles of high temperature soaks. Oxidation damage like this can be identified in the same way as the corrosion damage using mass loss estimates. Alternatively, oxidation causes a loss in density on the surface of the component. The speed of propagating waves changes when the density changes (recall Equation (2.30)); therefore, the change in the speed of propagating waves can also be used as an indicator of the amount of oxidation that has occurred.

3.1.6.1 Fiber Pullout and Fiber Breakage

In composite material components, loads are carried primarily by the fibers or filaments. For example, both of the images in Figure 3.12 contain composite materials with fibers providing the bulk of the mechanical resistance to load. Figure 3.12(a) shows an image of a composite material with fibers missing due to *fiber pullout*. This type of damage can occur during manufacturing or operation as high tensile loads along the fiber cause the fiber to debond and slide along the matrix. The region from which fibers have slid out is weaker than the bulk material leading to potential cracking in the plies. Similarly, Figure 3.12(b) shows an impact zone in a composite cylinder. In this zone of the material, fibers have broken and layers have

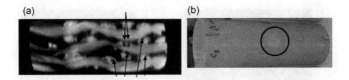

Figure 3.12 (a) Fiberpull out at various locations in woven composite (courtesy Dr S. Walsh, Army Research Laboratory) and (b) S2 glass composite canister impact zone with delamination and fiber breakage

delaminated leading to a high potential for cracks in the plies and propagation of those cracks to failure when the cylinder is pressurized.

In order to model fiber pullout and fiber breakage, the fibers are removed from composite constitutive relationships when calculating the effective elastic and shear moduli. For example, consider an orthotropic lamina with fibers along the x-direction. The constitutive law for this lamina consisting of composite materials under plane stress conditions is given by

$$
\left\{
\begin{array}{c}
\sigma_{xx} \\
\sigma_{yy} \\
\sigma_{xy}
\end{array}
\right\}
=
\begin{bmatrix}
\dfrac{E_{xx}}{1 - v_{yx}v_{xy}} & \dfrac{v_{yx}E_{xx}}{1 - v_{yx}v_{xy}} & 0 \\[2mm]
\dfrac{v_{yx}E_{yy}}{1 - v_{yx}v_{xy}} & \dfrac{E_{yy}}{1 - v_{yx}v_{xy}} & 0 \\[2mm]
0 & 0 & G_{xy}
\end{bmatrix}
\left\{
\begin{array}{c}
\varepsilon_{xx} \\
\varepsilon_{yy} \\
\varepsilon_{xy}
\end{array}
\right\},
\tag{3.8}
$$

where the subscripts denote the directionally dependent stresses, strains, moduli and Poisson's ratios (Soedel, 1994). The vector on the left-hand side contains the two normal stresses and shear stress, and the vector on the right-hand side contains the two normal strains and shear strain. The moduli can be calculated as a function of the amount of filament that is used relative to the amount of matrix that binds the filaments together. Specifically, the two moduli of elasticity along the x- and y-axes of the lamina and corresponding Poisson's ratio can be estimated as follows (Soedel, 1994):

$$
E_{xx} = E_f V_f + E_m V_m
\tag{3.9a}
$$

$$
E_{yy} = E_m \frac{1 + \alpha V_f}{1 - V_f}
\tag{3.9b}
$$

$$
v_{xy} = v_f V_f + v_m V_m
\tag{3.9c}
$$

$$
\text{with } \alpha = \frac{E_f/E_m - 1}{E_f/E_m + 1} \text{ and } V_f + V_m = 1
\tag{3.9d}
$$

where the subscripts 'f' and 'm' denote the properties of the filament and matrix, respectively. These expressions show that a composite lamina can be thought of as an orthotropic material with effective moduli and Poisson's ratio determined by the ratio of filament to matrix volumes in the material. In other words, materials with more filament behave essentially as if they were made entirely of that filament material.

Given the relationships in Equation (3.9), changes in modulus in each direction of a composite laminate material can be calculated when fibers are damaged or missing. For

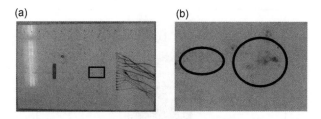

Figure 3.13 (a) Laminated composite sheet and (b) close-up of matrix cracking on surface

example, Equation (3.9a) indicates that if fibers become broken resulting in a lower damaged volume fraction V_{d-f} of filament, then E_{xx} decreases by the fraction:

$$\frac{E_f V_{d-f} + E_m(1 - V_{d-f})}{E_f V_f + E_m(1 - V_f)}.$$ (3.10)

This estimate can be used in conjunction with measurements of vibration and wave propagation response data based on the modeling concepts discussed previously.

3.1.7 Matrix Cracking

Matrix cracking is relatively common in composite materials. Figure 3.13(a) shows a composite plate with 20 plies of S2 glass fibers within an industrial matrix. A close-up picture in Figure 3.13(b) shows that certain regions of the plate surface have cracking in the matrix (area in between the woven fibers). Although this type of damage is less critical than broken fibers, matrix cracking is important and can be modeled using the method presented in the previous section. Composite elastic properties are a function of the fraction of matrix present in the material. If some of the matrix fails due to cracking, then the elastic properties weaken according to the formulae in Equation (3.9).

3.1.8 Microstructural Changes

Microstructural changes in materials are largely responsible for crack initiation and propagation in structural components. The word 'microstructural' indicates the properties of a material at the level of grains and grain boundaries in metals and alloys, for instance. As an example, Figure 3.14(a) shows a component that contains a friction stir weld between two aluminum lithium plates. This type of weld does not require that the joined materials be heated to their melting points. Instead, the weld is produced by plastically stirring the interfaces of both plates and applying pressure to join them. This process produces three distinct types of microstructure in the component. Figure 3.14(b) shows a micrograph of the material outside the welded zone. This microstructure shows the typical healthy laminar microstructure of a metal alloy such as Al–Li. The grains in this microstructure are regular and there are no clusters of dislocations that can lead to cracking. Figure 3.14(c) shows the microstructure in the material approaching the weld. The grains bend toward the weld due to the stirring operation applied to join the plates. This area is called the thermomechanically affected zone because it experiences elevated

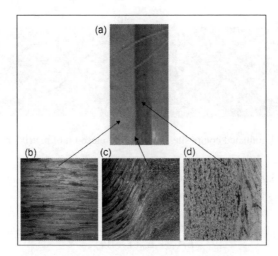

Figure 3.14 (a) Three areas of friction stir weld in (b) showing host material grain structure (c) thermomechanical heated zone and (d) weld nugget (Ackers *et al.*, 2006, DEStech Publications, Inc.)

temperatures during stirring. Finally, the welded zone (nugget) is shown in Figure 3.14(d), which exhibits a porous microstructure that is more likely to initiate cracks.

This type of variation in microstructure is difficult to detect and model because the changes from one microstructure to another do not cause major changes in the modulus, density and other material parameters that can be measured nondestructively. The most significant effects of microstructural changes are on failure criteria (e.g. fatigue life). However, if the changes are significant enough, they can be modeled using the same methods discussed in the previous sections (i.e. impedance, modulus, density, etc.).

3.2 DYNAMIC MODELS FOR DAMAGE

3.2.1 Phenomenological Models

In order to predict the growth of existing damage, there are numerous laws (formulae) available. Most of these laws are based on physics but are empirical in nature having been derived from experimental data. These are sometimes called *phenomenological laws.* The fundamental idea behind all damage laws is that both damage and load information is needed to predict the growth of damage (Grandt, 2004). There are fatigue crack growth laws that predict the rate of change in crack length with cycles of loading, creep laws that predict the strain rate of materials subjected to static stresses at high temperatures and corrosion laws that describe the rate at which corrosion spreads as a function of the environmental conditions and time. A sample of these laws is summarized here. The reader is referred to Jata and Parthasarathy (2006) for more details.

For example, Paris' law for crack growth in metals and metal alloys is given by

$$\frac{dc}{dN} = C\Delta K^m, \tag{3.11}$$

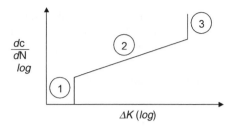

Figure 3.15   Plot of rate in crack length per cycle versus cyclic stress intensity factor

where c is the crack length, N is the number of loading cycles, C and m are empirical constants and K is the *stress intensity factor*. K combines the stress σ, crack length c and component geometry factor β to describe the susceptibility of a given fatigue crack to growth:

$$K = \sigma\sqrt{\pi c}\beta \qquad (3.12)$$

Large K values lead to more crack growth. ΔK is the difference between the maximum and minimum stress intensity factor in the loading cycle. The law in Equation (3.11) is based on the theory of *linear elastic fracture mechanics*. Figure 3.15 illustrates the typical format in which Paris' law (and other similar laws) is presented. Both axes are shown in logarithmic format per convention. The region marked '2' is where Paris' law applies. Regions '1' and '3' correspond to the threshold and fracture stress intensity factors at which cracks do not grow or cracks lead to runaway crack growth and fracture, respectively. Additional effects due to plasticity at the crack tip and other factors can also be incorporated into Paris' law (Grandt, 2004).

Damage laws are also available for creep. For example, under high stress and high temperature,

$$\frac{d\varepsilon}{dt} = A \, \exp\left(-\frac{Q}{RT}\right)(\sigma - \sigma_{th})^{n}, \qquad (3.13)$$

where ε is the creep strain, t is time, A is a constant for a specific material, Q is the activation energy for creep, R is the universal gas constant, σ is the applied stress and σ_{th} is a threshold stress state that depends on the microstructure of the material (Hertzberg, 1995). Similar damage laws for corrosion, oxidation, fiber breakage, ply cracks, etc. damage mechanisms are available in the literature. Many of these laws can be difficult to apply in practice through health monitoring because the constants are unknown and, therefore, estimates of the parameters must be used leading to uncertainty. The important aspect of these damage laws is that they incorporate the loading and damage variables that govern the rate at which damage accumulates.

3.2.2   Generalized Damage Growth Models

The alternative to phenomenological damage growth models are generalized models. As in Section 3.2.1, these models are based on data but motivated by physics. Generalized models aim to describe trends in damage growth and to use these trends for predicting future damage levels. Similar to phenomenological models, generalized models track damage and loading levels in order to make predictions. Unlike phenomenological models, generalized models do not require localized damage and loading data. For example, Paris' law relating the rate of change of crack length c to applied stress generalizes to other damage parameters that grow due to fatigue loading. If a damage parameter (or index) D varies with cyclic loading of magnitude ΔL, then

$$\frac{dD}{dN} = C\Delta L^m D^n, \tag{3.14}$$

where C, m and n are constants determined by data from the application of interest.

Equation (3.14) can be used in practical health-monitoring applications to relate loads to the rate of damage growth when damage parameters are extracted using the models presented in this chapter. For example, estimations of reductions in stiffness from embedded sensitivity models (Section 2.3.2.1) or virtual forces due to damage (Section 2.3.2.2) can be utilized along with restoring force curves (Section 2.3.1.2) or direct stress measurements to predict future damage levels. This technique will be applied in later chapters.

The given application determines which generalized method is suitable for damage prediction. For example, a polynomial model that indicates the rate at which a given damage indicator grows may be sufficient in some applications. Curve fitting using linear, quadratic and higher-order polynomials to predict the rate of change in damage indicators can be effective when loading is constant (time invariant). These methods are applied in later chapters to illustrate data-driven prediction methodologies.

3.3 FAILURE MODELS

The damage prediction models described above can be used to predict failure in structural material components as a function of usage. Failure models are available in the literature for metallic, composite, ceramic and other classes of materials. For example, in metals and metal alloys, there are three primary modes of failure: (i) deformation, (ii) fracture and (iii) material loss. Table 3.2 summarizes these failure modes for metals.

Deformation failure modes are reviewed in some detail here to provide an overview of typical failure models in structural materials and components. There are two broad classes of metallic materials: *ductile* and *brittle*. Ductile materials can reach more than 5% elongation at failure whereas brittle materials cannot. Deformation failure modes in ductile materials occur due to yield under high stresses or creep under a combination of high stresses and elevated temperatures. Consider the failure of healthy components due to yield under high stress. Failure due to yield is predicted based on the yield strength of the component σ_y. Several failure models under yield are used in engineering design:

$$\text{Max stress model}: \qquad \sigma_{max} = \sigma_y \tag{3.15a}$$

$$\text{Max strain model}: \qquad \varepsilon_{max} = \sigma_y/E \tag{3.15b}$$

$$\text{Max shear stress model}: \quad \tau_{max} = \sigma_y/2 \tag{3.15c}$$

$$\text{Energy distortion model}: \quad (\sigma_1 - \sigma_2)^2 + (\sigma_2 - \sigma_3)^2 + (\sigma_3 - \sigma_1)^2 = 2\sigma_y^2 \tag{3.15d}$$

Table 3.2 Failure modes in metals and metal alloys (courtesy of Jata and Parthasarathy (2006))

Failure mode	Damage mechanism	Description	Variables
Deformation	Yield	Athermal plasticity	Stress
			Temperature,
	Creep	Thermally activated plasticity	time, stress
	Static	Athermal/acyclic fracture	Stress
Fracture	Creep rupture	Thermally activated caviation leading to fracture	Temperature, time, stress
	Dynamic fatigue	Cyclic plasticity leading to nucleation and growth of cracks	Stress, time (cycles)
Material loss	Corrosion	Athermal loss of material	Time, environment
	Oxidation	Thermally activated loss of material	Time, environment oxygen

where σ_y is called the yield strength, σ_{max} and τ_{max} are the maximum normal stress and shear stress, ε_{max} is the maximum strain and σ_k are the principal stresses (Young, 2001). The yield strength of different materials is available in handbooks for engineering design. In health monitoring, it is important to note that σ_y is function of the yield strength of a single crystal of a given material, σ_o, as well as the grain size, d, in the material:

$$\sigma_y = \sigma_0 + kd^{-n}, \tag{3.16}$$

where k is the Hall–Petch coefficient and n is usually equal to 0.5. This relationship shows that as the microstructure of the material degrades, the yield strength changes.

Fracture modes of failure in components that are already cracked are encountered when the applied stress reaches a critical value causing existing cracks to quickly propagate in the component. In static fracture, the critical value of stress, σ_c, is related to the crack length through the fracture toughness K_c:

$$K_c = \sigma_c \sqrt{\pi c} \beta. \tag{3.17}$$

This fracture toughness relationship between the applied stress and crack length for a given geometry component is not valid for very small cracks (Grandt, 2004). Fracture toughnesses for various materials are also available in engineering design handbooks. For dynamic fracture, Paris' law can be used to determine if a component with a given crack length and applied cyclic stress is susceptible to runaway crack growth and failure via fracture (see region '3' in Figure 3.15).

In the case of material loss modes of failure and creep deformation and rupture, similar models apply. For example, when a material component corrodes or oxidizes (refer to Section 3.1.6), the effective thickness of the component is reduced leading to lower fracture toughness in either static or dynamic loading scenarios. Higher temperatures worsen the rate of loss in material leading to a greater potential for fracture when the component is exposed to operating loads. In creep-driven failure modes, changes in cross-sectional area and length introduced by creep (refer to Section 3.1.5) can also lower the component's ability to withstand loads. Combined failure modes including stress-corrosion cracking and creep fatigue generally accelerate failure when different loading variables such as temperature and mechanical stress occur simultaneously.

Although composite materials are highly heterogeneous unlike metals and alloys, criteria for static (fracture) and dynamic (fatigue) modes of failure are similar to those for metals. For example, maximum stress criteria for laminated composites indicate that if the principal stresses are greater than or equal to the tensile strengths (yield strengths), then the material will fracture (see Equation (3.15a)). Maximum strain criteria draw conclusions about imminent failure using the ultimate tensile strains instead of tensile strengths (see Equation (3.15b)). Other failure criteria including Hill–Tsai and Tsai–Wu are also available for use in laminated composite materials. Sun (2002) provides a rigorous review of failure criteria for composite structural components. For more details on damage mechanics and failure in homogeneous and heterogeneous materials, Lemaitre (1996) and Talreja (1994) have provided a rigorous summary of physics-based and empirical theories of failure.

3.4 PERFORMANCE MODELS

The dynamic damage models discussed above provide the means to predict potential failure modes given the current damage state and anticipated component loading and environmental conditions. However, the performance of a component often reaches substandard levels long before the component fails. By combining loading models like those discussed in Section 2.4 with static damage models discussed in Section 3.1, the dynamic response of a component can be predicted. Then this prediction can be used to determine if the component should continue to operate.

For example, Figure 1.6 illustrated that loading and damage information can be used to predict the growth of damage in the connector of a gas turbine wire harness. Suppose that the connector is determined to be loose either because it has not been screwed in all the way or is partially corroded. This reduction in stiffness can be incorporated into the connector dynamic vibration model, which can then be used to predict the level of deflection possible in future operation of the engine during hard landings or normal operational vibrations. If it is determined that the deflection will be excessive for this level of loosening especially at elevated temperatures, then the component is predicted to have failed even before an open circuit occurs at the connector.

3.5 SUMMARY

Models of damage for use in quantifying damage levels and predicting the rate of change of damage have been described in this chapter. Models for common types of failure have also been reviewed. A summary of key points in the chapter is given below:

- *Static damage models* of fasteners and joints, cracks, plastic deformation, erosion, separation, creep, buckling, corrosion and oxidation, fiber breakage and matrix cracking are required for health monitoring to interpret measurement data. Static damage models include geometric models (changes in area), material parameter models (changes in modulus or density, percent fraction of filament) and impedance models.

- *Dynamic damage models* for fatigue crack growth, creep, material loss and other damage mechanisms are used to predict the rate at which damage accumulates. The general dynamic damage law incorporates both the internal load to a component in the damaged area and the damage level.
- *Failure models* for metals, composites and other materials are based on maximum stress, maximum strain and other criteria.
- *Performance models* are needed in health monitoring because components reach the end of useful life long before they fail.

REFERENCES

Bickford, J.H. (1990) *An Introduction to the Design and Behavior of Bolted Joints*, Marcel Dekker, Inc., NY.

Jiang, J., Zhang, M., Park, T.-W. and Lee, C.H. (2003) 'A study of early stage self-loosening of bolted joints', *ASME J. Mech. Design*, **125**, 518–526.

Gere, J. M. and Timoshenko, S.P. (1990) *Mechanics of Materials*, 3rd edition, PWS-KENT Publishing Company, Boston, MA.

Tse, F.S., Morse, I.E. and Hinkle, R.T. (1978) *Mechanical Vibrations*, 2nd edition, Prentice Hall, Inc., New York, NY.

Adams, D.E. (2005) 'Prognosis applications and examples', Chapter 18 in *Damage Prognosis*, Inman, D. and Farrar, C. (Eds), John Wiley & Sons Ltd., Chichester, West Sussex, England.

Roemer, M. (2005), 'Rotating machine applications', *Damage Prognosis*, Inman, D. and Farrar, C. (Eds), John Wiley & Sons. Ltd., Chichester, West Sussex, England.

Soedel, W. (1994) *Vibrations of Plates and Shells*, Marcel Decker, New York, NY.

Sundararaman, S., Haroon, M., Adams, D. E., Jata, K. (2004) 'Incipient damage identification using elastic wave propagation through a friction stir welded Al–Li interface for cryogenic tank applications', *Proceedings of the European Workshop on Structural Health Monitoring*, pp. 525–532.

Grandt, A. (2004) *Fundamentals of Structural Integrity*, John Wiley & Sons, New York, NY.

Jata, K. and Parthasarathy, T. (2006) 'Physics of failure', *Proceedings of the NASA Workshop on Health Management*.

Hertzberg, R. (1995) *Deformation and Fracture Mechanics of Engineering Materials*, John Wiley & Sons, New York, NY.

Sun, C. (2002) 'Mechanics of composite materials and laminates', Lecture Notes for A&AE 555, School of Aeronautics and Astronautics, Purdue University, West Lafayette, IN.

Lemaitre, J. (1996) *A Course on Damage Mechanics*, 2nd edition, Springer Verlag, New York, NY.

Talreja, R. (1994) *Damage Mechanics of Composite Materials*, Elsevier Science, New York, NY.

Young, W. (2001) *Roark's Formulas for Stress and Strain* 7th Edition, McGraw-Hill, Hightstown, NJ.

PROBLEMS

(1) Consider the unmanned aerial vehicle in Figure 3.16(a). If the stabilizer strut becomes cracked, then the simplified model in Figure 3.16(b) can be used to model the crack with a reduction in thickness at the crack location. If the strut length is L with modulus E, estimate the change in stiffness for an aerodynamic load P applied at the end. Recall the deflections due to a point load and moment, respectively, at the end of a cantilever beam are $PL^3/3EI$ and $ML^2/2EI$, and the slopes caused by a point load and moment are $PL^2/2EI$ and ML/EI.

(2) Consider the fluid pipe shown in Figure 3.17. This pipe can become corroded leading to a loss of mass around the outer surface. This loss in mass can be modeled as either a change

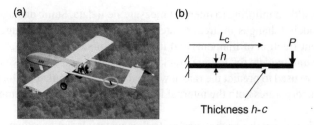

Figure 3.16 (a) Unmanned aerial vehicle with crack in stabilizer strut and (b) simplified model of strut

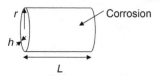

Figure 3.17 Pipe for fluid transfer subject to corrosion

in density ρ or decrease in effective thickness h if the corrosion affects the stiffness of the pipe in addition to its mass. The expression below can be used to calculate the natural frequencies of the cylinder for large m and n, where m and n describe the number of nodal lines in the circumferential and length directions. Plot the change in natural frequency with density and thickness. Use a modulus of 190 GPa, Poisson's ratio of 0.27 and density of 7850 kg/m^3. Comment on these two ways to model corrosion damage.

$$\omega_{mn}^2 = \frac{E}{12\rho(1-v^2)r^2} \left(\frac{h}{r}\right)^2 \left(\left(\frac{\pi mr}{L}\right)^2 + n^2\right). \tag{3.18}$$

(3) A composite windmill rotor blade (Figure 1.23(a)) is illustrated below in Figure 3.18. The fibers are S-glass ($E = 85$ GPa) and the binding matrix is epoxy ($E = 3.5$ GPa). Assume the blade is a symmetric laminate to avoid coupling between in-plane and bending motions; therefore, the blade can be treated like an isotropic BernoulliEuler beam with thickness h and modulus E_{xx} from Equation (3.9a). Assume that $V_f = 0.5$ and $V_m = 0.5$ in the undamaged rotor. Use $h = 0.02$ m, $w = 0.06$ m and $L = 2.5$ m. Plot the transverse propagating wave phase velocity as a function of frequency ω as V_f decreases from 0.5 to 0.3 due to damaged fibers.

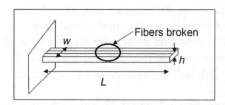

Figure 3.18 Composite motor blade in windmill station

4

Measurements

Data must be acquired and analyzed to identify loads and monitor the health of structural materials and components. The key issue to address in measurements is that *variability and errors in measurements affect data analysis leading to uncertain health-monitoring information*. Figure 4.1 illustrates that variability and other errors can distort measurements used for health monitoring. Then when data is analyzed, uncertainties due to these measurement errors arise in the diagnosis; i.e. damage might not be detected in the midst of the variability. This chapter addresses measurement issues. The goal in measurements is to reduce the effects of variability in raw data as much as possible before it is analyzed. Data analysis methods are discussed in the next chapter. In data analysis, the goal is to avoid the areas of data that are most influenced by variability and to detect, locate and quantify loads and damage.

4.1 MEASUREMENT NEEDS

Frequent measurements must be taken to identify and track changes in component health. Measurements used to identify loading and damage are almost always *indirect*; in other words, there are not many sensors designed to directly measure loads and damage (corrosion sensors are one of the few types of sensors that do directly measure damage). Instead, strain, acceleration and other types of sensors measure variables that can be used to infer load and damage information. Measurements can be made offline or online; however, health-monitoring methods ideally use *online* measurements, which are acquired in real time or near real time as components are operating. Online measurements suffer from variability due to the *measurement environment* and errors in the *measurement channel*. A measurement channel consists of *all the hardware and data acquisition settings utilized to make a measurement*. Figure 4.2(a) shows the components in a typical dynamic measurement channel. Figure 4.2(b) shows the physical form of these components in a health-monitoring system for a sandwich metallic aircraft panel. The components in this measurement channel and some key issues are as follows:

- The panel being monitored is exposed to loads such as impacts and sound pressure in operation. Transducers are attached to the rear of the panel in order to *avoid damage to the instrumentation* because one essential attribute of a health-monitoring system is that the instrumentation must supply *reliable measurements* for long-term use. Instrumentation must also be calibrated regularly; therefore, it should be *accessible*.

Health Monitoring of Structural Materials and Components: Methods with Applications   D. Adams
© 2007 John Wiley & Sons, Ltd

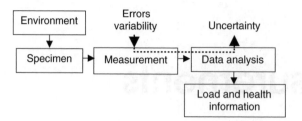

Figure 4.1   Issues in measurement and data analysis

- *Multiple passive sensors* called accelerometers are used to measure the acceleration of the panel in the transverse direction normal to the rear surface. Other response directions could also provide useful data for health monitoring. Active devices called *actuators* are used to excite the panel with user-defined vibrations. The forces supplied by these actuators are measured using *load cells* also pictured in Figure 4.2(b). By measuring the response and the excitation force, *actively sensed* input–output data is obtained to help reduce measurement variability and quantify damage.

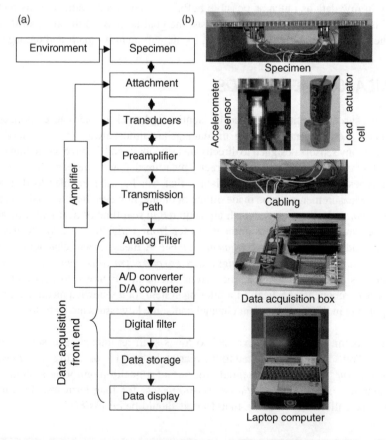

Figure 4.2   (a) Typical dynamic measurement channel and (b) physical example of portable measurement channel for online data acquisition

- The accelerometer and load cell in this system both use *preamplifiers* to condition signals before transmitting them along cables. This conditioning of the signals at the source reduces the effects of noise when transmitting the data.
- *Cables* are used in this panel instead of *wireless* transmitters. Cables add weight to the measurement system but are a reliable means of transmission because they provide power for preamplification of signals and provide electromagnetic shielding. Wireless transmission requires a transmitter and receiver, both of which add weight, and wireless transmission can experience noise with less power for preamplification.
- *Filters* are used to remove operational, environmental and sensor noise from the data when possible. For example, the accelerometer in this application does not measure accelerations below 5 Hz. Therefore, an *analog* high-pass filter is used to remove this content from the signal. *Digital* filters can also be used to filter sampled data.
- *Analog-to-digital and digital-to-analog converters* are used to sample or reconstruct analog signals, respectively. For example, digital signals are converted to analog (sometimes called continuous) signals for driving the actuators to produce vibration. Errors occur in this conversion process because analog data cannot be sampled infinitely fast nor can the data be sampled for infinitely long periods of time.
- Data must be *stored* for later use in health monitoring and prognosis. For example, baseline signatures in healthy data from a component help to detect damage in future data. Raw data is not stored because it requires too much memory; instead, processed data called *features*, such as damage indices and loading parameters, is stored.
- Data can be monitored online if a *display* is used as in the laptop system in Figure 4.2(b). Small amounts of data can also be stored in the black box (see figure).

Different aspects of dynamic measurement channels are described in the following sections. The emphasis in these discussions is on understanding data environments and making proper selections in the measurement channel for health monitoring.

4.2   DATA ENVIRONMENT

Before selecting transducers for acquiring health-monitoring data, the data environment must be identified. The examples in Chapter 2 demonstrated that operational loading affects damage in different ways depending on the type and location of the damage mechanism. For example, it was observed that the response data for the two-DOF system model in Figure 2.2(a) was sensitive to damage in certain frequency ranges but insensitive to damage in other frequency ranges. Similarly, the changes in friction at the interface of a loosened bolt were more evident in certain amplitude ranges (see Figure 2.16). Therefore, the types of sensors to be selected and how these sensors should be positioned to identify damage are in part determined by the data environment. Table 4.1 lists characteristics of the data environment and how these characteristics influence sensing for health monitoring.

4.2.1   Amplitude and Frequency Ranges

The anticipated amplitude levels in data are important when selecting measurement hardware. For example, all dynamic sensors are sensitive to *shock* (impulsive) loading. Piezoelectric

Table 4.1 Characteristics and influence of health-monitoring data environments

Characteristic	Description	Influence
Amplitude range	Excitations at different levels cause different response levels	• High levels of response can overload or burst sensors • Low levels of response can result in poor data • Certain response levels may not expose damage in data • High levels of response in one frequency range can mask the response in other ranges
Frequency range	Excitations in different frequency ranges produce different response frequencies and deflection patterns in a structural component	• Narrowband data contains short frequency bandwidths • Lower frequencies tend to be less sensitive to small damage • Certain frequencies excite damage better than others • Traveling waves combine with vibrations to accentuate damage in certain locations
Nature of data	Excitations and responses are stationary with constant amplitude, frequency and phase or nonstationary with time-varying characteristics	• Stationary response data is more repeatable requiring less data for diagnostics • Stationary data also tends to be cyclic and sometimes does not expose damage in data • Nonstationary response data is not as repeatable requiring more averaging of data • Nonstationary data tends to be transient exciting a broader frequency range to expose more types of damage
Temperature range	Temperatures fluctuate while components are operating	• Temperature shifts cause changes in sensor calibration • Prevents sensors from being positioned in certain locations • High/low temperatures can also cause failure of sensors and attachment mechanisms
Acoustic excitation	Air pressure fluctuations cause vibration and wave responses.	• Sensor housings respond directly to acoustic excitations
Electromagnetic interference	Electric and magnetic fields are created when sensors are used to convert a measurand to an electrical signal	• Shielding is needed in sensing (e.g. coaxial cables.) • Preamplification of signals is needed to minimize noise

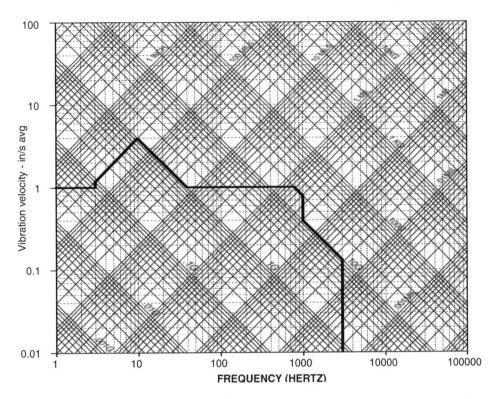

Figure 4.3   Sinusoidal vibration spectrum specification for wire harness and connector panel

elements can burst if the shock loading is too high. Typical piezoelectric accelerometer shock ratings for common applications are 2000–4000 g. The electronics and packaging in these accelerometers are also sensitive to shock loads, which can break soldered connections. Sensor cable connectors are also sensitive to shock and cyclic loading. Designers of sructural components such as the gas turbine engine wire harness and connector panel in Figure 1.6 usually specify the vibration spectrum that the component must endure. Figure 4.3 shows the specified sinusoidal velocity versus frequency spectrum for the wire harness. This plot is read along the two diagonal sets of axes for g level (left to right) and displacement (right to left). For example, from 10 to 40 Hz, the specification is 1 g of acceleration whereas from 1000 to 3000 Hz, the specification is 10 g of acceleration. Any measurement hardware inserted in this environment must withstand these same amplitude levels of acceleration.

The vibration spectrum measured while a structural component operates also gives a rough indication of component damage mechanisms that will be exposed in operating data and mechanisms that will be hidden in the data. For example, the measured operational vibration spectrum for the wire harness and panel is plotted in Figure 4.4. The amplitude of this spectrum is given in units of auto-spectral density (g^2/Hz, recall Equation (2.86)). This information is useful for determining if *passive measurements* are adequate for damage identification. Passive measurements do not require control of the vibration (or wave propagation) input to a system. If passive measurements are not sufficient, then the operating spectrum can also be useful for designing *active measurement* hardware. Active measurements use auxiliary actuators to cause components to vibrate (or waves to propagate) in a more desirable

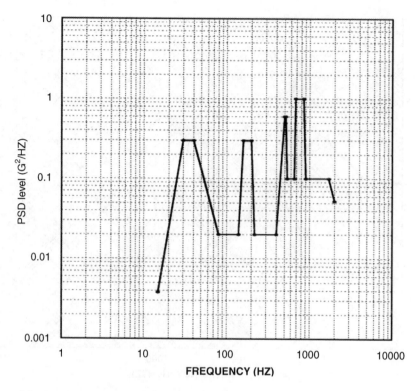

Figure 4.4   Power spectral density for engine mounted test on wire harness and connector panel

manner to detect damage (refer to Section 4.4). The screw connector in Figure 1.6 can experience loosening, which causes a reduction in bending stiffness across the connector. This change might be evident in the vibration response from 30 to 40 Hz because bending stiffness effects would be most evident in this frequency range. The high amplitudes from 30 to 40 Hz would cause the connector to bend thereby highlighting any loosening of the connector in the vibration data.

Conversely, small amounts of damage to the connector in the form of broken pins or corrosion may not be evident in operating vibration data because the vibration amplitude drops severely beyond 1000 Hz. Generally speaking, small changes in a component due to localized damage do not cause changes in low frequency vibration data. When a component is damaged in one small location, the vibrations must be high enough in frequency to cause that local area to deform in order to excite the damage. Recall that modal deflection shapes in vibrating systems do not exhibit much local deflection at low frequencies. In order to detect these types of damage, actuators that incite the component to vibrate at higher frequencies would need to be utilized in an active vibration measurement. When processing the data obtained from active measurements, it is still important to understand how lower frequency vibrations affect the measurements. For example, if the connector is vibrating, then the transmission of higher frequency waves through the threads to damaged areas inside the connector will change due to attenuation.

In summary, amplitude and frequency data such as the auto-spectral density in Figure 4.4 should be used to design measurement hardware and data analysis algorithms for damage

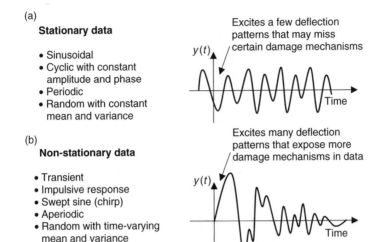

(a)
Stationary data

- Sinusoidal
- Cyclic with constant amplitude and phase
- Periodic
- Random with constant mean and variance

Excites a few deflection patterns that may miss certain damage mechanisms

$y(t)$

Time

(b)
Non-stationary data

- Transient
- Impulsive response
- Swept sine (chirp)
- Aperiodic
- Random with time-varying mean and variance

Excites many deflection patterns that expose more damage mechanisms in data

$y(t)$

Time

Figure 4.5   (a) Stationary and (b) nonstationary types of data with characteristics

detection. Frequency ranges over which the amplitude is high can be measured passively to detect certain types of damage. Usually, small damage mechanisms that are localized in one area of a component can be detected using only active measurements.

4.2.2 Nature of Data

There are two basic types of data. One type, called *stationary* data, has properties that do not change with time. For example, sinusoidal signals with constant amplitude and phase are stationary. The second type of data, called *nonstationary* data, does change with time. Impulsive data with decaying amplitude and time-varying frequency is nonstationary. In passive response measurements, which were defined in Section 4.2.1, stationary data is desirable because it is repeatable. Fewer averages are required to analyze this type of data. Figure 4.5(a) lists a few types of stationary data including periodic data and random data with constant statistical properties (e.g. mean and variance). In rotating machines, for instance, stationary response data is obtained if the machine is rotating at a constant speed and torque load. Averages for this type of data can be calculated once per rotation leading to more confidence in the diagnosis. The disadvantage of stationary data is that it usually only consists of a few response deflection patterns, which may not exercise all damage mechanisms in the component.

To demonstrate the advantages and disadvantages of stationary data, consider the nine-DOF system model illustrated in Figure 4.6. Each DOF is connected to its neighbors by springs and dampers that resist transverse (vertical) deflections. Four supports at DOFs 1, 3, 7 and 9 support the system. This model could represent the panel in Figure 4.2. It is assumed that an operational excitation, $f_1(t)$, is applied at DOF 1, and the response, $x_1(t)$, is measured passively at DOF 1. In passive measurements, only the responses are measured as in this example. The excitation is unknown. The FRF model between the excitation and response at DOF 1 was

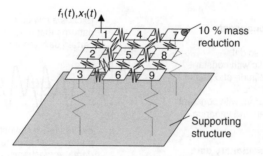

Figure 4.6 Nine-DOF model with transverse springs and dampers acting between the elements (Ackers *et al.*, 2006, Society for Experimental Mechanics, Inc.)

calculated using the simulation code 'ninedofdata.m' and is plotted in Figure 4.7. The FRF was calculated for the original mass at DOF 7 (__) and then for a 10 % reduced value of mass at that DOF (...). FRFs can help to identify what type of data is required to detect damage in different areas of a component even when the excitation is not measured. Note that the FRF does not change throughout most of the frequency range when the mass change occurs due to damage at DOF 7. The only visible changes in the FRF are between 1800 and 2500 Hz. Relatively little change is observed due to the mass reduction at DOF 7 because this local change in mass is exercised for only particular modal deflection shapes of the panel.

The implications of this limited sensitivity of the FRF to damage at DOF 7 are important when deciding what health-monitoring data to measure for this system. For example, Figure 4.8(a) shows an excitation with 100 Hz and 1000 Hz components. This excitation could arise due to a rotating machine (engine) next to the panel causing fixed amplitude and frequency forcing. When this stationary excitation is applied to the nine-DOF model, the steady-state response in Figure 4.9(a) is obtained before (__) and after (...) the mass at DOF 7 is changed. Note that there is no visible change in the measured response because the excitation is applied at frequencies where the FRF is insensitive to the simulated damage

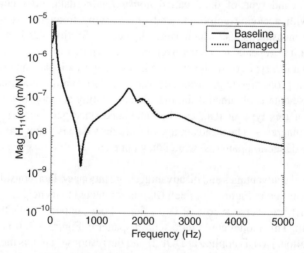

Figure 4.7 Frequency response function $H_{11}(\omega)$ for nine-DOF system with (...) and without (—) 10 % reduction in mass M_7 simulating oxidation

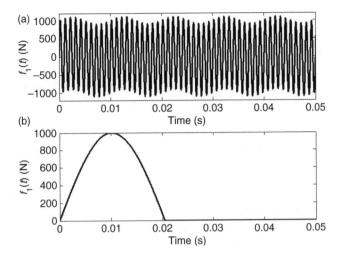

Figure 4.8 $f_1(t)$ excitation time histories with (a) stationary sinusoidal excitation at 100 Hz and 1000 Hz and (b) nonstationary pulse excitation

at DOF 7. This inability of cyclic data with constant amplitude and frequency to reveal localized damage is generally a disadvantage of stationary data.

On the contrary, the pulse in Figure 4.8(b) is a nonstationary type of excitation. Figure 4.5(b) lists a few types of nonstationary data including impulsive and swept sine signals with time-varying frequencies and amplitudes. Nonstationary data is often more revealing than stationary data because *transients highlight localized damage by exercising more deflection patterns in components*. For example, Figure 4.9(b) shows the response obtained for the

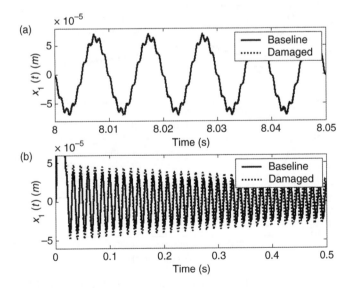

Figure 4.9 $x_1(t)$ time responses of nine-DOF system to (a) stationary sinusoidal excitation at 100 Hz and 1000 Hz and (b) nonstationary pulse excitation

pulse-type excitation using the code 'ninedofdata.m'. The difference between the baseline and damaged cases is evident in this response data. Nonstationary data is more difficult to analyze than stationary data because more averages are required. Also, more frequency components in the data lead to more complexity in data analysis algorithms. Before utilizing passive operational vibration measurements for damage detection, the nature of the data should be thoroughly understood.

4.2.3 Environmental Factors

Environmental factors due to temperature shifts, humidity variations, sound pressure fluctuations and other changes cause component response data to vary in ways similar to damage. These changes must be accounted for in data analysis algorithms in order to avoid false positive (and false negative) indications of damage. Also, these environmental factors cause measurement hardware such as sensors and sensor attachments to degrade over time causing drifts in measurements. In order to understand the effects of some of these variations, the nine-DOF system model in Figure 4.6 is considered.

The primary way to distinguish changes in measurements due to damage from changes due to environmental factors is that *environmental factors cause global changes in mechanical properties whereas damage causes more localized changes*. To demonstrate this rule of thumb, the simulation code 'ninedofenviron.m' is used to calculate changes in the FRFs for the nine-DOF system model in Figure 4.6 due to two environmental factors. First, a change in humidity is considered. It is assumed that an increase in humidity causes the panel to absorb moisture leading to an increase in panel mass. Panels made from composite materials do tend to absorb moisture in this way. Figure 4.10(a) shows the FRF $H_{11}(\omega)$ before (__) and after (...)

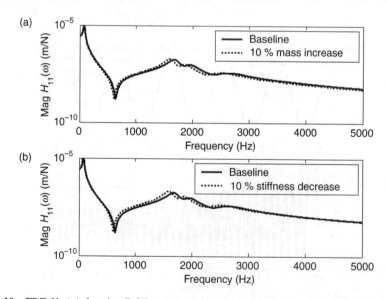

Figure 4.10 FRF $H_{11}(\omega)$ for nine-DOF system before (—) and after (...) (a) 10% increase in mass throughout system due to moisture absorption and (b) 10% decrease in stiffness due to temperature increase

a 10 % increase in mass due to moisture absorption. All peaks in the FRF shift downward due to the global mass addition; this result is expected because natural frequencies are proportional to $1/\sqrt{M}$. Compare this result to the FRFs in Figure 4.7, which change only in a short frequency bandwidth because the loss of mass in that example was localized at DOF 7.

Similarly, a change in temperature is considered by decreasing the stiffness 10 % throughout the nine-DOF panel model. The stiffness of a material decreases when the temperature increases because of the decrease in elastic modulus (refer to Section 2.4.2). Figure 4.10(b) shows the FRFs before and after the 10 % decrease in stiffness is introduced. The FRF again shifts throughout the entire frequency bandwidth suggesting that a global environmental change has occurred as opposed to localized damage.

4.3   TRANSDUCER ATTACHMENT METHODS

The impedance model in Section 2.2.5 was used to determine how FRFs of a component change when a sensor is attached to the component. In addition to the sensor mass, the effects of the attachment stiffness and damping should also be considered. Various methods are used to attach sensors and actuators (collectively referred to as *transducers*) including screw attachments and adhesives like epoxy or glue. Each of these attachment mechanisms introduces measurement limitations. Figure 4.11 illustrates the key issues to consider when attaching sensors. Physical examples of these issues are also pictured in the figure. Each of these issues is described in detail below.

4.3.1   Durability

Transducer attachments must be durable; otherwise, measurement quality degrades with time. The attachment method should be at least as resistant to shock loading as the transducer being attached. *Studs* are the most durable attachment mechanism because they provide a threaded hole into which sensors are screwed. However, it is not always possible or even desirable to drill and tap components to accept health-monitoring sensors. For example, strain sensors cannot be attached through studs because threaded holes would severely distort the local strain field of a component by introducing stress concentrations. When possible, it is preferable to attach sensors in hidden areas that are not exposed to debris or other causes of damage. For example, the sensors and actuators mounted on the rear of the panel in Figure 4.2 are positioned to avoid high temperatures and acoustic levels on the exterior surface. *Embedded* transducers are sometimes used to acquire health-monitoring data. These devices are placed inside of materials or structural components. For example, piezoceramic washers have been used by Yang and Chang (2006) to acquire response data at joints to detect loosening. Piezowafer sensors have also been embedded into laminated composite and filament wound materials to detect delamination and other types of damage (Islam and Craig, 1994).

Adhesives are often used to attach transducers. One measure of durability of adhesives is obtained by measuring the shear strength. Figure 4.11(a) shows several nuts attached to a metal plate with various adhesives including super glue and dental cement. A digital torque wrench was then used to apply a torque to the nut and measure the breakaway torque for each type of adhesive. Table 4.2 lists these breakaway torque values in decreasing order of shear strength. Concrete epoxy was the strongest and beeswax (commonly used for quick

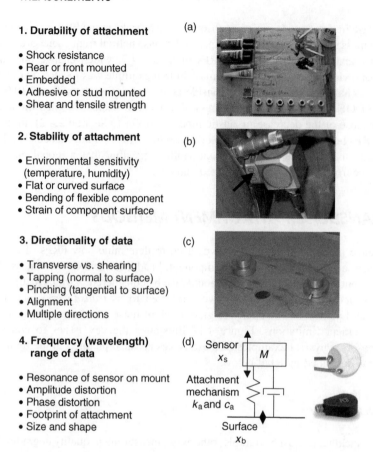

1. Durability of attachment (a)

- Shock resistance
- Rear or front mounted
- Embedded
- Adhesive or stud mounted
- Shear and tensile strength

2. Stability of attachment (b)

- Environmental sensitivity
  (temperature, humidity)
- Flat or curved surface
- Bending of flexible component
- Strain of component surface

3. Directionality of data (c)

- Transverse vs. shearing
- Tapping (normal to surface)
- Pinching (tangential to surface)
- Alignment
- Multiple directions

4. Frequency (wavelength) (d)
range of data

- Resonance of sensor on mount
- Amplitude distortion
- Phase distortion
- Footprint of attachment
- Size and shape

Figure 4.11 Issues to consider in attachment mechanism with (a) adhered bolts for breakaway torque test (b) magnetic mount to curved surface (c) tapping actuators and strain sensors and (d) resonance of sensor on attachment with two sensor shapes

measurements in a laboratory setting) was the weakest. Note that these measurements give the shear strength; however, in many types of sensors, the measured variable is vertical to the adhesive bond instead of tangential to it. The most relevant strength parameter in these cases is the tensile strength of the bond.

Table 4.2 Breakaway torque measured for different adhesives

Adhesive	Breakaway Torque (lb in)
Loctite™ concrete epoxy	238
Crazy glue	213
Loctite™ quickset epoxy	170
Epoxy	147
Loctite™ superglue	161
Dental cement	64
Bees wax	59

4.3.2 Stability

The stability of transducer attachment mechanisms is also important in health-monitoring measurements. A stable attachment is one which does not fluctuate in damping or stiffness with time or due to changes in the environment. The discussion in Section 4.2.3 demonstrated that environmental factors cause components to respond differently. Similarly, transducer attachment mechanisms that vary with environmental conditions cause measurements to change.

For example, superglue has a tendency to continuously cure for hours resulting in an increasingly stiffer bond. Figure 4.12(a) shows superglue being applied to an accelerometer (PCB 333B50), which is then positioned on the flange of a wheel as shown in Figure 4.12(b). The glue is applied to a small nut that is screwed into the accelerometer instead of applying glue directly to the accelerometer case to avoid damaging the case. The glue is then allowed to cure for an hour. Figure 4.12(c) shows the measured FRF between an excitation force in pounds applied to the front of the wheel and the acceleration response in g's on the back of the wheel. Three measurements are shown that were taken at 30-min intervals. Note that as the glue continues to cure, it becomes stiffer resulting in increasing FRF magnitude. This attachment is not stable with time. Superglue is also sensitive to changes in temperature and humidity.

Figure 4.13 depicts a more stable attachment mechanism. The accelerometer is shown screwed into a hole in a vinyl strap in Figure 4.13(a). The strap is then used to preload the sensor against the flange of the wheel in Figure 4.13(b). This strap is far more stable with time and is not nearly as sensitive to changes in environmental conditions as the glue. Figure 4.13(c) shows a comparison between the standard deviation in the FRF magnitude for measurements made with the glue relative to those made with the strap. The strap exhibits lower variation than the glue, but the strap is not effective at high frequency.

The other stability issue to consider in transducer attachments is the *geometry* of the component surface to which a sensor is mounted. When the profile of the attachment does

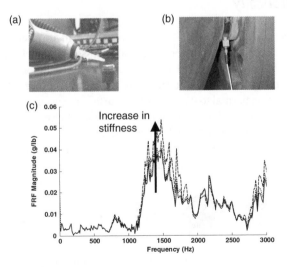

Figure 4.12 (a) Application of superglue to sensor (b) placement of sensor and (c) measured FRF at three different times showing change due to continuous curing

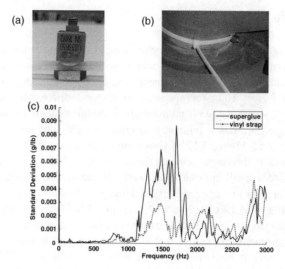

Figure 4.13   (a) Vinyl strap for attaching sensor (b) placement of sensor around wheel and (c) standard deviation in measured FRFs for glue (—) compared to vinyl (...) (Ackers *et al.* 2006, SPIE)

not match the profile of the mounting surface, the sensor can roll or pitch on the surface causing errors in the measurement. Figure 4.11(b) shows a square accelerometer with a flat mounting surface attached to a curved flange. This magnetic mount can be used on a flat surface out to approximately 1500 Hz; however, on a curved surface, the sensor can wobble causing measurement errors. To correct these errors, one solution is to machine a mounting adapter to match the contour of the curved surface.

4.3.3 Directionality

Different transducers are designed to actuate and sense in different directions. In some applications where components undergo lateral stresses and relatively small bending moments, sensors that measure strain in the plane of the component are used. In other applications where components exhibit relatively large out-of-plane displacements, sensors that measure acceleration normal to the component surface are used. It is important to properly align sensor axis directions when attaching sensors. Otherwise, the measured signal may contain off-axis data and be smaller (or larger) than desired.

Figure 4.14(a)–(e) shows five examples of acceleration and strain sensors used to make dynamic response measurements in structural components. Some sensors measure only one

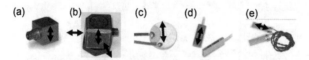

Figure 4.14   (a) Single-axis accelerometer (PCB 333A32) (b) triaxial accelerometer (PCB 356A08) (c) transverse piezoceramic disk transducer (APC850-51D) (d) piezoelectric strain sensors (PCB 740B02) and (e) piezoceramic strain sensors (Measurement Specialties LDT1-028K)

direction of response such as the accelerometer in Figure 4.14(a). Other sensors measure multiple response directions as in the triaxial accelerometer in Figure 4.14(b). These two accelerometer sensors are packaged by the manufacturer, who aligns the sensing elements to minimize *cross talk* errors between the desired measurement direction and off-axis directions. The user is responsible for attaching the sensor in the proper measurement direction. For example, too much adhesive applied to the sensor attachment surface in certain locations can cause tilting of these accelerometers and errors in the measurements. Other sensors like the piezoceramic disk in Figure 4.14(c) also measure the response motions normal to the surface of a component. This device can also serve as an actuator to provide a force per unit area (pressure) applied to the component surface. Figure 4.11(c) shows a second example of this type of transverse actuator that 'taps' on the surface of a steel plate. The two large disk-shaped devices in Figure 4.11(c) are piezoelectric actuators, which are also equipped with additional masses to increase the force they supply to the component.

The two sets of sensors in Figure 4.14(d) and (e) measure strain in the component tangential to the component surface. Figure 4.11(c) shows the first type of strain gauge installed on a steel plate to measure strain in various locations. These sensors measure dynamic strain only along the axes shown in the figure. A pair of gauges is generally used to provide a more complete measurement of strain at a point in the component. When attaching these sensors, it is important to avoid tilting due to uneven distribution of adhesive along the attachment interface. In these sensors, it is also important to avoid large amounts of adhesive at the interface due to the poor shear transfer obtained through thick, and usually uneven, adhesive layers.

4.3.4 Frequency Range (Wavelength)

Ideal attachments should transfer the response amplitude and phase of a component at a point directly to the sensor. However, attachments are imperfect causing changes to desired measurement variables. There are two basic types of filtering provided by attachments. Both types of filtering cause changes in amplitude and phase of measurements as the response frequency changes.

The first type of filtering is caused by the *mechanical properties of the attachment* (stiffness, damping or even mass). Softer attachments are less stiff with more damping resulting in lower response bandwidth. Consider the simplified attached sensor arrangement in Figure 4.11(d). The FRF relating the amplitude and phase in $X_s(\omega)$ transmitted to the sensor of mass M relative to the component response amplitude and phase, $X_b(\omega)$, is given by

$$\frac{X_s(\omega)}{X_b(\omega)} = \frac{j\omega c_a + k_a}{-M\omega^2 + j\omega c_a + k_a} \tag{4.1}$$

where c_a and k_a are the attachment damping and stiffness and M is the mass of the sensor. In practice, *thin adhesive layers result in higher stiffness and lower damping*. If both the numerator and denominator are divided by the stiffness, then the result is

$$\frac{X_s(\omega)}{X_b(\omega)} = \frac{j2\zeta\frac{\omega}{\omega_n} + 1}{-\left(\frac{\omega}{\omega_n}\right)^2 + j2\zeta\frac{\omega}{\omega_n} + 1} \tag{4.2}$$

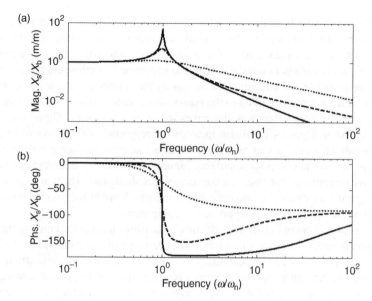

Figure 4.15 (a) Magnitude of component transmission to sensor and (b) phase for different values of ζ (——) 0.01, (- - -) 0.1 and (...) 0.7

where $\zeta = c_a/2\sqrt{k_a M}$ is the damping ratio associated with the attachment and $\omega_n = \sqrt{k_a/M}$ is the undamped natural frequency. This relationship shows that the amplitude and phase of the sensor response are nearly the same as those of the component when the frequency of response is much less than ω_n (resonance of attachment). When the response frequency increases beyond ω_n, the response amplitude and phase of the sensor become distorted versions of the component response. Note that similar relationships hold for strain and force transmission through attachment interfaces.

The simulation code 'attachtrans.m' was used to plot the magnitude and phase of Equation (4.2) in Figure 4.15 for three different values of ζ. These plots describe the response motion of the sensor relative to the component response. For $\omega \ll \omega_n$, the sensor and component responses are nearly identical; in other words, the attachment is ideal and can be neglected for $\omega \ll \omega_n$. Two kinds of distortion set in as ω approaches ω_n:

- *Amplitude distortion* results in a higher response amplitude at the sensor than at the component for ω approaching ω_n. This elevated response amplitude at the sensor prevents the use of factory calibration data when making dynamic measurements. When ω increases beyond ω_n and continues to increase, the sensor response amplitude decreases. In other words, the attachment prevents all of the response amplitude of the component from passing through the attachment mechanism. *Amplitude distortion is highest when the damping ratio is small.*
- *Phase distortion* results in a phase lag of the sensor response relative to the component response for ω approaching ω_n. In other words, the sensor response is slower than the component response. This delay makes it difficult to use wave propagation data because delays introduce errors in estimated damage locations. *Phase distortion is highest when the damping ratio is large.*

Table 4.3 Transducer mounting characteristics (Reproduced by permission of Dr. R. Allemang, University of Cincinnati, 1999)

Method	Frequency range (Hertz)	Main advantages	Main disadvantages
		Transducer mounting methods	
Hand-held	20–1000	Quick look	Poor measurement quality for long sample periods
Putty	0–200	Good axis alignment, ease of mounting	Low frequency range, creep problems during measurement
Wax	0–2000	Ease of application	Temperature limitations, frequency range limited by wax thickness, axis alignment limitations
Hot glue	0–2000	Quick setting time, good axis alignment	Temperature sensitive transducers (during cure)
Magnet	0–2000	Quick setup	Requires magnetic material, axis alignment limitations, bounce problem with impact excitation, surface preparation
Adhesive film	0–2000	Quick Setup	Axis alignment limitations, requires flat surface
Epoxy cement	0–5000	Mount on irregular surface, good axis alignment	Long curing time
Dental cement	0–5000	Mount on irregular surface, good axis alignment	Medium curing time, brittle
Stud mount	0–10 000	Accurate alignment if carefully machined	Difficult setup, requires drill and tap

In summary, thin adhesive layers provide high stiffness but low damping. The result of this trade-off is high resonant bandwidth of the attachment, which is desirable, and large amounts of distortion, which is undesirable. Usually, it is best to strike a balance by designing adhesive layers that are thin enough to provide reasonably high resonant frequency and also thick enough to provide sufficient damping on the order of $\zeta = 0.3$. Table 4.3 provides approximate frequency ranges over which different attachment methods can be utilized to acquire dynamic measurements.

The second type of filtering is caused by *geometric properties of the attachment* (length, width, radius). In general, *large attachment footprints cannot be used to measure short wavelength responses*. Regardless of whether an attachment is used to measure standing waves (vibration) or traveling waves (wave propagation), this limitation holds. Designers of health-monitoring measurement systems must be aware of this limitation because different sensors require different attachment footprints. For example, two sensors are shown in the far right of Figure 4.11(d). The first is circular and the second is pear shaped. Different attachments must be used for these two sensors resulting in differences in the way these attachments filter the component response.

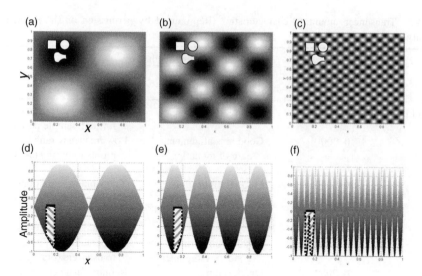

Figure 4.16 Different sensor footprints overlayed on top of (a) long wavelength (b) medium wavelength and (c) short wavelength deflection patterns; and (d)–(f) corresponding deflection profiles along the x direction relative to the square sensor

Figure 4.16 is a qualitative illustration of how different attachment footprints filter component responses leading to measurement errors. The code 'twoDshapesensor.m' was used to generate these patterns. Figure 4.16(a) is a plot of the modal deflection of a square component at a lower frequency with a longer wavelength. The x and y directions in the plane of the component are indicated in the plot. The component can be thought of as a flat plate (with simply supported boundary conditions) vibrating with amplitude given by the colors in the plot; high amplitudes are in white and low amplitudes are in black. Three attachment shapes are overlayed on top of the response pattern in Figure 4.16(a). The three attachments are all positioned in an area of low response amplitude. Figure 4.16(d) shows a different perspective of the response pattern in Figure 4.16(a). Only the x direction in the plane of the component is shown along with the response amplitude. The square-shaped attachment is shown in profile with a projection indicating how the square shape filters the response amplitude.

If the motion at one end of the sensor attachment is y_1 and the motion at the other end is y_2, then the square attachment transmits the average of these two amplitudes to the sensor, $(y_1 + y_2)/2$. As long as the sensor width is small relative to the wavelength of the motion, this average value is an accurate measure of the component motion. The attachment also filters component motions along the y direction. It is also often safe to assume that the base of the sensor is rigid compared to the component to which the sensor is attached. The analysis for different attachment footprints such as the circular and teardrop shapes changes somewhat from the analysis for a square attachment.

In Figure 4.16(b), the wavelength of the component motion is smaller. For this motion, there is an even greater difference between the motions on either side of the square attachment (see Figure 4.16(e)). This greater difference across the interface results in more measurement error due to in-plane strain of the attachment. When the wavelength of the motion is even shorter as in Figure 4.16(c) with approximately two wavelengths equal to the width of the attachment, the ends of the attachment move by the same amount but the center does not move as shown in

Figure 4.16(f). Depending on the stiffness of the attachment, this type of deflection profile produces significant amplitude measurement error. The general rule of thumb is that the attachment should be no more than a one-fourth wavelength of the response motion of interest to prevent measurement error.

4.4 TRANSDUCERS

Different transducers can be used to measure vibration, wave propagation and other response variables in structural components. There is no perfect sensor. 'Perfect measurements are possible only when the sensor is not there' (Tse and Morse, 1989). Section 2.2.5 already demonstrated that the mass of a sensor changes the response of a component. The previous section demonstrated that the attachment stiffness and damping also cause measurement errors. Figure 4.17 illustrates some measurement errors to which all sensors are susceptible. For the triaxial accelerometer shown in the figure (PCB 356B18), the sensor is intended to measure acceleration along the x, y and z axes. Potential measurement errors due to sensor issues include the following:

- Sensor misalignment can cause errors in the three different measurement directions;
- Acoustic sound pressure fluctuations can cause the accelerometer casing to vibrate, which does not relate directly to the component vibration that the sensor is intended to measure;
- Temperature fluctuations can cause shifts in the calibration of the sensing element or its internal electronics;
- The sensor can exhibit cross talk between the three measurement directions due to internal misalignment, which is usually quite small;
- Motions of the cable pull on the sensor causing it to respond, and these cable motions are not related to the desired measurement of the component motion;
- The screw nut attachment used in this sensor changes the local stiffness of the component material possibly leading to measurement errors; and
- The sensor element can become damaged if shock loads are above the capacity of the sensor (typically 1000–4000 g shock rating).

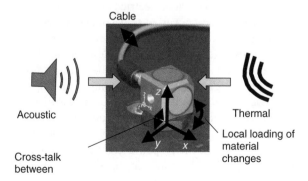

Figure 4.17 Typical sensor exposed to the environment including the desired measurands due to vibrations of the panel surface

Types of transducers and ways to minimize errors are discussed in the following sections:

4.4.1 Overview of Sensors and Actuators

Transducers convert one type of variable into another type of variable. Transducers take one of two forms in health-monitoring applications. A *sensor* converts mechanical motion or force, temperature, etc. into an electrical, magnetic or other measurable signal. Sensors measure the response of a component. On the contrary, an *actuator* works in the opposite direction to convert an electrical, magnetic or other user-defined signal into a mechanical motion or force. There are many types of sensors and actuators. Some examples of transducers are listed in Table 4.4 with references. Transducers are constructed of materials that can accomplish the conversion between forces/motions and signals for sensing and actuation. These materials are referred to as *smart materials*.

For example, the piezoelectric accelerometer in Figure 4.17 contains a piezoelectric material. This type of material strains when a charge is applied to it, so this material can serve as an actuator. A piezoelectric material also produces a charge when a strain is applied to generate a sensor signal. Piezoelectric materials can measure acceleration, strain, pressure and even temperature depending on which material is chosen and how the material is packaged by the manufacturer. For example, *quartz* is a piezoelectric material that is not sensitive to temperature changes; on the contrary, *piezoceramic* materials are quite sensitive to temperature variations.

Some smart materials are simply common materials combined in ways that provide useful transduction properties. For example, a *thermocouple* consists of two metallic materials with different thermal expansion coefficients. When combined in a thermocouple sensor, these two materials produce a measurable change in voltage when the temperature changes due to the Seebeck effect. Different combinations of metals enable measurements over different temperature ranges. For example, Tungsten and Tungsten alloys with different percentages of Rhenium are often used in thermocouples. Figure 4.18 gives a summary of different smart materials and their transduction properties.

There are two classes of sensing in health monitoring. *Passive sensing* uses sensors like the *acoustic emission* (AE) probes (PAC S9225) shown in Figure 4.19 to measure the response of a component to operational conditions. AE sensors measure pressure due to the acceleration of the component surface. Piezoelectric elements inside the sensor convert this pressure into current, which is then converted to a voltage signal for measurement. The word 'passive' does not suggest that the AE sensors are not powered to provide a signal output. On the contrary, most piezoelectric devices use preamplifiers as mentioned in Figure 4.2. AE sensors are called *non-self-generating* because an auxiliary source is used to amplify and condition the charge produced by the component response prior to transmitting the signal. In passive operation, AE sensors are primarily used to monitor crack initation and propagation events. Each crack event produces a transient pulse in the material that is detected by the AE sensor. In passive modes of operation, AE sensors rely on the growth of cracks to produce sufficient response for the measurement. In some cases, operational excitations also produce response in the same frequency band as the crack events causing low signal to noise ratios in the measurements.

Active sensing involves the use of an actuator that forces a component to respond in addition to a sensor that measures the response (or one device serving as both an actuator and a sensor).

Table 4.4   Types of transducers for health monitoring

Sensor	Description	Reference
Piezoelectric Piezoceramic	High frequency impedance transducers for detecting damage	Park *et al.*, 2000
	Uses electromechanical impedance to detect small cracks in steel beams	Giurgiutiu and Zagrai, 2001
	High frequency ultrasonic array	Giurgiutiu and Bao, 2002
	Uses propagating lamb waves to scan aluminum plates, pipes and concrete	Guo and Kundu, 2000; Jung *et al.* 2000; Villa *et al.*, 2001
	Built-in piezoceramic 'Smart layer' array for diagnosing damage	Choi *et al.*, 1994
Optical laser	Generates lamb waves and uses noncontact measurement with AE array	Mizutani *et al.*, 2003
Thermal imaging	Imaging of material loss in boiler water-wall tubing with scanning NDE	Cramer and Winfree, 2000
	Uses thermal transfer function to make thermal NDE more robust	Shepard *et al.*, 2000
Acoustic emission	Acoustography uses acoustic signals to develop ultrascan images quickly	Sandhu *et al.*, 2000; Sandhu *et al.* 2001
	Uses portable acoustic emission modules for fatigue crack characterization	Komsky, 2001
	Assess corrosion in airplane fuselage with multiple layers using acoustic emission	Komsky and Achenbach, 1996
Fiber Optic	Fiber optic health monitoring of civil infrastructure	Todd *et al.*, 1999; Bergmeister and Santa, 2001; Inaudi, 2001
	Distributed fiber Bragg grating (FBG) sensors were used to detect defects	Chang *et al.*, 2001
	Uses FBG sensors to monitor health of graphite epoxy missile motor casings	Heaton *et al.*, 2004
	Fiber optic AE sensors for harsh environmental SHM	Borinski *et al.*, 2001
Eddy current Electromagnetic	Detect cracking in concrete for civil infrastructure	Galleher *et al.* 2000
	Detect and monitor cracks using surface mounted sensors	Goldfine *et al.*, 2001
Conductive polymers	Senses moisture over extended surface areas in order to detect corrosion	Schoess, 2001
MEMS	Embedded strain microsensors with integrated signal conditioning	Walsh *et al.*, 2001
	Wireless remotely readable and programmable microsensors for SHM	Varadan and Varadan, 2000
	Sensor tags used to monitor physical variables of interest in TPS	Milos *et al.*, 2001
	Technology assessment of MEMS for PHM applications	Matzkanin, 2000
All	Review articles for civil and aero applications	Chong *et al.*, 2003; Ikegami, 1999
	Wireless architecture for civil infrastructure	Farrar *et al.*, 2001

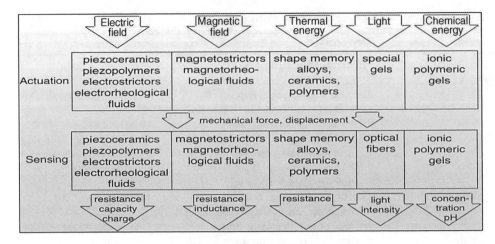

	Electric field	Magnetic field	Thermal energy	Light	Chemical energy
Actuation	piezoceramics piezopolymers electrostrictors electrorheological fluids	magnetostrictors magnetorheological fluids	shape memory alloys, ceramics, polymers	special gels	ionic polymeric gels
			mechanical force, displacement		
Sensing	piezoceramics piezopolymers electrostrictors electrorheological fluids	magnetostrictors magnetorheological fluids	shape memory alloys, ceramics, polymers	optical fibers	ionic polymeric gels
	resistance capacity charge	resistance inductance	resistance	light intensity	concentration pH

Figure 4.18 Transduction for actuation and sensing in different types of materials (courtesy Dr D. J. Inman, Smart Materials and Structures, Short Course)

Actuators for health monitoring using vibration and wave propagation measurements are often constructed of piezoelectric materials, that is materials that strain when a voltage is applied across them. Figure 4.19 shows a piezoceramic actuator plate (Polytec PI PIC255) used to generate propagating elastic waves and vibrations, which are then measured by the AE sensors described above. One advantage of active sensing is that the actuator can be driven with frequencies that fall outside the operating frequency range of the component. A second advantage of active sensing is that the excitation can be measured providing a means to quantify damage among other things. A third advantage is that point-to-point measurements, which enable damage to be located, can be obtained using active sensing.

For example, consider the set of measurements illustrated in Figure 4.20. These measurements were taken using a laser to measure the surface displacement normal to a sheet with two different types of cracks. A piezoelectric device was used to generate a packet of propagating waves that interfere with the cracks. From left to right, the images show the displacement patterns measured at different points in time after a wave pulse is transmitted into the cracked region of the sheet. When the propagating waves interfere with the fatigue crack, there are significant reflections in a spherical pattern because the crack serves as a point source of the reflections. This occurs because the crack width is small relative to the wavelength of the propagating waves. However, the sharp crack produces plane waves when it reflects the propagating waves because this crack is long compared to the wavelength of the waves. This

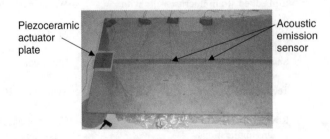

Piezoceramic actuator plate

Acoustic emission sensor

Figure 4.19 Passive acoustic emission sensors along with a piezoceramic actuator

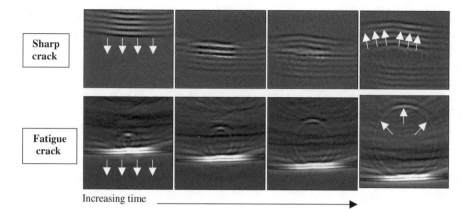

Figure 4.20 Laser ultrasound images for two different kinds of cracks showing change in wave propagation patterns normal to the material surface as a function of time (courtesy Dr J. Blackshire, AFRL, Materials and Manufacturing Directorate)

example demonstrates why active sensing is useful for health monitoring. The number of waves in the wave front and wavelength of the waves can be controlled with active devices in order to fully characterize damage in components. Without these active devices, response measurements would be limited to the waves that are produced by the cracks themselves.

Sensing can either be accomplished using contact sensors like those pictured in Figure 4.19 or noncontact sensors like those pictured in Figure 4.21. Contact sensors have the disadvantage that they require an attachment directly to the component being monitored. This attachment can cause the natural component response to change. Also, for high-temperature applications or applications involving shock loading, contact sensors can become damaged due to high mechanical and thermal transmission to the sensor.

Noncontact sensors like the *laser ultrasound sensor* shown in Figure 4.21(a) do not directly touch the component so the measurement is unobtrusive and more robust to severe environmental conditions. On the contrary, the laser source must be mounted in a noise-free environment to ensure high-quality measurements. It is not always possible to shelter noncontact sensor instrumentation from noise in the environment. This laser ultrasound measurement system was utilized to produce the images shown in Figure 4.20.

The *acoustic intensity probe* in Figure 4.21(b) measures the sound intensity (sound pressure and velocity) being radiated from a surface. This probe is also a noncontacting sensor. This

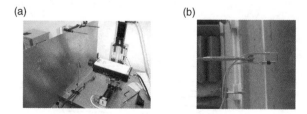

Figure 4.21 Noncontact sensors including (a) laser ultrasound measurement system (courtesy Dr James Blackshire, AFRL, Materials and Manufacturing Directorate) and (b) acoustic intensity probe (Hundhausen *et al.*, 2005, SAGE Publications, Inc)

device must also be mounted in a location where the intensity probe is not disturbed by background noise sources. Sources of noise might include vibration of the probe fixture or external sources of fluctuating sound pressure other than the sound produced by the component of interest (see Section 5.7.2 for application of this sensor).

Before inserting new transducers in a system to monitor components, all existing sensors in that system should be evaluated to determine if they may already provide the desired data. For example, a pressure sensor in an electrohydraulic actuator can be used to measure oil temperature indirectly. Likewise, a linear variable displacement transducer (LVDT) or force sensor in an air ride truck suspension system can be used to gather data on the dynamic response of the suspension without adding more sensors. Control systems and safety interlocks are other sources of sensing that sometimes already exist in structural systems.

4.4.2 Passive Sensors

Models of two types of passive sensors, a resistance strain gauge and a piezoelectric accelerometer, are presented in this section to demonstrate how models can be utilized to understand sensor performance.

4.4.2.1 Resistance Strain Gauge Model

Resistance strain gauges use changes in the resistance of a conductor to measure the local strain of a component (Hannah and Reed, 1992). If the wire (or foil) in the gauge shown in Figure 4.22 has cross-sectional area A, length L and resistivity ρ, then the resistance of the wire is

$$R = \frac{\rho L}{A}. \tag{4.3}$$

For example, a circular wire has $A = \pi d^2/4$ and a foil gauge has $A = wt$, where w is the width and t is the thickness of the foil. In order to utilize a foil strain gauge as a sensor, the differential change in resistance must be calculated (Tse and Morse, 1989):

$$
\begin{aligned}
dR &= \frac{\partial R}{\partial \rho}d\rho + \frac{\partial R}{\partial L}dL + \frac{\partial R}{\partial w}dw + \frac{\partial R}{\partial t}dt \\
&= \left(\frac{L}{A}\right)d\rho + \left(\frac{\rho}{A}\right)dL + \left(-\frac{\rho L}{w^2 t}\right)dw + \left(-\frac{\rho L}{wt^2}\right)dt.
\end{aligned}
\tag{4.4}
$$

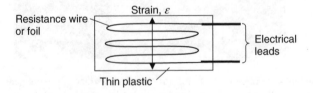

Figure 4.22   Resistance (or foil) strain gauge

This differential change in resistance of the strain gauge is usually normalized by the nominal resistance to calculate the sensitivity of the foil resistance to strain:

$$\frac{dR}{R} = \frac{d\rho}{\rho} + \frac{dL}{L} - \frac{dw}{w} - \frac{dt}{t}. \tag{4.5}$$

Then the strain, $\varepsilon = dL/L$, can be related to the foil Poisson's ratio if it is assumed that the foil is isotropic, $\nu = -(dw/w)/\varepsilon = -(dt/t)/\varepsilon$. With this assumption, Equation (4.5) can be written in the following form after dividing by the strain across the gauge:

$$\frac{dR/R}{\varepsilon} = \frac{d\rho/\rho}{\varepsilon} + 2\nu + 1. \tag{4.6}$$

This relationship indicates that the percentage change in resistance of the foil relative to strain is a function of Poisson's ratio of the foil and the sensitivity in resistivity of the foil material for changes in strain. In other words, the sensitivity of the foil is a function of both the *mechanical properties* and the *electrical properties* of the gauge material. Typically, $(dR/R)/\varepsilon$ is somewhere between 2 and 4 (Tse and Morse, 1989). Common materials used in strain gauges include constantan (copper–nickel alloy with $(dR/R)/\varepsilon = 2$) and platimum–tungsten $((dR/R)/\varepsilon = 4)$ (Dally *et al.*, 1993). Note that the foil shown in Figure 4.22 is not straight but curved leading to a lower sensitivity for the gauge than for a straight foil conductor of the same material, cross-sectional area and length.

Although it is desirable to have a high sensitivity, *there is a trade-off between sensitivity and robustness in sensing*. Higher sensitivity sensors can detect smaller signals and differences in signals between healthy and damaged components. However, higher sensitivity sensors are also more sensitive to environmental sources of variability. For example, ν and $d\rho/\rho\varepsilon$ are both functions of temperature in a strain gauge. If these parameters are selected to be large, then changes with temperature will also be large. In other words, *the signal-to-noise ratio drops as the sensitivity goes up*. The *threshold* of the sensor is important for health monitoring as well. The threshold is the resolution of strain that the sensor can detect. Resolution is only a function of $d\rho/\rho\varepsilon$.

If the input to the sensor is taken to be the strain and the output is taken to be the percent change in resistance, then the plot in Figure 4.23 is obtained relating the sensor input to the output. Over the operating range of the sensor, the output is linearly related to the input. The constant of proportionality given in Equation (4.6) is called the *sensitivity* of the sensor or the *static calibration constant*. When the strain becomes large, the calibration changes with the strain input. If the strain becomes too large, the foil can plastically deform, the plastic can crack or the bond can tear away from the component.

Other variables can also cause the characteristic in Figure 4.23 to change. For example, changes in temperature cause changes in resistivity, which consequently cause the sensitivity to vary. Circuits for measuring changes in resistance are designed to minimize sensitivity to temperature variations using Wheatstone bridges and other active electronic devices. Foil and wire strain gauges have *low output impedances* on the order of 100–500 Ω; therefore, there are usually no issues with accurately measuring the resistance change. For a Wheatstone bridge with equivalent resistors in three branches and a strain gauge in the fourth branch, the output voltage v_0 from the bridge is

$$v_0 = G\varepsilon, \tag{4.7}$$

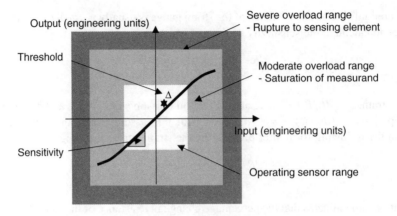

Figure 4.23   Calibration characteristics and terminology

where G is the *strain gauge sensitivity*. G is usually on the order of $1–10\ \mu V/\mu$strain[1] (Hannah and Reed, 1992).

4.4.2.2   Piezoelectric Accelerometer Model

Piezoelectric materials are sources of electric charge (Jaffe *et al.*, 1971). When a piezo-electric material is squeezed (or stretched), it produces a charge because positive and negative ions move to the top and bottom surfaces. Piezoelectric materials are normally modeled as charge sources with a capacitance that stores charge and some leakage resistance as well that loses charge. A piezoelectric accelerometer measures the charge produced by the inertial force experienced when a piezoelectric element accelerates. Two steps are required to understand the operation of a piezoelectric accelerometer. First, the mechanical behavior of the accelerometer must be modeled to understand how the acceleration of a seismic mass within the sensor relates to the component acceleration to which the sensor is attached. Second, signal conditioning of the charge produced by the piezo-electric element when a stress is applied to it must be understood in order to ensure an accurate measurement of the acceleration.

Consider the accelerometer illustration shown in Figure 4.24. This accelerometer is called a *shear mode* accelerometer. The advantage of this design is that it does not exhibit a measure-ment error due to bending of the accelerometer housing at the attachment interface. The accelerometer works as follows. The component surface moves with displacement $x_b(t)$. This displacement causes the rigid post to move. The post is attached to a cylindrical piezoelectric element with conductor plates on the inner and outer surfaces of the element. A seismic mass is then pressure fit around the piezo element. When the component displaces dynamically, the seismic mass resists this motion due to its inertia. Therefore, the outer surface of the piezo element moves with displacement $x_s(t)$ resulting in a shear stress on the piezo. This stress across the piezo can be represented by the relative motion (shear motion) $z(t) = x_s(t) - x_b(t)$.

---

[1] μ represents 1e-6 units, so $1\ \mu V = 1$e-6 V.

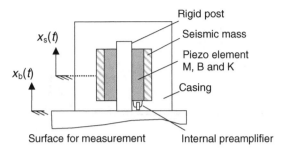

Figure 4.24 Shear mode accelerometer sensor model

The FRF relating the strain across the piezo to the component response acceleration is given by

$$\frac{Z(\omega)}{-\omega^2 X_b(\omega)} = \frac{-M}{K - M\omega^2 + j\omega B}, \tag{4.8}$$

where M is the effective piezo mass, B is the piezo's effective viscous damping coefficient and K is its shear stiffness. After dividing the numerator and denominator by M, this expression becomes

$$H_p(\omega) = \frac{-1}{\omega_n^2 - \omega^2 + j2\zeta\omega_n\omega}, \tag{4.9}$$

where the undamped natural frequency ω_n and damping ratio ζ are defined as in Section 2.2. $H_p(\omega)$ will be used to represent the *mechanical* FRF of the piezoelectric element.

For a resonant frequency of $\omega_n = 1$ rad/s, the magnitude of Equation (4.9) is plotted in Figure 4.25 for three values of ζ: 0.01 (__), 0.1 (- - -) and 0.7 (. . .). This plot was generated using the MATLAB® simulation code 'piezoaccelfrf.m'. An accelerometer is manufactured with a single calibration constant to measure g's of acceleration (1 g = 9.81 m/s²) throughout an operating frequency band. Therefore, accelerometers are usually operated far below their resonant frequencies where the *amplitude and phase distortion* are negligible. As in the attachment FRF (Figure 4.15), Figure 4.25 shows that a smaller value of ζ causes more distortion in the relative response magnitude between the piezo element strain and component vibration. Also, larger damping ratios cause distortions in the relative phase between the strain output and the acceleration input to the accelerometer. Consequently, accelerometers are usually designed with $\zeta = 0.3$ or greater to strike a balance between minimum amplitude and phase distortion.

The FRF in Equation (4.9) relates the relative displacement between the outer part of the piezoelectric element and the component vibration for the accelerometer in Figure 4.24. This relative displacement strains the piezoelectric element causing it to produce a charge, which is preamplified before transmitting a voltage to the other end of a cable. Typical sensitivities of piezoelectric accelerometers are 0.1–10 pC/g². The section below describes how this charge

---

[2]pico represents 1e-12 units, so 1 pC = 1e-12 C (Coulomb).

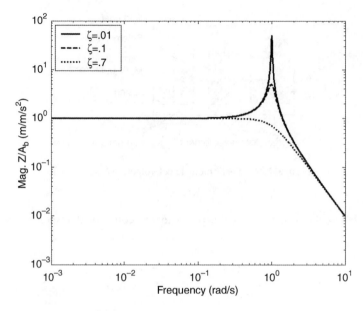

Figure 4.25   FRF magnitude relating piezoelectric element strain to acceleration of component for three values of damping ratio and $\omega_n = 1$ rad/s

sensitivity to acceleration is converted into voltage sensitivity to acceleration. Typical sensitivities of piezoelectric accelerometers are 10–1000 mV/g.

The analysis above dealt only with the acceleration input to the piezoelectric element from the base of the accelerometer. However, different types of piezoelectric materials are also sensitive to other inputs. For example, Figure 4.26 shows the percent deviation in the calibration constant for a piezoelectric sensor operating at temperature T. The plot shows that for the triaxial accelerometer pictured in Figure 4.14(b), the deviation is small as a function of temperature up to around 200 °F, which is the maximum temperature for quartz sensors. The high-temperature piezoelectric sensor shows slightly more deviation in

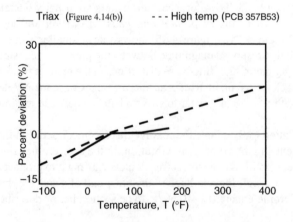

Figure 4.26   Piezoelectric change in calibration with temperature

calibration out to higher temperatures. It is important to remember that inputs other than acceleration cause the output voltage from an accelerometer to change. If the temperature is measured, then deviations in sensor calibration can be corrected in health-monitoring applications (refer to Section 5.9).

4.4.2.3   Transmission Models (Cable, Amplifier and Power Supply)

The previous section described the physics of how piezoelectric accelerometers operate. However, if the charge produced by the accelerometer is not carefully transmitted to the data acquisition system, then the measurement is compromised. Cables introduce capacitance and resistance that can degrade the quality of signals by filtering them. Power supplies and data acquisition systems also have capacitance and resistance, which can both compromise the quality of a measurement. For example, small input impedance data acquisition systems draw too much current from the measurement circuit causing attenuation and slow response in the measurement signal.

All of these potential sources of measurement error can be attributed to the inherently *high output impedance* of piezoelectric sensors. Unlike piezoelectric sensors, many sensors like the resistance strain gauges described in Section 4.4.2.1 have low output impedances. These types of sensors do not pose major measurement challenges because cabling, amplifiers and other data acquisition measurement devices do not electrically load the sensor. In other words, these transmission elements do not have a tendency to change the way the sensor performs. Consider Figure 4.27(a), which illustrates a sensor channel with impedance R_s that produces voltage v_s. The data acquisition device (DAQ), which could be a voltmeter or more sophisticated instrument, possesses an input impedance R_d and provides a measurement v_d of the sensor voltage. The FRF between v_d and v_s as a function of signal frequency is a measure of the potential for errors in the measurement due to loading of the sensor:

$$\frac{V_d(\omega)}{V_s(\omega)} = \frac{R_d}{R_s + R_d}. \tag{4.10}$$

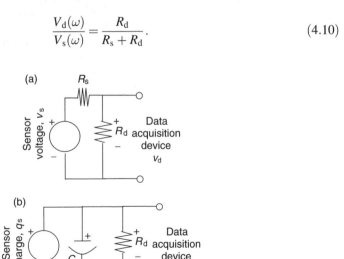

Figure 4.27   (a) Low output impedance sensor and (b) high output impedance sensor

This relationship shows that (a) the measured voltage is approximately equal to the sensor voltage when $R_d \gg R_s$, and (b) this requirement is constant as a function of frequency. The conclusion is small values of R_s lead to more accurate measurements.

Now consider Figure 4.27(b). This schematic shows a piezoelectric sensor that produces charge q_s, which is related to the sensor capacitance C_s as follows:

$$v_s = \frac{q_s}{C_s}. \tag{4.11}$$

This equation indicates that lower capacitance results in higher voltage sensitivity in piezoelectric sensors. Therefore, low capacitance on the order of 10 pF is desirable in piezoelectric sensing. On the contrary, the impedance relating the voltage v_s to the current i_s as a function of frequency on the sensor side of the circuit in Figure 4.27(b) is

$$\frac{V_s(\omega)}{I_s(\omega)} = \frac{1}{j\omega C_s}. \tag{4.12}$$

This equation indicates that low capacitance results in higher output impedance of the sensor on the order of several kΩ. The FRF relating v_d to v_s as a function of signal frequency is

$$\frac{V_d(\omega)}{V_s(\omega)} = \frac{j\omega R_d C_s}{j\omega R_d C_s + 1}. \tag{4.13}$$

This equation is the FRF for a *high-pass filter*. The magnitude of this FRF is shown in Figure 4.28 (generated using 'piezosensorfrf.m'). This plot shows that piezoelectric voltage

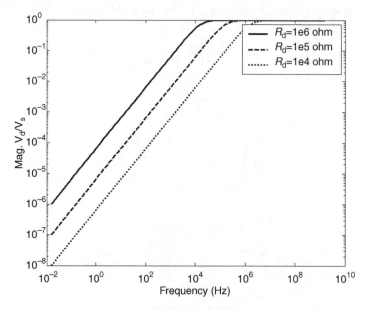

Figure 4.28 FRF magnitude relating measured voltage to piezoelectric voltage for three values of R_d showing attenuation of response due to low sensor output impedance

signals with frequencies above the cutoff frequency, $1/R_dC_s$, are passed to the DAQ. However, signals with frequencies below the cutoff are attenuated. The high cutoff frequency is primarily due to the high output impedance of the piezoelectric sensor, which usually possesses a very small capacitance of around 10 pF. The FRF also indicates that (a) the measured voltage is approximately equal to the sensor voltage when $j\omega R_dC_s \gg 1$, and (b) this requirement varies as a function of frequency. The conclusion is that small values of C_s and low frequencies lead to less accurate measurements. Also, the piezoelectric sensor in Figure 4.27(b) is very sensitive to capacitance changes in the transmission path due to cables, power suppliers and other sources. To ensure that the signal transmission path does not load piezoelectric sensors, preamplifiers are used to condition the piezoelectric sensor output.

In many piezoelectric sensors such as the accelerometer shown in Figure 4.17, there is an integrated *voltage amplifier* that conditions the charge produced by the piezoelectric element to avoid errors in the measurement due to the high piezo output impedance. The accelerometer shown in Figure 4.17 uses this type of voltage amplifier and is called an Integrated Circuit Piezoelectric (ICP®) sensor[3]. The preamplifier converts the charge into a voltage that is measured at the other end of a cable. In other types of accelerometers, *charge amplifiers* are inserted in line with the cable to amplify the measured charge from the piezoelectric element. There are advantages to both types of signal amplification. For example, high-temperature sensing can result in damage to ICP circuitry; therefore, charge amplification is preferred in that case. Charge amplifiers can be placed down line from the sensor where temperatures are not so high. A comparison between these two types of signal conditioning methods is given in Table 4.5.

Consider the typical piezoelectric sensing block diagram shown in Figure 4.29. This diagram shows a piezoelectric sensor with capacitance C_s. The sensor produces a charge output, which is preamplified by a metal oxide semiconductor field effect transistor (MOSFET). This transistor is integrated into the housing of the sensor. The transistor is powered over a coaxial cable, which also possesses some capacitance. Fortunately, the capacitance of the cable and the capacitance of the piezoelectric element can be ignored when the MOSFET amplifier is used because it dictates the overall impedance of the sensor. The constant current power supply with 18–30 V DC (direct current) is used to power the MOSFET device. This power supply has a capacitance C_d to remove DC voltage (i.e. drift) from the output of the signal conditioner. Assume that the coupling capacitor is 10 μF. The

Table 4.5 Voltage and charge amplification approaches to signal conditioning

Voltage amplifier	Charge amplifier
High frequency ($> 1\,\text{MHz}$)	Limited frequency ($< 100\,\text{kHz}$)
Low cost	More costly
Small size	Larger size
Used with quartz	Used with ceramic materials
	Low noise

---

[3]Refer to www.pcb.com for technical drawings and specifications of ICP® sensing infrastructure.

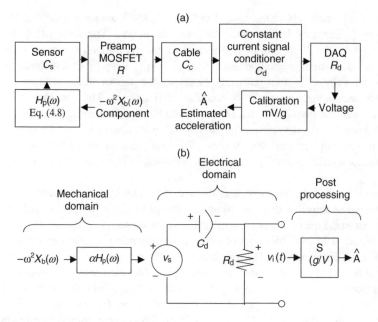

Figure 4.29 (a) Piezoelectric sensor and signal preamplification block diagram and (b) simplified electromechanical circuit diagram

voltage output from the signal conditioner is measured with a data acquisition system, which has input impedance R_d. Assume that the input impedance is 1 MΩ. As long as the frequency range of the signal is in the sensor's operating range, the voltage can then be divided by the calibration constant to produce a measurement with engineering units.

The total FRF relating the estimated acceleration $\hat{A}(\omega)$ from the shear mode accelerometer in Figure 4.24 to the actual component acceleration is given by

$$\frac{\hat{A}(\omega)}{-\omega^2 X_b(\omega)} = S \times \frac{jR_d C_c \omega}{jR_d C_c \omega + 1} \times \alpha H_p(\omega), \tag{4.14}$$

where α is a constant that relates the relative motion of the seismic mass to the charge and ultimately the voltage produced by the piezoelectric element, and S is the sensor calibration in units of g/V. Typical values of S are 100–1 g/V. This FRF is a composite of the FRFs in Figure 4.25 for the mechanical response of the accelerometer and Figure 4.28 for the electrical response of the accelerometer. This same block diagram approach can be utilized to model other types of dynamic sensors and their signal conditioning circuitry.

Other factors may influence the performance of piezoelectric accelerometers (and other piezo sensors). For example, the constant *current power source* illustrated in Figure 4.29 may become damaged. If the current drops below 2 mA, piezoelectric sensors with internal preamplifiers like those described above will not function properly. Sensors like these cannot just be plugged into voltmeters; these sensors must be externally powered. *Electromagnetic interference* (EMI) can also influence measurements. Piezoelectric accelerometers are used with coaxial cables, which possess a signal-carrying cable surrounded by a cylindrical grounding plane. These cables shield the desired measurement from sources of EMI along the cable length. Also, *ground loops* can affect piezoelectric sensor measurements. Ground loops occur if the component to which sensors are attached is not properly grounded. Ground loops result in loading of the measurement that floats with the component as its voltage fluctuates. In these cases, a nonconducting attachment mechanism can be used to isolate the sensor housing from the component. *Faraday cages* are also sometimes used for this purpose.

4.4.3 Active Piezoelectric Transducers (Actuators)

Active piezoelectric transducers are used to apply forces to components so that sensor measurements can be used to characterize damage in those components. Active transducers are often called *actuators*. The simplest piezoelectric actuator is the *modal impact hammer* (Figure 4.30(a)), which is used for off-line testing to impart and measure broadband excitations. Impact hammers can be used to apply forces at any accessible location. A nonpiezoelectric type of actuator is the *eddy current* transducer (Hellier, 2001). This device is based on the use of actuators that generate a magnetic field, which senses changes in the permeability of materials that are damaged. The advantages of active sensing were described in Section 4.4.1. There are also disadvantages to active sensing. For example, a power source is required to drive actuators. Power sources are often too heavy for use in applications. Also, active sensing often requires additional measurement hardware such as digital-to-analog converters (refer to Figure 4.2).

As mentioned before, when a voltage is applied to a piezoelectric material, the material contracts (or extends). Figure 4.30(b) shows an example of an automatic piezoelectric actuator. This actuator is a *stack* actuator consisting of 10 piezoelectric crystals in series. Figure 4.31(a) is a picture of a packaged stack actuator. Each of the crystals is L_0 thick and is polarized in the vertical direction to provide the largest piezoelectric action when energized in that direction. The strain across this stack actuator produced by a voltage V applied across each crystal for n crystals is given by the ratio of the total change in length ΔL of the stack of crystals to the original length of the stack:

$$\varepsilon_{\text{stack}} = \frac{\Delta L}{n\,L_0} = \frac{V\mathrm{d}}{L_0}. \tag{4.15}$$

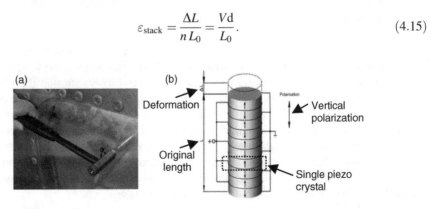

Figure 4.30   (a) Modal impact hammer (PCB Piezotronics model 086C03) and (b) piezoelectric stack actuator (10 crystals) for longitudinal deformation (courtesy Physik Instrumente (PI) GmbH & Co. KG)

(a)  (b)

Figure 4.31 (a) PI P-010.00P (1000 V, 129 kHz and 4 gm) stack actuator and (b) APC 850 actuator with 38 g forcing mass (10 mm diam. by 2 mm thick APC 850) alongside a 10 kHz PCB 333C65 accelerometer (Ackers *et al.*, 2006, Mr. Jonathan White, Purdue University)

The constant d is called the *piezoelectric constant* with units of m/V (Jaffe *et al.*, 1971). The force amplitude produced at a given frequency as a consequence of this strain is given by

$$
\begin{aligned}
f_{\text{stack}} &= \omega^2 M_a \frac{\Delta L}{2} \\
&= \frac{1}{2}\omega^2 M_a n L_0 \varepsilon_{\text{stack}} \\
&= \frac{1}{2}\omega^2 M_a n V d.
\end{aligned}
\tag{4.16}
$$

The parameter M_a in this equation is the effective mass of the moving parts of the actuator. This equation indicates all of the following:

- Higher mass actuators produce more mechanical force; this is why additional masses are often attached to actuators as shown in Figure 4.31(b).
- Higher frequencies lead to larger actuation forces for constant applied voltage.
- Larger numbers of crystals in the stack produce larger actuation forces.
- Piezoelectric materials with higher piezoelectric constants d produce larger actuation forces.
- Larger applied voltages produce larger actuation forces; voltages as high as 1 kV are not uncommon (Figure 4.31(a)).

In addition to these operating characteristics, it is also important to note that piezoelectric crystals perform best in compression. Many types of actuators and sensors (e.g. piezoelectric load cells) are therefore designed with preloaded crystals to ensure that the piezo device behaves linearly in compression and tension. Actuators are designed for either high voltage/low current operation or vice versa. It is important to utilize voltage amplifiers or current amplifiers based on the type of actuator. If the current is not adequate to drive an actuator, it will not perform to the specifications provided by the manufacturer. Also, actuators exhibit dynamic response characteristics just like the piezoelectric sensor that was modeled in Section 4.4.2.2. When generating signals that will be used to drive actuators, it is necessary to account for the limited bandwidth of the actuator dynamic response.

For example, Figure 4.32(a) shows a PI P-8410.20 stack actuator hanging freely. The actuator is attached to a load cell at its base to measure the force supplied by the actuator when it is energized. Figure 4.32(b) shows a triangular-shaped drive voltage along with the corresponding force measurement for this actuator. The triangular-shaped pulse provides a sudden (transient)

(a)

(b)

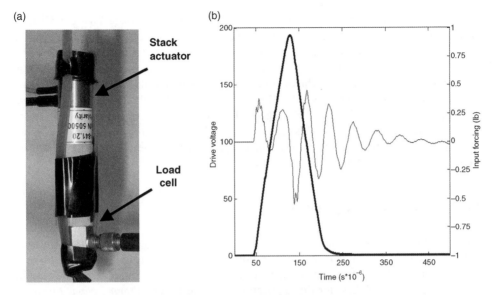

Figure 4.32   (a) PI P-8410.20 stack actuator attached to PCB 209C01 load cell for force measurement and (b) transient force measurement compared to driving voltage (Reproduced by permission of Jonathan White, Purdue University, 2006)

voltage to the actuator around 50 μs. This transient voltage causes the actuator to 'ring'. In other words, the transient force excites the free response of the stack actuator at the natural frequency of the piezoelectric crystals (refer to Section 2.2.3). Because there are many crystals within the stack, there are many natural frequencies that can become excited by such transient drive signals. Instead of using a quickly changing triangular pulse to drive the actuator, a slowly changing pulse should be used to avoid exciting the piezo resonant frequencies.

It is also important to verify that the actuator (in conjunction with its attachment mechanism) is providing the actuation force that is desired for damage detection in health monitoring. Actuation force can be measured using a force sensor like the load cell shown in Figure 4.32(a). Figure 4.33 shows an example of how an actuator and force sensor can be

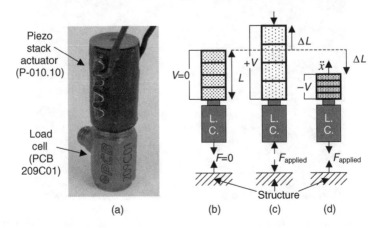

Figure 4.33   (a) PI P-010.10 piezoelectric stack actuator bonded to a PCB 209C01 load cell for measuring force input (b)–(d) illustration of operation of this actuator–sensor pair (Reproduced by permission of Mr. Jonathan White, Purdue University, 2006)

utilized in practice. Figure 4.33(a) shows the two devices in series and Figure 4.33(b)–(d) show the basic principle of operation. When the actuator extends, it compresses the load cell resulting in a pushing force on the structural component. When the actuator contracts, it pulls on the load cell, which pulls on the component. By measuring the force applied by the actuator, various data analysis techniques can be used to more fully characterize component damage as described in Section 5.11.3. Also, the integrity and calibration of the actuator can be verified continuously by measuring the FRF between the voltage drive signal and the measured force at the load cell. This type of *self-calibration* whereby actuators and sensors help to identify any changes in their measurement performance is needed in health monitoring to ensure that measurements remain error free.

4.4.4 Other Types of Sensors

Appendix B provides information on many other types of sensors that can be used for health monitoring. Displacement, velocity, acceleration, force, temperature and pressure sensors are reviewed. Each type of sensor has advantages and disadvantages so the user must carefully select the sensor that is best for the application of interest. For example, *capacitive accelerometers* can measure DC acceleration, which may be of interest in aircraft or vehicle dynamics applications; however, capacitive accelerometers are heavier than piezoelectric accelerometers because of the extra signal conditioning required. Appendix B also provides information on several other types of piezoelectric actuators that can be used for health monitoring.

Also, *wireless* sensors such as the Oceana two-channel device shown in Figure 4.34 are available for use in health monitoring. The transmitter for this two-channel device supplies the piezoelectric accelerometers shown in the figure with power for signal conditioning. The transmitter sends the voltage signal measured from the sensors to a receiver and data

Figure 4.34 Two-channel wireless transmitter attached to wheel of tire for use with piezoelectric sensors in rotating reference frame of the tire

acquisition system, which are located in a fixed reference frame. Wireless sensing is attractive in applications like this one where the response variables of interest occur in a rotating reference frame. Wireless sensing is also useful in weight-sensitive applications where large numbers of cables are prohibited. However, wireless sensing is not as convenient for use with active sensing because transmitters do not generally provide enough power for actuation. Wireless sensing can also be susceptible to interference due to barriers or external communication systems.

4.4.5 Transducer Placement and Orientation

Sensors and actuators must be positioned in locations and in directions that are indicative of the loads and damage of interest in health monitoring. There are two basic methods for placing and orienting transducers: (a) physics-based techniques and (b) data-driven techniques. Physics-based techniques utilize a first-principle model of the component to select the transducer configuration that is most sensitive to the loads and damage to be identified. Data-driven techniques treat the component like a 'black box', which produces data. Indicators are calculated through trial-and-error for use in maximizing the information content in the data.

The basic rule in positioning sensors is that the loading and damage mechanisms of interest should be *observable* with the sensors. An observable quantity is one that can either be directly measured or indirectly measured through the use of sensors and a model of some kind. Several considerations for placing sensors are illustrated in Figure 4.35. There are always trade-offs in performance associated with any sensor configuration.

In Figure 4.35(a), a cantilevered plate is shown undergoing displacement in the transverse direction only. First, consider measurements along the surface of the plate. Strain gauge SG1 in the location and direction shown provides a good measure of the response because the strain is high at this point of high curvature. The strain measurement at SG1 is referred to as a *measurement degree of freedom*. Each measurement DOF consists of a sensor location and

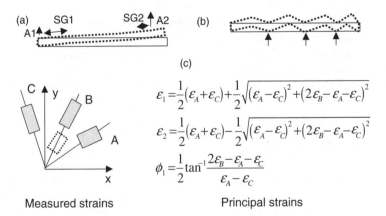

$$\varepsilon_1 = \frac{1}{2}(\varepsilon_A + \varepsilon_C) + \frac{1}{2}\sqrt{(\varepsilon_A - \varepsilon_C)^2 + (2\varepsilon_B - \varepsilon_A - \varepsilon_C)^2}$$

$$\varepsilon_2 = \frac{1}{2}(\varepsilon_A + \varepsilon_C) - \frac{1}{2}\sqrt{(\varepsilon_A - \varepsilon_C)^2 + (2\varepsilon_B - \varepsilon_A - \varepsilon_C)^2}$$

$$\phi_1 = \frac{1}{2}\tan^{-1}\frac{2\varepsilon_B - \varepsilon_A - \varepsilon_C}{\varepsilon_A - \varepsilon_C}$$

Measured strains   Principal strains

Figure 4.35 Sensor placement considerations in structural components (a) strain in one direction (b) modal vibration creates nodes and antinodes and (c) indirect estimation of principal strains using strain Rosette

orientation. If this same gauge is oriented into the page, very little strain is measured because the loading does not produce large strains in that direction (except through Poisson's effect). SG2 also measures small strains because the boundary conditions prevent large strains at the free end. Next, consider measurements normal to the surface of the plate. An accelerometer located and oriented at A1 does not measure much acceleration because the boundary condition for the component prevents motions at that point in that direction. In contrast, accelerometer A2 measures high accelerations. This example illustrates the need for considering the point and direction of applied loads when positioning sensors. The preferred positions for strain and acceleration measurements are often different.

The other issue to consider is that sensors in locations and directions of high response amplitude are most likely to distort the component response. For example, the A2 accelerometer measurement amplitude just described in Figure 4.35(a) is high relative to the A1 measurement. Consequently, the mass loading due to accelerometer A2 (see Section 2.2.5) is much higher than for A1. The durability of sensors can also be affected by the positions in which they are installed. Higher strains at SG1 provide a better measure of the component response than SG2, but measurements with SG1 are more susceptible to variability due to degradation in the strain gauge and bond.

Figure 4.35(b) illustrates another consideration for sensor placement. A component is shown undergoing a vibration displacement shape with three *nodal lines* along the length of the plate. If sensors are placed in any of the locations indicated with arrows, no transverse response is measured. In order to determine where these nodes of vibration are located, models can be used as described in Chapter 2. If the frequency range of the operational excitation excites modes of vibration with nodes in certain locations, then vibration sensors should not be positioned in those locations. If actuators are used, then frequencies that excite vibration shapes with nodes at sensor positions should be avoided. These issues are of less concern in wave propagation response data because traveling waves do not exhibit nodes in the same sense as standing vibration responses. However, the displacement shapes associated with different types of traveling waves must be understood because different shapes are less or more sensitive to certain type of damage.

Sensors provide raw data, but health-monitoring information can be extracted only by using models to process that data. If a quantity is observable with a given set of sensors, then it is possible to estimate the quantity of interest indirectly. For example, Figure 4.35(c) shows a strain Rosette with three strain gauges positioned along directions A, B and C. The strain measurements in these three directions are sufficient to calculate principal strains in the x, y and shear direction (ϕ) as shown in the equations listed in Figure 4.35(c). If the sensor along the A direction in the strain Rosette is replaced with a second sensor along the B direction (indicated with dotted lines), then there is not sufficient data to calculate the principal strains. In this situation, the principal strains are said to be *unobservable* with the raw strain data.

The concept of observability can be extended to actuators as well. If active sensing is used in health monitoring, then an actuator must be positioned and oriented to produce an auxiliary source of standing or traveling waves that is measurable with the sensors. If it possible to excite the vibration or wave propagation responses needed for damage identification using a certain actuator position and orientation, then the health-monitoring damage identification problem is said to be *controllable*. For example, if it is desirable to excite the vibration response shown in Figure 4.35(b) to detect certain types of damage, then an actuator placed at one of the arrows normal to the surface of the plate should not be used. An actuator in one of these locations and directions cannot excite this vibration pattern because the component does not move at those locations, so there is no way to impart energy to the component at this frequency of vibration.

The nine-DOF system model in Figure 4.6 can be used to examine these issues in more depth. The simulation code 'ninedofselect.m' calculates the modal frequencies and deflection shapes of the nine-DOF system using the technique described in Section 2.2.2. The nine modal deflection shapes normalized with respect to the maximum deflection are plotted in Figure 4.36(a). The 'o' symbols indicate the deflection of each DOF for the nine real normal modes for which there is either no damping or proportional damping. To aid in interpreting these shapes, the modal deflection shape for mode 5 at 2517 Hz is illustrated in Figure 4.36(b). The normalized deflection shape for mode 5 is plotted with solid lines overlaid on top of the static shape of the panel. First consider this particular mode shape. The DOF at the center of the component is the only node in this mode of vibration. All other DOFs displace either up or down in this deflection shape. In this mode of vibration, the only poor choice for a sensor location measuring transverse motion is at DOF 5 situated in the center because this mode of vibration response is not observable from this sensor location. The location at DOF 5 is also a

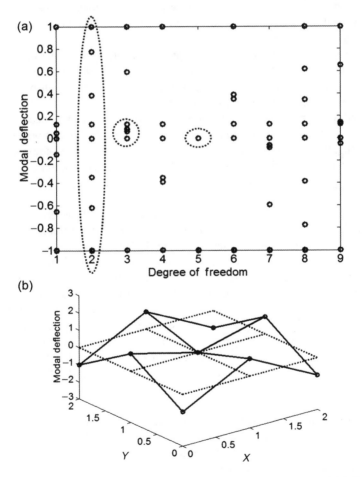

Figure 4.36 (a) Modal deflection patterns for nine-DOF system model and (b) modal deflection pattern for mode 5 at 2517 Hz

poor location to place an actuator because it does not impart any energy at this frequency lacking the motion to do so. A strain sensor, on the contrary, does measure strain at DOF 5 for the modal deflection shape at mode 5 because the curvature of the component is high at this position in this mode of response.

The deflection shape at 2517 Hz represents one possible response mode of vibration that can become excited by the operational loading spectrum or a user-defined excitation delivered by actuators. In order to understand how transducer positioning is accomplished when all of these modes are excited by a broadband forcing spectrum, consider the composite plot of all modal deflection shapes in Figure 4.36(a). The following observations can be made in this plot:

- Transducers should not be positioned at DOF 5 to measure or impart transverse vibration at the center of the component. This DOF is a node of vibration for several modal deflection shapes making much of the motion unobservable using a sensor and uncontrollable using an actuator at this DOF. It is generally true that transducers placed along lines of symmetry result in less observability and controllability of the dynamic response of a component.
- On the contrary, if it is desirable to isolate the vibration response at one mode of vibration (mode 1 or mode 9), then a sensor positioned at DOF 5 is ideal because it measures only motions at these two modes.
- DOF 3 appears to be a node of vibration for approximately three deflection shapes, 1, 2 and 8; however, the motions of DOF 3 at several other modes of vibration are small. This DOF would, therefore, be a relatively poor location at which to position sensors to measure the response of the panel over a wide frequency range.
- DOFs 2 and 8 are the ideal locations to place sensors and actuators because these DOFs appear to be nodes for only one mode of vibration, mode 4. The component response is, therefore, highly observable over a broad frequency range in data acquired from sensors positioned at DOFs 2 and 8. The same can be said of the controllability achieved when actuators are placed at these DOF locations.

Figure 4.37 illustrates a practical example of selecting transducer configurations. Figure 4.37(a) shows a fiber-reinforced polymer beam used to reconstruct bridge decks. The beam cross section is highlighted in the figure. Figure 4.37(b) shows a top view of the beam schematic with the damaged region (2 in × 7 in) of interest highlighted. In this damage detection problem, three accelerometer sensors are used to measure transverse vibration of the beam in the locations circled (O) in Figure 4.37(b). A modal impact hammer with a force load cell to measure input force is used to excite the beam in the transverse direction at locations I1 through I24. This type of actuation method is appropriate for offline use to monitor the health of this beam at certain inspection intervals. This combination of 24 input measurement DOFs and three output measurement DOFs provides $3 \times 24 = 72$ input–output FRF measurements. This set of actuator and sensor measurement DOFs are chosen for the following reasons:

- The three sensor measurement DOFs are on different lines along the width of the beam; therefore, *bending* motions as well as *torsional* motions are observable in the data provided by these sensors. By exciting both bending and torsional motions, the damage is exercised in both bending and torsion to enhance changes in vibration data.
- The three sensors are also positioned at different positions along the length of the beam to ensure that they do not all lie on the same *nodal lines*.

(a)

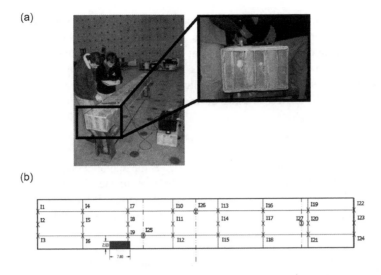

(b)

Figure 4.37 (a) Fiber reinforced polymer beam used in bridge deck reconstruction and (b) input locations (X), output locations (O) and damaged area shaded

- The 24 actuator measurement DOFs span the width and length of the beam to ensure that bending and torsional modes are exercised and to avoid nodal lines.

These concepts in observability and controllability can be made more concrete by defining quantitative metrics for comparing one transducer configuration to another. Two criteria are defined below for this purpose. The first criterion is to determine the observability of a given set of sensors. The second criterion is to determine the controllability of the component's response for a given set of actuators. Friedland (1986) gives a detailed derivation of these criteria.

4.4.5.1 Observability Criterion

Consider the state–space model for a structural component with n DOFs where the [A], [B] and [S] matrices are defined as follows (refer to Section 2.2.1):

$$\frac{d}{dt}\left\{\begin{matrix}\{x\}\\\{\dot{x}\}\end{matrix}\right\}_{2n\times1} = \left[\begin{matrix}[0] & [I]\\-[M]^{-1}[K] & -[M]^{-1}[C]\end{matrix}\right]_{2n\times2n}\left\{\begin{matrix}\{x\}\\\{\dot{x}\}\end{matrix}\right\}_{2n\times1} + \left[\begin{matrix}[0]\\[I]\end{matrix}\right]_{2n\times n}\{f(t)\}_{n\times1} \quad (4.17a)$$

$$= [A]\left\{\begin{matrix}\{x\}\\\{\dot{x}\}\end{matrix}\right\} + [B]\{f(t)\} \quad (4.17b)$$

$$\{y\}_{m\times1} = [S]_{m\times2n}\left\{\begin{matrix}\{x\}\\\{\dot{x}\}\end{matrix}\right\}_{2n\times1}. \quad (4.17c)$$

The physical meaning of the matrix [S] is that it defines the sensor set being evaluated. The *observability test matrix* is,

$$[O]_{2n\times2nm} = [S^T \ (A^T)^1 S^T \ \cdots \ (A^T)^{2n-1} S^T]. \quad (4.18)$$

If the rank[4] of this matrix is less than $2n$, which is the order of the system, then all of the dynamics are not observable using the chosen set of sensors defined by [S].

In order to evaluate the rank of [O] in all but the simplest applications, numerical methods are used. One method for evaluating the rank is to compute the eigenvalues of the normal matrix, $[O]^T[O]$. These eigenvalues are called the *singular values*. The *singular value decomposition* of [O] is expressed using these singular values as follows:

$$[O]_{2n \times 2nm} = [U]_{2n \times 2n} [\Sigma]_{2n \times 2n} [V]^T_{2n \times 2nm}, \tag{4.19}$$

where [U] is the matrix of *right singular vectors*, [Σ] is the diagonal matrix of *singular values* and [V] is the matrix of *left singular vectors*. The rank of the observability test matrix is determined by examining the size of successive singular values. If the singular values drop by a threshold amount, then the sensor measurements will not provide data from which all dynamics are observable.

Consider the nine-DOF system model discussed earlier in Section 4.2.2 (see Figure 4.6). The simulation code 'ninedofselect.m' is used to calculate the singular values of the observability test matrix for this model. Several different sensor sets are considered:

Case 1: Displacement sensor at DOF 1, $[S] = [1\ 0\ 0\ 0 \ldots\ 0]$
Case 2: Displacement sensor at DOF 2, $[S] = [0\ 1\ 0\ 0 \ldots\ 0]$
Case 3: Displacement sensor at DOF 3, $[S] = [0\ 0\ 1\ 0 \ldots\ 0]$
Case 4: Displacement sensor at DOF 4, $[S] = [0\ 0\ 0\ 1 \ldots\ 0]$
Case 5: Displacement sensor at DOF 5, $[S] = [0\ 0\ 0\ 0\ 1 \ldots\ 0]$
Case 6: Displacement and velocity sensors at DOFs 1–9, $[S] = [I]$, where [I] is the 18 by 18 identity matrix

The 18 singular values for each of these cases are shown in Figure 4.38. Different symbols are used for each case, and the singular values are normalized by the maximum singular value for each case. Case 6, for which all DOF displacement and velocity variables are measured as outputs, is indicated with a diamond shape (◇). This case is always optimum; if all states of the system are measured, then all of the dynamics are observed in those measurements. Note that the size of the singular values is shown in decreasing order. There are no large drops in the singular values for this case indicating that [O] has rank 18. However, it is not practical to measure all of the displacement and velocity responses. Usually, only one or at most several DOF responses can be measured.

For example, consider Case 5 in which the displacement of DOF 5 is measured (□). The modal deflection shapes in Figure 4.36(a) show that DOF 5 is a node of vibration for six out of the nine modal deflection shapes upon closer inspection ('ninedofselect.m'). Only modes 1, 5 and 9 can be observed from the displacement measurement at DOF 5. It is, therefore, expected that the rank of the observability test matrix for Case 5 will not be 18, i.e. [O] will be *rank deficient*. The singular values for Case 5 are seen to drop by nearly 20 orders of magnitude at the solid arrow. This large drop in the singular values indicates that the rank of [O] is 5. The rank is equal to 5 in this case because only two of the modes of vibration are fully observed using a sensor at DOF 5. Each mode that is observable is associated with two singular values.

---

[4]The rank of a matrix is the minimum number of independent columns or rows in the matrix.

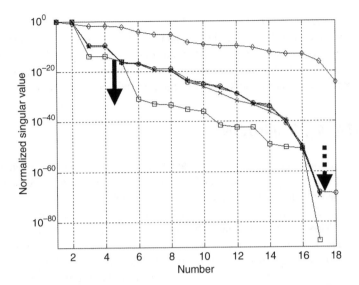

Figure 4.38 Normalized singular values for (($\Diamond$) all 18 states measured, ($\bullet$) displacements of DOF 1, (o) DOF 2, (x) DOF 3, (+) DOF 4 and ($\Box$) DOF 5

Mode 6 is actually not numerically observable because the modal deflection at DOF 5 is only 0.002 units; therefore, mode 6 is associated only with one significant singular value.

Next consider the other cases, 1–4. The singular values for these cases are indicated by the ($\bullet$), (o), (x) and (+) symbols, respectively. For the most part, the singular values trend together for all of these cases. In particular, when the singular values for DOF 5 drop down at the solid arrow, the singular values for Cases 1–4 do not. Only at the 17th singular value do these cases show a large jump down indicated by the dotted arrow. This result suggests that the rank of the observability matrices for Cases 1–4 is 16. Based on the earlier discussion of the modal deflection shapes for the nine-DOF system, it is expected that one mode of vibration will not be observable. This one mode of vibration is associated with two insignificant singular values (a loss in rank of 2). The differences in singular values for Cases 1–4 are due to the previously mentioned differences where DOFs 1–4 sit relative to the nodes of vibration for the various modes. For instance, a sensor positioned at DOF 3 indicated with the (x) symbol shows lower singular values than sensors at the other DOFs because DOF 3 is nearly a node of vibration for several modes (2 and 6) whereas the others are not.

4.4.5.2 Controllability Criterion

Consider the state-space model for a structural component with n DOFs where the [A] and [B] matrices are defined as above in Equation (4.17). The *controllability test matrix* is given by

$$[E]_{2n \times 2n^2} = [B \ A^1 B \ \cdots \ A^{2n-1} B]. \tag{4.20}$$

If the rank of this matrix is less than $2n$, which is the order of the system, then all of the dynamics are not controllable using the chosen set of actuators defined by [B]. Note that in some applications, the number of columns in [B] is chosen to be less than n for convenience in writing the input vector, $[f(t)]$, in Equation (4.17). The criterion is the same regardless of the number of actuators used. For example, the panel in Figure 4.2 is instrumented with four actuators resulting in four columns in [B]. On the contrary, the beam in Figure 4.37 is excited by 24 modal impacts resulting in 24 columns in [B]. In both cases, however, the condition for controllability requires that the rank of [E] is equal to twice the number of DOFs (modes).

4.5 DATA ACQUISITION

The previous sections discussed issues related to measurement data and sensing. Sensors deliver data to DAQ. If the hardware configuration and settings in the DAQ are not chosen properly, good sensor data can be ruined. There are many different types of DAQ systems. Figure 4.39 shows three types of dynamic DAQ systems. Figure 4.39(a) is a 16-channel LDS Genesis system. This system is semiportable and uses a laptop to control the DAQ. It is too large for many applications but is useful in industrial settings. Figure 4.39(b) is a two-channel system with a sound card and preamplifier running through the PCMCIA and USB slots, respectively. A laptop is again used as the interface to control the sound card settings and store and process data. This system is highly portable for field use but with only two channels, it can provide only local health-monitoring capability for a specific component. Figure 4.39(c) is an eight-channel system that uses two four-channel National Instruments 9233 cards for acquiring input signals. Four output channels are also available for driving actuators. The 'front end' of this DAQ consisting of the input and output channels communicates with the laptop through the USB interface. This system is highly portable for field use and provides a sufficient number of channels to monitor the health of large components through both passive and active sensing.

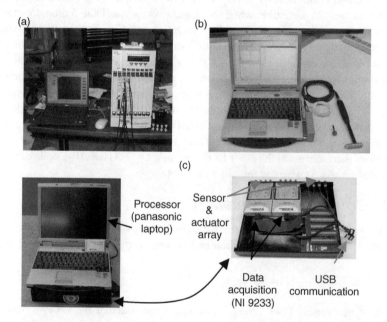

Figure 4.39   (a) Large channel data acquisition system (b) two-channel system and (c) multichannel system with breakout of components (Reproduced by permission of Jonathan White, Purdue University, 2006)

Regardless of which DAQ is used, *all DAQ devices impose limitations on the measurement data*. For example, each of the systems in Figure 4.39 *digitally samples* data at a certain rate for a certain period of time. Unlike analog signals, which are continuous functions of time, digitally sampled signals are discrete functions of time. Both of these digital sampling *limitations introduce errors in the data*, so it is fair to ask 'Why not just use analog data?' Digital signal acquisition is preferred to analog methods because digital data can be processed using sophisticated algorithms running on microprocessors. Computers represent data in digital form. Analog data can be processed only using complicated arrays of analog filters.

This section discusses many of the errors introduced by data acquisition limitations. First, errors associated with the amplitude of a signal are discussed. Then errors associated with the frequency of a signal are described. Data acquisition methods that overcome certain limitations are also described including filtering and multiplexed signal acquisition. Vendors who sell data acquisition equipment can be helpful when selecting the proper hardware and settings; however, there is no substitute for a good personal understanding of data acquisition limitations and methods for overcoming these limitations.

4.5.1 Common Errors

When analog signals are represented digitally, the conversion illustrated in Figure 4.40 is performed by an *analog-to-digital converter* (ADC) circuit (Dally *et al.*, 1993). The input signal amplitude is drawn on the abcissa of this plot, and the digital output code is drawn on the ordinate. This conversion is limited in several ways, and these limitations do cause errors in the acquired data. The basis for all these limitations is *sampling*. Analog signals are converted to digital format by sampling signals a certain number of time steps.

To begin with, the input signal is restricted to a certain amplitude range ($-V$ to $+V$). Usually, the input signal to the ADC is preset to accept a certain amplitude range (e.g. -10 to $+10$ V or -5 to $+5$ V). If the input signal amplitude goes beyond these limits, then the ADC circuit is overloaded. Figure 4.41 illustrates the typical digital output signal obtained when an ADC circuit is overloaded. The curve in the top is drawn in units of Volts as a function of time. The curve in the bottom is a zoomed version of the signal in the top. Although this curve looks like a continuous signal, it is discrete but is usually drawn as a continuous line in most data acquisition software because otherwise it is difficult to see trends in the data. The range for the ADC used in this data acquisition system is -5 to 5 V. The signal is seen to surpass this range causing the ADC circuit to output an exponentially decaying response superimposed on the actual dynamic response measured by the sensor. This type of error makes it difficult to use

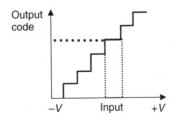

Figure 4.40   Analog-to-digital conversion showing bit roundoff (quantization) error

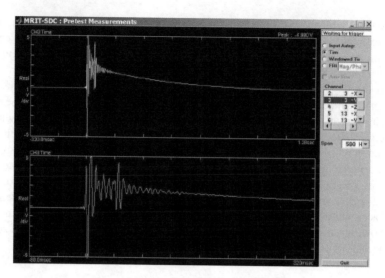

Figure 4.41 Overload in ADC followed by exponential decay with full signal and zoomed version of signal

the data for health monitoring. When setting amplitude ranges for data acquisition systems, some padding should be applied on either end of the anticipated range to avoid overloads.

The conversion illustrated in Figure 4.40 is also limited by the number of digital output levels into which it converts the analog signal amplitude range. These levels are drawn as steps in the figure. For example, an 8 bit ADC converts a −5 to 5 V analog signal into a digital series of 256 (2^8) values. More bits in the ADC translate into more digital amplitude levels. A 16-bit ADC converts the same −5 to 5 V signal into a series of 65 536 digital values. As of the year 2007, most digital data acquisition systems use at least a 16-bit ADC and some high-end systems use 24 bits. As explained above, an ADC always has amplitude limitations. These limits are set, and then available bits of amplitude resolution for converting the analog signal are distributed evenly across the signal range. Therefore, it is important in digital data acquisition to optimize the range settings of the ADC to match the signals being monitored. For instance, one should not use an ADC with a −10 to 10 V range to acquire a signal with a −0.5 to 0.5 V range.

These limitations on the amplitude range and number of bits of resolution of the analog signal can introduce other subtle errors in measurement data. The simulation code 'adcerrors.m' was used to generate the plots shown in Figures 4.42–4.44. Figure 4.42(a) shows a 500 Hz signal sampled at a certain time step. A range of −5 to 5 V is sufficient to acquire this data. However, if there is 60 Hz noise added to this signal as in Figure 4.42(b), the ADC will experience an overload with a ±5 V range. Sixty hertz noise is commonly observed in AC power lines and is also caused by fans running in HVAC (heating, ventilating and air conditioning) equipment as well as other types of rotating equipment. In this case, the range must be increased to ±10 V. This increase in range avoids an overload and also reduces the number of bits available for converting the signal into digital format. For example, Figure 4.43(a) shows two different sets of quantized data points for the 500 Hz signal contaminated with 60-Hz noise. Four bits of resolution is assumed in this example to emphasize the errors involved in the analog-to-digital conversion. The (o) indicates the results

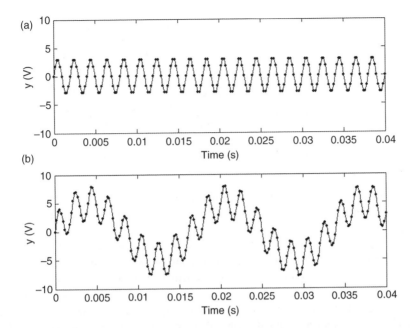

Figure 4.42 (a) Measured digital signal at 500 Hz and (b) same digital signal with 60 Hz noise showing overload requiring increase in range and decrease in resolution

for the ±5 V range and the (x) indicates the results for the ±10 V range. The dotted line indicates the analog (true continuous) signal for reference. Note that the ±5 V range quantization provides a good estimate of the analog signal whereas the ±10 V quantization does not because there are insufficient bits in the 20 V peak-to-peak range. One solution to this problem

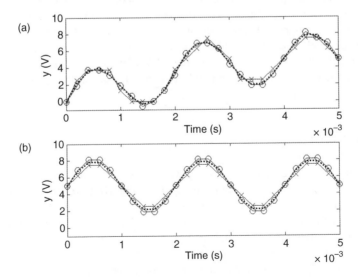

Figure 4.43 (a) Quantized signal at 500 Hz with 60 Hz noise component for ±5 V (o) and ±10 V (x) range with 4 bits of resolution and (b) same comparison of 500 Hz quantized signal with 5 V DC offset

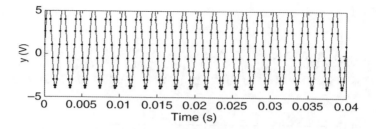

Figure 4.44   500 Hz signal with DC component

is to insert an analog high-pass filter (or band-pass filter) to remove the 60 Hz noise. When using a filter, it must be kept in mind that filters will distort the amplitude and phase in the data somewhat.

A similar case of contaminated 500 Hz data is shown in Figure 4.44. A DC component of 3 V is shown superimposed on the 500 Hz signal causing an overload of the ± 5 V range. DC bias is common when using piezoelectric sensors, which are powered by a current source (see Figure 4.29). To avoid this overload, a ± 10 V range on the ADC can be used. Figure 4.43(b) shows two different sets of quantized data points for the 500 Hz signal contaminated with DC noise. ± 5 V range results correspond to the (o) symbols and ± 10 V results correspond to the (x) symbols. As in the case of 60 Hz noise, the increase in range reduces the number of bits available for converting the signal into digital format resulting in errors. These errors are called *quantization errors*. Errors like this one may not seem significant but when detecting damage, all errors are significant because they accumulate leading to false positive and negative indications of damage.

There are many other errors associated with quantization. *Bit dropout* can occur if the ADC hardware malfunctions. If one or more bits fail, then there are even fewer levels into which analog signals are quantized leading to measurement errors. ADC circuits use reference voltage sources, and if these voltages vary, that can also lead to variations in the ADC digital outputs. A classic example of quanitization error is shown in Figure 4.45 for an empty measurement channel. The data should be zero for all time points. This *digitizer noise* occurs because of drift in the reference voltage of the ADC and environmental factors that lead to nonzero digital outputs that jump from one level to another according to the available bits of resolution.

4.5.2   Aliasing

ADC can sample only at a finite rate usually with a fixed time step. For example, the 10 Hz signal shown in Figure 4.46(a) is being sampled at 1000 Hz (refer to the simulation code 'aliasdemo.m' for details). At this high *sampling frequency*, the digital output has so many points that it is difficult to see them in the figure. The sampling frequency is 100 times the signal frequency providing 100 digital data points per cycle. It is usually not possible to sample all of the frequency components in a signal this fast. In Figure 4.46(b), the signal is being sampled at 100 Hz. It is easy to see the sampled data at this rate with 10 digital data points per cycle. The digital output is still representative of the 10 Hz analog signal when it is sampled at 10 times the signal frequency.

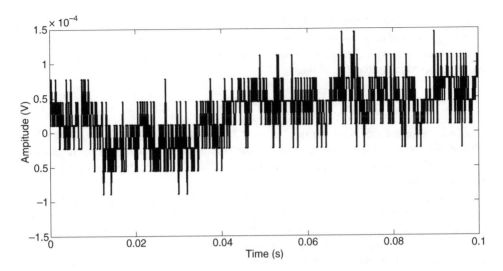

Figure 4.45   Measurement channel with nothing plugged in showing digitizer noise

In Figure 4.46(c), the digital output begins to lose accuracy. The signal is being sampled at 20 Hz, which is only two times the frequency of the signal. Two different starting points are used to sample the signal. The (*) symbol and the (+) symbol are two digital representations of the 10 Hz signal at a 20 Hz sampling frequency. The (*) digital data points do not represent the analog signal well due to an unlucky choice of where to begin sampling. The (+) data points give a much clearer picture of the 10 Hz signal. When the signal is sampled at less than 20 Hz as in Figure 4.46(d), it does not matter where the signal is sampled; the digital output will register a DC frequency (0 Hz).

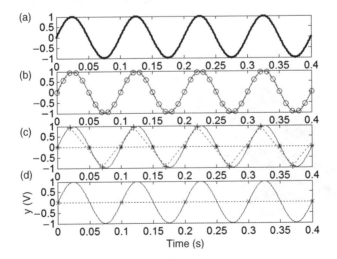

Figure 4.46   Sampling of 10 Hz response signal at (a) 100 points per cycle (1000 Hz) (b) 10 points per cycle (100 Hz) (c) 2 points per cycle (20 Hz) and (d) 1 point per cycle (10 Hz) showing aliasing of high frequency into lower frequencies in (c) and (d)

This example demonstrates that the choice of sampling frequency is critical when converting analog to digital data. *Shannon's Sampling Criterion* requires that the sampling frequency F_{samp} is more than twice the maximum signal frequency F_{max} to guarantee that the true signal can be observed using the digital data:

$$F_{samp} > 2F_{max}. \tag{4.21}$$

If a signal is sampled that is greater than one half the sampling frequency, then the digital data will appear to be at a lower frequency than it actually is. This error is called *aliasing*. Books on digital signal processing like Oppenheim and Shafer (1989) describe these and other issues in digital data representation.

For example, Figure 4.47(a) shows a speaker cabinet with an electromechanical diaphragm undergoing acceleration that is measured on the cone. The response amplitude is plotted using grayscale colors in Figure 4.47(b). In this *spectrogram*, the amplitude is shown as a function of the input frequency and digitally measured output frequency. A 4000 Hz sampling frequency is used. The lowest, thickest diagonal line in the plot is the response to the input frequency of the voltage source used to drive the speaker. The output frequency is seen to match the input frequency as expected based on the frequency response models described in Section 2.2.4. However, speaker diaphragms possess flexures with nonlinear stiffness that result in response frequencies at harmonics of the input frequency as evident in Figure 4.47(b). A *second harmonic* (twice the input frequency) and *third harmonic* (three times the input frequency) are seen in the data. The third harmonic reaches 2000 Hz (one half the sampling frequency) in the circled region. Up until that point, the output frequency correctly captures the third harmonic response frequency. As the third harmonic frequency increases beyond 2000 Hz, the spectrogram output frequency incorrectly shows that the third harmonic frequency begins to decrease. This error is due to aliasing. The problem with aliasing is illustrated in the spectrogram as the third harmonic frequency continues to increase causing the aliased frequency to mix with the second harmonic frequency and then the response at the input frequency.

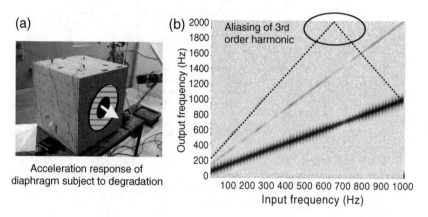

Figure 4.47   (a) Speaker with diagphragm undergoing sweep excitation and (b) spectrogram showing signal amplitude as a function of output frequency versus input frequency with aliasing of higher harmonic response components

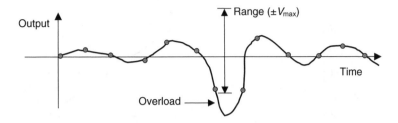

Figure 4.48 Signal not sampled fast enough to observe overload on ADC

As Figures 4.46(c) and 4.47(b) demonstrate, the fact that the sampling frequency is twice the signal frequency does not guarantee that the signal can be observed using digital data points. A sampling frequency four to five times the signal frequency is often chosen to avoid aliasing. Also, analog low-pass filters are usually used in data acquisition systems to filter out frequencies above one half the sampling frequency. If used properly, these filters help to avoid major errors in the conversion of analog to digital measurement data. In addition to preventing aliasing, Figure 4.48 shows another such error that can be avoided. An output signal is shown as a function of time. Digital data points are shown at certain time steps. Although the sampling frequency is high enough to observe the analog signal, the overload on the ADC is not detected. This overload produces serious errors in the sampled data (see Figure 4.41). An analog low-pass filter inserted before the ADC will avoid errors of this type in the measurement. Figure 4.49 is the circuit schematic for a *single-pole* low-pass filter. This filter provides 20 dB of attenuation per decade increase in frequency beyond the cutoff frequency, $1/R_f C_f$. The frequency response function relating the input and output of this filter is

$$\frac{V_d(\omega)}{V_s(\omega)} = \frac{1}{j\omega R_f C_f + 1}. \tag{4.22}$$

The magnitude and phase of this FRF are plotted in Figure 4.50. The magnitude is plotted in dB (1 dB = $20\log_{10}$(magnitude)). Note that for each value of the resistance, there is a different cutoff frequency. Beyond this cutoff frequency, the slope of the magnitude plot is -20 dB per decade in frequency. The phase indicates that even frequencies below the cutoff frequency are affected by the filter. These phase errors should be avoided by setting the cutoff frequency at a factor of 1.5 above the maximum frequency of interest. Also, the slope of the filter magnitude characteristic is not steep enough to prevent frequencies just above the cutoff frequency from entering into the data to cause aliasing. Therefore, a cutoff frequency set at $1.5F_{max}$ will ensure that both the magnitude and phase of the data up to F_{max} is relatively error free.

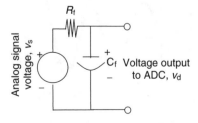

Figure 4.49 Analog low-pass filter

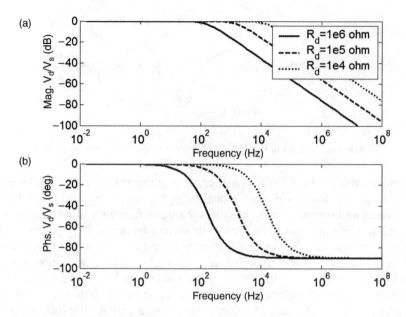

Figure 4.50 (a) Magnitude in dB of low-pass filter for three different cutoff frequencies (159 Hz, 1591 Hz and 15 915 Hz) and (b) phase

4.5.3 Leakage

Frequent measurements must be taken to identify and track changes in component health. Long measurements are not taken because they make it difficult to pinpoint when a change in the data occurs. Limitations in data acquisition systems also prevent the acquisition of long time histories. Therefore, signals are sampled only for finite periods of time. This limitation leads to another type of error. Figure 4.51 illustrates two response signals drawn with different line types: (—) and (- - -). The signals have nearly the same frequency. The dashed box indicates the length of the data block acquired by the data acquisition system. Note that for this small amount of data, it is difficult to distinguish the two signals from one another. There is not sufficient time in the block of data acquired for the small difference in frequency to generate a large difference in phase between the two signals. This inability to distinguish closely spaced frequencies is due to *leakage* error. Leakage is introduced because a finite length of data is measured.

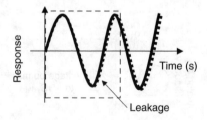

Figure 4.51 Finite length time of signal is acquired leading to finite frequency resolution error called leakage

Rayleigh's criterion relates the time length of the data block measured, T, to the smallest frequency difference Δf which can be resolved in that measurement:

$$\Delta f = \frac{1}{T}. \tag{4.23}$$

Leakage can be interpreted in another way. Leakage involves energy from analog signal frequencies leaking into adjacent frequencies during the process of converting analog data to digital format. More will be said about leakage and ways to reduce it later in the book when signal processing approaches to filtering of time data are described.

4.5.4 Channel Limitations in Data Acquisition

When multiple channels of data are needed for load and damage identification, data acquisition systems must be designed to enable these measurements. In many cases, the number of measurement channels is limited to four, eight, etc. If the number of measurement channels is too limited, it can be difficult to identify the loading and damage characteristics needed for health monitoring of components. There are usually more limitations on the number of *source channels* than measurement channels. Source channels are needed for active sensing. Devices called *function generators* are used to drive actuators with user-defined signals. The signals generated using the function generator are selected based on the principles discussed in Section 4.2. After the actuators excite the component, sensors measure the responses.

To overcome limitations on the number of channels, *multiplexers* are used. A multiplexer is a routing circuit that increases the number of effective data acquisition source and measurement channels without increasing the number of physical channels in the hardware. Figure 4.52 illustrates a typical multiplexer arrangement in a DAQ system. The components

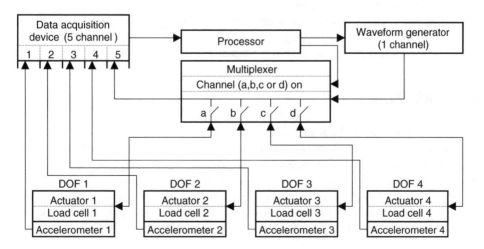

Figure 4.52 Multiplexer acquisition system including processor, waveform generator, multiplexer, sensor/actuator pairs and DAQ (Reproduced by permission of Mr. Jonathan White, Purdue University, 2006)

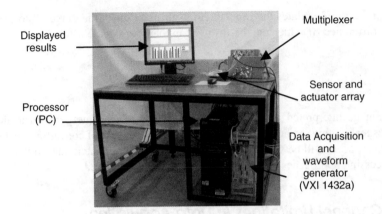

Figure 4.53   Physical hardware corresponding to the flowchart in Figure 4.52 (Reproduced by permission of Jonathan White, Purdue University, 2006)

of this system are the processor, waveform generator, data acquisition device and pairs of sensors and actuators. Figure 4.53 shows an actual set of measurement hardware corresponding to the flowchart given in Figure 4.52. The sequence of operations in this system is as follows:

(1) The processor creates a desired drive waveform and transmits this wave to the waveform generator.
(2) The processor initializes the DAQ by running a series of diagnostic tests.
(3) The processor uses the multiplexer to select an actuator to emit a vibration excitation (set by channels a–d of the multiplexer).
(4) The drive signal from the waveform generator causes one of the actuators to vibrate, and these vibrations are measured with a force load cell, which is connected to channel 5 of the data acquisition system.
(5) The vibration response is measured by four accelerometers, which are connected to channels 1–4 of the data acquisition system.
(6) The sampled signal from the data acquisition system is transferred to the processor where algorithms are used for analyzing the data.
(7) This process is repeated for each of the actuators.

4.6 SUMMARY

This chapter has reviewed basic measurement issues that affect the accuracy and repeatability of health-monitoring data. Different elements of the measurement channel were discussed including the type of data being measured (stationary or nonstationary), characteristics of the data such as the amplitude and frequency range, attachment mechanisms for transducers and the effects of attachment properties on measured data, various transducers for passive and active sensing, hardware modeling for data acquisition, transducer placement and errors in the data acquisition process. Key issues in the chapter are highlighted below:

• The goal in measurements is to reduce the effects of variability in raw data as much as possible before it is analyzed.

- Measurements used to identify loading and damage are almost always *indirect*.
- Before selecting transducers for acquiring health-monitoring data, the data environment must be identified.
- Passive measurements do not require control of the vibration (or wave propagation) input to a system. Active measurements use auxiliary actuators to cause components to vibrate (or waves to propagate) in a more desirable manner to detect damage.
- Small changes in a component due to localized damage do not cause changes in low frequency vibration data.
- The disadvantage of stationary data is that it usually consists of only a few response deflection patterns, which may not exercise all damage mechanisms in the component.
- Nonstationary data is often more revealing than stationary data because *transients highlight localized damage by exercising more deflection patterns in components.*
- Environmental factors due to temperature shifts, humidity variations, sound pressure fluctuations and other changes cause component response data to vary in ways similar to damage.
- *Environmental factors cause global changes in mechanical properties whereas damage causes more localized changes.*
- In addition to the sensor mass, the effects of the attachment stiffness and damping should also be considered.
- Softer attachments are less stiff with more damping resulting in lower response bandwidth.
- *Amplitude distortion is highest when the damping ratio is small. Phase distortion is highest when the damping ratio is large.*
- *Large attachment footprints cannot be used to measure short wavelength responses.*
- Although it is desirable to have a high sensitivity, *there is a trade-off between sensitivity and robustness in sensing. The signal-to-noise ratio drops as the sensitivity goes up.*
- Many potential sources of measurement error can be attributed to the inherently *high output impedance* of piezoelectric sensors.
- The basic rule in positioning sensors is that the loading and damage mechanisms of interest should be *observable* with the sensors.
- Digital sampling *limitations introduce errors in the data.*
- *Shannon's sampling criterion* requires that the sampling frequency F_{samp} is more than twice the maximum signal frequency F_{max}.
- *Rayleigh's criterion* relates the time length of the data block measured, T, to the smallest frequency difference, Δf, which can be resolved in that measurement.

REFERENCES

Ackers, S., Kess, H., White, J., Johnson, T., Evans, R., Adams, D.E. and Brown, P. (2006) 'Crack detection in a wheel spindle using modal impacts and wave propagation', *Proceedings of the SPIE Conference on Nondestructive Evaluation and Smart Structures and Materials*, Chulha Vista, CA, Vol. 6177, Paper #61770B.

Allemang, R. (1999) *Vibrations: Experimental Modal Analysis*, University of Cincinnati, Cincinnati, Ohio.

Bergmeister, K. and Santa, U. (2001) 'Global monitoring concepts for bridges', *Proc. SPIE*, **3995**, 312–323.

Borinski, J.W., Boyd, C.D., Dietz, J.A., Duke, J.C. and Home, M.R. (2001) 'Fiber optic sensors for predictive health monitoring', *AUTOTESTCON Proceedings*, IEEE Systems Readiness Technology Conference, pp. 250–262.

Chang, K.-C., Lin, T.-K., Lin, Y.-B. and Wang, L. (2001) 'FBG sensors for structural health monitoring', *Proc. SPIE*, **4335**, 35–42.

Choi, K., Keilers, C.H., Jr. and Chang, F.-K. (1994) 'Impact damage detection in composite structures using distributed piezoceramics', *AIAA/ASME/ASCE/AHS/ASC Structures, Structural Dynamics and Materials Conference*, No. 1, pp. 118–124.

Chong, K.P., Carino, N.J. and Washer, G. (2003) 'Health monitoring of civil infrastructure', *Smart Mat. Struct.*, **12**, 483–493.

Cramer, K.E. and Winfree, W.P. (2000) 'Thermographic imaging of material loss in boiler water-wall tubing by application of scanning line source', *Proc. SPIE*, **3995**, 600–609.

Dally, J., Riley, W. and McConnell, K. (1993) 'Instrumentation for engineering measurements', 2nd edition, John Wiley & Sons, Inc., New York.

De Villa, F., Roldan, E., Tirado, C. and Mares, R. (2001) 'Defect detection in thin plates using S_0 lamb wave scanning', *Proc. SPIE*, **4335**, 121–130.

Farrar, C.R., Sohn, H., Fugate, M.L. and Czarnecki, J.J. (2001) 'Integrated structural health monitoring', *Proc. SPIE*, **4335**, 1–8.

Friedland, B. (1986) *Control System Design: An Introduction to State-Space Methods*, McGraw-Hill, Inc., New York.

Giurgiutiu, V. and Bao, J. (2002) 'Embedded-ultrasonics structural radar for nondestructive evaluation of thin-walled structures', *Proceedings of IMECE*, Paper # 39017, pp. 1–8.

Giurgiutiu, V. and Zagrai, A. (2001) 'Electro-mechanical impedance method for crack detection in metallic plates', *Proc. SPIE*, **4335**, 131–142.

Guo, D. and Kundu, T. (2000) 'A new sensor for pipe inspection by lamb waves', *Mat. Eval.*, **58**(88), 991–994.

Hannah, R. and Reed, S. (1992) *Strain Gage: User's Handbook*, Elsevier Applied Science, London.

Heaton, L. C., Kranz, M. and Williams, J. (2004) 'Embedded fiber optics for structural health monitoring of composite motor cases', *Proc. SPIE*, **5393**, 178–185.

Hellier, C. (2001) *Handbook of Nondestructive Evaluation*, McGraw-Hill, New York.

Ikegami, R. (1999) 'Structural health monitoring: assessment of aircraft customer needs', *Struct. Health Monit.*, 12–23.

Inaudi, D. (2001) 'Application of optical fiber sensor in civil structural monitoring', *Proc. SPIE*, **4328**, 1–10.

Islam, A. and Craig, S. (1994) 'Damage detection in composite structures using piezoelectric materials (and neural net)', *Smart Mat. Struct.*, **3**, 318–328.

Goldfine, N.J., Zilberstein, V.A., Schlicker, D.E., Sheiretov, Y., Walrath, K. and Washabaugh, A.P. (2001) 'Surface-mounted periodic field eddy current sensors for structural health monitoring', *Proc. SPIE*, **4335**, 20–34.

Galleher Jr., J.J., Stift, M.T., Mergelas, B. and Stine, G.P. (2000) 'Detecting damaged prestressed concrete cylinder pipe using remote field eddy current/transformer coupling technology', Environmental and Pipeline Engineering, 60–69.

Jaffe, B., Cook, W. and Jaffe, H. (1971) 'Piezoelectric ceramics', Academic Press Limited, R.A.N. Publishers, Marietta, OH.

Jung, Y.C., Kundu, T. and Ehsani, M. (2000) 'Detection of internal defects in concrete panels by lamb waves' *Proceedings of the AIP Conference*, **509**(1), 1669–1676.

Komsky, I.N. (2001) 'Application of portable modules for fatigue crack characterization', *Proceedings of SPIE*, **4335**, 290–299.

Komsky, I.N. and Achenbach, J.D. (1996) 'Ultrasonic imaging of corrosion and fatigue cracks in multilayered airplane structures', *Proceedings of SPIE*, **2945**, 380–388.

Matzkanin, G.A. (2000) 'Technology assessment of MEMS for NDE and condition-based maintenance', *Proceedings of SPIE*, **3994**, 80–89.

Milos, F.S., Watters, D.G., Pallix, J.B., Bahr, A.J. and Huestis, D.L. (2001) 'Wireless subsurface microsensors for health monitoring of thermal protection systems on hypersonic vehicles', *Proceedings of SPIE*, **4335**, 74–82.

Mizutani, Y., Morino, Y. and Takahashi, H. (2003) 'Acoustic emission monitoring of CFRP tank and ambient and cryogenic temperatures', *44th AIAA/ASME/ASCE/AHS/ASC Structures, Structural Dynamics, and Materials Conference*, **1937**, 1–10.

Oppenheim, A.V. and Shafer, R.W. (1989) 'Discrete time signal processing', Prentice Hall, Englewood Cliffs, NJ.

Park, G., Cudney, H.H. and Inman, D.J. (2000) 'Impedance-based health-monitoring of civil structural components', JInfrastructure Systems, **6**(4), 153–160.

Sandhu, J.S., Sincebaugh, P.J., Wang, H. and Popek, W.J. (2000) 'New developments in acoustography for fast full-field large-area ultrasonic NDE', *Proceedings of SPIE*, **3993**, 191–200.

Sandhu, J.S., Wang, H., Popek, W.J. and Sincebaugh, P.J. (2001) 'Acoustography: it could be a practical ultrasonic NDE tool for composites', *Proceedings of SPIE*, **4336**, 129–134.

Schoess, J.N. (2001) 'Conductive polymer sensor arrays: a new approach for structural health monitoring', *Proceedings of SPIE*, **4335**, 9–19.

Shepard, S.M., Ahmed, T., Lhota, J.R. and Rubadeux, B.A. (2000) 'Systems-based approach to thermographic NDE', *Proceedings of SPIE*, **3993**, 14–18.

Todd, M.D., Johnson, G.A., Vohra, S.T., Chang, C.C., Danver, B. and Malsawma, L. (1999) 'Civil infrastructure monitoring with fiber bragg grating sensor arrays', *Proceedings of the 2nd International Workshop on Structural Health Monitoring*, 534–539.

Tse, F. and Morse, I. (1989) 'Measurement and instrumentation engineering', Marcel Dekker, Inc., New York.

Varadan, V.K. and Varadan, V.V. (2000) 'Microsensors, microelectromechanical systems (MEMS), and electronics for smart structures and systems', in Smart Materials and Structures, **9**, 953–972.

Walsh, S.M., Butler, J.C., Belk, J.H. and Lawler, R.A. (2001) 'Development of a structurally compatible sensor element', *Proceedings of SPIE*, **4335**, 63–73.

White, J. (2006) 'Impact and thermal damage identification in metallic honeycomb thermal protection system panels using active distributed sensing with the method of virtual forces', Masters Thesis, School of Mechanical Engineering, Purdue University, West Lafayette, Indiana.

White, J. Adams, D. E., Jata, K., 'Damage Identification in a Sandwich Plate Using the Method of Virtual Forces,' 2006, Proc. of the International Modal Analysis Conference, St. Louis, MO, Paper #67.

Yang, J., and Chang, F., 'Detection of Bolt Loosening in C-C Composite Thermal Protection Panels: I. Diagnostic principle,' 2006, Smart Materials and Structures 15, pp. 581–590.

PROBLEMS

(1) Suppose that a loading variable f is related to a measurement variable x in the following way:

$$f = \alpha x^3 + \beta x^2 + \chi, \tag{4.24}$$

where α, β and χ are constants containing material and geometrical parameters. The mean value of x (assumed to be normally distributed) is 2 and the variance is $\sigma^2 = 0.5$. Generate the probability density function for f. What are the mean value and variance of the estimated value of f?

(2) Consider the rolling tire shown in Figure 1.21. Based on Table 4.1, describe the data environment issues involved with measuring the response and identifying potential damage in this tire.

(3) Based on the measured vibration spectrum shown in Figure 4.3, what is the vibration level in g's of acceleration for the wire harness and connector at 1000 Hz? What is the vibration velocity in in/s for a 1 g acceleration at a displacement amplitude of 0.2 in?

(4) Consider the nine DOF system model shown in Figure 4.6. Suppose there is 10 % mass reduction in DOFs 2, 3 and 6. In what frequency range does the driving point FRF, $H_{11}(\omega)$, exhibit the most significant changes due to this damage? In what frequency range does $H_{22}(\omega)$ exhibit the largest changes? Use the simulation code 'ninedofdata.m' as a starting point for the analysis.

(5) Suppose the nine DOF system model is excited by aerodynamic forcing applied in the form of a uniform pressure to all nine DOFs shown in Figure 4.6. If the bandwidth of the forcing frequency is from 0 to 2000 Hz, in which coupling stiffnesses in the model could damage be detected by measuring the response at DOF 1? Use the simulation code 'ninedofdata.m' to answer this question.

(6) How do the results in Figure 4.7 for a 10 % reduction in the mass at DOF 7 change if the stiffness of the supports of the panel model are increased by 50 %? What if the stiffness of the supports is decreased by 50 %? Use the simulation code 'ninedofdata.m' to answer this question.

(7) For a nonstationary excitation force with amplitude 1000 N and time period 0.05 s shaped like the transient pulse in Figure 4.8(b), replot the results in Figure 4.9(b) and comment on

the results relative to those obtained in Figure 4.9(b). Use the simulation code 'nine-dofdata.m' to answer this question.

(8) Suppose that the nine-DOF system model in Figure 4.6 undergoes a temperature change leading to a 10 % global decrease in modulus. Consider the transmissibility function $T_{16}(\omega)$ equal to $H_{16}(\omega)/H_{66}(\omega)$. In what frequency range and for what percent level of oxidation damage in the center of the panel at DOF 5 (modeled as a reduction in mass) does the difference between the baseline transmissibility and damage transmissibility measurements clearly distinguish between the change in temperature and damage? Use the simulation code 'ninedofenviron.m' to answer this question.

(9) Describe the issues listed in Figure 4.11 related to sensor attachment on the spindle of the wheels in the mining truck shown in Figure 1.2(a). Assume that bending and torsional loads are of interest, the damage mechanism is due to cracking, and fatigue failures as well as overloads are possible.

(10) If the uniaxial accelerometer shown in Figure 4.14(a) is attached with an angle of inclination of 3 % on DOF 1 of the two DOF panel model in Figure 2.2(a), plot the difference between the acceleration FRFs $-\omega^2 H_{12}(\omega)$ with and without this angle included. In what frequency range is the error due to inclination small? This scenario can occur when health-monitoring sensors are renewed and replaced. Use the simulation code 'twodoffrf.m' to answer this question.

(11) Suppose an epoxy adhesive possessing shear modulus G, thickness h and length L is used to attach the strain gauges shown in Figure 4.14(d) to a component with modulus E that undergoes a local strain of S. Assume the cross-sectional area of the component is A_c and the planar area of the adhesive is A_a. Show that the ratio of measured strain to actual strain is $1 - 2A_c Eh/A_a GL$.

(12) What value of ζ leads to less than 1 % amplitude distortion error in $X_s(\omega)$ transmitted to the sensor relative to those of the component, $X_b(\omega)$, in Figure 4.15? What value of ζ leads to less than 1 % phase distortion error? Use the code 'attachtrans.m' to conduct the necessary analysis.

(13) If the planar area of an adhesive sensor attachment is A, the modulus in the direction normal to the component is E and the thickness is h, what is the stiffness k_a of the attachment?

(14) What is the maximum frequency modal deflection pattern for which the displacement can be accurately measured in Figure 4.16 with a 0.01 m × 0.03 m sensor? What location and orientation should be chosen for the sensor? Use the code 'twoDshapesensor.m' to answer the question.

(15) Describe all possible sources of error in the noncontact acoustic sound intensity sensor shown in Figure 4.21(b). Refer to Figure 4.17 for an example of measurement errors.

(16) If a high-temperature accelerometer is used to measure the acceleration of the two-DOF system model shown in Figure 2.2(a), plot the maximum and minimum measured transmissibilities $T_{12}(\omega)$ and FRF $H_{12}(\omega)$ corresponding to a temperature change of 0–350 °F in the two-DOF system model. Assume the stiffness parameters of the two-DOF system model decrease by 5 % over this range and use the data shown in Figure 4.26 for the change in sensor calibration.

(17) Plot the error in the measured voltage v_d relative to the sensor voltage v_s as a function of the input impedance R_d of a typical data acquisition device (see Figure 4.27(b)). For what range of R_d relative to R_s is the error less than 1 %?

(18) Plot the third modal deflection pattern for the nine-DOF system model in Figure 4.6 as was done in Figure 4.36(b) for the fifth mode. Use the simulation code 'ninedofselect.m' to extract and plot the deflection pattern. What is the natural frequency of this mode of vibration? Which sensor locations would be more ideal from a standpoint of observability based on this deflection pattern?

(19) Calculate and plot the singular values of the observability test matrix, [O], for the three DOF system model shown in Figure 2.12 assuming that displacement measurements are taken at DOF 1, DOF 2 and DOF 3 individually, and then at all three DOFs. What conclusions can be drawn about the optimal position(s) for sensors? After some modifications, the simulation codes 'ninedofselect.m' and 'threedoftrans.m' can be used to solve this problem.

(20) Suppose that no low pass filter is used with an analog-to-digital sampling frequency of 1000 Hz. Frequency components at 10, 50 and 200 Hz appear in the sampled data. It is known that the 10 and 200 Hz frequencies are due to aliasing. What are the actual frequencies corresponding to these two aliased frequency components?

5

Data Analysis

As mentioned at the beginning of the previous chapter, health monitoring measurement data is almost always indirectly related to component health. *Models are the means by which load and damage information are extracted from raw health-monitoring data.* There are many different types of models. Some models are based only on the data. For example, a model of the healthy frequency domain displacement response signal of a component can be used to detect changes to the signal in order to identify damage. Other types of models are based more on the physics of the component. For example, the embedded sensitivity method in Chapter 2 utilized a lumped parameter model of a component to estimate the change in stiffness due to damage based on frequency response function (FRF) data. All models used to extract component health information from measurement data augment variability in the raw data in some manner. The data analysis process should be designed in order to minimize the variability in health-monitoring information.

5.1 DATA ANALYSIS NEEDS AND FRAMEWORK

The framework for data analysis illustrated in Figure 5.1 is discussed in this section. Each of the steps in this framework is needed to identify loads and damage and then predict the growth of damage. Note that none of these steps can correct for poor measurement data. Errors in measured data propagate through the data analysis process.

5.2 FILTER DATA

A filter is *any weighting function used to emphasize certain parts of the data and de-emphasize other parts of the data.* Filters can be applied to data in either the time domain or the frequency domain. Spatial filters can also be applied to emphasize certain response patterns in data acquired at different locations and in different directions. The utility of these filters is demonstrated in the following sections.

Health Monitoring of Structural Materials and Components: Methods with Applications   D. Adams
© 2007 John Wiley & Sons, Ltd

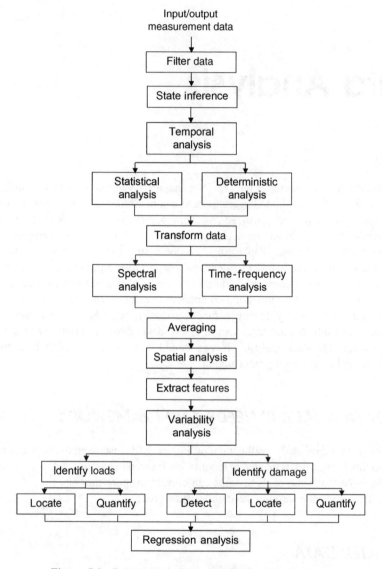

Figure 5.1   Data analysis framework for health monitoring

5.2.1 *Time Domain Filters*

One type of filter applied in the time domain simply deletes blocks of data that are undesirable in some way. For example, Figure 5.2(a) shows a response time history. At 0.2 s, the time history exhibits an overload at 5 V where the signal saturates. This saturation is usually due to the data acquisition analog-to-digital converter (ADC), which is limited to a preset amplitude range (refer to Section 4.5.1). A simple filter can be applied to remove this data from 0 to 0.5 s

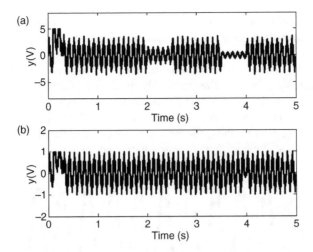

Figure 5.2 (a) Time history with overload and several segments of low amplitude response and (b) filtered data to normalize response amplitude

from consideration. In data analysis algorithms, this time domain filter would check the signal amplitude and discard data that is beyond a certain amplitude range.

Figure 5.2(a) also exhibits two segments of data with low amplitude. In data analysis for health monitoring, it is usually desirable to have consistent amplitude measurements for damage detection and other purposes. For example, digital data acquisition systems have limited bits of amplitude resolution as described in Section 4.5.1. When the amplitude is a fraction of the peak amplitude range, the bits are not utilized fully leading to imprecise measurements. In this instance, the data can be discarded as in the case of the overload. However, if there are no quantization errors, then the data can be normalized to enable more straightforward data analysis in some cases. If it assumed that this measurement corresponds to the response of a structural component in healthy condition, then a filter can be applied to the data to normalize it. Figure 5.2(b) shows the results of applying such a normalization filter ('filterdemo.m'). This filter calculates the maximum amplitude of every 50 signal data points. Then the filter divides the amplitudes of those 50 data points by the maximum amplitude. The variance can also be used to normalize data. The result is a time history with more consistent amplitude values, which can be used to establish a baseline condition of the component response.

Various window functions are also regularly applied to time domain data to improve the ability to process the data. Recall from Section 4.5.3 that one limitation in measurements is that data can only be acquired for a limited period of time. This limitation leads to leakage errors in the data. Leakage causes frequency components in the data to be misinterpreted. Consider the time histories shown in Figure 5.3(a) and (c). The time history shown in Figure 5.3(a) is at 10 Hz and is sampled for 10 complete periods of the signal. The signal in Figure 5.3(c) is at 6.7 Hz and is sampled for a noninteger number of periods. The code 'windowdemo.m' is used to demonstrate the effects of different filters applied to this data. Both of the time histories are measured for 1 s. For the signal in Figure 5.3(a), which is sampled an integer number of signal periods, the frequency spectrum of the block of data is plotted in Figure 5.3(b). The frequency spectrum contains the Fourier series coefficients plotted as a function of frequency (refer to Equation (2.83)). In this plot, only the magnitudes of the Fourier series coefficients are shown. The only nonzero coefficient occurs at 10 Hz. The amplitude is one because the signal has unity amplitude. All other coefficients are zero (10^{-16} order of magnitude). This signal is said to be *completely observed* because it provides a complete and accurate picture of the frequencies in the data, which is sampled an integer number of periods. Completely observed signals are not contaminated by truncating the data.

In contrast, the 6.7 Hz signal is not completely observed. The signal is effectively cut off before the period of oscillation ends. The frequency spectrum of the 6.7 Hz signal is shown in Figure 5.3(d). Note that although there is only one frequency component in the data at 6.7 Hz,

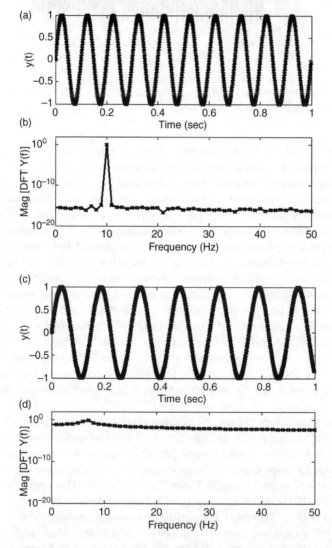

Figure 5.3 (a) 10 Hz sinusoidal signal sampled for exactly 10 periods (b) frequency spectrum of 10 Hz signal (c) 6.7 Hz signal sampled for noninteger number of periods and (d) frequency spectrum of 10 Hz signal showing leakage error

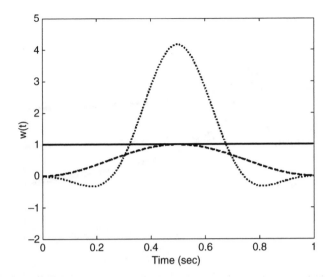

Figure 5.4   Boxcar time domain window (___), Hanning window (- - -) and flattop (P301) window (. . .)

the energy from this frequency has *leaked* throughout the entire frequency band. This leakage makes it difficult to pick out the 6.7 Hz frequency. If higher frequencies were present in the data with lower amplitudes than the 6.7 Hz component, then those higher frequency components would be masked by the leakage error.

To correct this error, time domain filters called *windows* are used. Windows are *weighting functions multiplied by the time data to reduce the effects of leakage*. Three different window functions often used in dynamic signal analysis are shown in Figure 5.4. The boxcar window (___) applies uniform weighting over the block of time data analyzed. This window was being used in the analysis performed in Figure 5.3. As mentioned above, this type of weighting allows the cutting off of the signal in Figure 5.3(c) to cause leakage errors. The sharp discontinuity at the end of the time history is perceived in subsequent analysis to have come from high-frequency components in the data.

In order to suppress this discontinuity at the end of the block of data, different windows can be used. The Hanning window (- - -) in Figure 5.4 is zero at the beginning and end of the time window and varies like a cosine in between. This window causes the signal to go to zero at the beginning and end of the data block. This suppression of the discontinuities at either end of the signal helps to reduce leakage errors. Likewise, the P301 (or flattop) window (. . .) shown in Figure 5.4 also causes the signal to go to zero at the ends helping to reduce leakage errors. The difference between the Hanning and P301 windows is that the Hanning window enables better frequency estimates in the data whereas the P301 window enables better amplitude estimates.

Figure 5.5 shows the frequency spectra for the 6.7 Hz signal in Figure 5.3(c) with these three different time domain windows applied. The boxcar window exhibits the most leakage (x). The spectra for the Hanning (o) and P301 (+) windows exhibit much less leakage. The spectrum for the P301 window gives a very good estimate of the signal amplitude; however, the broad peak makes it difficult to identify the 6.7 Hz frequency. On the contrary, the Hanning window clearly identifies the 6.7 Hz frequency, but the amplitude estimate obtained with this window is less accurate than for the P301 window.

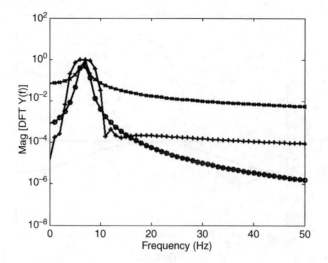

Figure 5.5　Frequency spectra for 6.7 Hz signal sampled for non-integer number of periods using boxcar window (x), Hanning window (o) and flattop window (+)

Time domain digital filters can also be applied to *decimate* or *interpolate* digital data sampled at a certain time step. For example, the signal in Figure 5.2 is shown in Figure 5.6(a) after interpolating midway between the original data points. The simulation code 'filter.m' utilizes the MATLAB® command 'interp' to produce this interpolated data. The interpolated data (o) is identical to the original data (x) at the original time points. Between the original points, a low-pass filter is used to interpolate so that no frequencies beyond half the sampling frequency appear in the interpolated data. Figure 5.6(b) shows the digital data decimated (o) to one half the original number of points (x). This type of filter applies a low-pass filter and then resamples the data. Decimation filters are often used to smooth out time history measurements.

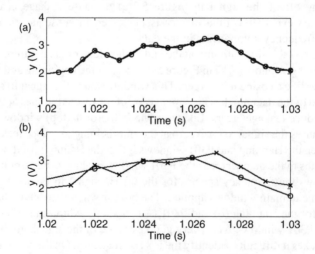

Figure 5.6　(a) Interpolated (ooo) and original (xxx) signals and (b) decimated (ooo) and original (xxx) signals

5.2.2 Frequency Domain Filters

Frequency domain filters are *weighting functions in amplitude and phase applied to either analog or digital data to emphasize certain frequency data and de-emphasize other data.* Analog low- and high-pass filters used during the measurement process were described in Section 4.5. If there are high-frequency components in analog data, then these components should be filtered out using a low-pass analog filter prior to digitally converting the data. However, digital data can then be filtered using two types of digital filters: *finite impulse response (FIR)* and *infinite impulse response (IIR)* filters.

FIR filters are one type of filter used to suppress certain frequency ranges in digital data. The FIR filter is expressed as an input–output difference equation model similar to those used in Section 2.3.1.3 of Chapter 2. The input to the filter, $u(n)$, at each time step n is related to the filtered output $y(n)$ as follows:

$$y(n) = b_0 u(n) + b_1 u(n-1) + \ldots + b_{M-1} u(n - (M-1)) = \sum_{i=0}^{M-1} b_i u(n-i) \qquad (5.1)$$

where M is the number of moving average filter coefficients b_i. For example, the simulation code 'filterdemo.m' was used to design a *bandpass* filter for digital data sampled at 5000 Hz. The bandpass filter was designed[1] to pass frequencies in the data from 500 to 1000 Hz and to suppress frequencies in the data below 300 Hz and above 1200 Hz. Below 300 Hz, input data to the filter is suppressed by 60 dB, and above 1200 Hz, data is suppressed by 80 dB. Sixty-three coefficients were used in the FIR. Figure 5.7 shows the magnitude (a) and phase (b) of the FRF of this FIR filter. Note that the filter does pass frequencies in the range 500–1000 Hz and suppress frequencies in the digital data outside that range. Low-pass, high-pass and bandstop filters can also be designed in this way (Hamming, 1989).

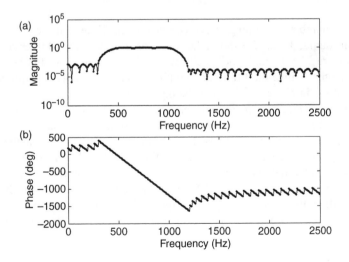

Figure 5.7 (a) Magnitude and (b) phase response of FIR digital filter with passband from 500 to 1000 Hz

---

[1]The MATLAB® filter design and analysis tool was used to design this equiripple FIR filter.

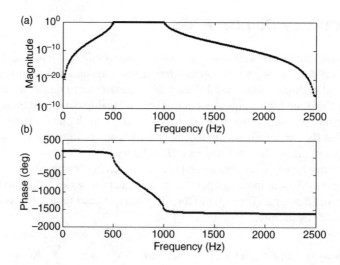

Figure 5.8 (a) Magnitude and (b) phase response of IIR digital filter with passband from 500 to 1000 Hz

IIR filters can also be used to filter digital data. The input and output of this filter are also related using a finite difference model as follows:

$$a_0y(n) + a_1y(n-1) + \ldots + a_{N-1}y(n-(N-1)) = b_0u(n) + b_1u(n-1) + \ldots$$

$$+ b_{M-1}u(n-(M-1)) \sum_{i=0}^{N-1} a_iy(n-i) = \sum_{i=0}^{M-1} b_iu(n-i) \tag{5.2}$$

where N is the number of autoregressive coefficients a_i. Note that the IIR filter involves feedback of the output. Figure 5.8 shows the equivalent result for an IIR filter compared to the FIR filter in Figure 5.7 with the same specifications. This particular IIR filter is a Chebyshev-type filter. Note that unlike the FIR filter, the phase characteristic is not straight but curved somewhat resulting in phase distortion of the input data.

The primary advantage of FIR filters is that they tend not to distort the phase of the input data as much as IIR filters do. This advantage is important when data is being compared between multiple measurement channels. The advantage of IIR filters is that they require less computational power to compute. For example, the FIR filter described above required 63 coefficients whereas the IIR filter required only 20 coefficients resulting in less computational demands when filtering data using the IIR finite difference model. Both of these filters are applied to the digitally sampled data. Other types of frequency-domain filters are applied directly in the frequency domain by modifying the magnitude and phase of the data.

5.2.3 Spatial Filters

Spatial filters *weight data from different measurements according to some rule that empha-sizes certain components of the data and de-emphasizes other components.* In vibration data

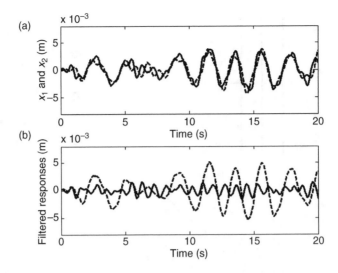

Figure 5.9 (a) Responses x_1 (___) and x_2 (- - -) of two-DOF system model in Figure 2.2(a) for uniform random excitation and (b) spatially filtered responses along two modal deflection shapes $[11]^T$ (- - -) and $[1-1]^T$ (___)

analysis, spatial filters are constructed using modal deflection shapes. When these shapes are used to filter the data, the responses of independent modes of vibration can be extracted. For example, consider the two-DOF system model in Figure 2.2(a). This model was excited with a uniformly distributed random excitation using the simulation code 'filterdemo.m'. The displacement responses of the model are shown in Figure 5.9(a). Each of the response variables comprised the responses at the two modes of vibration. The two modal deflection shapes of these modes are $[11]^T$ and $[1-1]^T$. In the first mode, the two DOFs move in the same direction in phase, and in the second mode, the two DOFs move directly out of phase. These two modal deflection shapes can be used to filter the responses of the system by multiplying each modal coefficient by the corresponding displacement and then adding the results for all DOFs:

$$y_1(t) = [1\ 1]\left\{ \begin{array}{c} x_1(t) \\ x_2(t) \end{array} \right\} \tag{5.3a}$$

$$y_2(t) = [1\ -1]\left\{ \begin{array}{c} x_1(t) \\ x_2(t) \end{array} \right\} \tag{5.3b}$$

The results of performing these operations are shown in Figure 5.9(b). Note that the low frequency response corresponding to the first mode of vibration is primarily isolated in the response y_1 (- - -), whereas the higher frequency response at the second mode of vibration is primarily isolated in the response y_2 (___). By filtering the response data in this manner, the responses of the individual modes of vibration have been obtained. In other words, the physical displacement variables have been filtered to produce the modal displacement variables of the system response.

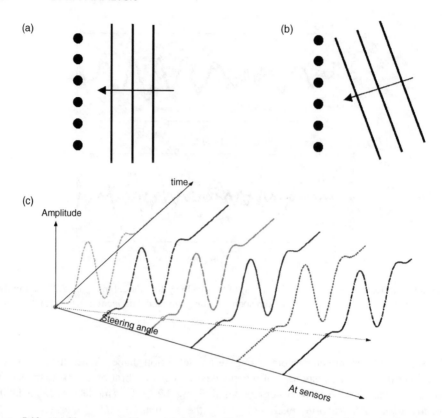

Figure 5.10   (a) Six sensors with wave arriving at all sensors simultaneously (b) wave arrives at sensors at different times and (c) measured signal at each sensor with time for situation in (b) showing the steering angle of the phased array

Another type of spatial filter for use with propagating waves is based on the application of *phased array* techniques. Phased arrays are *sets of omnidirectional sensors (or actuators) that function together as one directionally sensitive sensor (or actuator)*. For example, consider the case where each of the six sensors pictured in Figure 5.10 measures the wave propagation response of a structural component. In Figure 5.10(a), the incoming transient propagating wave reaches all sensors simultaneously. However, in Figure 5.10(b), the propagating wave reaches the top sensor first, followed by each of the sensors along the array. Figure 5.10(c) shows a typical set of measured propagating wave responses (e.g. vertical displacements) for the situation in Figure 5.10(b). Note that this series of measurements indicates that the wave is approaching from a particular *steering angle*. In phased sensor arrays, spatial filters are produced when *sensors receive data, and then the data is weighted and phase shifted to increase the sensitivity to waves approaching from specific directions*. The primary motivation for implementing spatial filters is to eliminate interference, or jamming, signals that cannot be removed through temporal or frequency domain filtering. Spatial filters can be designed to 'look' toward the signal of interest thereby removing the interference.

Spatial filters can also be designed in the reverse direction to transmit propagating waves in certain directions using an array of actuators instead of sensors. For example, Figure 5.11

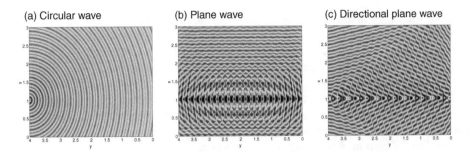

Figure 5.11 Radiation from (a) single source (b) 23 sources with no phase delay and (c) 23 sources with phase delay resulting in a directional plane wave (Reproduced by permission of Shankar Sundararaman, Purdue University, 2003)

shows three distinct wave propagation patterns achieved using actuators. The dark colors signify peak particle displacements in propagating elastic waves. Figure 5.11(a) is a circular elastic wave (omnidirectional) transmitted from a single actuator. When 23 of these actuators are lined up and activated simultaneously, the plane wave pattern in Figure 5.11(b) is generated. When this same set of actuators is activated with a fixed time delay between each successive drive signal from right to left in the row of actuators, the plane wave is steered in a direction dictated by the chosen time delay.

A *beamformer* is one type of spatial filter used for propagating waves. Beamformers were originally developed in the communications industry using antennae dishes as continuous spatial filters to filter out interference from transmitted propagating electromagnetic waves. Many applications of beamforming involving radar, sonar, imaging, geophysical exploration, astrophysical exploration and medical diagnostics are discussed by Van Veen and Buckley (1988). The work of pioneers such as Widrow and Sterans (1985), Frost (1972), Griffiths and Jim (1982) and others are summarized in Van Veen and Buckley. It is also interesting to note that orb web spiders use beamforming as a means for locating insects that become ensnared in fibrous webs.

Consider the general geometrical arrangement of point sensors shown in Figure 5.12 following the development in Dietrich (2000). An incident wave, assumed to be a plane

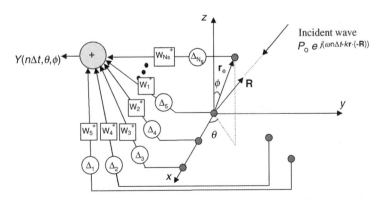

Figure 5.12 A beamformer setup with weights and delays applied to each sensor (Reproduced by permission of SAGE Publications, Inc.)

wave for simplicity, arrives along the vector $-\mathbf{R}$, and the individual sensor location vectors with respect to the origin of the array are $\mathbf{r}_e$, where $e = 1, 2, \ldots, N_s$ and N_s is the number of sensing elements. A Cartesian coordinate system is used to define the configuration of the array although a cylindrical system is often useful to simplify the algebra in some applications. The azimuth steering angle in the x–y plane is denoted by θ, and the elevation angle is denoted by ϕ. Subscripts e and o are used to denote the angular coordinates of the eth sensor location and direction of arrival of the plane wave(s) (single in this case), respectively. The unit vector triad $(\hat{\imath}, \hat{\jmath}, \hat{k})$ is used to define the orthogonal coordinate directions; o is also used to denote the desired 'look' direction of the array. These vectors are given by

$$\mathbf{r}_e = |\mathbf{r}_e| \sin \phi_e \cos \theta_e \hat{\imath} + |\mathbf{r}_e| \sin \phi_e \sin \theta_e \hat{\jmath} + |\mathbf{r}_e| \cos \phi_e \hat{k} \qquad (5.4a)$$

and

$$\mathbf{R} = \sin \phi_o \cos \theta_o \hat{\imath} + \sin \phi_o \sin \theta_o \hat{\jmath} + \cos \phi_o \hat{k}. \qquad (5.4b)$$

A complex representation for the propagating plane wave has been used in Figure 5.12 because it is the most general form of an oscillating spatio-temporal function, and the complex algebra avoids cumbersome trigonometry; the real and imaginary parts of the complex exponential can be extracted when needed. If the incident plane wave is traveling at a speed c(m/s) with circular frequency ω(rad/s), then the wave number, $k = 2\pi/\lambda$ (rad/m), determines the delay in phase that occurs due to the propagation of the wave front from point to point in the medium. The medium is assumed to behave as an elastic solid. The difference in phase, ς_e (rad), of arrival for the plane wave at arbitrary locations, $\mathbf{r}_e$, and at each sensor location with respect to the origin of the array at $\mathbf{r}_e = 0$ is found by multiplying the dot product of $\mathbf{r}_e$ and $\mathbf{R}$ k:

$$\varsigma_e(\theta_o, \phi_o) = k\mathbf{r}_e \bullet \mathbf{R} = k(x_e \sin \phi_o \cos \theta_o + y_e \sin \phi_o \sin \theta_o + z_e \cos \phi_o), \qquad (5.5)$$

where the Cartesian coordinates of the eth sensor are given by (x_e, y_e, z_e).

A simple linear beamformer for single-frequency waves can be created by weighting and then summing the contributions from all of the sensor readings to create the array output, $Y(n\Delta t, \theta, \phi)$, for arbitrary pairs of azimuth and elevation angles, (θ, ϕ), where $n\Delta t$ denotes the nth time sample. The superscript * denotes the complex conjugate. The beamformer in Figure 5.12 can then be expressed as an inner (dot) product:

$$Y(n\Delta t, \theta, \phi) = \sum_{e=1}^{N_s} w_e^* P_o e^{j(\omega n\Delta t + \varsigma_e(\theta,\phi))} = \mathbf{w}_e^H \mathbf{A}_r P_0 e^{j\omega n\Delta t}, \qquad (5.6)$$

where $\mathbf{w}_e^H = [w_1^* \; w_2^* \; \ldots \; w_{N_s}^*]$ is the vector of complex conjugate weights, which possess both magnitude and phase; P_0 is the amplitude of the wave, and $\mathbf{A}_r$, the *array response vector* with respect to sensor $e = 1$, is

$$\mathbf{A}_r(\theta, \phi) = [1 \; e^{-j\varsigma_2(\theta,\phi)} \; \cdots \; e^{-j\varsigma_{N_s}(\theta,\phi)}]^H \qquad (5.7)$$

The formation of the inner product between the weighting vector and the array response vector defines the directional sensitivity of the beamformer. When these two vectors are parallel, the output response is large, but when the vectors are perpendicular, the output response is small.

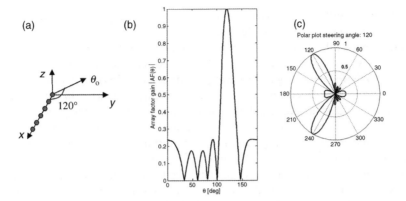

Figure 5.13 (a) Sensor array (b) array factor gain, $|A_{fp}(\theta)|$, for $N_s = 6$ and $d = \lambda/2$ and (c) corresponding polar plot showing steered direction of sensitivity at $120°$ in the upper half-space (Reproduced by permission of Taylor & Francis)

By selecting the weighting vector appropriately, the directional sensitivity of the beamformer can be optimized. For example, the simplest kind of beamformer steers the sensitivity of the array to certain desired azimuth and elevation reception angles, $\theta = \theta_o$ and $\phi = \phi_o$, by selecting the weights to be equal to the array response vector at the angle of arrival of a certain plane wave of interest, $\mathbf{w}_e = \mathbf{A}_r(\theta_o, \phi_o)$. Any other waves that arrive at arbitrary angle pairs, (θ, ϕ), are attenuated by the filter for this choice of $\mathbf{w}_e$ because their respective array response vectors are nearly orthogonal to $\mathbf{w}_e$. Other choices for $\mathbf{w}_e$ lead to additional beneficial characteristics in the beamformer output.

The performance of a beamforming spatial filter will now be examined using a simplified linear array along the x axis with element 1 at the origin (see Figure 5.13(a)). To simplify the analysis, the incident plane wave is assumed to have zero elevation ($\phi_o = 0°$). For this phased sensor array, the phase delay at each element due to its position in the array is $\varsigma_e(\theta_o, 90°) = 2\pi x_e \cos\theta_o/\lambda = \omega x_e \cos\theta_o/c$. It will also be assumed that the sensors are uniformly spaced such that $x_e = (e-1)d$ for $e = 1, 2, \ldots, N_s$, where d is the spacing between the sensors and that the amplitude of the propagating plane wave is $P_0 = 1$.

With these specifications and with weights of unity magnitude, the steered output of the array becomes

$$Y(n\Delta t, \theta) = \sum_{e=1}^{N_s} w_e^* P_0 e^{j(\omega n\Delta t + \varsigma_e(\theta, 90°))} = e^{j\omega n\Delta t} \sum_{e=1}^{N_s} e^{j\frac{2\pi(e-1)d}{\lambda}(-\cos\theta_0 + \cos\theta)} = e^{j\omega n\Delta t} \times A_{fp}(\theta)$$

(5.8)

if *phase scanning* is used and

$$Y(n\Delta t, \theta) = e^{j\omega n\Delta t} \sum_{e=1}^{N_s} e^{j(-\frac{\omega(e-1)d}{c}\cos\theta_0 + \frac{2\pi(e-1)d}{\lambda}\cos\theta)} = e^{j\omega n\Delta t} \times A_{ft}(\theta)$$

(5.9)

if *time scanning* is used. $A_{fp}(\theta)$ and $A_{ft}(\theta)$ are called the *array gain factors*. Both methods yield the same results for narrowband operation, but in practice the time scanning method is more appropriate because there are usually multiple frequency components in wave propagation data. Phase scanning fails for multiple frequencies.

The magnitude of the array factors for both of these beamsteering arrays at one frequency with $d = \lambda/2, N_s = 6$ and a steering angle of $\theta_0 = 120°$ is shown in Figure 5.13. The plot in Figure 5.13(b) shows the array gain factor magnitude as a function of the scan angle, θ, and the plot in Figure 5.13(c) shows the polar envelope of the array gain factor. In both plots, the $120°$ steering angle is evident and implies that this array with weighting vector $\mathbf{A}_r(\theta_0, \phi_0)$ emphasizes plane waves in the $120°$ direction and attenuates waves from other directions by no less than a factor of 0.23; this attenuation for off-axis reception is desirable.

Next, the effects of changes in the number of sensing elements, N_s, and the spacing, d, between the sensors are examined. The simulation code 'beamsteer.m' can be used to animate the array gain factor for various 'look' directions and numbers of sensors. Changes in the array factor for three values of N_s (6, 4 and 2) are shown in Figure 5.14(a) and (b). As the number of elements is decreased, the width of the main lobe of the spatial filter increases and the amplitudes of the side lobes increase. In other words, the array becomes less directionally sensitivity for fewer sensors. This leakage of sensitivity to adjacent azimuth angles is analogous to leakage described in Section 5.2.1 and is not desirable in beamforming. In contrast, an increase in the number of sensing elements sharpens main lobes and reduces side lobes.

Other analogies with temporal filtering also translate to beamforming arrays including spatial aliasing in which too large a spacing between sensing elements creates what are called *grating lobes*. Spatial aliasing is illustrated in Figure 5.14(c) and (d) in which d has been increased from $\lambda/2$ to λ and then to 2λ. An increase in the sensor element spacing creates grating lobes at $60°$ for $d = \lambda$ and then at $0°$ for $d = 2\lambda$. In effect, the array has not sampled the region from which plane waves propagate with adequate spatial resolution, so waves from other directions can pass through the filter in addition to the desired waves from $120°$. Wooh and Shi (1999) showed that the sensor spacing must satisfy $d < \lambda/(1 + \sin\theta_{max})$ to avoid grating lobes where θ_{max} is the maximum scan angle. They also proved that the sensor element dimension, a, should be chosen such that $a < \lambda/4$ and/or $|\theta_{max}| < 30°$ in order to avoid

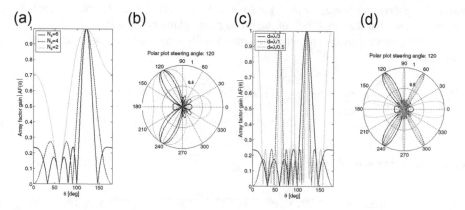

Figure 5.14 (a) Array factor gain for three numbers of elements (___) 6, (- - -) 4 and (...) 2 with $d = \lambda/2$ (b) polar plot with widening of main lobe (c) array factor gain for three interelement spacings (___) $d = \lambda/2$, (- - -) $d = \lambda/1$ and (...) $d = \lambda/0.5$ with $N_s = 6$ and (d) polar plot with grating lobes (Reproduced by permission of Taylor & Francis)

distortion in array gain factors due to sensor size. Finite sensor widths attenuate the array factor near $\theta = 0°$.

5.3 ESTIMATION OF UNMEASURED VARIABLES (STATE INFERENCE)

It is usually not possible to measure all of the response variables of interest for health monitoring of structural components. Therefore, measured variables must be used to infer (estimate) unmeasured variables. Models like those presented in Chapter 2 must be used for this purpose. In some cases, state inference may be a static process; in other words, the unmeasured variable is algebraically related to the measured variables. For example, the use of strain Rosettes to calculate principal strains as illustrated in Figure 4.35 is an example of state estimation with algebraic relationships. In this example, there are no dynamics between the measured and unmeasured variables. *In many applications, the unmeasured variables have different dynamics than the measured variables.* In order to estimate unmeasured variables in these applications, *state observers* are required.

Consider the sandwich panel shown in Figure 5.15. This panel represents a thermal barrier for aerospace vehicles. It is constructed of low thermal conductivity and high melting point (1330 °C) metallic material. Two face sheets bracket a honeycomb core in the panel, which has a cross-sectional area A. The temperature in the gas flowing past the exterior face sheet is T_0. The airframe protected by the panel is at temperature T_a. The coefficients of thermal conductivity for the panel layers and air gap beneath the panel are listed in the figure. Contact resistances between the layers are ignored for simplicity. The layers also possess thermal capacitances, which are given by c_1 and c_2. The temperature at each interface in the panel is different because of the thermal resistive and capacitive elements. Heat transfer through the small struts that attach the panel to the airframe are neglected. The coefficient of convective heat transfer at the exterior face sheet is h. By applying the first law of thermodynamics (energy conservation), the dynamic equations of state for this lumped parameter thermal system are found to be

$$
\begin{Bmatrix} \dot{T}_a \\ \dot{T}_{12} \\ \dot{T}_{23} \\ \dot{T}_{34} \end{Bmatrix} =
\begin{bmatrix}
-\frac{1}{c_a R_g} & \frac{2}{c_a R_g} & -\frac{2}{c_a R_g} & \frac{2-b}{c_a R_g} \\
\frac{1}{c_1 R_g} & -\frac{2}{c_1 R_g}-\frac{2}{c_1 R_1} & \frac{2}{c_1 R_g}+\frac{4}{c_1 R_1} & \frac{b-2}{c_1 R_g}+\frac{2b-4}{c_1 R_1} \\
0 & \frac{2}{c_2 R_1} & -\frac{4}{c_2 R_1}-\frac{2}{c_2 R_2} & \frac{4-2b}{c_2 R_1}+\frac{4-2b}{c_2 R_2} \\
0 & 0 & \frac{2}{c_1 R_2} & \frac{2b-4}{c_1 R_2}-\frac{b}{c_1 R_{conv}}
\end{bmatrix}
\begin{Bmatrix} T_a \\ T_{12} \\ T_{23} \\ T_{34} \end{Bmatrix}
$$

$$
+ \begin{Bmatrix}
-\frac{a}{c_a R_g} \\
\frac{a}{c_1 R_g}+\frac{2a}{c_1 R_1} \\
-\frac{2a}{c_2 R_1}-\frac{2a}{c_2 R_2} \\
\frac{2a}{c_1 R_2}+\frac{1-a}{c_1 R_{conv}}
\end{Bmatrix} T_0(t)
\tag{5.10}
$$

$$\{\dot{x}\} = [A]\{x\} + \{B\}u$$

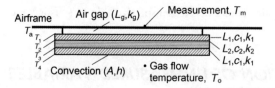

Figure 5.15   Panel subject to thermal loading by external temperature

where the mean temperatures of the layers serve as the states and are defined by

$$T_{12} = \frac{T_1 + T_2}{2} \tag{5.11a}$$

$$T_{23} = \frac{T_2 + T_3}{2} \tag{5.11b}$$

$$T_{34} = \frac{T_3 + T_4}{2} \tag{5.11c}$$

and the constants a and b are determined by the following relationships:

$$a = \frac{1/R_{conv}}{\dfrac{1}{R_{conv}} + \dfrac{2}{R_1}} \qquad b = \frac{2/R_1}{\dfrac{1}{R_{conv}} + \dfrac{2}{R_1}} \tag{5.12a, b, c}$$

$$T_4 = aT_0 + bT_{34}$$

State–space notation has been used in Equation (5.10) (refer to Section 4.4.5). The resistance and capacitance coefficients are defined in terms of the component geometrical and thermal parameters:

$$R_g = L_g/k_g A \quad R_1 = L_1/k_1 A \quad R_2 = L_2/k_2 A \quad R_{conv} = 1/hA,$$
$$c_a = \rho_a c_{sp-a} V_a \quad c_1 = \rho_1 c_{sp-1} V_1 \quad c_2 = \rho_2 c_{sp-2} V_2 \tag{5.13a – g}$$

where c_{sp-x} denotes the respective *specific heats* of the system components.

The simulation code 'thermalobserver.m' was used to generate the simulated data shown in Figure 5.16. A $0.46\,\text{m}^2$ area sandwich panel ($L_1 = 5$ mm; $L_2 = 2$ cm) of high-temperature metal is mounted with an $L_g = 1$ cm airgap above an aluminum airframe ($L_a = 5$ mm). The mechanical and thermal properties of the panel are given in 'thermalobserver.m'. The exterior temperature (input) is shown in (a). T_0 increases linearly from 20 to 1000 °C and then decreases to simulate a spacecraft reentry event. Figure 5.16(b) shows the temperature responses of the airframe (__) and the various panel layers (T_{12} (- - -), T_{23} (...) and T_{34} (-.-)). The airframe temperature does not experience much of a temperature change (0–42 °C). Radiation effects have been ignored leading to an even smaller temperature change in the airframe. On the contrary, T_{34} experiences a large temperature change. From a health-monitoring perspective, the peak temperature would be of interest. This peak temperature can be estimated using a *dynamic observer*.

The model in Equation (5.10) can be used to calculate the dynamic changes in temperature of the panel as well as the steady-state temperatures. Suppose initially that the exterior

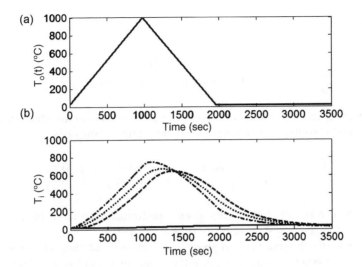

Figure 5.16 (a) Exterior temperature $T_0(t)$ of hot gas flowing past panel and (b) T_a (__), T_{12} (- - -), T_{23} (...) and T_{34} (-.-) of panel

temperature $T_0(t)$ of the hot gas flow is known and that the temperature on the airframe T_m is measured:

$$T_m = T_a \text{ with } \{y\} = [1\ 0\ 0\ 0]\begin{Bmatrix} T_m \\ T_{12} \\ T_{23} \\ T_{34} \end{Bmatrix} = [S]\{x\}. \tag{5.14}$$

This measurement can be used to estimate the other temperature states of the panel. For example, the temperature T_{34} would be of particular interest because unusually high values would trigger automatic replacement of the panel. The block diagram in Figure 5.17 is the framework for a linear dynamic observer. The basic approach is to use the panel model along with the measured response to estimate the unmeasured state variables. By driving the error between the estimated and measured response variable(s) to zero, the unmeasured states are estimated with accuracy. The equation for the observer states is derived based on the structure

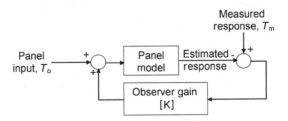

Figure 5.17 Dynamic observer model designed to estimate unmeasured state variables

in Figure 5.17:

$$\{\dot{\hat{x}}\} = [A]\{\hat{x}\} + \{B\}u + [K](\{y\} - [S]\{\hat{x}\})$$
$$= ([A] - [K][S])\{\hat{x}\} + \{B\}u + [K]\{y\} \tag{5.15}$$

Then this observer is simulated in real time along with the system for which states are to be estimated using the following combined set of state variable equations:

$$\left\{\begin{array}{c}\{\dot{\hat{x}}\} \\ \{\dot{x}\}\end{array}\right\} = \left[\begin{array}{cc}[A] - [K][S] & [K][S] \\ [0] & [A]\end{array}\right]\left\{\begin{array}{c}\{\hat{x}\} \\ \{x\}\end{array}\right\} + \left\{\begin{array}{c}\{B\} \\ \{B\}\end{array}\right\}u. \tag{5.16}$$

The matrix $[K]$ can be selected to give specific performance in terms of how quickly the observer tracks the actual system behavior (Friedland, 1986).

Figure 5.18 shows the results of an observer designed to estimate all of the temperature states of the panel system given the temperature input shown in Figure 5.16(a) and T_m. An observer gain matrix of $[K] = [0.001\ 0.001\ 0.001\ 0.001]^T$ was used to generate these results. The initial temperatures of the observer states were chosen to be 0 °C. After only a short period of time, the estimated temperatures for T_{12}, T_{23} and T_{34} are approximately equal to the actual system states. The estimated temperature for the airframe takes a longer time to reach the actual measured temperature due to the low conductivity of the panel and airgap. These estimated temperature states can then be used to estimate the maximum temperature reached by the outer face sheet, T_{34}.

Of course, it is often the case that the input (T_0) is unknown. In these cases, two approaches can be taken. First, the methods presented in Section 5.10 for identifying

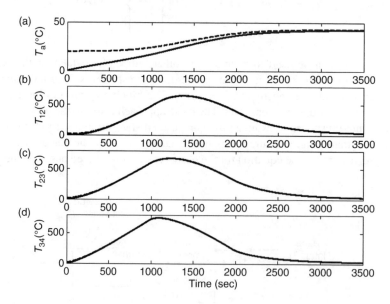

Figure 5.18   Temperature responses of actual system (- - -) and observer (___) with (a) T_a, (b) T_{12}, (c) T_{23} and (d) T_{34}

external loads using models can be applied to estimate the input. Then the method just described for state estimation can be used. Second, a different observer can be defined that includes the input as an unknown 'metastate' variable. In either case, the quality of the estimated states depends on the quality of the system model and the measured data. If the measurements do not make the system states observable (as defined in Section 4.4.5), then it is not possible to estimate all of the states. If there is significant noise in the measured data, it may also not be possible to estimate the states accurately. For example, the scenario above used the temperature of the airframe to estimate the panel temperatures. This scenario might lead to errors in practice because the panel temperature varies only 20 °C making precise measurements over that range difficult. Kalman filters and other robust methods for state estimation with noise in the model and measurements can be used to overcome some of these limitations. Friedland (1986) provides examples of setting up and using these different methods of state estimation.

The methodology in Figure 5.17 can be applied using any type of model other than the state variable model just described. For example, if an FRF model is measured for a dynamic system (refer to Section 2.2.4), then the observer for a system described by this FRF model can be expressed as follows:

$$\{\hat{X}(\omega)\} = [H(\omega)]\{F(\omega)\} + [K](\{Y(\omega)\} - [S]\{\hat{X}(\omega)\}). \tag{5.17}$$

In this expression, the output matrix $[S]$ determines which of the response variables is measured. The actual measurements are denoted by $\{Y(\omega)\}$. The observer gain matrix $[K]$ is selected to provide a good match between the actual and estimated responses.

5.4   TEMPORAL ANALYSIS

The advantage of analyzing time history data directly is that *no transforms are required leading to fewer errors*. The disadvantage of analyzing time histories is that *loading and damage characteristics may not be as easily identified nor interpreted using raw time data*. Two basic classes of time history analysis methods are described below. The first class of methods analyzes the basic statistics of the data. The second class of methods analyzes the deterministic characteristics of measured variables and relationships between different variables. The two-DOF system model in Figure 2.2 and the rod model in Figure 2.23 are used to demonstrate many techniques in this section as well as frequency and time–frequency analysis techniques in the subsequent sections.

5.4.1   Statistical (Nondeterministic) Analysis

Statistical analysis uses a variety of parameters and functions to characterize the temporal nature of one or more measured time histories *regardless of whether or not those measurements can be modeled in any way*. There are many parameters and functions from which to choose. Table 5.1 lists some of the choices available. Some of these will be immediately recognizable such as the *sample mean* and *variance*. However, others such as the *mean time* and *kurtosis* may be less familiar. To demonstrate the meaning of these various parameters and functions, first consider the two-DOF system model in Figure 2.2. This

Table 5.1 Statistical parameters/functions for time history analysis

Term	Expression
Sample mean	$\mu = \bar{y} = \sum_{n=1}^{N} y(n)/N$
Mean time	$\mu_t = \bar{t} = \sum_{n=1}^{N} t_n y^2(n)/N$
Root mean square (rms)	$\mu_{\mathrm{rms}} = \bar{y}_{\mathrm{rms}} = \sqrt{\sum_{n=1}^{N} y^2(n)/N}$
Variance	$\sigma^2 = \sum_{n=1}^{N} (y(n) - \bar{y})^2/N$
Temporal variance	$\sigma_t^2 = \sum_{n=1}^{N} (t_n - \bar{t})^2 y^2(n)/N$
Sample variance	$\bar{\sigma}^2 = \sum_{n=1}^{N} (y(n) - \bar{y})^2/(N-1)$
Skewness	$g = \dfrac{N \sum_{n=1}^{N} (y(n)-\bar{y})^3}{[\sum_{n=1}^{N} (y(n)-\bar{y})^2]^{3/2}}$
Kurtosis	$\mathrm{Ku} = \dfrac{N \sum_{n=1}^{N} (y(n)-\bar{y})^4}{[\sum_{n=1}^{N} (y(n)-\bar{y})^2]^2}$
Crest factor	$\mathrm{CF} = (|y(n)|_{\max})/\mu_{\mathrm{rms}}$
Difference data	$y_{\mathrm{d}}(n) = y_1(n) - y_2(n)$
rms difference mean	$\mu_{\mathrm{drms}} = \sqrt{\sum_{n=1}^{N} (y_{\mathrm{d}}(n))^2/N}$
Joint sample variance	$S_{\mathrm{P}}^2 = \dfrac{(N_1-1)S_1^2 + (N_2-1)S_2^2}{N_1+N_2-2}$
Sample correlation	$R_{y_1 y_2} = \sum_{n=1}^{N} y_1(n) y_2(n)/N$
Sample covariance	$C_{y_1 y_2} = \sum_{n=1}^{N} (y_1(n) - \bar{y}_1)(y_2(n) - \bar{y}_2)/N$
Auto-correlation	$R_{yy}(n) = \dfrac{1}{N} \sum_{i=1}^{N} y(i) y(i+n)$
Cross-correlation	$R_{y_1 y_2}(n) = \dfrac{1}{N} \sum_{i=1}^{N} y_1(i) y_2(i+n)$

system was simulated using the code 'twodoftemporal.m' with a sinusoidal excitation ($f_1(t) = \sin(2t) + \sin(5t) + \sin(8t)N$) combined with Gaussian white noise of unity variance[2]. Five time histories of length 1e7 points were simulated, and the parameters listed in Table 5.2 were computed for each of the time histories. The average of these five sets of parameters for a coupling stiffness of 20 N/m is listed in the column labeled 'Baseline'. Then the five experiments were repeated for a coupling stiffness of 18 N/m (a 10 % reduction in stiffness), and the list of parameters in the column labeled 'Damaged' was calculated. The subscripts '1' and '2' denote x_1 and x_2 responses.

An example of the x_1 time history is shown in Figure 5.19 before and after the coupling stiffness is reduced. Although the two time histories look somewhat different, it is not

---

[2]The 'seed' of the random number generator must be reset before each simulation to obtain random data.

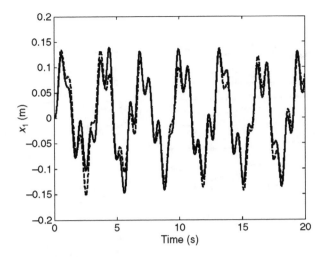

Figure 5.19 x_1 time history response for $K_2 = 20$ N/m (___) and $K_2 = 18$ N/m (- - -)

immediately obvious that the two signals are actually different from a statistical point of view. In other words, random changes in the excitation (and random errors in the measurements) could be responsible for the differences seen in the figure. The data in Table 5.2 is helpful for analyzing the statistical differences between the two time histories. These numbers represent averages across the five simulations run for each of the two stiffness cases. Figure 5.20 shows the eleven statistical parameters for all 10 simulations with baseline (___) and damaged (- - -) experiments indicated using different linetypes. It is clear from these comparisons that many of the statistical parameters do not indicate a clear difference between the two sets of responses. For example, the skewness parameters (g_1 and g_2) do not show a consistent difference in the two stiffness cases. Three of the parameters ($\bar{\sigma}^2_1$, $\mu_{1\mathrm{rms}}$, and $\mu_{d\mathrm{rms}}$) do show a consistent difference between the baseline and simulated damage cases. These three parameters show the most consistent difference because x_1 experiences larger motion on average with a lower coupling stiffness.

Table 5.2 Statistical parameters for a two-DOF system model with and without change in coupling stiffness

Parameter	Baseline ($K_2 = 20$ N/m)	Damaged ($K_2 = 18$ N/m)
μ_1	−0.0001	−0.0001
μ_2	−0.0001	−0.0001
$\mu_{1\mathrm{rms}}$	0.0745	0.0734
$\mu_{2\mathrm{rms}}$	0.0670	0.0667
$\bar{\sigma}^2_1$	0.0055	0.0054
$\bar{\sigma}^2_2$	0.0045	0.0044
g_1	−0.3436	−0.1093
g_2	−8.1100	−4.9500
$\mu_{d\mathrm{rms}}$	0.0542	0.0521
R_{12}	0.0036	0.0036
C_{12}	0.0036	0.0036

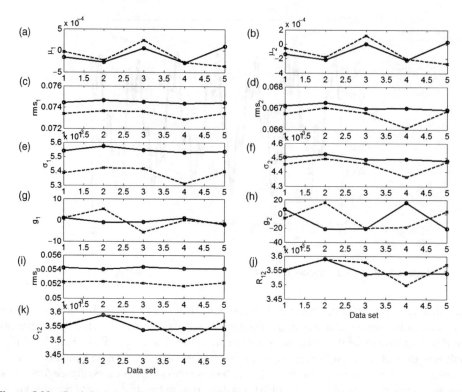

Figure 5.20 Statistical parameters for $K_2 = 20$ N/m (___) and $K_2 = 18$ N/m (- - -) with (a) μ_1, (b) μ_2, (c) $\mu_{1\text{rms}}$, (d) $\mu_{2\text{rms}}$, (e) $\bar{\sigma}^2_1$, (f) $\bar{\sigma}^2_2$, (g) g_1, (h) g_2, (i) μ_{drms}, (j) R_{12} and (k) C_{12}.

This example demonstrates that it is usually necessary to calculate a number of different parameters to distinguish healthy from damaged time history data. The example also shows that certain measurements may be more sensitive to changes due to damage in a component. In this example, x_1 is more sensitive to changes in stiffness than x_2. Lastly, the variation in statistical parameters in Figure 5.20 emphasizes that it is usually necessary to analyze multiple time histories in order to develop confidence that one time history is different from another due to damage.

Next consider the rod model in Figure 2.23. Longitudinal waves propagate along this rod. Recall from Section 2.2.8.2 that a finite element model can be used to simulate the response of the rod to applied forces. It was also demonstrated in that section that waves propagate differently in rods with localized changes in density, modulus or geometry due to damage (e.g. cracks). These differences must be identified using data analysis. Time history analysis methods are considered here based on the statistical functions listed in Table 5.1.

The differences between baseline and damaged responses at various DOFs along the rod were already plotted in Figure 2.26 for a crack located at the midpoint of the rod. The crack is modeled using a localized change in stiffness (modulus or cross-sectional area). Although the statistical parameters calculated above for the vibrating panel are informative in that application, parameters that highlight the traveling nature of the response are more

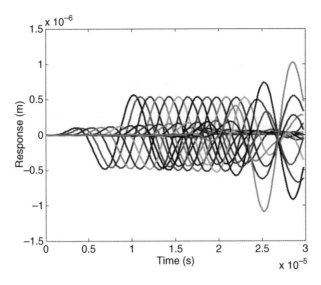

Figure 5.21 Displacement responses for all 18 DOFs of the rod subject to a pulse excitation force entering at the left end of the rod

appropriate for analyzing propagating waves. Figure 5.21 shows the longitudinal displacement responses of the rod at all of its DOFs when the rod is excited by a pulsed force entering at the left end. The traveling nature of the wave is observed with each DOF measuring the passage of the wave from left to right. Statistical parameters like the mean time and temporal variance (see Table 5.1) are useful for identifying changes in these types of traveling wave response measurements. The auto-correlation and cross-correlation functions are also indicative of changes in propagating wave response due to the presence of damage along the rod. The simulation code 'rodtemporal.m' is used to demonstrate the application of statistical methods for temporal data analysis of propagating waves in the rod.

The mean time and temporal variance of the first 150 time points for the 18 displacement measurements along the rod are shown in Figure 5.22(a) and (b). For the first few DOFs of the rod, there are no noticeable differences in the mean time and temporal variance before (__,o), and after (- - -,x) the stiffness change is introduced in the rod. However, DOFs 6, 7 and 8 indicate a significant difference in both of these parameters. The measurement at DOF 8 indicates the largest change in these two parameters because the stiffness change that simulates damage is located at DOF 8 in this example. When the traveling wave reaches the damaged region, a portion of the wave's energy reflects backward due to the impedance change in this region (refer to Section 3.1.2) and the remainder of the wave passes through the lower stiffness region of the rod. Consequently, a few DOFs to the left of DOF 8 experience changes where the bulk of the response is located along the time axis resulting in a change in the mean time and temporal variance parameters.

Figure 5.22(c) shows the autocorrelation function, $R_{11}(n)$, for the measurement at DOF 1. *The auto-correlation function indicates how a signal is related to delayed versions of itself as a function of the delay.* For example, the auto-correlation of an impulse is another impulse. Figure 5.22(d) shows the cross-correlation function, $C_{12}(n)$, for the measurements at DOFs 1 and 2 (side by side). As with the auto-correlation function, the cross-correlation function

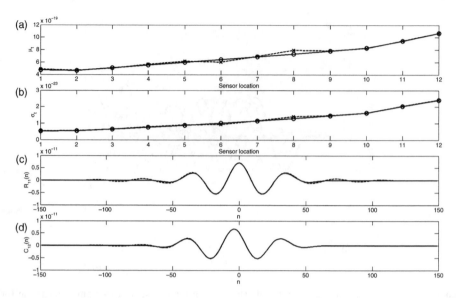

Figure 5.22 Statistical parameters and functions for $K_2 = 20$ N/m (—) and $K_2 = 18$ N/m (- - -) with (a) μ_t, (b) σ_t, (c) $R_{11}(n)$ and (d) $C_{12}(n)$

indicates how a signal is related to delayed versions of a second signal. Both of these functions are computed using the 'xcorr' function in MATLAB® in the simulation code 'rodtemporal.m.' The functions are computed for positive and negative values of n (refer to Table 5.1). $R_{11}(n)$ in (c) indicates a difference between the autocorrelation function for the DOF 1 measurement before (—) and after (- - -) the stiffness at DOF 8 is changed. The primary difference occurs for positive and negative delays in the signal. $C_{12}(n)$ in (d) also indicates a difference between the wave propagation responses of the baseline and rod with simulated damage at DOF 8. In the cross-correlation function, the primary differences occur for negative delays because the damage is located at DOF 8 and the first measurement in the cross-correlation (DOF 1) is located to the left of the second measurement (DOF 2). There are many more possible auto-correlation and cross-correlation functions that provide more indications of changes in the rod. By processing the data using these types of statistical analysis techniques, damage indicators can be extracted using the methods presented later in the chapter.

5.4.2 Deterministic Analysis

The statistical temporal analysis techniques presented in the previous section for analyzing time history data assumed that there were no underlying predictable characteristics in the data. Deterministic temporal analysis techniques utilize *time series models often used for prediction to analyze time history data*. Several of these models were described in Chapter 2. Temporal analysis methods involving restoring force models and autoregressive models are applied in this section to the vibration and wave propagation examples discussed in the previous section. These two methods are combined to describe a third method using time delay embedding techniques to analyze time histories.

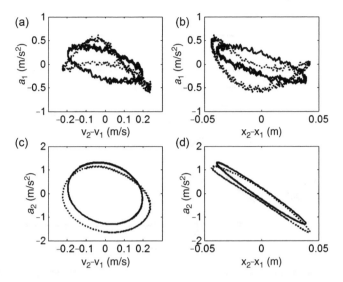

Figure 5.23   Restoring force projections in two-DOF system for $\mu_2 = 0$ N/m^2 (___) and $\mu_2 = 150$ N/m^2 (...) simulating buckling for (a), (b) a_1, and (c), (d) a_2 acceleration data

Restoring force models relate the acceleration in one particular DOF of a component to the displacement and velocity variables that result in damping and stiffness forces acting upon that DOF (Section 2.3.1.2). The response of the two-DOF system model in Figure 2.2 was simulated using 'twodoftemporal.m' with and without a nonlinear spring inserted between DOFs 1 and 2. Recall from Section 2.3.1.2 that this spring represents the effects of buckling damage to the component. The model was excited with a force applied to DOF 1 at a frequency of 5 rad/s with unity amplitude ($f_1(t) = \sin(5t)$N) in addition to a 0.05 RMS amplitude Gaussian signal with variance 0.25 to simulate the effects of other broadband excitations in real-world applications. Restoring forces are most useful for analyzing sinusoidal-type data with one or a few frequency components. Restoring force projections were then constructed as described earlier relating acceleration variables at the two inertias to relative velocity and displacement variables between the two DOFs. These restoring force projections are shown in Figure 5.23.

Restoring force projections for damping and stiffness on DOF 1 are shown in Figure 5.23(a) and (b). These projections undergo significant changes after the nonlinear spring is inserted to simulate damage. Restoring forces for damping and stiffness on DOF 2 are shown in Figure 5.23(c) and (d). These projections also exhibit significant changes after the stiffness is inserted. At 5 rad/s, the restoring forces for DOF 1 are more jagged than those for DOF 2 because of the nature of the system dynamics at this frequency. Because they are geometrical objects, these restoring force projections can be analyzed using geometrical methods. For example, the areas of the two projections in (d) are calculated in 'twodoftemporal.m' to be 0.018 m^2/s^2 (___) and 0.022 m^2/s^2 (...). The larger area value indicates the insertion of the nonlinear stiffness. By calculating this area, the damage due to buckling can be detected and quantified. The areas of the ellipses in (c) can also be computed in this manner providing similar results; a larger area is obtained for the case involving nonlinear coupling stiffness between the two DOFs.

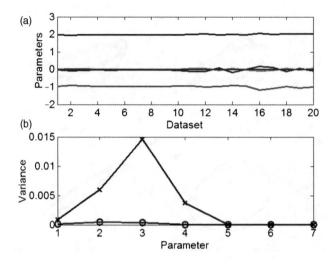

Figure 5.24 (a) Discrete time model coefficients for $\mu_2 = 0$ N/m² (1–10) and $\mu_2 = 150$ N/m² (11–20) in two-DOF simulation and (b) sample variance in parameters in healthy (ooo) and simulated damage (xxx) cases

Discrete time models can also be used to analyze time history data. These models relate one or more input forcing functions to one or more response variables using time delays of the input and response measurements (Section 2.3.1.3). The two-DOF system model in Figure 2.2 is again used to demonstrate temporal analysis with discrete time models. The simulation code 'twodoftemporal.m' was used to simulate the response of the two-DOF system to the same excitation used for the restoring force analysis presented above. The response was simulated with and without the nonlinear stiffness also used in the restoring force analysis. Ten time histories were generated for each condition. The linear input-output model in Equation (2.60) was used to estimate seven moving average (b_i) and autoregressive coefficients (a_i). The random variations in the excitation force influence each time history leading to variations in the estimated discrete time model coefficients. Although the model is linear, it can be used to analyze time histories produced by nonlinear vibrating systems like the two-DOF system with nonlinear coupling stiffness. Figure 5.24(a) shows the estimated coefficients for datasets 1–10 with $\mu_2 = 0$ N/m² and 11–20 with $\mu_2 = 150$ N/m². After the nonlinear stiffness is introduced, the coefficients exhibit more variation from dataset to dataset. This increased variance is plotted in (b) for the seven coefficients. The (x) symbols correspond to the simulated damage case and show larger variance as expected. The variance increases because of the amplitude dependence of the nonlinear spring.

The same approach was applied for a change in linear coupling stiffness, K_2, from 20 to 18 N/m, a 10 % reduction in stiffness. The coefficients for 10 datasets simulated in each of these two conditions are plotted in Figure 5.25(a). The coefficients once again exhibit more variation after the stiffness is decreased than in the baseline stiffness condition as plotted in (b). The variance in coefficients increases in this case because the response amplitude increases with decreased stiffness. The mean values of the coefficients also change somewhat.

These two examples demonstrate that discrete time models can be curve fitted to time history data by estimating the model parameters. These parameters and the variance in

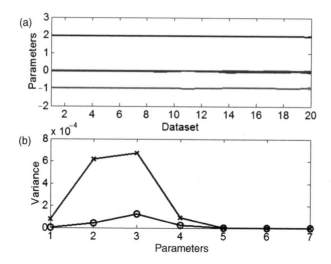

Figure 5.25   (a) Discrete time model coefficients for $K_2 = 20$ N/m (1–10) and $K_2 = 18$ N/m (11–20) in two-DOF simulation and (b) sample variance in parameters in healthy (ooo) and simulated damage (xxx) cases

parameters can then be used to detect changes in time histories due to damage. The prediction error of the model, called the *residual*, can also be calculated and used to analyze changes in time histories with loading and damage.

As demonstrated above, restoring force projections can be used to construct damping and stiffness characteristics when multiple response variables are measured. However, if only one response variable is measured, then the discrete time modeling techniques just described can be combined with the notion of restoring forces to produce *time delay embeddings* (Stark *et al.*, 1997). Time delay embeddings are sets of delayed measurements that relate one variable to time-delayed versions of that variable. Recall this same kind of relationship was expressed in the discrete time model (refer to Equation (2.60)). The result of this group of delayed measurements is a geometric object similar to the restoring forces described above. For instance, the phase–plane diagram presented in Figure 2.34 related the velocity and displacement variables of M_2; velocity leads the displacement by 90°. Similarly, diagrams of time delay embeddings, called *return maps*, relate delayed variables to one another (Packard *et al.*, 1980). As with restoring force analysis, return maps are most informative for sinusoidal-type data of low *dimension*. There are quite a few different 'dimension' parameters used in time-series analysis. Regardless of which parameter is used, the dimension determines the number of delays needed to reconstruct the dynamics, or alternatively the *phase space*, of a system from which a time history is measured. Takens theorem (Takens, 1981) states that the return map of response data points, $x(n), x(n-1), \ldots, x(n-(m-1))$, reconstructs the total phase space of a dynamic system if the *embedding dimension, m*, is equal to $2d + 1$, where d is the dimension of the underlying dynamics. In practice, an iterative approach is used to determine what m should be. The embedding dimension is increased and the return maps are analyzed. The return map is repeatedly redrawn until all crossings are eliminated. When the return map becomes disentangled, the embedding dimension is said to be found (Strogatz, 1994).

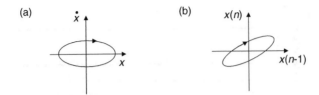

Figure 5.26   (a) Phase–plane diagram of a mass-spring oscillator and (b) return map constructed using time delay embedding with embedding dimension $m = 2$

For example, the phase–plane diagram of an unforced single-DOF mass-spring oscillator is shown in Figure 5.26(a). Instead of plotting displacement versus time, time is removed from the plot, and the velocity versus the displacement is plotted. The resulting planar elliptical object is the fingerprint so-to-speak of this dynamic system. The analytical expression of the phase–plane diagram is given by

$$\frac{\dot{x}^2}{1/M} + \frac{x^2}{1/K} = \text{const} \tag{5.18}$$

where const is a constant obtained from the initial conditions through integration. The lengths of the major and minor axes are equal to the inverse square roots of the stiffness and mass parameters of the system. If a forward difference relationship is used to approximate the velocity variable (refer to Section 2.3.1.3), then the phase–plane diagram can be reconstructed using only $x(n)$ and $x(n - 1)$. Therefore, the embedding dimension of this unforced single-DOF system is $m = 2$. The return map constructed using the measurement and one delay is illustrated in (b). It is a tilted and squeezed version of the original phase–plane diagram, but it does represent the system dynamics.

The shape of the return map depends on what delay is used in the embedding sequence. For example, Figure 5.27(a), (c) and (e) in the top row of plots shows the return map for the simple

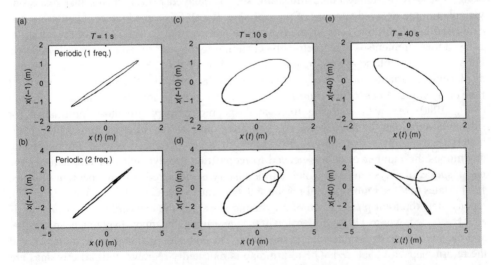

Figure 5.27   Return maps with one ((a), (c) and (e)) and two ((b), (d) and (f)) frequency components for ((a) and (b)) 1 s delay, ((c) and (d)) 10 s delay and ((e) and (f)) 40 s delay

harmonic forced response of a single-DOF system at different embedding delays (1 s, 10 s and 40 s from left to right). The delayed variable versus the undelayed variable is plotted in these maps. Note that regardless of the delay used, the shape of the map is elliptical. The embedding dimension of this time history is $m = 2$; therefore, only two terms in the embedding sequence are needed to disentangle the maps. On the contrary, Figure 5.27(b), (d) and (f) in the bottom row of plots shows the return map for the harmonic forced response of a single-DOF system to two sinusoidal excitation frequencies. The shape of the map again changes depending on the delay used. In addition, the map exhibits crossings for all delays indicating that the embedding dimension is greater than two, $m > 2$. If an additional delayed variable is added to the sequence ($m = 3$), then the return map becomes three dimensional and does not exhibit any crossings.

Return maps can be analyzed for vibrating systems like the two-DOF system in Figure 2.2. The case where nonlinear stiffness (μ_2 in the model from Section 2.2.7) is added to simulate buckling of a component is again considered. After the two-DOF system was excited by two sinusoidal frequencies, $f_1(t) = \sin(2t) + \sin(8t)N$, the return maps shown in Figure 5.28 were constructed for the DOF 1 displacement using the simulation code 'twodofembed.m.' Figure 5.28(a) shows the results for two delays at 0.15 and 0.30 s before (black line) and after (gray line) the nonlinear stiffness was inserted. There are no crossings in either of the return maps indicating that the embedding dimension is $m = 3$. There are also clear differences in the two maps due to the simulated damage from the nonlinear spring. When the delays are increased to 0.47 and 0.94 s, the shape of the map changes as shown in (b). However, the primary characteristics of the map remain the same; there are no crossings, and the maps before and after the nonlinear stiffness is introduced exhibit significant changes in shape.

As in the case of the restoring force, return maps constructed using time delay embedding can be analyzed to distinguish baseline maps from the maps produced after components become damaged. For example, the two sets of diagrams in Figure 5.28 were used in 'twodofembed.m' to calculate the *Euclidean distance* between corresponding points on the baseline and damage return maps, $\sqrt{x_d^2(n) + x_d^2(n-1) + x_d^2(n-2)}$, where x_d is the

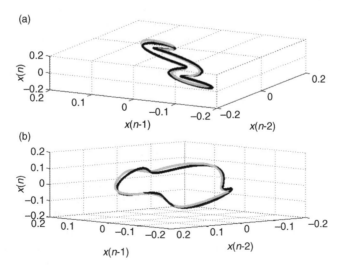

Figure 5.28 Time delay embedding phase diagrams of x_1 for $\mu_2 = 0$ N/m² (dark line) and $\mu_2 = 100$ N/m² (gray line) in two-DOF simulation with (a) 0.15 and 0.30 s delays and (b) 0.47 and 0.94 s delays

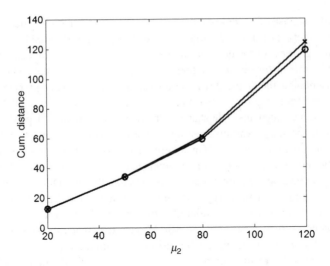

Figure 5.29 Cumulative Euclidean distance between baseline and damaged x_1 return maps for two-DOF system with $\mu_2 = 20$, 50, 80 and 120 N/m² for (x) 0.15 and 0.30 s delay and (o) 0.47 and 0.94 s delay showing increase with level of nonlinear stiffness

difference data listed in Table 5.1. These distances were then summed over a certain number of vibration cycles. This parameter was then calculated for different values of the nonlinear stiffness parameter, μ_2, that defines the damage. The results are shown in Figure 5.29 for the two different sets of embedding time delays. Note that the cumulative Euclidean distance increases as μ_2 increases because the return map for increasing nonlinear stiffness deviates significantly from the baseline return map.

The return map obtained through time delay embedding is often reduced into simpler geometric objects when analyzing time history measurements. Return maps quickly become complicated even for two-DOF systems, let alone real-world structural components with many more DOFs. *Poincare sections* are generated by sampling the return map at certain points in time (Virgin, 2000). By sectioning the return map in this manner, more readily identifiable geometrical objects can be constructed. Simpler geometries are consequently easier to analyze when detecting changes in mechanical loading and damage. Poincare sections are examined in the end-of-chapter homework problems.

5.5 TRANSFORMATION OF DATA

Temporal analysis of data is limited because all of the data is analyzed at once. Temporal analysis results are easiest to interpret when time histories are narrowband in nature (i.e. sinusoidal). Time history data can be transformed into other forms that separate the components in data as a function of frequency, time and other variables. *The advantage of transformed data analysis is that broadband data can be analyzed more readily in addition to narrowband data.* The disadvantages of transforming data are that errors are introduced by the transformation process. Errors can be reduced through the use of windows (refer to Section 5.2.1) and other means (like synchronous averaging described later). Frequency analysis is described first followed by higher order frequency domain techniques. Then time–frequency transforms including spectrograms and wavelets are described.

5.5.1 Spectral (Frequency) Analysis

Frequency analysis of data is often called *spectral analysis*. The basic idea of spectral analysis is that time history signals are decomposed into sinusoidal-type components. Each component has a frequency of oscillation, amplitude, phase and sometimes an attenuation coefficient. These components individually or in groups can be analyzed to identify loading and damage in structural components. For example, modes of vibration as described in Section 2.2 can be analyzed through spectral analysis by extracting the frequencies associated with those modes from response data.

Fourier series (Equation (2.83)) and Fourier transforms (Equation (2.84)) are used to model and analyze *continuous* signals that are defined for all points in time. Spectral analysis can be performed on both stationary and nonstationary types of data. Examples of these various types of data were given in Section 2.4.1. Fourier series are generally applied to periodic signals[3]. Periodic data repeat every T_0 seconds, the *period*; different frequency events have different periods of repetition. The frequency, ω_0, of a signal component in rad/s is related to the period, T_0, by $\omega_0 = 2\pi/T_0$. Fourier transforms are used for both periodic and aperiodic signals. For example, the frequency content in transient data that does not repeat can be found using the Fourier transform.

Sinusoidal signals are transformed into one pair of a_n and b_n coefficients in the case of Fourier series and one set of impulses in the case of Fourier transforms as illustrated in Figure 5.30. Figure 5.30(a) shows a sinusoidal signal of period T_0 with unity amplitude and zero phase. For a Fourier series with a fundamental period equal to T_0, the Fourier series coefficients from Equation (2.83) are illustrated in (b). The Fourier transform of the sinusoid is illustrated in (c), which contains two impulses of amplitude $\pm j/2$. Both positive and negative frequency components are needed in the transform to represent real signals by virtue of Euler's formula: $\sin \omega_0 t = (e^{j\omega_0 t} - e^{-j\omega_0 t})/2j$. Note that the Fourier series and Fourier transform both transform a *continuous signal with one frequency into a frequency spectrum with only one frequency component*.

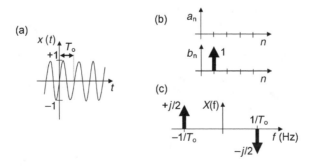

Figure 5.30 (a) Continuous sinusoidal, periodic signal with period T_0 with (b) Fourier series coefficients and (c) Fourier transform of that sinusoid

---

[3]Fourier series can represent any continuous signal exactly over a finite segment of time but will fail to describe the signal outside this segment of time if the signal is aperiodic.

Digital data can also be analyzed using frequency spectra; however, the data must first be truncated and then sampled as described in Section 4.5 before calculating the frequency transform. Figure 5.31 shows the process for calculating the *discrete Fourier transform* of a digitized sinusoidal signal. Figure 5.31(a) shows the truncation step in which the sinusoidal signal is truncated at the end of a block of time, T. This multiplication is equivalent to a *convolution* of the Fourier transform of the continuous sinusoid with the Fourier transform of the boxcar window (Brigham, 1988). The convolution, $c(x)$, of two signals, $s_1(x)$ and $s_2(x)$, with independent variable x (time or frequency) is calculated with the following formula

$$c(x) = s_1(x) * s_2(x) = \int_{-\infty}^{+\infty} s_1(\tau)s_2(x - \tau)d\tau. \tag{5.19}$$

As shown in Figure 5.31(b), the Fourier transform of the boxcar window is a *sinc* function, and the Fourier transform of a continuous sinusoid is two impulses at positive and negative

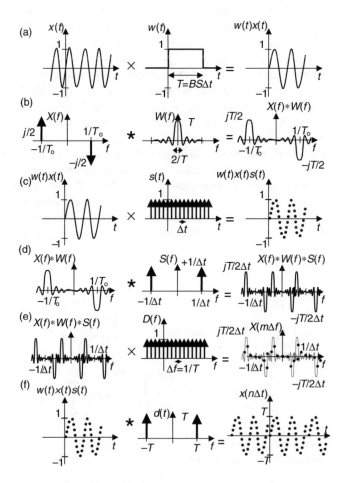

Figure 5.31 Discrete Fourier transform process for sinusoid with ((a) and (b)) truncation, ((c) and (d)) time domain sampling and, ((e) and (f)) frequency domain sampling (Brigham 1988, Pearson Education)

frequencies. Convolution of a function with an impulse is equivalent to drawing the function centered at each impulse. Therefore, the convolution of these two frequency transforms is as shown in the far right of (b). This result suggests that cutting off the sinusoid with a finite length observation window introduces the potential for leakage errors because the sinc function smears the frequency content of the true Fourier transform of an infinite sinusoid. In other words, discontinuities at the beginning and end of the observed signal in the far right of (a) introduce other frequencies in the transform of the cutoff sinusoid. One way to reduce this leakage error is to utilize time domain filters called windows, which were described in Section 5.2.1. Windows lessen the effects of leakage by modifying the sinc function shown in (b).

The process of computing the discrete Fourier transform (DFT) continues in Figure 5.31(c) with the sampling of the cutoff sinusoid. The sampling function, $s(t)$, is a row of impulse functions (refer to Section 2.4.1.1) separated by Δt. This time step is related to the observation period, T, used in the first step in the process: $T = BS\Delta t$, where BS is the number of time steps in the observation period. When multiplied by the windowed sinusoid, a discrete series of BS points is generated as shown in the far right of (c). The equivalent convolution operation is illustrated in (d). Note that the Fourier transform of $s(t)$ is also a row of impulses separated by $1/\Delta t$, which is the sampling frequency, F_{samp}. Convolution of the Fourier transform of the windowed sinusoid with the Fourier transform of $s(t)$ produces the result in the far right of (d). The modified sinusoidal transform has been replicated at each impulse centered at multiples of the sampling frequency in the convolution process. By replicating the frequency content of the sinusoids in this manner, convolution introduces the potential for aliasing. Aliasing was described in Section 4.5.2. It occurs when the sampling frequency is not at least twice the maximum frequency of interest in the digitized signal. The convolution in (d) explains the source of Shannon's criterion.

The last step in the DFT process is shown in Figure 5.31(e). Just as the time signal is digitized to produce a discrete series of data points, so too must the Fourier transform be digitized after the previous convolution operations. The sampling in this case is done in the frequency domain with a train of impulses separated by $1/T$. Recall from Rayleigh's criterion in Section 4.5.3 that the frequency resolution, Δf, is equal to $1/T$. Sampling produces the discrete series of frequency domain data points shown in the far right of (e). The equivalent convolution operation in the time domain is performed on the sampled and truncated sinusoid. The result of this final convolution is a periodic train of sampled and truncated sinusoids. It can be concluded that the DFT inherently assumes that the data acquired digitally is periodic in nature even if the data is not periodic.

All of the steps in Figure 5.31 can be summarized in one expression to calculate the DFT $X(m/BS\Delta t)$ at frequency steps $m/BS\Delta t$ of a digitized time domain signal $x(n\Delta t)$ sampled at time points $n\Delta t$, where m and n are integers. By applying the rectangular rule for integration to the Fourier transform integral given in Equation (2.84), the following formula for the DFT is found (Brigham, 1988):

$$X\left(\frac{m}{BS\Delta t}\right) = \sum_{n=0}^{BS-1} x(n\Delta t)e^{-j\frac{2\pi nm}{BS}} \text{ for } m = 0, 1, 2, \ldots, BS - 1. \tag{5.20}$$

A scale factor, Δt, must be included in front of the summation to make the DFT equivalent to the Fourier transform. As in the case of the Fourier transform and the Fourier series, the DFT calculates the projection of a time domain signal onto sinusoidal and cosinusoidal functions. In MATLAB[®], the DFT is computed using the *fast Fourier transform* (FFT) algorithm. Note that a scale factor of $2/BS$ must be applied to the DFT to obtain the coefficients for the corresponding Fourier series[4]. The DFT is complex; the real part represents the cosinusoidal components of $x(n\Delta t)$, and the imaginary part represents the sinusoidal components of $x(n\Delta t)$. The DFT provides an exact approximation of the true Fourier transform when the following four conditions are met:

1. $x(n\Delta t)$ is periodic (recall the concluding assumption of periodicity in Figure 5.31(f))
2. $x(n\Delta t)$ is bandlimited meaning that the amplitudes of its frequency components go to zero beyond a certain frequency, the signal bandwidth (otherwise, aliasing has the potential to occur no matter how high one samples the data as shown in Figure 5.31(d))
3. The sampling frequency, $1/\Delta t$, must satisfy Shannon's sampling criterion, Equation (4.21) (aliasing can occur if one samples to slowly)
4. The observation period, T, must span exactly an integer multiple of the period of $x(n\Delta t)$ (otherwise, leakage can occur as shown in Figure 5.31(b) and (e)).

In many applications, the DFT is calculated using FRFS as described in Section 2.3.2, and then the DFT is transformed back into a time domain signal. The *inverse discrete Fourier transform* is calculated using the following formula:

$$x(n\Delta t) = \frac{1}{BS} \sum_{m=0}^{BS-1} X\left(\frac{m}{BS\Delta t}\right) e^{j\frac{2\pi nm}{BS}} \text{ for } n = 0, 1, 2, \ldots, BS - 1. \quad (5.21)$$

To demonstrate the utility of the DFT, consider the two-DOF system model in Figure 2.2(a) before and after K_2 is changed to simulate buckling damage. The excitation force from Section 5.4.1 with frequency components at 0.32, 0.80 and 1.3 Hz in addition to a Gaussian random component was used in the simulation code 'twodofdft.m' to generate the spectrum of the response time history data x_1. The magnitude of the complex DFT spectrum is shown in Figure 5.32(a). The spectrum is shown in certain frequency ranges in Figure 5.32(b)–(d). As described above, the DFT is discrete in nature; it only has values at multiples of $1/T$, where T is the length of the time history. Therefore, the discrete DFT values are indicated with (o) for $K_2 = 20$ N/m and (x) for $K_2 = 18$ N/m. In some cases, the discrete values of the DFT are not directly indicated in plots but are always implied (e.g. see Figure 2.37(b)). The units of the DFT are meters or whatever the units of the measured variable. When a multiplication by T is used in Equation (5.20) to make the DFT equivalent to the Fourier transform, the units of the DFT are m/Hz. The 'fft' command in MATLAB® does not use this factor so the results are only scaled by BS.

Consider the shape of the spectrum in Figure 5.32(a). The frequency components of the excitation are evident as are the resonant frequencies of the two-DOF system (0.5 and 1.1 Hz). No differences are evident before and after the stiffness is changed in the first frequency range (a) because the modal deflection shape in that range does not exercise the damaged element. However, differences are apparent in two of the three frequency ranges ((c) and (d)). This example demonstrates that *DFT analysis is useful for distinguishing between frequency*

---

[4]The results in this book apply a factor of $2/BS$ to the positive frequency components in the DFT to indicate the actual amplitude of time history data as a function of frequency. Negative frequency components are not shown but can be plotted using the command 'fftshift' in MATLAB®.

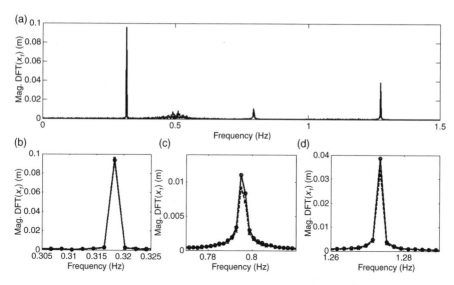

Figure 5.32 x_1 DFT magnitudes for two-DOF system with $K_2 = 20$ N/m (__) and $K_2 = 18$ N/m^2 (...) simulating buckling with (a) 0–1.5 Hz range and (b)–(d) for short frequency ranges

components that are sensitive to changes in the component and frequencies that are insensitive to changes.

Damage due to buckling in the two-DOF system can also be modeled by introducing the nonlinear stiffness, μ_2, as was done in some of the previous sections. The DFT spectrum before and after μ_2 is inserted is shown in Figure 5.33(a). The most obvious

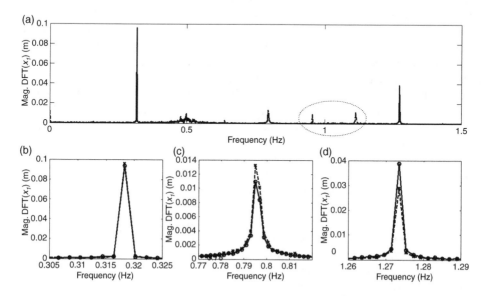

Figure 5.33 x_1 DFT magnitudes for two-DOF system with $\mu_2 = 0$ N/m^2 (__) and $\mu_2 = 100$ N/m^2 (...) simulating buckling with (a) 0 to 1.5 Hz range indicating harmonics and (b)–(d) for short frequency ranges

difference between the DFT for the nonlinear stiffness and the previous case with linear stiffness is the presence of additional response frequencies just below and just above 1 Hz. These frequencies are due to harmonic distortion caused by the nonlinear quadratic spring. If the sum of two sinusoidal frequencies is squared, *harmonic distortion* occurs because two frequency components interact as shown in the following trigonometric sequence:

$$
\begin{aligned}
[\sin(\omega_1 t) + \sin(\omega_2 t)]^2 &= \sin^2(\omega_1 t) + \sin^2(\omega_2 t) + 2\sin(\omega_1 t)\sin(\omega_2 t) \\
&= \frac{1 - \sin(2\omega_1 t)}{2} + \frac{1 - \sin(2\omega_2 t)}{2} + 2\sin(\omega_1 t)\sin(\omega_2 t) \\
&= 1 - \frac{1}{2}[\sin(2\omega_1 t) + \sin(2\omega_2 t)] + \cos((\omega_1 - \omega_2)t) \\
&\quad - \cos((\omega_1 + \omega_2)t)
\end{aligned}
\tag{5.22}
$$

For example, the frequency at 0.98 Hz is from the frequency combination 1.3–0.32 Hz (last term in Equation (5.22)). The frequency at 1.12 Hz is from the frequency combination 0.32 + 0.8 Hz (second to the last term in Equation (5.22)). The first term, 1, in Equation (5.22) is at 0 Hz and this frequency component can also barely be seen in Figure 5.33(a). These response frequency components can be used for damage detection and other purposes. Other differences in the DFT before and after the nonlinear stiffness is introduced to model buckling are also evident in Figure 5.33(c,d). The DFT spectrum in the second and third frequency ranges in these plots exhibits a greater change when the nonlinear stiffness is introduced than when K_2 was changed. As in the case where K_2 was used to model the damage, there is no change to the DFT in the lowest frequency range.

DFT spectra can be further manipulated to construct other types of spectra. For example, the transmissibility between two measurements can be calculated by taking DFT ratios. Figure 5.34 shows the results of taking the ratio between the DFT of x_1 and x_2 for the two-DOF system before (___) and after (- - -) the nonlinear spring is inserted. The magnitude of the transmissibility is plotted from 0 to 1.5 Hz in (a) and in short frequency ranges in (b)–(d). The Gaussian random excitation signal causes small response components whose ratio produces large amounts of variance in the transmissibility plot. In the vicinity of the excitation frequencies (0.32, 0.8 and 1.3 Hz), the variance is smaller as indicated in Figure 5.34(b)–(d). A Hanning window can be applied as described in Section 5.2.1 to improve upon these results. Recall that this time domain weighting function reduces the amount of leakage in the DFT. The result in Figure 5.35 is an improved transmissibility plot for the case in which $\mu_2 = 0$ N/m². By comparing this transmissibility magnitude to that for the case when $\mu_2 = 100$ N/m², major differences (variance) are apparent due to nonlinear harmonic distortion in the response.

FRFs can also be formed by calculating the ratio of a response measurement DFT to an excitation measurement DFT. However, the random component in the excitations used in the examples presented above introduces large amounts of variance if the quotient is computed directly without any averaging. Figures 5.34 and 5.35 showed that the variance is high for the transmissibility ratio of the two response DFTs. The same is true of the FRF. Averaging will smooth the DFTs and improve the estimate of amplitude and phase amidst random variations in the excitation and response time histories. Averaging techniques are described later in Section 5.6.

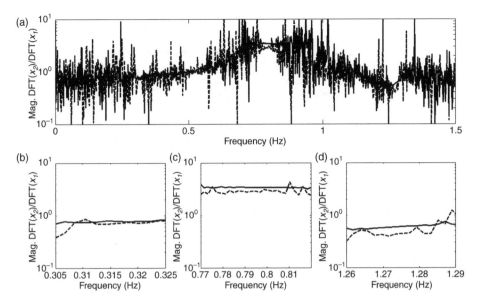

Figure 5.34 Transmissibility ratio of DFT magnitudes for two-DOF system with $\mu_2 = 0$ N/m^2 (___) and $\mu_2 = 100$ N/m^2 (...) simulating buckling with (a) 0–1.5 Hz range and (b)–(d) for short frequency ranges near excitation frequencies

5.5.2 Higher-Order Spectral Analysis

The DFT is useful for analyzing the response of structural components that are linear in nature. Linear structural components respond only at the frequencies of the excitation forces and the system resonances. DFT analysis decomposes time history measurement data into frequency components, some of which are more sensitive to damage than others. None of these frequency components interact in a linear structural component. However, when components respond in a nonlinear manner after becoming damaged or even prior to becoming damaged, frequency components do interact. Higher-order spectral analysis is used to detect these

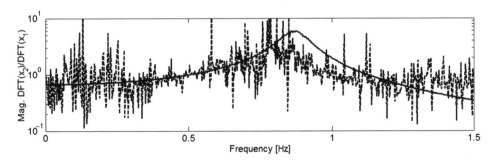

Figure 5.35 Transmissibility ratio of DFT magnitudes for two-DOF system with $\mu_2 = 0$ N/m^2 (___) and $\mu_2 = 150$ N/m^2 (...)

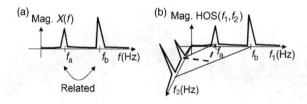

Figure 5.36 (a) Magnitude of DFT spectrum with response at two frequencies, and (b) multi-dimensional spectrum showing interaction between various frequency components

interactions and identify their source (i.e. the damage). Nikias and Petropulu (1993) give a detailed treatment of higher-order correlation and spectral functions.

The basic idea behind *higher-order spectra* (HOS) is that interactions between various frequency components can be detected by analyzing the spectral content of the data along more than one frequency axis. Figure 5.36(a) shows a typical DFT spectrum plotted as a function of frequency. Two frequency components, f_a and f_b, in the response are found. However, by examining this information alone it is unknown whether or not the two frequency components are related in some way. For example, damage such as that in the previous section involving a nonlinear coupling stiffness could be introducing harmonic distortion causing interactions between the two frequency components (refer to Equation (5.22)). In order to detect these interactions, the approach illustrated in (b) can be applied. A higher-order spectrum is plotted as a function of two frequencies instead of one. By examining how this spectrum varies with both these frequencies, interactions between f_a *and* f_b *can be detected.*

HOS are calculated by first calculating the DFT (Section 5.5.1). Table 5.3 lists the formulas for calculating two types of HOS in addition to the power (or auto) spectrum previously defined in Equation (2.85). The function $X(\omega)$ represents the DFT of the time history $x(m\Delta t)$, or the Fourier transform of $x(t)$. The $\lessdot$ symbol indicates the angle of the complex argument. The primary difference between the power spectrum, which only involves one frequency variable, and the *bispectrum* and *trispectrum*, which involve two

Table 5.3 Formulas for higher-order spectra of $x(n\Delta t)$.

Spectrum	Formula
Power (auto) spectrum	$G_{xx}(\omega) = X(\omega) \times X^*(\omega) = |X(\omega)|^2$
Bispectrum	$G_{xxx}(\omega_1, \omega_2) = X(\omega_1) \times X(\omega_2) \times X^*(\omega_1 + \omega_2)$ so $|G_{xxx}(\omega_1, \omega_2)| = |X(\omega_1)| \times |X(\omega_2)| \times |X(\omega_1 + \omega_2)|$ and $\lessdot G_{xxx}(\omega_1, \omega_2) = \lessdot X(\omega_1) + \lessdot X(\omega_2) - X(\omega_1 + \omega_2)$
Trispectrum	$G_{xxxx}(\omega_1, \omega_2, \omega_3) = X(\omega_1) \times X(\omega_2) \times X(\omega_3) \times X^*(\omega_1 + \omega_2 + \omega_3)$ so $|G_{xxxx}(\omega_1, \omega_2, \omega_3)| = |X(\omega_1)| \times |X(\omega_2)| \times |X(\omega_3)| \times |X(\omega_1 + \omega_2 + \omega_3)|$ and $\lessdot G_{xxxx}(\omega_1, \omega_2, \omega_3) = \lessdot X(\omega_1) + \lessdot X(\omega_2) + \lessdot X(\omega_3) - \lessdot X(\omega_1 + \omega_2 + \omega_3)$

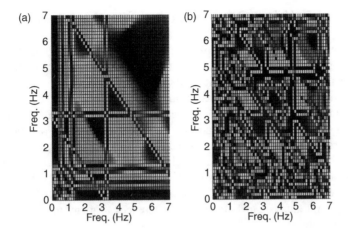

Figure 5.37   Bispectra phase plots of x_2 for two-DOF system with resonances at 0.5 and 1.2 Hz with (a) $\mu_2 = 0$ N/m^2 and (b) $\mu_2 = 100$ N/m^2 for an excitation at 3.1 and 8.0 Hz

and three frequency variables, respectively, is that the power spectrum is entirely real. In other words, *there is no phase information in the power spectrum*. On the contrary, *the bispectrum and trispectrum contain phase information*. This information is what enables HOS to detect interactions among frequency components.

Consider the two-DOF system used in the previous section to demonstrate spectral analysis techniques using the DFT. The two-DOF system possesses resonant frequencies at 0.5 and 1.2 Hz. The simulation code 'twodofhos.m' was used to excite the system with an excitation force, $f_1(t) = \sin(20t) + \sin(50t)N$, with frequency components at 3.1 and 8.0 Hz. Then the bispectrum was calculated for x_2. To calculate the bispectrum, the DFT was calculated first and then the formula in Table 5.3 was applied. Figure 5.37(a) shows the phase of the bispectrum before the nonlinear stiffness, μ_2, is introduced to simulate damage due to buckling. The phase is used instead of the magnitude because as mentioned above the phase is what distinguishes the bispectrum from the power spectrum. The two frequencies are along the x and y axes, and the phase is plotted in color.

At the two resonant frequencies and the two excitation frequencies, the bispectrum phase of the response exhibits a change along each of the frequency axes indicated by the brighter colors. In addition, any pairing of two frequencies along these axes that sums to one of the four frequencies listed above leads to a change in the bispectrum. This result suggests that there are no interactions between the four response frequencies. After the nonlinear spring is introduced, the bispectrum phase plot in Figure 5.37(b) is obtained. Note all of the additional diagonal, horizontal and vertical lines on the plot indicating that the frequency components are interacting in the response time history x_2. The same type of analysis can be performed on the magnitude of the bispectrum.

Although the bispectrum can be used to analyze frequency interactions due to quadratic-type nonlinearities as in the previous example, the trispectrum can be applied to analyze cubic-type nonlinearities. The trispectrum is generally not analyzed all at once because there are three frequency variables (see Table 5.3); instead, projections of the trispectrum are analyzed one frequency at a time along the third frequency axis.

5.5.3 Analysis Using Other Spectral Transformations

There are many other spectral transformations and methods for analyzing dynamic measurement data. For example, the *Hilbert transform* (Worden and Tomlinson, 1990) can be applied to Fourier spectra to analyze nonlinearity and time-varying mechanical properties due to the effects of damage. *Cepstral analysis* can also be applied to analyze frequency spectra. A cepstrum is the DFT of the DFT of a time history (two applications of Equation (5.20) back to back). The cepstrum can be used to identify frequencies that are harmonically related by detecting patterns in the DFT spectrum. The *power spectral density* (PSD) is another form of the DFT. The PSD is calculated using averaging, as are many spectral functions (including some of those already presented in this book). For example, the MATLAB® function 'psd' uses Welch's method to calculate the PSD. Refer to Bendat and Piersol (1993) for a complete overview of spectral analysis techniques with references to many more techniques used in engineering applications.

5.5.4 Time-Frequency Analysis

Time history analysis and spectral analysis examine how certain amplitude, phase, and other characteristics of measurement data change as a function of time or frequency. In many health-monitoring applications, it is more useful to analyze how measurement characteristics change as a function of both time and frequency. For example, if a 10 Hz component is detected in the data, spectral analysis could determine *if* it is present; however, spectral methods cannot identify *when* it is present. By including both time and frequency as independent variables in the analysis, changes in frequency content can be detected as a function of time and vice versa. This approach to data analysis is called *time–frequency analysis*. Cohen (1995) provides an intuitive and rigorous description of time–frequency analysis.

The most widely used time–frequency analysis technique is the *Short Time Fourier Transform (STFT)*. The STFT is sometimes called the *spectrogram*. Spectrograms were used in Section 4.5.2 on aliasing to demonstrate what happens when the analog-to-digital sampling rate is not high enough. Figure 5.38 illustrates how spectrograms are computed. A linear sweep (chirp) signal is shown in (a). The expression for this chirp signal, $y(t)$, is,

$$y(t) = \sin(2\pi(0.02t)t). \tag{5.23}$$

The frequency of this signal varies from 0 to 2.6 Hz[5]. The spectrogram is computed by taking the Fourier transform of small time segments of the signal, one at a time. For example, Figure 5.38(b) shows the first time segment of data. The original chirp is windowed so that the Fourier transform applies only to the first segment of the signal. In this example, a P301 window (see Section 5.2.1) is applied to each 8 s segment. The third segment of the signal is brought into focus in (c) through windowing and the fifth segment is brought into focus in (d). It is clear at least from (b) and (c) that the signal frequency in each successive time segment increases. Then the DFT is calculated for each of these segments of data. The magnitude of the

---

[5]The frequency of a sinusoidal signal with phase $\phi(t)$ is $d\phi(t)/dt$; e.g., from Equation (5.), $\phi(t) = 0.04\,\pi t^2$ and $\omega(t) = 0.08\pi t = 2\pi \times 0.04t$ leading to a 0 to 2.6 Hz sweep in frequency.

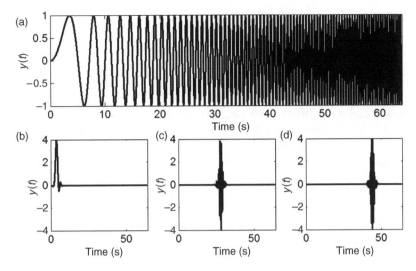

Figure 5.38   (a) Linear chirp from 0 to 2.6 Hz, (b) window applied in first time segment, (c) third time segment and (d) fifth time segment

resulting spectrogram is plotted in Figure 5.39 in two formats. Figure 5.39(a) shows a perspective view of the three-dimensional spectrogram. This plot shows that the amplitude of the signal is unity for all frequencies. The spectrogram in (b) shows that the frequency content in the signal varies linearly from 0 to 2.6 Hz. These plots were generated with the code 'timefreqanalysis.m' along with the MATLAB® function 'specgram,' which computes the spectrogram of a signal automatically.

The expression for calculating the spectrogram is obtained by formalizing the graphical approach presented in Figure 5.38. If a time domain window function, $w(t)$, is used to analyze

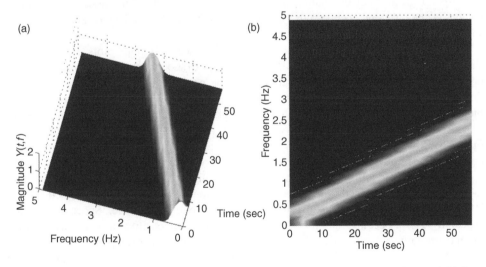

Figure 5.39   (a) Spectrogram of linear chirp from 0 to 2.6 Hz from perspective view, and (b) from above view

the signal, $y(t)$, then the spectrogram is defined by

$$Y(t, \omega) = \int\limits_{-\infty}^{+\infty} w(\tau - t)y(\tau)e^{-j\omega\tau}d\tau. \tag{5.24}$$

The window function is shifted to time t as required before performing the Fourier transform per Equation (2.84). For digitized data, the DFT of each windowed segment of sampled time data is calculated to obtain the spectrogram. For example, the spectrogram in Figure 5.39 was calculated using windowed blocks of data 2048 samples long. The spectrogram is a function of two variables, t and ω, which are evident in Figure 5.39. The major disadvantage of the spectrogram is that it mixes the spectrum of the signal together with the spectrum of the windowing function. This mixing is not always a problem, but there are alternative time–frequency transformations that avoid the problem (e.g. Wigner distribution; Cohen (1995)). Although this book analyzes the frequency content at a given time using a narrow time domain window function, the time content *at a given frequency* can also be analyzed. For this latter type of analysis, a broad time domain window function is chosen to provide a narrow frequency-domain window. This transformation is then called a *Short Frequency Time Transform (SFTT)*.

The way in which the chirp signal in Figure 5.38(a) is windowed in Figure 5.38(b)–(d) is reminiscent of the way elastic waves propagate in components (refer to Figure 5.21). Waves travel from point to point in a component similar to how the spectrogram window $w(t)$ travels along a time history measurement to analyze the change in frequency content with time. For example, consider longitudinal wave propagation in a rod with a crack. The rod is modeled using finite element methods as described in Chapter 2. The crack is located in the eighth element from the left hand side of the rod. A longitudinal force is applied to the left end to analyze the propagating waves using spectrograms. This example was described in Section 5.4. The simulation code 'timefreqanalysis.m' was used to simulate the response and analyze the response data. After computing the spectrograms for each of the nodal displacement responses, the difference between the spectrogram in the cracked and uncracked simulations was calculated.

Figure 5.40 shows 16 of the 18 spectrogram difference magnitude plots. Figure 5.40(a) shows the difference magnitude for node 1, Figure 5.40(b) shows the difference magnitude for node 2 and so on. As in the time history plots, the presence of a crack causes a nonzero difference in the plots. All of the plots indicate that a change has occurred in the rod. The plots also suggest that the crack is located in the eighth element from the left (Figure 5.40(h)). The very first occurrence in time of a difference around 20 µs in any of the plots is in Figure 5.40(h) (indicated by an arrow). By summing the difference magnitudes at various points in time, a quantitative measure indicating where damage is located can be obtained.

Another use of spectrograms is in identifying nonlinear types of damage. Several other methods already described in previous sections including restoring force projections and HOS were also useful for identifying nonlinearity due to damage. Spectrograms are especially useful because they identify the effects of nonlinearity at a particular moment in time providing a means for analyzing nonstationary data. One example of this capability was presented in Section 4.5.2 where a loudspeaker exhibited harmonic distortion, which caused aliasing errors in the digitally sampled data. As another example, consider the two-DOF system with nonlinear stiffness inserted between DOFs 1 and 2 to simulate damage

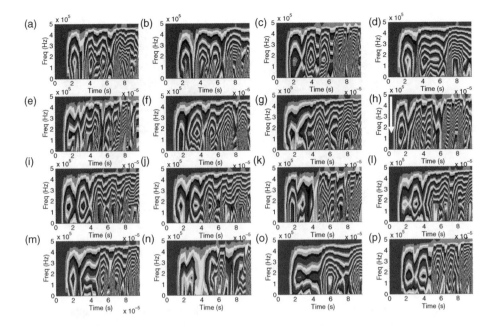

Figure 5.40 Magnitude of differences in spectrograms before and after crack is inserted at element 8 of rod undergoing longitudinal wave propagation for nodal displacements (a) 1, (b) 2, (c) 3, (d) 4, (e) 5, (f) 6, (g) 7, (h) 8 (with arrow indicating damage is located here), (i) 9, (j) 10, (k) 11, (l) 12, (m) 13, (n) 14, (o) 15 and (p) 16

due to buckling. This example has been used in several previous sections of this chapter (refer to Section 5.4.2). Instead of using a stationary excitation with fixed frequencies of excitation as was done in all previous examples, consider an excitation with time-varying frequencies of the form, $f_1(t) = \sin(2(1 + 0.005t)t) + \sin(5(1 + 0.005t)t) + \sin(8(1 + 0.005t)t)$N with an additional 1 N RMS amplitude Gaussian random excitation component with unity variance. The excitation is applied for 500 s (8.3 min). It is anticipated that this excitation will cause harmonic distortions due to frequency interactions such as those expressed in Equation (5.22).

Figure 5.41 shows the spectrograms for the measured time history x_2 before (Figure 5.41(a)) and after (Figure 5.41(b)) the nonlinear spring was inserted. The magnitude scales of these plots are on a log scale (base 10). Prior to inserting the nonlinear spring, the only frequencies evident in the data are the excitation frequencies from the above expression and the two resonant frequencies of the two-DOF system (0.5 and 1.1 Hz). After the nonlinear stiffness is inserted to simulate damage due to buckling, all of these same frequencies are still evident in the spectrogram in addition to many other frequencies resulting from harmonic distortion from the stiffness nonlinearity. These harmonics can be extracted from the spectrogram and used for damage detection and other purposes in health monitoring.

The spectrogram time–frequency transformation expressed in Equation (5.24) has two primary limitations. First, spectrograms use sinusoids and cosinusoids to decompose measured signals into frequency components. The exponential function, $e^{-j\omega\tau} = \cos\omega\tau - j\sin\omega\tau$, in Equation (5.24) is called the *kernel* of the transformation. As long as a measurement looks

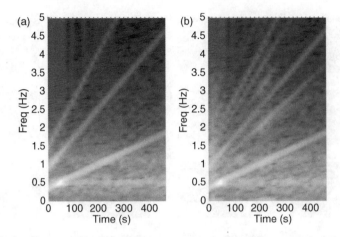

Figure 5.41 Magnitude of spectrogram before (a) and after (b) nonlinear stiffness is inserted ($\mu_2 = 80\,\text{N/m}^2$) in two-DOF system model showing harmonic distortion

like the sinusoidal kernel used in the spectrogram formula, spectrograms are appropriate for decomposing the measurement. However, if the measurement is broadband in nature with transient characteristics, then a kernel that looks more like the measured data should be used. A sinusoidal kernel function does not look much like an impulse; therefore, the spectrogram must use a broad range of frequencies to decompose an impulsive signal making it difficult to decipher anything specific about the impulse. Second, the *localization* properties of the spectrogram are constant with time and frequency. In other words, at 10 Hz, the frequency resolution of the spectrogram is Δf and at 100 Hz, the frequency resolution is also Δf. When analyzing broadband data with transient characteristics, it is often beneficial to reduce the frequency resolution as the frequency increases. On the contrary, it is beneficial to increase the time resolution as the frequency increases to better localize transient events.

These two limitations can be overcome by using *wavelet* analysis. Wavelets are functions that can be stretched and shifted to represent a variety of signals. All of the beneficial properties of wavelet time–frequency transformations are derived from the way in which the wavelet kernel function is used. Figure 5.42 illustrates how wavelet analysis works in comparison to spectrogram analysis. Figure 5.42(a) shows a time history with low frequency content and a transient event midway through the measurement. In Figure 5.42(b), spectrogram analysis is used to decompose this signal into a series of low- and high-frequency sinusoidal components at each point in time. The problem with this approach is that the sinusoids into which the signal is decomposed are not transient leading to smearing of the transient pulse across all frequencies. Also, the ability to localize the occurrence of the transient pulse along the time axis is compromised by the lowest frequency level, which requires a relatively long block of data to calculate the contribution of the lowest frequency component to the signal.

The wavelet analysis approach illustrated in Figure 5.42(c) is different than the spectrogram approach. Wavelet analysis uses an increasingly greater number of time intervals at higher frequencies leading to a better ability to localize high-frequency (transient) components in the signal. In addition, it uses a closer frequency spacing at lower frequencies leading to a better ability to resolve closely spaced low frequencies. In other words, *the wavelet kernel shortens*

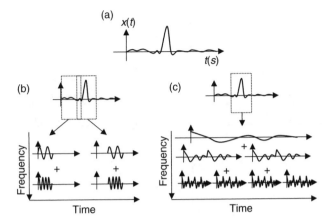

Figure 5.42 (a) Time history with transient event, (b) decomposition using spectrogram analysis as a function of time and frequency, and (c) decomposition using wavelets

the time interval for analyzing high-frequency signal components and lengthens the time interval for analyzing low frequency signal components. In contrast, the spectrogram has a fixed length time interval.

The wavelet transform, $Y_w(b, a)$, of a continuous signal, $y(t)$, is calculated using the wavelet kernel, $\psi(t)$, which is called the *basis* or *mother wavelet* (Chiu, 1992):

$$Y_w(b, a) = \sqrt{a} \int_{-\infty}^{+\infty} y(\tau)\psi^*\left(\frac{\tau - b}{a}\right)d\tau \quad \text{with} \quad \int_{-\infty}^{+\infty} \psi(\tau)d\tau = 0. \tag{5.25}$$

The continuous wavelet transform is a function of two independent parameters. b is the *time shift* parameter, which allows the wavelet transform to localize events in time. a is the *dilation* parameter, which squeezes and stretches the basis wavelet function to increase the frequency resolution at low frequencies and decrease resolution at high frequencies. The basis wavelet should be chosen to match the signal of interest as well as possible and there are several constraints for constructing basis wavelets including the zero mean constraint indicated in Equation (5.25). If the signal looks like a combination of sines and cosines, then *harmonic wavelets* should be chosen because they look like sines and cosines. On the contrary, if the signal looks like an impulse or sharp transient, then *Daubechies wavelets* are a better choice (Newland, 1993). In vibrations and wave propagation data analysis, harmonic wavelets are often used.

Figure 5.43 shows the time domain wavelet basis functions and Fourier transforms associated with these wavelets. For example, Figure 5.43(a) is the Daubechies wavelet using two coefficients (Newland, 1993) with the Fourier transform in Figure 5.43(b). If more coefficients are added to generate the Daubechies wavelet, the basis function becomes smoother but still retains its transient nature. Figure 5.43(c) is the harmonic basic wavelet function (real and imaginary parts) and Figure 5.43(d) is its Fourier transform. These are the types of wavelets that are utilized in Equation (5.25) to decompose signals using wavelet

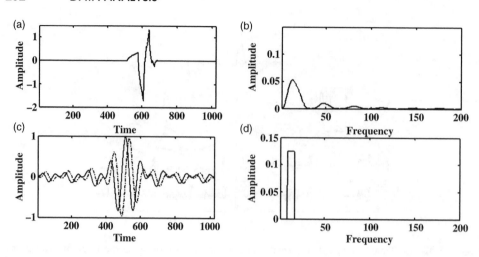

Figure 5.43 (a) Daubechies two wavelet, and (b) its Fourier transform, (c) real (___) and imaginary (-.-) part of harmonic wavelet and (d) its Fourier transform

time–frequency analysis. Consider the harmonic wavelet as an example. The Fourier transform of this basis wavelet is

$$\Psi(\omega) = \begin{cases} \dfrac{2^{-a}}{2\pi} e^{-\frac{j\omega b}{2^a}}, & \text{for } 2\pi 2^a \leq \omega < 4\pi 2^a \\ = 0 & \text{elsewhere} \end{cases} \tag{5.26}$$

and the corresponding time domain basis wavelet function is,

$$\psi(2^a t - b) = \frac{e^{j4\pi(2^a t - b)} - e^{j2\pi(2^a t - b)}}{j2\pi(2^a t - b)} \quad \text{for } 2\pi 2^a \leq \omega < 4\pi 2^a. \tag{5.27}$$

It is clear from Equation (5.26) that each harmonic wavelet at a given dilation, a, has constant amplitude in the scaled frequency range, $2\pi 2^a \leq \omega < 4\pi 2^a$. Outside this frequency range, the wavelet has zero spectral content. Therefore, the result in Figure 5.43(d) is obtained.

When analyzing digitally sampled data, the discrete wavelet transform (DWT) is used as in the case of the previously discussed discrete Fourier transform. A discrete time signal, $x(n\Delta t)$, is decomposed into a series of wavelet components with the basis wavelet function expanded using the dyadic[6] a and time–shift parameter[7] b as follows:

$$x(n\Delta t) = c_0 + \sum_a \sum_b c_{a,b} \psi(2^a n\Delta t - b). \tag{5.28}$$

---

[6]Discrete wavelet transforms are calculated assuming an even number of samples, 2^N.

[7]The same parameter, a, is used to denote dilation in continuous wavelet transforms and wavelet level in discrete wavelet transforms for consistency; however, rigorously speaking the two parameters are different.

The number of wavelet levels a is determined by the number of sampled data points, 2^N. There are $N+1$ wavelet levels, which are integers that are labeled starting at -1 and extending through $N-1$. The level -1 corresponds to a DC component.

There are numerous algorithms for calculating the coefficients, $c_{a,b}$, of the DWT (e.g. Mallat's tree algorithm). The MATLAB® wavelet toolbox can be used to compute these coefficients. The results in this textbook utilize Newland's algorithms (Newland, 1993). Once the coefficients in the DWT are computed using Equation (5.28), they are usually plotted in the form of a map as in the case of the spectrogram. The coefficients are sometimes squared and summed to generate a wavelet contour map or *mean square map*. In terms of the coefficients, $c_{a,b}$, the mean square map, $M_{a,b}$, is given by

$$M_{a,b} = c_0^2 + \sum_b \sum_a c_{a,b}^2 \left(\frac{1}{2^a}\right). \tag{5.29}$$

The *scalogram* is also commonly used to present the coefficients obtained using wavelet analysis. The scalogram differs from the mean square map in that the scalogram plots the coefficients in terms of the frequency rather than the wavelet level.

Consider the sweep signal in Equation (5.23), which was analyzed using the spectrogram in Figure 5.39. Figure 5.44(a) shows the chirp with an added transient midway through the signal. It is transient events like this one that inspire the use of wavelet analysis. Spectrograms tend to smear this transient event across a wide range of times unless unreasonably small window functions are used to process the data leading to poor frequency resolution in the spectrogram. Wavelets overcome this tradeoff between frequency and time localization. The mean square maps for the harmonic basis wavelet (Figure 5.44(b)) and the Daubechies wavelet with 10 coefficients (Figure 5.44(c)) are both shown in the figure. In the harmonic wavelet map (Figure 5.44(b)), wavelet level 13 corresponds to the 64–128 Hz bandwidth, wavelet level 12 corresponds to 32–64 Hz and so on. The frequency sweeps from level 1

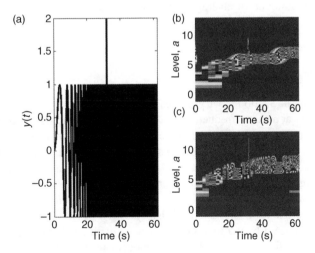

Figure 5.44   (a) Swept signal from Equation (5.23) with transient added midway through the signal, (b) mean square map using harmonic basis wavelet and (c) mean square map using Daubechies wavelet with 10 coefficients

(0.015–.030 Hz) to level 7 (2–4 Hz) as expected. Midway through the map, a narrow spike occurs at the location of the transient and there is very little smearing of this spike along the time axis of the map. The same is true of the mean square map for the Daubechies wavelet in Figure 5.44(c). In this case, the content of the map is not confined to the same frequency ranges as in the previous case because the Daubechies wavelet is not designed to span octave frequency bands.

5.6   AVERAGING OF DATA

Many of the previous examples dealt with simulated measurement data that contains random variations. For example, the transmissibility functions calculated in Section 5.5.1 using DFTs were difficult to interpret because only one sample of the random data was used to form the ratio between the two response measurements. Random variations were also observed in the statistical parameters calculated in Section 5.4.1. In order to improve the accuracy with which time, frequency and time-frequency data can be analyzed, blocks of measurement data must be *averaged*. Two basic types of averaging can be used in health monitoring.

Synchronous averaging is used if there is a reference signal either from an actuator applied to actively sense the component response or from some other sort of trigger (e.g. tachometer pulse measured once per revolution of a rotating component). This type of averaging is always preferable because it eliminates the randomness associated with the phase of blocks of data that are averaged. *Asynchronous averaging* is used when there is no reference signal with which the averaging can be performed. Both types of averaging are discussed and demonstrated below. Note that no amount of averaging can eliminate serious errors in measurement data. Only *random* variations in data can be reduced through averaging. *Bias errors* cannot be removed. For example, if a linear FRF model is being used to relate an excitation measurement to a response measurement but the system from which these measurements are acquired is nonlinear, then no amount of averaging will eliminate the errors in the FRF due to nonlinearity.

5.6.1   Cyclic Averaging

The DFT formulated in Section 5.5.1 provides information about the frequency content of a sampled time history at discrete frequencies, $m/N\Delta t$. Cyclic averaging is a form of asynchronous averaging that aims to reduce random variations and bias errors due to leakage at the discrete set of frequencies in the DFT (Allemang, 1999). In a measured time history with N_t total points, the basic idea is to select a blocksize BS that emphasizes signals with certain periods and de-emphasizes signals with other periods. *Blocksizes that contain an integer number of periods of the signal(s) of interest are optimum for cyclic averaging.*

For example, Figure 5.45(a) shows a time history with a 10 Hz sinusoidal component with unit amplitude covered up by additive Gaussian noise with an RMS amplitude of two and variance of four. It is difficult to see the 10 Hz component because of the noise. The simulation code 'averagingdemos.m' was used to perform two types of block averaging on this data. Figure 5.45(c) shows the result of asynchronous averaging using a block size of 683 data points corresponding to a time period of 2.7 s. The original time history with 2^{16}

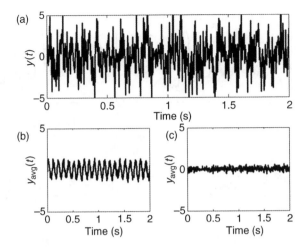

Figure 5.45 (a) 10 Hz sinusoidal signal corrupted with Gaussian random noise, (b) cyclic averaging with 4 s blocksize, and (c) asynchronous averaging with 2.7 s blocksize showing loss of signal

(65 536) data points was divided into 95 blocks, the blocks were averaged and the result is plotted in Figure 5.45(c). Note that the 10 Hz signal has a period of 0.1 s; therefore, 2.7 s is not an integer number of periods of the desired signal. The result is an averaging away of the 10 Hz signal because the data blocks do not synch up with the signal. In contrast, Figure 5.45(b) shows the results of cyclic synchronous averaging using a blocksize of 1024 points corresponding to 4 s of data. After 64 averages of these contiguous blocks of data, the 10 Hz signal is extracted with approximately the correct amplitude. Two different types of averaging lead to two entirely different conclusions about the information in the data.

Cyclic averaging using a block of data $BS\Delta t$ time points in length improves the DFT accuracy at frequency points $m/BS\Delta t$ (refer to Equation (5.20)). Therefore, the blocksize for averaging should be selected to coincide with the frequencies of most interest for analysis. Consider the two-DOF system model from Figure 2.2(a) used in the previous sections. The excitation force used in this example contains three sinusoidal components, $f_1(t) = 10\sin(2\pi(1.25)t) + 100\sin(2\pi(1.8)t) + 100\sin(2\pi(2.5)t)$N plus a 20 N RMS amplitude component of Gaussian random noise with variance 400. The MATLAB® code 'averagingdemos.m' was used to simulate the response of the two-DOF system to the excitation force. Two simulations were run with $K_2 = 20$ N/m to observe the effects of the noise component on the response. Then two more simulations were run with $K_2 = 20$ N/m and $K_2 = 10$ N/m to simulate the effects of damage due to buckling stiffness. The data was processed using two different blocksizes to average the DFT of x_1.

Figure 5.46(a) shows the results of processing two baseline datasets with no simulated damage using a blocksize of 4 096. This blocksize corresponds to a frequency resolution of 0.0625 Hz. The two DFTs are nearly the same with a few small exceptions. All of the response frequencies due to the excitation are correctly identified. The two resonant frequencies at 0.5 and 1.1 Hz are also identified. A Hanning window (Section 5.2.1) was used when computing the DFT to optimize the accuracy of the frequency values. Figure 5.46(b) shows the results for a baseline case (__) and simulated damage case (- - -) with $K_2 = 10$ N/m. The most noticeable change in the DFT is that the resonant frequency formerly at 1.1 Hz drops in frequency to

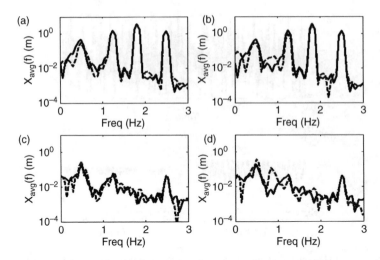

Figure 5.46 DFT of x_1 in two-DOF system subject to sinusoidal and random excitations where (a) BS = 4096 without damage for two different datasets, (b) *BS* = 4096 with $K_2 = 20$ N/m (___) and (- - -) $K_2 = 10$ N/m, (c) BS = 5154 without damage for two different datasets, and (d) *BS* = 5154 with $K_2 = 20$ N/m (___) and (- - -) $K_2 = 10$ N/m

0.87 Hz. This change is repeatable for new datasets generated using different Gaussian random components with the same amplitude and variance. The other noticeable change occurs at the 1.25 Hz frequency response, which drops in amplitude noticeably when K_2 is decreased. These two changes together suggest that a permanent change in the component has occurred.

Figure 5.46(c) shows the results of processing two baseline datasets with no damage using a blocksize of 5154. This blocksize corresponds to a frequency resolution of 0.0497 Hz. For this blocksize, only two of the frequency components in the data are correctly identified (0.5 and 2.5 Hz). The other frequencies are missed because they are not integer multiples of the frequency resolution provided by this blocksize. The results for a lower coupling stiffness, $K_2 = 10$ N/m, are shown in Figure 5.46(d). Although the 0.87 Hz peak is detected, poor resolution in the DFT due to asynchronous averaging with the wrong blocksize makes it difficult to interpret the change at 0.87 Hz. Unlike the result for the 4096 blocksize in Figure 5.46(b), the frequency component at 1.25 Hz cannot be resolved in Figure 5.46(d) to confirm that a permanent change has occurred in the component.

5.6.2 Frequency Response Function Estimation

An FRF relates the relative amplitude and phase of an excitation (input) variable and a response (output) variable. The FRF relationship assumes that the response is proportional to the excitation and is, therefore, linear in nature. Section 2.2.4 provided the analytical derivation for FRFs in a multi-DOF system model. In order to estimate FRFs using measured data, averaging must be used. Measurements can contain noise from electromagnetic, thermal, ambient vibration and other sources leading to errors in estimated FRFs. The right kind of

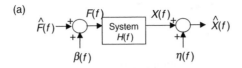

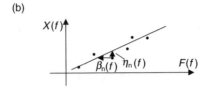

Figure 5.47 (a) Frequency response measurement and estimation framework, and (b) graphical interpretation of estimation with measurement noise

averaging can alleviate these errors depending on the type of noise contained in the data. Synchronous averaging is preferred for the reasons already described above. Synchronous averages can be performed by using excitation measurements to synchronize the phase of the input and output data.

Figure 5.47(a) shows the framework for estimating the FRF between one input spectrum measurement and one output spectrum measurement when both measurements are corrupted by noise. The true excitation and response spectra are denoted by $\hat{F}(f)$ and $\hat{X}(f)$, respectively. The measured excitation and response spectra are denoted by $\hat{F}(f)$ and $\hat{X}(f)$. Additive noise spectra corrupting the excitation and response measurements are denoted by $\beta(f)$ and $\eta(f)$, respectively. The FRF of the system is $H(f)$, which relates the true response spectrum to the true excitation spectrum. Figure 5.47(b) illustrates this straight line relationship. The objective is to estimate $H(f)$ as well as possible using measured spectra. Averages are performed on multiple measurements of these spectra, $\hat{F}_n(f)$ and $\hat{X}_n(f)$ with noise components $\beta_n(f)$ and $\eta_n(f)$, where n is the measurement number. Each of these measurements is indicated by a data point in Figure 5.47(b). The noise components are indicated by arrows in the figure.

The FRF relationship for each measurement can be written as follows:

$$X_n(f) = H(f)F_n(f)$$
$$\hat{X}_n(f) - \eta_n(f) = H(f)(\hat{F}_n(f) + \beta_n(f))$$

(5.30a, b)

This relationship can be written in vector-matrix notation for N_{avg} measurements:

$$\left\{ \begin{array}{c} \hat{X}_1(f) \\ \hat{X}_2(f) \\ \vdots \\ \hat{X}_{N_{avg}}(f) \end{array} \right\} - \left\{ \begin{array}{c} \eta_1(f) \\ \eta_2(f) \\ \vdots \\ \eta_{N_{avg}}(f) \end{array} \right\} = H(f) \left\{ \begin{array}{c} \hat{F}_1(f) \\ \hat{F}_2(f) \\ \vdots \\ \hat{F}_{N_{avg}}(f) \end{array} \right\} + H(f) \left\{ \begin{array}{c} \beta_1(f) \\ \beta_2(f) \\ \vdots \\ \beta_{N_{avg}}(f) \end{array} \right\}.$$

(5.31a, b)

$$\{\hat{X}_n(f)\} - \{\eta_n(f)\} = H(f)\{\hat{F}_n(f)\} + H(f)\{\beta_n(f)\}$$

In order to estimate $H(f)$, the *method of least squares* is used. Several different approaches can be taken to solve for the estimate, $\hat{H}(f)$, depending on the nature of the measurement noise. Different assumptions about the noise are made to estimate $H(f)$.

NOISE ON RESPONSE MEASUREMENT

If there is no noise on the excitation measurement, then $\beta_n(f) = 0$. The complex conjugate transpose (called the *Hermitian transpose*) of the vector of excitation measurements is then premultiplied on both sides of Equation (5.31b):

$$\{\hat{F}_n(f)\}^H \{\hat{X}_n(f)\} - \{\hat{F}_n(f)\}^H \{\eta_n(f)\} = H(f)\{\hat{F}_n(f)\}^H \{\hat{F}_n(f)\}. \qquad (5.32)$$

If it is assumed that the excitation measurements are *uncorrelated* with the response measurement noise[8], then $\{\hat{F}_n(f)\}^H \{\eta_n(f)\} = 0$ for an adequate number of measurements. Therefore, the estimate of $H(f)$ is:

$$\hat{H}(f) = \frac{\{\hat{F}_n(f)\}^H \{\hat{X}_n(f)\}}{\{\hat{F}_n(f)\}^H \{\hat{F}_n(f)\}} = \frac{G_{XF}(f)}{G_{FF}(f)}, \qquad (5.33)$$

where $G_{XF}(f)$ is the *cross-power spectrum* and $G_{FF}(f)$ is the *auto-power spectrum*[9]. This estimate is usually referred to as the H_1 FRF estimate. This solution minimizes the sum of the squared errors in the FRF expression written for all of the measurements.

NOISE ON EXCITATION MEASUREMENT

If there is no noise on the response measurement, then $\eta_n(f) = 0$. If it is assumed that the response measurements are *uncorrelated* with the excitation measurement noise, then $\{\hat{X}_n(f)\}^H \{\beta_n(f)\} = 0$. Therefore, the estimate of $H(f)$ is:

$$\hat{H}(f) = \frac{\{\hat{X}_n(f)\}^H \{\hat{X}_n(f)\}}{\{\hat{X}_n(f)\}^H \{\hat{F}_n(f)\}} = \frac{G_{XX}(f)}{G_{FX}(f)}. \qquad (5.34)$$

This estimate is usually referred to as the H_2 FRF estimate. H_2 is not used as often as H_1 because $G_{FX}(f)$ often goes to zero at certain frequencies where the excitation and response become uncorrelated.

NOISE ON EXCITATION AND RESPONSE MEASUREMENTS

If there is noise on both the excitation and response measurements, then $H(f)$ is estimated by forming the matrix (Allemang, 1999):

$$\begin{bmatrix} G_{FF}(f) & G_{FX}(f) \\ G_{XF}(f) & G_{XX}(f) \end{bmatrix}, \qquad (5.35)$$

---

[8]Two vectors of signals (or spectra) are uncorrelated if their dot product is zero, i.e., the two vectors are orthogonal.
[9]Recall Eq. (2.85) which provided the auto power spectrum for a continuous spectrum with no noise. Equation (5.33) is valid for discrete spectra and incorporates averaging to reduce the effects of measurement noise.

and then solving for the eigenvalues and eigenvectors. The FRF estimate, $\hat{H}(f)$, that minimizes the sum of the squared errors across all of the measurements is found by normalizing the eigenvector associated with the *minimum* eigenvalue as follows:

$$\left\{ \begin{array}{c} \hat{H}(f) \\ -1 \end{array} \right\}. \tag{5.36}$$

ERROR IN ESTIMATION

The FRF estimates are not equal to the true FRF unless the measurement noise is zero. In order to determine the random estimation error in the FRF magnitude and the standard deviation in the phase estimate, the following expressions derived by Bendat and Piersol (1993) for the normalized magnitude error and deviation in phase can be used:

$$\varepsilon_{MAG} = \frac{1}{\sqrt{2N_{avg}}} \times \frac{\sqrt{1 - \hat{\gamma}^2(f)}}{|\hat{\gamma}(f)|} \tag{5.37a, b}$$

$$\sigma_{PHS} = \sin^{-1}(\varepsilon_{MAG}),$$

where $\hat{\gamma}(f)$ is the *ordinary coherence function* given by:

$$\hat{\gamma}^2(f) = \frac{|G_{FX}(f)|^2}{G_{FF}(f)G_{XX}(f)}. \tag{5.38}$$

The symbol $|\cdot|$ denotes the magnitude of a complex number. Equation (5.37a) shows that the error in the FRF magnitude estimate decreases as the number of measurements (i.e. N_{avg}) increases. The magnitude error is zero when $\hat{\gamma}(f) = 1$ and the standard deviation in phase is also zero. Conversely, the magnitude error is infinite when $\hat{\gamma}(f) = 0$. It can be concluded that if the coherence function is close to unity, then fewer averages are required to estimate the magnitude with a given level of confidence.

Consider the two-DOF system (Figure 2.2(a)) subject to a Gaussian random excitation, $f_1(t)$, with a 20 N RMS amplitude and 400 N variance. The code 'frfestimation.m' was used to simulate the response of the system to this excitation force. Then the FRFs between the force and the two response displacements were computed using the code 'rfrf.m.' The same code can be used to estimate the FRFs for non-random excitation and response measurements; however, only the FRF results at frequencies contained in the excitation are valid when considering narrowband excitations. A blocksize of 5140 time points was used to synchronously average the data to estimate the H_1 FRF from Equation (5.33). Fifty averages were used. A 50 % *overlap* of the datablocks was also used. An overlap is used because when a window function, like a Hanning window, is applied to the measurement data, the front and back ends of the data block are lost. To recover the information contained at the ends of one block, the subsequent block of data is selected to overlap with the surrounding blocks. The results are shown in Figure 5.48.

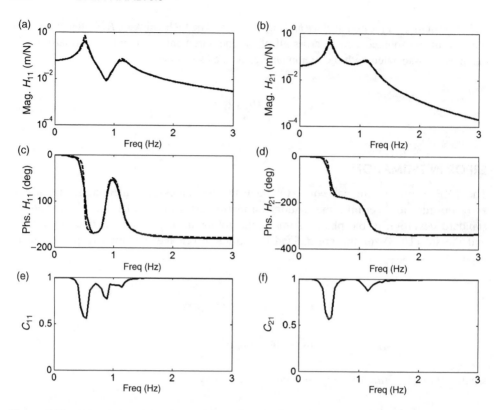

Figure 5.48   ((a) and (b)) Magnitude and phase ((c) and (d)) of $H_{11}(f)$ and $H_{21}(f)$ estimated (___) and true (- - -) FRFs (H_1 estimator) for overlap averaging with 5140 blocksize and 50 averages and ((e) and (f)) ordinary coherence functions $C_{11}(f)$ and $C_{21}(f)$

Figure 5.48(a) and (b) shows the magnitudes of the FRF estimates (___) and analytically computed magnitudes using matrix inversion (- - -). The estimated magnitude matches the analytical calculation well except for near the resonant peaks and valleys. At these frequencies, the frequency resolution is not small enough to prevent leakage errors in the DFT leading to errors in the estimated FRFs. These leakage errors are apparent in the ordinary coherence functions in Figure 5.48(e) and (f). Recall from Equation (5.37a) that a low coherence value translates into more random error in the magnitude estimate. Leakage errors are also seen in the phase estimates in Figure 5.48(c) and (d) at the resonances and between the resonances where the phase transitions are sharp. When the blocksize is increased to 23 831 and the number of averages are reduced to 10, the magnitude estimates and coherence functions shown in Figure 5.49 are obtained. By improving the frequency resolution in the DFT, the leakage error is reduce. The coherence functions also improve significantly (Figure 5.49(c,d)).

5.6.3 Averaging of Data in Rotating Systems

As described in Section 5.6.1, cyclic averaging is performed by selecting a data block with a period that is an integer multiple of the signal period of interest for analysis. In components

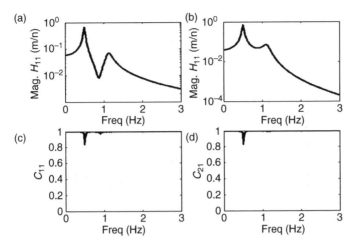

Figure 5.49 ((a) and (b)) Magnitude of $H_{11}(f)$ and $H_{21}(f)$ estimated (__) and true (- - -) FRFs for overlap averaging with 23 831 blocksize and 10 averages and ((c) and (d)) ordinary coherence functions $C_{11}(f)$ and $C_{21}(f)$

that are near or contained within rotating systems, cyclic averaging is performed by triggering the start of a new block of data each time the rotor rotates. By averaging with these one-per-revolution data blocks, the accuracy of amplitude and phase information at frequencies that are integer multiples of the rotational period is improved. A trigger signal is required to synchronize averages for data acquired from rotating systems. This trigger can be obtained using optical tapes that reflect light off of a reflective surface on the rotating component. The trigger can also be obtained using magnetic pickups that detect changes in magnetic flux due to a reference tooth or hole in the rotating component.

For example, consider the rolling tire in Figure 1.21. The accelerometer voltage signal measured in the vertical direction on the spindle of this rotating tire is shown in Figure 5.50 for 500 rotations. 1200 data points were acquired for each rotation of the tire. The data clearly varies from rotation to rotation. It is difficult to detect changes to the response due to tire damage

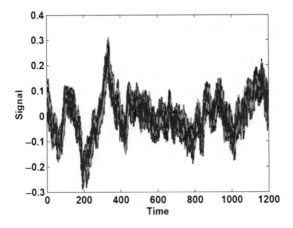

Figure 5.50   Measurements of spindle accelerometer voltage for 500 rotations of tire

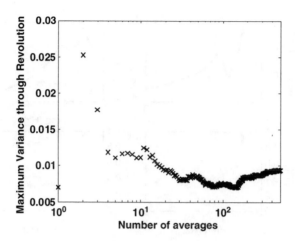

Figure 5.51   Maximum variance in measured signal for all rotational positions of the tire as a function of number of averages

using this data because of the significant variability. The objective with averaging is to accurately approximate the mean value of the signal at each position of the tire. As more averages are taken, the mean value of the averaged data is approached assuming the data is stationary. Theoretically, the variance should also approach a constant value as more averages are acquired. To determine if the mean value is being approximated increasingly better with more averages, the maximum variance at any position through the rotation of the tire was calculated for increasing numbers of averages. The results that were obtained are shown in Figure 5.51. As the number of averages increases to around 200, the variance converges to approximately 0.007. However, as the number of averages is further increased, the variance begins to rise. This situation is typical in experimental data. There is usually an optimum number of averages beyond which nonstationary effects in the data cause the mean and variance to drift. For this set of data, approximately 150 averages would be the preferred number to accurately approximate the mean acceleration signal at each position of the tire rotation.

5.7   SPATIAL DATA ANALYSIS

The time history, spectral and time–frequency analysis methods presented in the previous sections provided various ways to *detect* changes in dynamic measurements of a component for use in health monitoring. In order to *locate* those changes in a structural material or component, spatial data analysis techniques are required. The spatial filtering techniques described in Section 5.2.3 for vibration and wave propagation data are a means of highlighting spatial characteristics in the data. This section describes four basic techniques for analyzing these spatial characteristics to locate loading and damage phenomena in structural components. Global vibration deflection patterns are considered first. Then point-to-point analysis methods are discussed using transmissibility and other techniques. Methods for analyzing multidirectional spatial data are also described. Lastly, triangulation of propagating elastic waves is discussed to identify loading and damage in two-dimensional components.

5.7.1   Modal and Operational Deflection Patterns

Section 2.2.2 discussed modal vibration natural frequencies and deflection shapes in structural components. Each natural deflection shape defines how the component DOFs move at a given natural frequency of oscillation. When a component is subjected to operating loads, then one or more modal deflection shapes are exercised causing the DOFs to move relative to one another. Section 2.3.3 showed that the linear forced response of a structural component is the superposition of the modes of vibration. Measured vibration data can be analyzed in various ways to identify the nature of component loading and damage by characterizing modal deflection patterns in the data.

Consider the nine-DOF system model in Figure 4.6. The simulation code 'ninedofspatial.m' was used to calculate the FRFs of this system for an excitation applied at DOF 1. These FRFs can be estimated in practice using the techniques presented in Section 5.6.2. The FRF magnitudes for all nine DOFs are plotted in Figure 5.52(a). These FRFs indicate that each DOF responds with a different amplitude at the resonant frequencies of the system. In other words, the modal deflection shapes exhibit flexibility. The only exception is at the first resonant frequency; all of the DOFs have approximately the same FRF magnitudes at this frequency suggesting that the first modal deflection shape is nearly a rigid body mode of the system. The shapes in Figure 5.52(b) and (c) confirm these initial indications. These two modal deflection shapes for modes 1 and 2 were estimated from the imaginary parts of the FRFs using the experimental modal modeling technique presented in Section 2.3.3. The dotted line images are the static shapes of the system and the solid lines are the deflection shapes. The deflection shapes have been normalized so that the maximum absolute deflection is unity.

The first modal deflection shape of the system (Figure 5.52(b)) exhibits very little flexibility. Only the supports shown in Figure 4.6 deform (see Figure 2.1 for physical nature of supports being modeled). Because the neighboring DOFs do not move relative to one another at this mode, it is not anticipated that the response at this mode will be sensitive to certain physical changes in the system (e.g. damping and stiffness). The first mode will be sensitive to mass

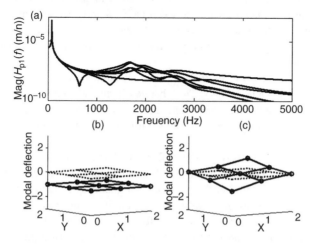

Figure 5.52   (a) Magnitude of FRFs for excitation at DOF 1 and modal deflections for (b) mode 1 and (c) mode 2 extracted from imaginary parts of FRFs

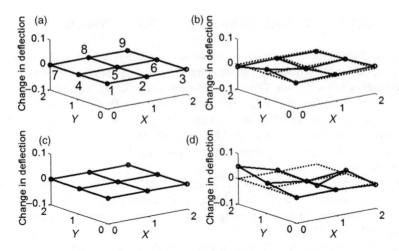

Figure 5.53 Difference in modal deflections for ((a) and (b)) modes 1 and 2 for damage between DOFs 4 and 5 and ((c) and (d)) modes 1 and 2 for damage between DOFs 6 and 9

changes due, for example, to corrosion because these changes affect the inertial response of the first mode. On the contrary, the deflection shape at mode 2 (Figure 5.52(c)) does exhibit flexibility suggesting that the response at the second mode will be sensitive to mass, damping and stiffness changes in the panel.

The code 'ninedofspatial.m' was used to reduce the stiffness acting between DOFs 4 and 5 and then recalculate the FRFs and estimate the modal deflection shapes. This stiffness reduction is meant to represent a separation or crack in the panel affecting the bending stiffness. Figure 5.53(a) and (b) show the changes in deflection shape at mode 1 and mode 2, respectively. As expected, the deflection shape for mode 1 does not change because that mode is inflexible and, therefore, insensitive to stiffness changes in the panel. The deflection shape for mode 2 does exhibit a change due to the reduction in stiffness between DOFs 4 and 5. Furthermore, the largest changes occur between DOFs 4 and 5 suggesting that the damage is located in that region of the system. This example demonstrates that damage can be detected and located using spatial analysis techniques based on the modal deflection shapes estimated from the FRFs of the system.

A similar result is obtained when the coupling stiffness is reduced between DOFs 6 and 9 (see Figure 5.53(a) for DOF labels). The changes in modal deflections for this simulated damage case are shown in Figure 5.53(c) and (d). The first mode does not indicate any changes in shape, but the second mode does indicate significant changes especially in the vicinity of DOFs 6 and 9.

A more formal way to analyze modal deflection shapes is to calculate the gradient and curvature of the shapes. For two-dimensional mode shapes, these calculations can be performed using MATLAB® 'gradient' and 'del2' commands. For simplicity, the calculations can be done along one dimension as well. The simulation code 'ninedofspatial.m' was used to calculate the gradient and curvature as a function of DOF. The results are plotted in Figure 5.54 for the gradient and Figure 5.55 for the curvature. In Figure 5.54, the stiffness change between DOFs 4 and 5 is evident in the gradient for mode 2 (b) but not in mode 1 (a) as discussed before. The stiffness change between DOFs 6 and 9 is evident in the gradients of

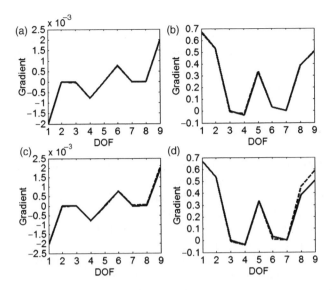

Figure 5.54 Gradient of modal deflection for ((a) and (b)) modes 1 and 2 for damage between DOFs 4 and 5 and ((c) and (d)) modes 1 and 2 for damage between DOFs 6 and 9

both modes in Figure 5.54(c) and (d). In the curvature plots (Figure 5.55), there is a less clear result because the one-dimensional curvature does not capture the two-dimensional curvature very well.

If the forced response of a vibrating system is measured without a measurement of the excitation, then the *operating deflection shapes* from the data can be extracted to help locate damage. Operating shapes are generally the sum of multiple modal deflection shapes;

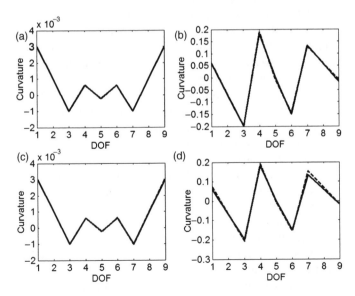

Figure 5.55 Curvature of modal deflection for ((a) and (b)) modes 1 and 2 for damage between DOFs 4 and 5 and ((c) and (d) modes 1 and 2 for damage between DOFs 6 and 9

therefore, operating shapes sometimes do not provide as clear an indication of the location of damage as do the pure modal deflection shapes. To estimate operating deflection shapes, one response measurement is treated as a reference for the others. Then the relative amplitude and phase of all DOFs are calculated with respect to that reference. This information is then used to estimate the operating shapes.

5.7.2 Transfer Path Analysis

The spatial analysis methods discussed in the previous section focus on the vibration deflection patterns that exist across the entire structural component. It is also useful to analyze how forces and responses transfer from one DOF to another. This type of analysis is called *transfer path analysis*. There are many types of transfer path analysis techniques. Each of these techniques is motivated by an underlying physical model of either propagating elastic waves or transmitted vibrations in structural components.

For example, Section 5.5.4 described one type of transfer path analysis for detecting and locating damage in a rod using longitudinal propagating waves in conjunction with spectrograms. An input force was applied at the left end of the rod and the responses at various DOFs along the rod were measured. Figure 5.40 showed that damage located at DOF 8 caused the response of DOF 8 to register the earliest change in its spectrogram. This approach is considered a *time-of-flight analysis*. The traveling wave that is reflected by the impedance change at DOF 8 affects the response at DOFs closest to the damaged DOF first because the scattered wave from the damage must travel farther to reach more distant DOFs.

Another type of transfer path analysis was considered in Section 2.2.6. *Transmissibility models* relating two response variables as a function of frequency were used to detect and locate damage by effectively 'trapping' it between the two response DOFs. The transmissibility function model revealed that local changes in the mass, damping and stiffness properties of a component affect the ratio of two response spectra more than the ratio of a response and an excitation spectra (i.e. an FRF). Equation (2.24) showed that transmissibility functions can be calculated using ratios of FRFs if the input excitation force is measured. If the input is not measured, the methods presented in Section 5.6.2 for estimating FRFs using averaging techniques can be applied to estimate transmissibility functions.

Consider the building frame structure shown in Figure 5.56(a). This frame was constructed by Farrar *et al.* (2002). The structure was intended for use in simulating the affects of earthquake seismic loads across joints. It is comprised of Unistrut® columns and aluminum floor plates. A side view of the 1.5 m tall structure is shown in Figure 5.56(b) (Allen *et al.* (2001)). The global x and y axes are in the plane of the base plate, and the z axis is in the vertical direction. The floors are labeled a, b and c from the top to the bottom floors. As shown in Figure 5.56(a), pairs of PCB 336C accelerometers (1 V/g sensitivity) were attached on the columns and plates in each corner labeled 1 through 4 in the clockwise direction. A Vibrations Test Systems, model VG100-6 100 lb electromagnetic shaker attached to the base was used to excite the frame at the location shown and in the direction indicated in Figure 5.56(a). A Techron model 5530, 1000 Watt amplifier was also used to amplify the drive signal by a factor of 20. A broadband random input with a 200 Hz cutoff frequency was used. The sample rate was 512 Hz and 8 s of acceleration data were acquired. An eight channel Hewlett Packard (HP) 35655A input module with an HP 35633A source module was used for data acquisition.

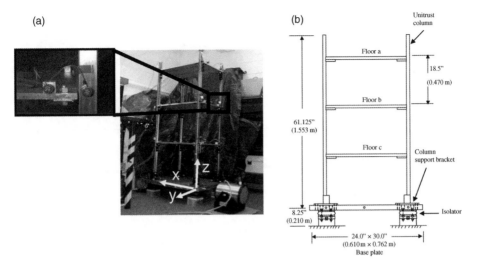

Figure 5.56 (a) Three-story building frame structure constructed by Farrar *et al.* (2002) and (b) side view of structure showing dimensions given by (Reproduced by permission of Society for Experimental Mechanics, Inc.)

Damage was simulated by loosening bolts on the brackets connecting the plates to the columns. The fully tight joint was torqued to 220 in·lbs. The three levels of damage were 15 in·lbs of torque (tight), 5 in·lbs of torque (hand tight) and no bolt. For calculating transmissibility functions, 4096 data points were segmented into blocksizes of 1024 with 50% overlap to provide seven asynchronous spectral averages (refer to Section 5.6). The transmissibility function magnitudes from 60–140 Hz for three different pairs of acceleration measurements are shown in Figure 5.57(a)–(c). The joints to which the transmissibility plots correspond from top to bottom in the figure are 2c, 2b, and 2a, which is where the joint was loosened. As discussed in Section 2.2.6, the transmissibility function across the damaged joint (c) changes much more significantly than the transmissibility across the undamaged joints for the three levels of damage (Figure 5.57(a) and (b)). There is not much change in the transmissibility at joint 2a when the bolt torque is lowered from 15 to 5 in·lbs. Note that from the perspective of spatial analysis using transmissibility functions, *locating damage is tantamount to detecting damage in different spatial pathways* from one measurement to another.

The method presented in Section 2.3.2.2 for modeling damage using virtual forces to represent changes in component mass, damping or stiffness is also considered a transfer path analysis technique. In that method, different pairs of input–output FRFs for the undamaged component are used to process FRF data to estimate the change in force due to damage. Similarly, the sensitivity modeling approach presented in Section 2.3.2.1 was also a transfer path technique. The sensitivity method also compares FRFs prior to a component becoming damaged with FRFs measured after damage occurs.

Another example of transfer path analysis is the use of transmitted sound measurements through panels to detect damage. This method uses sound measurements to identify how damage affects components that vibrate due to externally applied dynamic pressure fields. For example, an aircraft fuselage is excited during flight by aerodynamic pressures. Spacecraft during launch and reentry are also excited by dynamic pressures. When sound pressure fields on one side of a component excite the panel to vibrate, the vibration radiates sound to the other

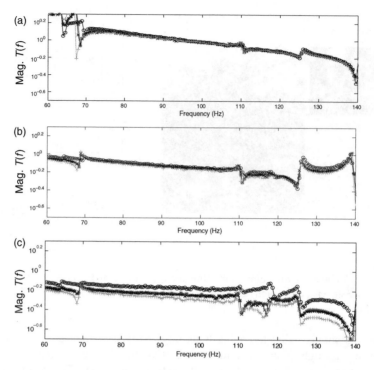

Figure 5.57 Transmissibility function magnitudes for joints (a) 2c, (b) 2b and (c) 2a for 220 in · lbs of torque (o), 15 in · lbs (x), 5 in · lbs (*) and no bolt (+)

side of the component. If the component becomes damaged, then the vibration changes as discussed in Chapter 2 subsequently causing the sound transmitted through the component to change as well. The vibration models presented in Chapter 2 are combined with acoustic analysis methods to develop this method (Jiang *et al.*, 2006).

Consider the baffled panel shown in Figure 5.58(a). The test panel used was an aluminum panel 1.2 m wide, 1.2 m long and 1.6 mm thick. Silicone sealant was used to seal all edges and joints in order to block any acoustic leakage. A uniform 9×9 grid was marked on the panel to help position the intensity probe and map the transmitted intensity on the rear of the panel. The test setup is illustrated in Figure (5.58)(c). A diffuse sound field was generated inside the reverberation room causing the test panel to vibrate. Then the vibrating panel radiated sound into the area behind the panel. The transmitted *sound intensity*[10] was measured behind the panel by using a noncontacting sensor – a B&K Intensity Probe type 3520 (see Figure 4.21(b)). The intensity probe contains two 0.5 inch B&K 4183 free-field, phase-matched microphones in a face-to-face configuration. The probe was connected to a pulse cross spectrum analyzer (B&K Pulse). During the measurement, the distance between the surface of the panel and the middle of the intensity probe was 3 in. An anechoic enclosure was used around the intensity probe to reduce the reflections from the surrounding environment at the rear of the panel.

---

[10]Intensity is the product of sound pressure and particle velocity.

(a)   (b)

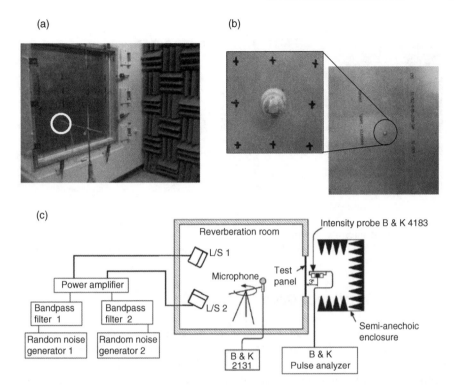

Figure 5.58   (a) Baffled aluminum panel with sound intensity probe, (b) 38 g mass attached at center of panel to simulate change in material density and (c) test setup (Reproduced by permission of SAGE Publications, Inc.)

The frequency range of interest was determined to be 300 – 7000 Hz based on data available from NASA reports from acoustic launch environments. To simulate damage due to oxidation, a small mass of 38 g weight was attached to the panel to introduce a local density change as shown in Figure 5.58(b). It is expected that there would be high transmitted intensity because an increase in mass at the attachment point causes the sound radiation to increase at this point. The transfer path-type measurement approach is to use transmitted sound intensity in the panel's resonant frequency range to identify where the density defect is located by visualizing high-amplitude points in the intensity maps.

Figure 5.59(a) shows the sound intensity map for the mass placed at location 5×5. The intensity is seen to be higher at point 5×5 than at other points. Figure 5.59(b) shows the intensity map measured when the mass was placed at location 7×3. The intensity is again higher at the point of attachment of the simulated mass damage. To examine a different form of damage due to a loss in material thickness due to corrosion, the panel was then damaged at location 7×3 as shown in Figure 5.60(a). Material in a square area of 60×60 mm was removed to create an average thickness of 0.8 mm, which was only half the thickness of the panel (1.6 mm). The intensity map in Figure 5.60(b) measured after damaging the panel again shows an increase in intensity at location 7×3.

In summary, any data analysis technique that considers how forces (including pressure) or motions (including sound particle velocity) transmit from one point to another point can be considered a transfer path analysis technique.

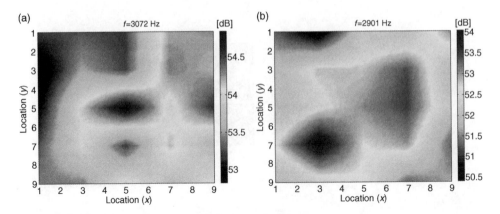

Figure 5.59 Sound intensity measurements at the rear of the panel at (a) 3072 Hz with simulated damage at location 5 × 5 and (b) 2901 Hz for damage at location 7 × 3 (Reproduced by permission of SAGE Publications, Inc.)

5.7.3 Multidirectional Data

In many applications, loads cause components to respond in multiple directions. For example, the rolling tire in Figure 1.21 responds differently in multiple directions when the bead area of the tire becomes damaged. Similarly, the panel in Figure 2.1 responds differently in multiple directions if one of the supporting struts becomes damaged. This multidirectional response data can be used to identify loading levels in various directions and also emphasize the changes due to damage in measurement data. Both vibration and wave propagation data exhibit multidirectional characteristics. For instance, the propagating waves presented in Section 2.2.9 travel outward from the source, which could be a crack growing or clapping due to fatigue loading. Propagating waves from a discrete source like

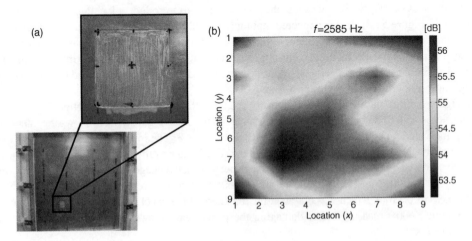

Figure 5.60 (a) 60 × 60 mm square area reduced to half thickness of panel to simulate corrosion damage and (b) sound intensity showing damage at location 7 × 3 (Reproduced by permission of SAGE Publications, Inc.)

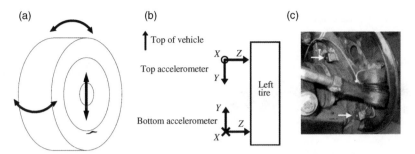

Figure 5.61   (a) Radial tire undergoing response due to asymmetric stiffness (b) measurement setup to capture response and (c) two sensors (Reprinted with permission from SAE Paper # 2006-01-1621® SAE International SAE)

this are spherical near the source and can be approximated as planar away from the source. As spherical waves travel, they excite a component in multiple directions. By measuring the response in these multiple directions, certain information about the location of the source (crack) can be determined.

As a specific example of multidirectional data analysis, consider the rolling tire shown in Figure 1.21. If this tire becomes cracked along the bead area where the tire meets the rim as shown in Figure 5.61(a), an asymmetric radial stiffness is formed at the corresponding circumferential location where the crack is located. Figure 3.7(b) illustrated one way in which this asymmetric stiffness can be modeled using a change in modulus in the radial direction. Each time the tire rotates, this asymmetric stiffness results in an imbalance excitation force that causes the tire to displace in an abnormal way vertically and to rock, twist and pitch to some extent as well. All of these motions, therefore, indicate that a crack has formed in the bead area of the tire. Suppose that two triaxial accelerometers are positioned within the suspension of the tire as shown in Figure 5.61(b). The physical locations of these sensors are pictured in Figure 5.61(c). Each sensor measures acceleration in the three directions indicated. The responses in each direction can be processed separately; however, if the crack is small, then the individual responses to the effects of asymmetric stiffness will be small as well.

It is desirable to process the six directional responses simultaneously using spatial analysis techniques to highlight the presence of small cracks in the tire. There are two basic techniques for combining the multidirectional response data. First, the measured response amplitudes in the various directions of motion can be *added*. Addition represents an OR operation; that is if only one of the motions is significant, then the sum of all motions is significant. Second, the measured response amplitudes can be *multiplied*. Multiplication represents an AND operation, that is all motions must be significant to generate a significant product. In some applications, OR operations on the data are preferred and in other applications, AND operations are preferred to combine multidirectional data.

Figure 5.62 illustrates a numerical example of these two different methods for combining multidirectional measurement data. Figure 5.62(a) shows a 2-unit amplitude event that is added to the six Gaussian noise signals in Figure 5.62(b) to produce the simulated measurement signals in Figure 5.62(c). The noise signals represent the underlying response of a tire prior to becoming damaged. The 2-unit amplitude event represents the passage of an imbalance force in one rotational position of the tire due to cracking. Note that this event

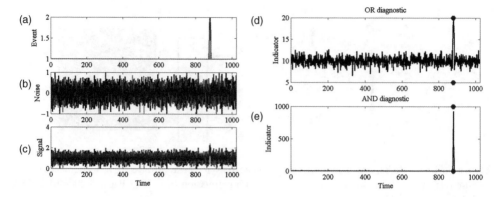

Figure 5.62 (a) 2-unit amplitude event at time point 880, (b) 6 channels of unity variance Gaussian random data, (c) sum of (a) and (b) across all channels of data, (d) summing of signals in (c) and (e) multiplication of signals in (c)

is clouded by the background response. Figure 5.62(d) is the sum of all six simulated measurement signals and Figure 5.62(e) is the product of these signals. In both of these combinations, the event is more clearly identified than in the individual signals. The AND operation provides an especially clear indication of the event because it only indicates events that are common to all of the six simulated pieces of data such as the event in Figure 5.62(a).

Next, consider the actual rolling tire shown in Figure 1.21. Multidirectional acceleration response measurements for the tire were acquired prior to damage and then with a crack present in the bead area (see Figure 5.61(a)). Data were acquired continuously as the tire-rolling test proceeded. The DFTs, $X_i(m\Delta f)$, of the response time histories collected for each rotation were then calculated (refer to Section 5.5.1). Synchronous averaging was then performed with N_{avg} rotations of data as described in Section 5.6.3 to produce averaged DFTs for each channel of response data:

$$\frac{1}{N_{\text{avg}}} \sum_{i=1}^{N_{\text{avg}}} X_i(m\Delta f). \tag{5.39}$$

Then the magnitudes of the averaged DFTs were calculated. The magnitudes in one frequency range, m_j from m_{1j} to m_{2j}, were then summed representing an OR operation:

$$\sum_{m_{1j}}^{m_{2j}} \left| \frac{1}{N_{\text{avg}}} \sum_{i=1}^{N_{\text{avg}}} X_i(m_j \Delta f) \right|. \tag{5.40}$$

This operation emphasizes events that occur consistently in a given channel of data in one frequency range. The summing operation (OR) was used instead of the multiplication operation (AND) because the asymmetric stiffness in the tire did not cause measurable response amplitudes at all frequencies in a given range. The response is usually tonal occurring more strongly at certain harmonic frequencies of the tire rotations. Then the summed magnitudes were combined by multiplying across many frequency ranges:

$$\prod_j \sum_{m_{1j}}^{m_{2j}} \left| \frac{1}{N_{avg}} \sum_{i=1}^{N_{avg}} X_i(m_j \Delta f) \right| \qquad (5.41)$$

where the $\prod$ symbol denotes a product. The goal in this example was to combine the data acquired for all six of the measurement channels pictured in Figure 5.61(c). This objective was accomplished by multiplying the results of Equation (5.41) for each of the measurement channels ($X1$, $Y1$, $Z1$, $X2$, $Y2$ etc.):

$$\prod_{\substack{X1,Y1,Z1 \\ X2,Y2,\dots}} \prod_j \sum_{m_{1j}}^{m_{2j}} \left| \frac{1}{N_{avg}} \sum_{i=1}^{N_{avg}} X_i(m_j \Delta f) \right|. \qquad (5.42)$$

This expression represents an AND operation that emphasizes events that simultaneously affect all of the response motions illustrated in Figure 5.61(a).

The results of performing this series of operations on the acceleration data in sequential rotations of the tire over a period of several days of testing are shown in Figure 5.63(a). The damage index given by Equation (5.42) is plotted as a function of data file number representing the test time. There is one event highlighted by the damage index. This event occurs when the tire becomes cracked during the test. This event is not sustained because after the tire cracks, the residual stress in the zone around the crack is alleviated as the tire continues to rotate. As the stress is alleviated, the asymmetric stiffness due to the crack diminishes. Figure 5.63(b) shows a second result for a different tire where only one triaxial accelerometer was used. Only three response channels (X, Y and Z) were combined to form this plot. This tire experienced two cracks at around data file 110 and data file 140. An example of one of these cracks or splits in the bead area of the tire is pictured in Figure 3.7(a).

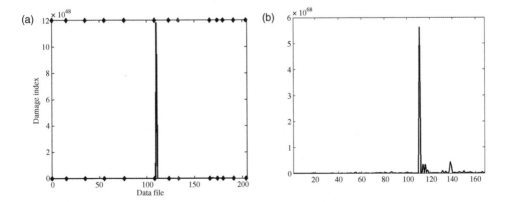

Figure 5.63 Damage index from Equation (5.42) for tire with (a) one and (b) two cracks

5.7.4 Triangulation

The previous section deals with measurements in multiple directions. In health-monitoring applications, measurements at multiple *locations* are also often acquired. These measurements provide multiple perspectives on the component response that can aid in locating sources due to loads or damage. *The process of locating some kind of source using distributed measurements is called triangulation.* For example, when propagating elastic waves travel from one location to another location in a plane, the point from which the waves originate can be located by identifying two directions (or distances) to the source from two different locations on a component. The intersection of the two different directions or radii determines the source of the waves.

Consider the plate shown in Figure 5.64(a). The plate is 1 m × 1 m with a 1 mm thickness. It is made from aluminum with $E = 7.1\text{e}10\,\text{N/m}^2$, $\rho = 2.7\,\text{e}3\,\text{kg/m}^3$ and $v = .33$. In order to demonstrate how triangulation works, this plate is modeled with four-noded thin plate finite elements. One hundred elements in a 10 × 10 grid are used along with 121 nodes. Each node possesses three DOFs (vertical translation, rotation in x direction and rotation in y direction). As in Section 2.2.8.4, a three-cycle sinusoidal pulse force is used to excite the plate. The 100 N amplitude force is applied at node 1 in the upper-left-hand corner. The frequency bandwidth of the force is from 60 to 180 Hz (see Figure 2.29 for an example of the spectrum of a pulse waveform).

The code 'platefemsim.m' was used to simulate the response of the plate at all 121 nodes. All of the vertical (transverse) nodal displacements are shown in Figure 5.65. The abscissa in each plot is time in seconds and the ordinate is displacement ($\times 10^4$ m). The plot is similar to the transverse wave propagation plots for a beam shown previously in Figure 2.30. Like the transverse waves in a beam, the plate waves are also flexural involving bending leading to dispersion as lower frequencies travel at different speeds than higher frequencies. It is clear from the plots that the wave travels from the forced node (node 1, upper-left-hand corner) to all of the nodes below and to the right of the forced node. The symmetry in waveforms in the lower-left-hand and upper-right-hand corner nodes is also evident due to the square dimensions of the plate. The phase velocity of this kind of flexural wave at frequency, ω, is given by:

$$v_{\text{flex},p} = \sqrt[4]{\frac{D\omega^2}{\rho h}}, \tag{5.43}$$

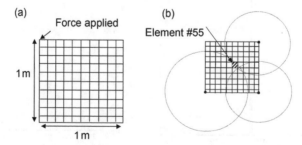

Figure 5.64 (a) Two-dimensional FEM of plate and (b) triangulation of change in modulus in element 55 using data in Figure 5.67

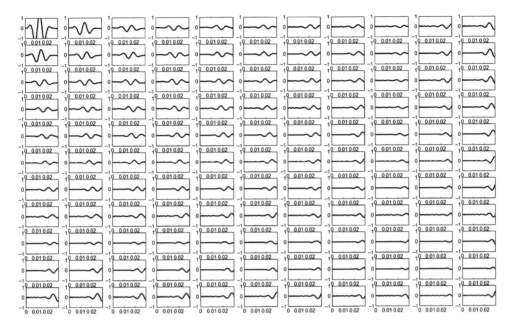

Figure 5.65 121 simulated nodal displacement responses of free-free plate subject to sinusoidal pulse at node 1 (displacements shown $\times 10^4$ and time in s)

and the corresponding group velocity is obtained from $d\omega/dk$ (as discussed in Chapter 2):

$$v_{\text{flex},g} = 2v_{\text{flex},p}. \tag{5.44}$$

Based on the simulation, the group velocity of the waveform is found to be 71 m/s, which is approximately the same as the 68 m/s result calculated from Equation (5.44) for a pulse applied at node 1 with a 119 Hz center frequency.

Next, the modulus in element 10 in the lower-left-hand corner of the plate is reduced by 50 %. The difference between the displacement responses before and after the local modulus change is plotted in Figure 5.66. As in the case of the longitudinal wave in a rod that is cracked (refer to Section 2.2.8.2), the first occurrence of a difference in the displacement is in the neighborhood of the modulus change in the plate. This result is obtained because the wave time of flight causes differences in the response in the vicinity of the damage before these differences propagate outward as illustrated in Figure 5.66. If displacement data for all nodes is available, then these differences clearly indicate the location of the localized change in modulus.

The approach just illustrated is a form of triangulation; however, a more practical example is illustrated in Figure 5.67. Only 4 of the 121 nodal displacement differences are shown. In this scenario, the modulus in element 55 is reduced by 50 %. With only these four measurements, it is not possible to pinpoint the location of the damage so easily. With limited numbers of measurements, triangulation can be accomplished by calculating the distance from each of the measurement locations to the location of the source that is transmitting the waveform.

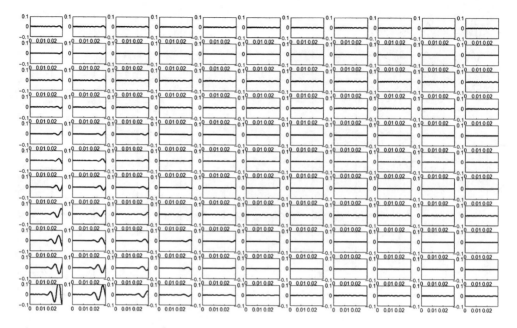

Figure 5.66 Difference between all 121 nodal displacement responses of plate with and without change in modulus in element 10 (lower left-hand corner)

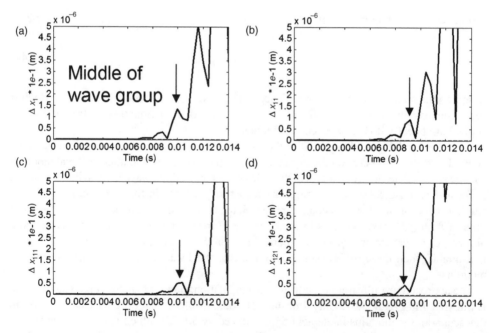

Figure 5.67 Magnitude of baseline (- - -) and difference (___) between nodal displacement responses of plate with and without change in modulus in element 55 (see Figure 5.64(b)) at (a) node 1, (b) node 11, (c) node 111 and (d) node 121

For example, Figure 5.67(b)–(d) indicate that the arrivals of the displacement difference wave group at these three locations are at 9.2, 11.8, and 8.8 ms, respectively. Given that the group velocity of the wave pulse is 68 m/s, the distances from the source to each of these three locations are 0.62, 0.80 and 0.60 m. These are sketched out as shown in Figure 5.64(b). The radii of each circle indicate the distance to the source from the respective nodal measurement location. Triangulation is then performed by identifying the intersection of these circles. This intersection is the approximate location of the source, which in this example is the change in modulus of element 55 in the plate model. There is some error in the estimated location of the simulated damage because the grid of elements is relatively coarse to begin with and because it is difficult to identify the first peak in the measured displacement differences shown in Figure 5.67(b)–(d). There are many different techniques for improving the accuracy of the estimated source location. Appendix B provides a summary of literature dealing with wave propagation-based methods for source localization.

Other examples of triangulation are described later in the book. For example, beamforming filters (described in Section 5.2.3) can be used to identify the direction of arrival of waves due to loads or damage in a component. Then the intersection of two or more directions of arrival with respect to different measurement reference locations can be found to identify the location of the source.

5.8 FEATURE EXTRACTION

Loads identification and damage identification in health monitoring are accomplished by analyzing measurement data using the various methods described in the previous sections. When analyzing raw data, important parameters are extracted, called *features*, from which conclusions can be drawn about the nature of loading and damage in a component. Farrar *et al.* (1999) describe health monitoring in the context of feature pattern recognition. Many examples of features have already been introduced in the previous sections and chapters. For instance, the example in Section 5.7.3 involving the rolling tire cracked around the bead area used features to combine multidirectional response data in several frequency ranges in various response directions. The features in that example were derived by rotationally averaging the data, summing signal amplitudes over each frequency range, and then multiplying the amplitudes over several frequency ranges and measurements. Also, features were used throughout Section 5.4.1 to analyze the basic statistics in measured time histories. Features instead of raw data must be extracted and stored in practice; otherwise, there would be excessive memory requirements and too much raw data for use later in diagnosing loads and damage.

There are two basic methods of feature extraction. *Model-based* methods use models of some kind to estimate parameters, which are then processed and used as features. The embedded sensitivity model presented in Section 2.3.2.1 is a source of features, which help to quantify component damage as discussed in that section. The features in that case are the estimated changes in mass, damping or stiffness in certain regions of a component. Section 2.3.2.2 introduced virtual force models, which can be used to extract virtual forces that quantify damage in components. The FRF models in Section 2.3.2 were also sources of model-based features to estimate loads applied to components. Other examples include the direct parameter time domain models introduced in Section 2.3.1.1 and the discrete time models introduced in Section 2.3.1.3. These discrete time (autoregressive) models were applied in

Section 5.4.2 to estimate model coefficients, which served as features for damage detection. In general, model-based features provide more physically meaningful information about loading and damage. However, model-based features are most suitable in applications where the user has *a priori* knowledge of the loading and damage. For instance, the embedded sensitivity method is most useful when the location of damage is known beforehand so that the proper measurements can be taken to quantify the damage level.

Signal-based methods extract features directly from signals without applying any sort of deterministic component model to the data beforehand. For example, the statistical data analysis techniques presented in Section 5.4.1 extract features directly from signals using formulas like those listed in Table 5.1. The spectrogram analysis method presented in Section 5.5.4 to locate damage in the rod using time-of-flight information extracted signal-based features (time–frequency maps). The use of raw DFTs computed in Section 5.5.1 to characterize damage due to linear and nonlinear stiffness changes is also a signal-based method of feature extraction. The spectral coefficients are the features in this case. In general, signal-based features provide less physically meaningful information about the component loading and damage. On the contrary, signal-based features are useful when the loading and damage are not fully comprehended before measurements are taken. For instance, DFT features were useful in Section 5.5.1 for detecting two types of damage; knowledge of the type of damage was not required beforehand in order to detect the damage simulated using linear and nonlinear stiffness changes. Other examples of signal-based feature extraction methods include *principal components analysis* and *multilayer perceptron* neural networks (Bishop, 1995).

Examples of model-based feature exraction for both loads and damage identification are provided in the sections below. The example in Section 5.7.3 involving the detection of tire bead area damage provides a demonstration of signal-based feature extraction. Issues in feature extraction relating to the dimensionality of features and statistical models of features are also addressed in the following sections.

5.8.1 Model-Based Feature Extraction (Damage)

Transmissibility function models can be used to extract features that are useful in locating damage for the reasons discussed in Section 2.2.6. For a given model, there are an infinite number of ways to extract features. Assume that an excitation with spectrum, $F_q(\omega)$, is applied to a structural component. The transmissibility between two response measurements, $X_{p1}(\omega)$ and $X_{p2}(\omega)$, can be computed in either of the following ways:

$$T_{p_1, p_2(q)}(\omega) = \frac{X_{p1}(\omega)}{X_{p2}(\omega)} = \frac{H_{p1,q}(\omega)}{H_{p2,q}(\omega)}, \tag{5.45}$$

where the first calculation is in terms of the response spectra and the second calculation is in terms of the FRFs if the excitation is measured. Note that this ratio is dependent on whether $X_{p1}(\omega)$ or $X_{p2}(\omega)$ is in the numerator (denominator). The *transmissibility power* is computed to avoid this dependence on the ordering of the two response measurements:

$$TP_{p1,p2(q)}(\omega) = \| \log T_{p_1, p_2(q)}(\omega) \|, \tag{5.46}$$

where the bars denote the magnitude of the natural logarithm of the transmissibility. To detect and locate damage, the *damage power* is then formulated by taking the difference between a baseline estimate of *TP* and the current estimate of *TP* for comparison:

$$DP_{p1,p2(q)}(\omega) = \frac{TP_{p1,p2(q)}(\omega)|_{current} - TP_{p1,p2(q)}(\omega)|_{baseline}}{TP_{p1,p2(q)}(\omega)|_{baseline}} \qquad (5.47)$$

If *DP* is zero, then there is no difference between the baseline and current measurements of *TP* leading to the conclusion that the component is undamaged. If *DP* is large in a narrow or broad frequency range, then the conclusion is that the component is damaged. Although it can be advantageous to consider the entire vector of values given in Equation (5.47), it is more convenient to reduce the vector into one number, the *transmissibility damage feature*, by summing the damage power magnitude over a given frequency range and taking the absolute value as follows:

$$DF_{p_1,p_2(q)} = |\sum_{\omega=\omega_1}^{\omega=\omega_2} DP_{p_1,p_2(q)}(\omega)|. \qquad (5.48)$$

It is understood in this equation that $DP(\omega)$ is actually a discrete spectrum, $DP(m\Delta\omega)$, as discussed earlier so the summation only occurs over these discrete frequency values.

As one application of this damage feature, consider the three-story building frame discussed in Section 5.7.2. The transmissibility functions were computed across the joints of this frame within the three stories, and then the damage powers were computed. Equation (5.48) was then applied over the frequency range from 0–255 Hz. In order to check the baseline condition to determine its level relative to the various damage levels, three sets of baseline data were acquired. The first two baseline datasets were used to calculate the baseline transmissibility powers given in Equation (5.47). The third baseline dataset was used to calculate the damage feature (*DF*) to characterize the baseline (or healthy) condition. The results of these calculations are listed in Table 5.4 when damage is imposed at joint 2a in three stages as described previously. The first column indicates the joint across which the transmissibility damage feature was computed (e.g. 2a indicates the second joint on floor a). The second column gives the direction of transmission across the joint.

The third column lists the values of *DF* when all bolts are fully tight. All of these features are much below 1 with the exception of the *DF* for joint 2b. This result could indicate an unusually large amount of measurement noise at that joint or perhaps a defective joint with bent connectors or bolts. This information is important because it suggests that the *DF* for joint 2b should be treated with caution in the data analysis process. The fourth column gives the damage features for the 15 in·lbs torque condition at joint 2a. The first three rows of the table indicate that all three of the transmissibility damage features in the three coordinate directions change by at least one order of magnitude. The largest changes occur in the *X* and *Z* directions. Other joints also register order of magnitude changes including joints 1a, 4a, 1b, 4b, 2c, 1c, 3c and 4c. However, the resultant damage features in these undamaged joints are less than 1. It can, therefore, be concluded that the *X* and *Z* direction transmissibility damage features are best for locating damage. Also, note that these damage features increase as the damage level increases in the fifth and sixth columns of Table 5.4. In this example, the transmissibility

Table 5.4 Transmissibility damage features for 0–255 Hz for baseline and damage cases for three-story frame (Reproduced by permission of Timothy Johnson, Purdue University, 2002)

Joint	Direction	No damage	Tight	Hand tight	No bolts
2a	X−	0.0363	2.34	2.6	2.91
2a	Y+	0.0242	0.119	0.146	0.218
2a	Z+	0.0917	8.11	8.71	10.8
1a	X−	0.0287	0.123	0.121	0.0441
3a	X+	0.025	0.0241	0.023	0.0582
4a	X+	0.00763	0.115	0.116	0.174
2b	X−	0.197	0.31	0.309	0.599
1b	X−	0.0244	0.108	0.108	0.14
3b	X+	0.0146	0.0608	0.0611	0.143
4b	X+	0.0426	0.22	0.231	0.19
4b	Y−	0.0701	0.343	0.272	0.105
2c	X−	0.0254	0.263	0.268	0.275
1c	X−	0.0394	0.296	0.298	0.364
3c	X+	0.00502	0.117	0.131	0.173
4c	X+	0.0719	0.135	0.0997	0.137

damage feature has the capability of detecting, locating and quantifying (on a relative scale) the damage.

5.8.2 Model-Based Feature Extraction (Loading)

The previous section identified damage using damage features extracted from transmissibility functions. Features can also be extracted from models to identify loading. For example, consider the damaged stabilizer bar in an automotive suspension system that was pictured in Figure 2.21(b) and is shown again with a notch (circled) and instrumented with accelerometer sensors (squares) in Figure 5.68(a). The accelerometers shown are triaxial making it possible to measure the motions in all three directions on either side of the stabilizer bar. This acceleration data can be used to characterize the forces that load the bar causing the damage to grow at the notch through fatigue crack growth. To this end, the suspension was excited using the two-post shaker test rig shown in Figure 5.68(b). The shakers were driven in the vertical direction from 0–15 Hz with a sweep rate of 0.025 Hz/s. An analog low-pass filter was used to attenuate data beyond 100 Hz and the data was sampled at 600 Hz. In order to identify the internal loads in the stabilizer bar, the acceleration data was integrated numerically. To reduce errors due to the initial conditions, a first-order function was fit to the acceleration data and then subtracted from the data to remove any trends that might emerge after the integration. Then the trapezium rule for integration was used ('cumtrapz' command in MATLAB®), and a digital high-pass filter was applied to remove any trends below 2 Hz at the end of the integration process.

Restoring force projection models were used to extract features for identifying the internal loads in the stabilizer bar as the notch in the bar grew a crack and the crack length subsequently increased during testing (see Sections 2.3.1.2 and 5.4.2). One cycle of data at 7.9 Hz was

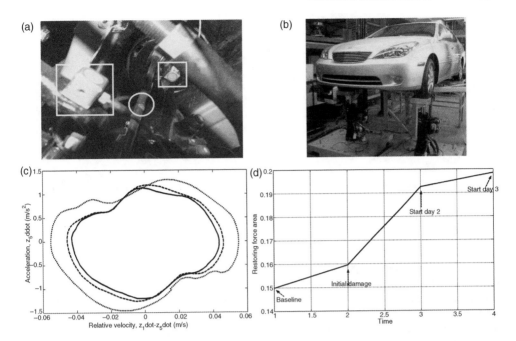

Figure 5.68   (a) Stabilizer bar link with notch, (b) two-post shaker to simulate vehicle response, (c) damping restoring force projection for stabilizer bar damage in suspension at 7.9 Hz with (__) baseline, (- - -) initial cut and (. . .) growth in crack within cut and (d) change in restoring force area damage feature over course of test

plotted as shown in Figure 5.68(c). The acceleration at the rightmost sensor in the transverse direction versus the relative velocity across the bar in the transverse direction is plotted in this figure. The three different curves indicate the baseline condition, (__) and two different levels of damage in increasing order, (- - -) and (. . .). The areas inside these curves represent the internal damping force acting across the bar causing the damage in the notch to grow. Note that the damping force increases as the notch damage grows. Although these curves are themselves a feature for loads identification, it is more convenient to extract one number to serve as a load feature. In this case, the area inside the restoring force projection is used as the load feature. The method for calculating this area was discussed in Section 5.4.2. The MATLAB® commands 'convull' and 'trapz' can be used to compute this area although the curves must be simply connected (do not exhibit crossings). The results of this area calculation at various times during the testing of the suspension are plotted in Figure 5.68(d). As expected, the area increases as the test progresses indicating that the internal loading due to dissipation of energy across the notch in the strut is increasing.

The change in restoring force projections shown in Figure 5.68(c) is linear in nature; in other words, the restoring force curves simply expand as the loading changes. This same methodology can be applied when the loading changes in a nonlinear manner. For example, consider the shock module clevis with a notch pictured in Figure 3.2(a). The strut was tested over a period of days to examine its durability. It was subjected to an accelerated durability profile used by the automotive industry to simulate driving loads.

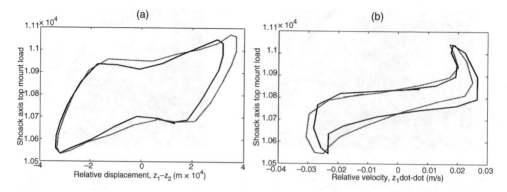

Figure 5.69 (a) Stiffness and (b) damping restoring forces for baseline (___) and intermediate (...) conditions showing increase in stiffness loading (area changes) and damping (slope changes)

Then at intermediate periods a sine sweep was used to excite the strut from 0–35 Hz with a 0.058 Hz/s sweep rate in order to generate the restoring force projections. A sinusoidal swept excitation was used because restoring force projections are much more useful in characterizing the loading carried by the clevis for a sinusoidal input. Accelerometers on either side of the strut were used to measure the accelerations. A load cell was also used to measure the load input to the shock module. The restoring forces were then generated at various damage levels over the course of testing. These restoring forces for stiffness (a) and damping (b) are shown in Figure 5.69. Note that the area of the stiffness restoring force increases over the course of the test as the damage increases. The damping restoring force also exhibits an increase in the damping because the slope of the damage curve (...) is larger than the slope for the baseline curve (___) in Figure 5.69(b). The nonlinear nature of the restoring force curve also changes as the damage increases. These changes are due to the nonlinear nature of the crack, which 'breathes' as the strut is forced to respond resulting in crack growth and eventual failure of the clevis.

This example demonstrates that *multiple* features are sometimes needed to fully character-ize the change in loading (or damage) in a component. In this scenario, both the stiffness restoring force and damping restoring force were needed to fully characterize the change in loading with increase in damage level.

5.8.3 Dimensionality of Feature Sets

The previous example involving loads identification in an automotive shock clevis with a crack demonstrated that multiple features are sometimes needed to fully characterize loading. The same is true for damage. For example, the modal curvatures shown in Figure 5.55 contained nine features – one for each measurement DOF in the panel. Without all of the nine features, it is not possible to locate the damage in a general case. Multiple features were also required to distinguish between oxidation damage and temperature changes in the nine-DOF model using measured FRFs. Unless the FRF magnitude is measured at several frequencies across the range shown in Figure 4.7 for simulated oxidation damage at DOF 7, local changes in the FRF due to this simulated damage cannot be distinguished from the more global changes shown in Figure 4.10(a) due to the change in temperature.

In other examples, a single feature is adequate for the purpose of health monitoring. For example, the feature calculated for the rolling tire was capable of detecting cracks in the bead area of the tire in Section 5.7.3. Likewise, the cumulative Euclidean distance between the return maps for the baseline and damaged component response in Figure 5.29 was sufficient to detect and quantify damage in the two-DOF system.

When selecting features for loads and damage identification, the health-monitoring objectives must first be considered. More features are needed in some applications and fewer features are required in other applications. Whenever possible, it is better to choose features that have a *lower dimension*. In other words, it is better from a measurement point of view to select features that consist of one or at most a few numbers as opposed to long vectors of numbers. The reason that *fewer features are preferred is because less measurement data is needed to identify loads and damage with confidence using fewer features*. More specifically, D^N is the number of data points needed to classify all possible feature values if a feature vector containing N values is used, and each feature is divided into D divisions (Bishop, 1995). This relationship between the number of data points required and the number of features is often called the *curse of dimensionality*.

The code 'ninedoffeature.m' was used to generate FRF measurements for the nine-DOF system model in Figure 4.6 for various simulated damage cases in order to demonstrate the advantages and disadvantages of higher dimensional features. In each of the nine simulated damage cases, the mass at one of the DOFs in the nine-DOF system model was reduced by 10 %. This reduction in mass simulates corrosion damage. Figure 5.70(a) shows one of the FRFs, $H_{11}(\omega)$, with 10 % correlated Gaussian noise added to the FRF. A feature vector with five elements was then extracted by, first, summing the FRF magnitudes over 500 Hz frequency bands from 0 to 2500 Hz. This summation can be thought of as an integration, or

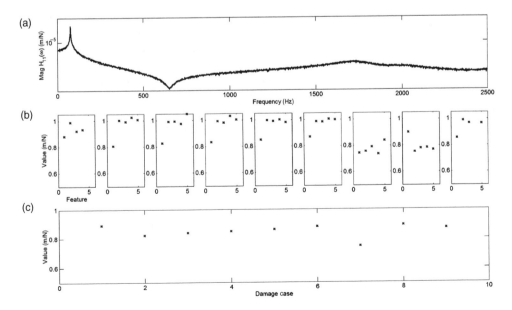

Figure 5.70 (a) $H_{11}(\omega)$ from nine-DOF system model, (b) set of five features in 500 Hz frequency bands for nine different damage cases and (c) single damage feature for nine damage cases

area under the FRF magnitude curve. Second, the summed values were divided by the values obtained for the healthy condition. These feature values are plotted in Figure 5.70(b) for the nine different damage cases. Several of the feature vectors are similar to one another such as for damage cases 2 through 6; however, others are quite different such as for damage cases 1, 8 and 9. In a practical application, all of these damage cases could not be measured before implementing a health-monitoring damage detection algorithm. Only a subset of these cases could be measured in addition to the case when there is no damage. This subset of data is referred to as the *training data*. Training data is used in health monitoring to establish the normal and abnormal condition for a component. *Thresholds* are then set up to differentiate between the normal and abnormal condition of the component. Statistical models used to differentiate between these two conditions are described in the next section.

Suppose that only damage cases 1 and 8 are measured. If these two sets of feature vectors were used to detect corrosion damage in general, the two features would fail to do so if the corrosion occurs at DOFs 2–6 because of the significant differences in the feature vectors for damage located at DOFs 2–6 and damage located at DOFs 1 and 8. It is concluded that more training data is required to define the abnormal (damaged) condition of the component. In contrast, consider the one-dimensional feature plotted in Figure 5.70(c) for the nine different damage cases. This feature was calculated by summing the magnitude of the FRF over the entire frequency band from 0–2500 Hz and dividing by the summation for the healthy condition[11]. Note that for all nine damage cases, the one-dimensional feature is lower than 1. Theoretically, only one damage case would need to be measured in order to define abnormal values for this low-dimensional feature in arbitrary damage cases if measurement variations were neglected (see Section 5.9).

Of course, the one-dimensional feature in Figure 5.70(c) is not helpful in locating the damage within the nine–DOF component. There is no pattern that relates the feature value to the location of the mass loss in the model. On the contrary, the set of five features plotted in (b) can be used to locate the damage. One simple way of locating the damage is to calculate a new one-dimensional feature equal to the dot product of the feature vectors for the training data with the measured feature vectors in data acquired after damage is thought to have occurred. The simulation code 'ninedoffeature.m' performs these dot product operations and plots the results in Figure 5.71(a)–(j). Each subplot presents the results of the dot product using the feature vector for that condition. For example, subplot (a) is the dot product of the feature vector for the undamaged condition (condition 1) with the feature vectors for the nine damage conditions. Likewise, subplot (j) is the dot product of the feature vector for damaged condition 10 with mass loss at DOF 10 with the feature vectors for the nine damage training vectors. In most cases, the results indicate the correct location. There are several cases where multiple damage cases are identified such as the result in subplot (j), which indicates damage could be located at either DOF 10 or DOF 5.

This example demonstrates that lower dimensional features are preferred for damage detection in health monitoring because fewer data points are needed to define the values of these features in normal and abnormal conditions. However, longer feature vectors are sometimes needed to characterize damage either by locating it or by classifying the type of damage (e.g. change in mass, stiffness etc.). The same is true in loads identification.

---

[11]This process of normalization was also carried out in Section 5.8.1 for transmissibility functions and is common in health-monitoring algorithms. By normalizing the feature with respect to the normal condition of the component, any deviation from 1 indicates abnormality (damage).

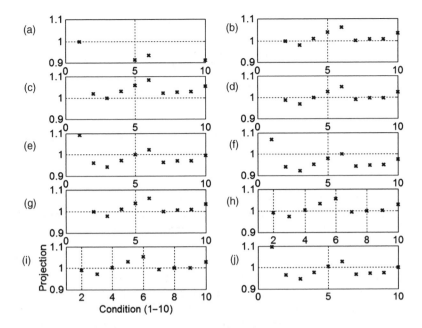

Figure 5.71 Projections to locate change in mass simulating corrosion in nine-DOF model for conditions 1–10 (a) healthy through (j) mass change at DOF 9

5.8.4 Statistical Models for Features

The previous section concludes that fewer features are usually preferred because less data is required for detecting damage with confidence. In addition to acquiring enough data to make a physical determination of whether features correspond to normal or abnormal conditions, it is also necessary to acquire enough data to make this determination even when there are deterministic or random variations in the measured data. To account for random variations in data, statistical models of features are used. Deterministic variations in measurement data are addressed in Section 5.9.

First, consider a feature, x, which can be treated like a *random variable* that is normally distributed about a mean μ with variance σ^2. The probability density function for this random variable is described by Equation (2.91). The probability distribution is the integral of the density as given in Equation (2.92). Instead of considering the random variable, x, it is more convenient to apply the coordinate transformation, $z = (x - \mu)/\sigma$, to produce the standard normal distribution. This transformation process is illustrated in Figure 5.72. Tables for the standard normal distribution such as found in Dougherty (1990) can then be used to calculate the probability that x lies in certain intervals (i.e. confidence intervals). For example, the probability that x lies in the range x_1 to x_2, $x_2 > x_1$, is,

$$P\left(\frac{x_2 - \mu}{\sigma}\right) - P\left(\frac{x_1 - \mu}{\sigma}\right), \tag{5.49}$$

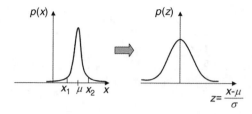

Figure 5.72 Transformation from normal distribution to standard normal distribution

where the values of $P(\cdot)$ can be found in tables for the standard normal distribution. For example, the probability that x will fall inside the interval $\mu \pm 2\sigma$ is 95 %. The probability that x will fall inside the interval $\mu \pm 3\sigma$ is 99 %. This formula is useful in both loads identification and damage identification.

In damage detection, the most common approach is to compare a healthy feature extracted from baseline measurements with the same feature measured at a later time after damage has occurred. This approach has been used in examples presented in the previous two chapters. The question that often arises when applying this approach is how to detect *statistically significant* changes in the feature. *Hypothesis testing* is the field of statistics concerned with choosing between two (or more) different conjectures about a population of features using a *test statistic*. In health monitoring, the two conjectures are usually that the component is either (0) healthy or (1) damaged. The first conjecture is called the *null hypothesis* and is denoted by H_0. The second conjecture is called the *alternative hypothesis* and is denoted by H_1. Common test statistics used in health monitoring are the sample mean and sample variance of any feature such as those presented in the earlier sections of this chapter.

For example, consider the nine-DOF system model examined in Section 5.8.3. The feature used in that section for damage detection was the cumulative sum of the magnitude of $H_{11}(\omega)$ over the frequency range from 0–2500 Hz normalized by the baseline value. Two hundred ninety nine simulations were run using the code 'ninedoffeature.m' to calculate $H_{11}(\omega)$. As in the previous section, 10 % correlated random noise with a normal distribution was added to $H_{11}(\omega)$ to simulate measurement variability. The resulting probability density from these simulations is shown in Figure 5.73(a). The sample mean and sample variance were then calculated and used to transform this probability density to the standard density function shown in Figure 5.73(b). Next, suppose the mass of DOF 4 is reduced by a factor of 10 % to simulate corrosion damage. The new probability density function for the damaged case (__) is shown in Figure 5.74 along with the density for the healthy case (...). The calculated sample means and standard deviations of these two density functions are $\bar{\mu}_1 = 1.02$, $\bar{\mu}_2 = 0.81$, $\bar{\sigma}_1 = 0.034$ and $\bar{\sigma}_2 = 0.018$. Bars over the statistics are used to indicate that these are only *estimates* of the means and deviations. Note that the sample mean, $\bar{\mu}$, is itself a random variable and has a mean and variance equal to μ and $\bar{\sigma}^2/N$, where N is the number of samples. It is evident from these values that the mean and variance have changed. However, the question of whether or not a statistically meaningful change has occurred must be answered.

The hypothesis test for this case is a *one-tailed test* because all that needs to be decided is whether or not $\bar{\mu}_2$ is actually less than $\bar{\mu}_1$ in a statistical sense. The sample mean can be used

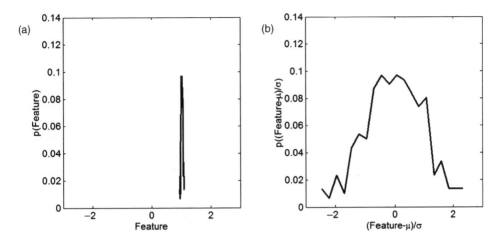

Figure 5.73 (a) Probability density of feature in healthy condition and (b) standardized density after applying coordinate transformation for normal random variable

because it is an unbiased estimate of the actual mean. The one-tailed hypothesis test to decide if damage has occurred is expressed as follows:

$$H_0 : \qquad \mu = 1.02$$
$$H_1 : \qquad \mu < 1.02. \tag{5.50}$$

If the sample mean lies in the *critical region*, defined by $\bar{\mu} < C$, then the null hypothesis will be rejected and damage will be detected. C is often called the *threshold for detection*.

The next step is to calculate the value of C. C is calculated by weighing the risk of missing damage in an unhealthy component (Type II error) with the risk of detecting damage when the

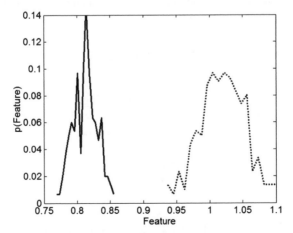

Figure 5.74 Probability density of feature in healthy condition (. . .) and damaged condition (___) with 10 % mass loss at DOF 4 in nine-DOF model

component is healthy (Type I error). These two types of detection error are called *false negative* and *false positive* alarms. The false positive alarms are found with the following probability expression:

$$P(\bar{\mu} \leq C; H_0) = \frac{\% \text{ False positives}}{100},$$
(5.51)

where the expression in parentheses indicates that the null hypothesis is mistakenly rejected (damage is detected) even though the component is healthy because the sample mean lies in the critical region. Likewise, the probability of obtaining false negative alarms is found from the following expression:

$$P(\bar{\mu} > C; H_1) = \frac{\% \text{ False negatives}}{100}.$$
(5.52)

Large % False positives are not usually acceptable because they lead to removal of healthy components. On the contrary, large % False negatives can be devastating leading to catastrophic failure of the component in operation.

Suppose that the desired % False positives is 1%. The sample mean test statistic can then be transformed into a standard normal random variable to produce the condition for calculating the threshold, C. When making this transformation, the variance, $\bar{\sigma}_1^2/N$, of the sample mean rather than just the variance of the feature, $\bar{\sigma}_1$, must be used, otherwise the test will overestimate how much spread there is in estimating the mean. The correct transformation of coordinates is $(\bar{\mu} - \mu)/(\sigma/\sqrt{N})$. The probability for false positives is then expressed using this standardized random variable:

$$P\left(z < \frac{C - 1.02}{\bar{\sigma}_1/\sqrt{299}}\right) = 0.01.$$
(5.33)

The corresponding value of the argument of $P(\cdot)$ in Equation (5.53) is found to be -2.325, which leads to the condition on C that:

$$\frac{C - 1.02}{\bar{\sigma}_1/\sqrt{299}} = -2.325 \Rightarrow C = 1.015.$$
(5.54)

The null hypothesis can therefore be rejected indicating that the component has become damaged when the sample mean is $\bar{\mu}_2 = 0.81$, which is much less than C. The probability of obtaining a false negative diagnosis of damage can also be calculated as:

$$P\left(z > \frac{1.015 - 0.81}{\bar{\sigma}_1/\sqrt{299}}\right) \ll 0.01.$$
(5.55)

It is seen that the probability of missing damage at this level of mass loss is very small. If the loss in mass is equal to 3 % instead of 10 %, then the probability of false negatives increases somewhat because the shift in the mean is smaller for less mass change. If only 10 samples of the feature are taken to estimate the sample mean in a case when there is only a 3 % mass change, then the probability of a false negative diagnosis of damage is equal to 46 %. In the limit, if only one sample is taken to estimate the sample mean, then $N = 1$ and the probability of obtaining a false negative diagnosis approaches 100 %. This example demonstrates the tradeoffs between false positive and false negative alarms in damage detection. In general

More samples result in more confidence in the sample mean and, therefore, the diagnosis. Less damage results in less confidence due to the variance in the sample mean.

The previous example utilized a lower one-tailed hypothesis test. In other words, damage was detected when the sample mean of the measured feature dropped below a threshold value. If damage occurs due to buckling, nitridation or other types of damage that cause the feature to increase instead of decrease, then an upper one-tailed hypothesis test would be used. And if both types of damage occur, then a two-tailed hypothesis test of the form:

$$H_0 : \quad \mu = 1.02$$
$$H_1 : \quad \mu \neq 1.02, \tag{5.56}$$

would be needed instead to detect both decreases and increases in the sample mean.

If the feature distribution is not a normal distribution, then there are two options for statistically modeling the feature and for hypothesis testing. First, the actual probability density and distribution functions displayed by a feature can be used as provided in Dougherty (1990) and other texts. For example, magnitude measurement errors often assume a *Rayleigh* distribution in FRF estimation. Features based on the use of FRF magnitude measurements would be affected by these errors. In this case, the Rayleigh distribution is used to model variations in features

$$\eta^{-2} x e^{-\frac{(x/\eta)^2}{2}} \quad \text{for} \quad x > 0 \quad \text{with} \quad \mu = \sqrt{\frac{\pi}{2}} \eta \quad \text{and} \quad \sigma = \frac{4 - \pi}{2} \eta^2. \tag{5.57}$$

Other examples include the uniform, gamma, chi-square, Weibull, and lognormal distributions.

Instead of using the formulae for the actual feature distribution, the second option for dealing with non-normal feature distributions is to utilize a relatively large number of samples when computing the sample mean and variance statistics. Then according to the *central limit theorem*, the probability density function of the standardized sample mean, $(\bar{\mu}-\mu)/(\sigma/\sqrt{N})$, of feature x approaches the standard normal density function. Then the previous analysis can be used as if the distribution was normal. To identify the type of distribution, histograms can be used in addition to other methods described by Bishop (1995). For example, the 'hist' command in MATLAB® can be used.

The example described above draws inferences about the sample mean of one measured feature that was extracted from a single population of random variables. The variance of the sample mean, $\bar{\sigma}_1^2/N$, for the baseline distribution of features was used in all of the calculations to calculate and apply the threshold of detection. If damage causes the variance to change significantly from that of the healthy component, then it is more appropriate to apply hypothesis tests designed to distinguish between the sample means of two different distributions of features. Also, sustained operational loading can cause features used for damage detection to change over time causing the variance to change. For example, the two-tailed hypothesis test to distinguish between the baseline mean value of a feature, μ_1, for new components and the measured mean value of that feature, μ_2, for components that have operated for a certain period of time is expressed as

$$
\begin{aligned}
H_0 : & \quad \mu_1 = \mu_2 \\
H_1 : & \quad \mu_1 \neq \mu_2.
\end{aligned}
\tag{5.58}
$$

The statistic for testing this hypothesis is the difference between the sample means, Equation (5.59a), and the sample variance of this statistic, Equation (5.59b), is calculated by summing both variances:

$$
\bar{\mu}_1 - \bar{\mu}_2 \quad \text{and} \quad \frac{\bar{\sigma}_1^2}{N_1} + \frac{\bar{\sigma}_2^2}{N_2}.
\tag{5.59a, b}
$$

There are many more techniques used to draw inferences about features using statistics. *Control charts* base decisions about normalcy on to what extent the sample mean of a feature varies from the mean of the feature for the healthy component. For example, *upper and lower control limits* for a feature that is acquired continuously as a component operates can be set at

$$
\text{Upper control limit} = \mu + \frac{3\sigma}{\sqrt{N}} \quad \text{and Lower control limit} = \mu - \frac{3\sigma}{\sqrt{N}}.
\tag{5.60a, b}
$$

If the sample mean falls outside this interval, there is a 99 % probability that the component from which the measurement was taken has fundamentally changed due to damage or some other factor (temperature, humidity, fault in sensor etc.). The reader is referred to Farrar *et al* (1999) for a detailed discussion of statistical modeling and classification of features for damage detection.

5.9   VARIABILITY ANALYSIS

Some sources of variability that are random in nature can be lumped together and dealt with using the statistical methods presented in the previous section. Other sources of variability should be analyzed (and eliminated) one by one from a physical point of view. Figure 5.75 illustrates the primary sources of variability in health-monitoring systems. All of these sources of variability combine to produce variations in the loading and damage indicators, which are

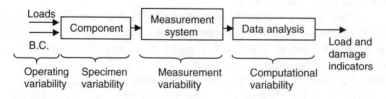

Figure 5.75   Variability entry points and propagation for health monitoring

generated using the modeling, measurement and data analysis techniques described in this book. The four primary sources of variability listed are (1) Operating variability, (2) Specimen variability, (3) Measurement variability and (4) Computational variability.

Operating variability consists of changes in the loading, environmental and boundary conditions of a component during operation. For example, suppose the frequency bandwidth of the dynamic loading on the nine-DOF system model in Figure 4.6 changes during operation. If some of the resonant modes of vibration are excited at one instant but not consistently, then it is more difficult to characterize the loading and damage affecting the component. The reason for this difficulty is that it is challenging to determine if the changes are due to the loading, the component or a combination of the two. In some applications, changes in loading lead to fundamentally different behaviors in the component response. For example, the nonlinear model of a panel described in Section 2.2.7 possessed stiffness and friction nonlinearities, which are both sensitive to changes in excitation amplitude. When the amplitude of the excitation changes, these nonlinearities are excited differently causing the component to undergo significant changes in its resonant frequencies, damping and other dynamic characteristics. Changes like these due to how nonlinearities in a healthy component are exercised by the operating loads are difficult to distinguish from changes due to damage.

One source of variability due to environmental conditions was discussed in Section 4.2.3 where changes in temperature causing shifts in the stiffness of components were considered. The changes in stiffness of the nine-DOF model due to temperature changes caused significant changes in the FRFs measured from the component. The effects of temperature changes were also considered in the panel model discussed in Section 2.4.2. Other sources of environmental variability include changes in humidity, chemical pH, ultraviolet radiation and electromagnetic interference. Each of these factors causes variations in component mechanical properties and response characteristics. For instance, humidity is absorbed by composite components like the cylinder in Figure 1.3(c) leading to a higher density material. Higher densities subsequently change the way waves propagate through these composite components. This type of variability was simulated in Section 4.2.3. Although environmental sources of variability often change with time making them dynamic, the changes are usually slow compared to component vibrations.

For both operating and environmental sources of variability, several methods can be used to detect the source and mitigate the effects of variability on data analysis algorithms. For example, changes in material modulus due to temperature were studied in Section 2.4.2. By measuring the temperature of a component, the variations in dynamic response due to changes in modulus can be distinguished from variations in the response due to geometric changes in the material. Also, the variations in response due to a temperature change in Section 4.2.3 were global in nature indicating that a temperature change caused the observed variations in the FRFs of the nine-DOF system model rather than a local change in the component due to damage. Finally, if the observed changes in the component characteristics are permanent, then it can be concluded that damage has caused the changes as opposed to a fluctuation in temperature.

Variations in component response due to changes in boundary conditions are also common. Component vibrations are quite sensitive to changes in boundary conditions. Consider the panel shown in Figure 2.1. The panel is supported by four struts, which serve as boundary conditions. If the temperature increases, then the panel expands outward causing the struts to deform. This deformation causes the stiffness of the panel boundary conditions to change leading to changes in natural frequencies and modeshapes. Likewise, the building frame in

Figure 5.56(a) experiences changes in its boundary condition damping and stiffness if the air bags are pressurized to different levels. These changes are meant to simulate changes in the soil into which a building foundation is poured. The aluminum sheet in Figure 5.58(a) meant to simulate an aircraft panel is shown simply supported. If clamps are used instead, the vibration characteristics change considerably. Another type of boundary condition variability is structureborne vibration through the component interfaces to the structural system. For example, the panel shown in Figure 2.1 might be mounted near an aircraft engine causing vibration response at harmonics of the engine rotating speed that vary with the speed and thrust of the engine. Boundary conditions can change during operation or from installation to installation so they are considered both static and dynamic sources of variability.

Specimen variability is due to changes in the makeup of a component due to manufacturing variability or differences in usage. For example, in order to detect cracks in the spindle illustrated in Figure 1.5(b), variability in the spindle response due to differences in the installation of the spindle in various wheels and vehicles must be taken into account. The spindle is surrounded by other components including gears that mate with the spline of the spindle. These interfaces can change slightly from wheel to wheel and vehicle to vehicle resulting in changed boundary conditions for the spindle. As mentioned above, changes in boundary conditions result in variations in the vibration response of components. Slight changes in the interfaces between the spindle and wheel components also cause substantial variations in the wave propagation through the spindle. Specimen variability is considered a static source of variability.

Numerous sources of measurement variability have been described earlier in the book. Some examples include changes in sensor calibration constants with temperature, changes in transducer attachment stiffness and damping properties and changes in cable length. Figure 5.76 illustrates several sources of measurement variability at the sensor level. Among the sources of variability illustrated are the sensor–cable connection and the orientation taken by the sensor when attached. Changes in attachments are considered both static and dynamic sources of variability, whereas many other sources of measurement variability are considered static.

Computational variability is the least obvious source of variation in health monitoring. Section 4.1 pointed out that the objective in taking health-monitoring measurements is to reduce variability as much as possible prior to analyzing the measured data. This objective

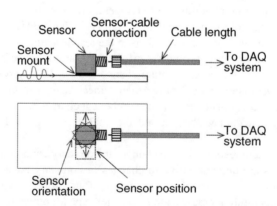

Figure 5.76 Examples of measurement sources of variability at sensor (Reproduced by permission of Harold Kess, Purdue University, 2006)

may give the false impression that nothing can be done in data analysis to deal effectively with variability that enters the measurement. On the contrary, algorithms with which one analyzes health-monitoring data can either (a) amplify variability or (b) attenuate variability in the features that are extracted for loads and damage identification. For example, consider the FRF estimation process described in Section 5.6.2. If there is measurement noise on the response measurement, then the H_1 algorithm is selected to estimate the FRF between the excitation force and response. If the H_2 FRF estimator is mistakenly selected, then the FRF will contain significantly larger amounts of variability. Note that the same measurement data leads to two drastically different amounts of variability (or error) in the estimated FRF. This difference in the way algorithms treat sources of variability in the measurement should be taken into account when selecting data analysis algorithms.

For example, consider the transmissibility function discussed in Section 2.2.6. Transmissibility is used to process FRF data for the purpose of detecting and locating damage; therefore, it is important to understand how the calculation of transmissibility is affected by variability in the FRFs. The transmissibility between two response measurement DOFs for an excitation applied at DOF k can be calculated by taking the ratio of the two FRFs for these response DOFs:

$$T_{p_1 p_2 (k)}(\omega) = \frac{H_{p_1 k}(\omega)}{H_{p_2 k}(\omega)}. \qquad (5.61)$$

The total variation in the transmissibility given a change in the variable, V, can be calculated using a Taylor series expansion of this function:

$$\Delta T_{p_1 p_2 (k)}(\omega) = \left(\frac{\partial T_{p_1 p_2 (k)}}{\partial H_{p_1 k}}\frac{\partial H_{p_1 k}}{\partial V} + \frac{\partial T_{p_1 p_2 (k)}}{\partial H_{p_2 k}}\frac{\partial H_{p_2 k}}{\partial V}\right)\Delta V + \text{h.o.t.}, \qquad (5.62)$$

where h.o.t. denotes higher order terms. The frequency, ω, has been removed from this equation for simplicity. Only the first term in the Taylor series expansion involving a finite (not infinitesimal) change in V was considered in Equation (5.62). V could represent a change in cable length or a change in some damage variable like crack length.

There are two types of terms in Equation (5.62). Terms like $\partial T_{p_1 p_2 (k)}/\partial H_{p_1 k}$ can be calculated directly from the transmissibility formulas in Equation (5.61). For example,

$$\frac{\partial T_{p_1 p_2 (k)}}{\partial H_{p_1 k}} = \frac{1}{H_{p_2 k}}$$

$$\frac{\partial T_{p_1 p_2 (k)}}{\partial H_{p_2 k}} = -\frac{H_{p_1 k}}{H_{p_2 k}^2}. \qquad (5.63a, b)$$

These two formulas indicate the relative sensitivity of the transmissibility to the two FRFs. It is clear from the expressions that there are significant differences as a function of frequency between Equations (5.63a) and (5.63b). It is also clear that both of these sensitivity functions depend on the excitation DOF and the response DOFs. When attempting to reduce the total change in transmissibility caused by the numerous sources of variability described above, the excitation and sensor measurement DOFs can be selected to reduce the sensitivity at frequencies where the variability is highest.

Table 5.5 Types of variability including static and dynamic sources (Reproduced by permission of Harold Kess, Purdue University, 2006)

Variability	Type		
	Static	Dynamic	Both
Operational:			
Forcing amplitude and frequency		x	
Environmental:			
Temperature, humidity and so forth.		x	
Boundary Conditions:			
Simply supported, clamped and so forth.			x
Structureborne vibration		x	
Specimen:			
Boundary conditions	x		
Geometrical and material parameters	x		
Measurement:			
Cable length	x		
Sensor attachment			x
Sensor remounting	x		
Sensor-cable connection	x		
Sensor/actuator position	x		
Sensor angular orientation	x		
Sensor manufacturing	x		
Sensor specifications	x		
Computational:			
Computational error	x		

The other type of term in Equation (5.62) like $\partial H_{p_1 k}/\partial V$ cannot generally be calculated beforehand in a general application. This term must be estimated as the measurement system is being developed and tested to determine the best course of action for installing sensors and analyzing data. For example, the cable length can be increased (ΔV) and the difference in measured FRFs ($\Delta H_{p_1 k}$) can be calculated due to this increase. The product of this term and the previous sensitivity term determines the total change in transmissibility as a function of the variability in V. This same approach can be applied to any feature extracted using models such as the embedded sensitivity and virtual force models described in Chapter 2. In some cases, variability affects the same frequency ranges in the data as the damage or loading of interest; however, in most cases, this process of decomposing the total variation of a given feature into two parts, *one due to the sensitivity of the data to the source of variation and the other due to the way in which the feature is calculated*, helps to identify and minimize computational variability.

Table 5.5 summarizes these sources of variability and denotes whether they are generally considered static, dynamic or both in nature. Many of these sources of variability are demonstrated next in an example.

To illustrate the sources of variability described above, consider the woven composite plate shown in Figure 5.77. The plate is 1 ft × 2 ft × 0.5 in with 22 plies of glass fibers bound with epoxy resin. This type of material is used in composite armor for military applications. The backside of the plate is shown in (a) with five single-axis accelerometers (PCB 333B32) attached along with eight soft springs to support the plate for testing. The accelerometers were connected to the data acquisition system using 10 ft cables (PCB 002C10). The

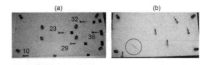

Figure 5.77 (a) Woven composite plate with spring supports and five sensors attached on backside and (b) plate with crack in lower-left-hand corner (Reproduced by permission of Harold Kess, Purdue University, 2006)

accelerometers were mounted with beeswax for this example. An 8-channel VXI (Agilent E1432A) system was used to acquire the data. In (b), the plate is shown with a crack in the lower-left-hand corner. The objective of this example is to demonstrate how sources of variability affect the transmissibility features used for damage detection. Also, the effects of the crack relative to these sources of variability will be considered to determine the best strategy for processing the data.

With five sensors, there were 10 possible transmissibility functions to consider between pairs of sensor measurements (i.e. 1–2, 1–3, 1–4, 1–5, 2–3 etc.). Only the transmissibility pathways between sensor locations 10 and 29 and 10 and 23 are considered in detail here (refer to Figure 5.77 for locations of sensors). Both of these paths span the area across which damage is introduced. In practice, one would want to select between these two transmissibility pairs to minimize the variation in transmissibility due to the previously mentioned sources of variation. A modal impact hammer (PCB 086C03) is utilized to excite the plate to vibrate in order to simulate the effects of loads that act on the plate while the vehicle to which the plate is attached operates. In practice, the excitation force would not be measured; however, this example measures the excitation force and the response accelerations to calculate FRFs between the input and output measurements. Then these FRFs are used to calculate the transmissibilities from sensor 10 to 29 and from sensor 10 to 23 for comparison.

In order to determine how each source of variability contributes to the overall variability in the transmissibility function, a series of tests were designed to isolate the effects of each source of variation. Table 5.6 lists these tests. The first test determined the variability in the baseline measurement prior to changing the measurement setup. Several sets of baseline data were acquired. Then static sources of variability were considered; the effects of small changes in impact location, remounting of the sensor and the other changes listed were studied. For example, Figure 5.78 illustrates the experiment for measuring variation in transmissibility as a function of the change where the excitation force is applied. Dynamic sources of variability including temperature changes of the sensors, temperature changes of the plate and changes in the boundary conditions of the plate were also examined. For example, Figure 5.79 shows the apparatus used to simulate changes in boundary conditions on the two ends of the plate. These changes in boundary conditions would occur in practice as the vehicle to which the plate is attached steers in different directions causing changes in static bending and in-plane loads to the plate. Although great care was taken to isolate individual sources of variability indicated by the (x) symbol in Table 5.6, tests sometimes involve secondary sources of variability as indicated by the (*) symbol.

In order to understand the effects of measurement and computational variability, consider the sensitivity functions shown in Figure 5.80(a) and (b). These are the two functions given in Equations (5.63a) and (5.63b). They indicate that each transmissibility pair leads to a different sensitivity to sources of variation as a function of frequency. The response DOFs are indicated in the legend of the plots ('TR1023' for response sensors 10 and 23) and the excitation DOF is

Table 5.6 Test matrix showing primary variables (x) and secondary variables (*) that influenced the measurement variability (Reproduced by permission of DEStech Publications, Inc.)

Variability	|	Testing sequence												
	1	2	3	4	5	6	7	8	9	10	11	12	13	14
Baseline	x													
Static														
Impact location		x												
Sensor remounting		x	*	*	*	*	*	*	*	*				
Sensor orientation					*	*	*	*	*	*	*	*		
Sensor position					x									
Sensor-cable conn.					x	*	*	*	*	*				
Cable length							x							
Sensor manufacturing							x							
Sensor sensitivity									x					
Sensor range										x				
Mounting material											x	*		
Dynamic														
Sensor temp.												x		
Global temp.													x	
Boundary condition														x

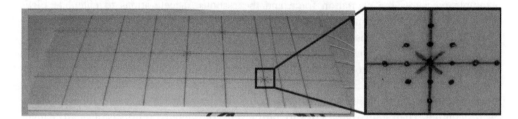

Figure 5.78 Test for examining variation in vibration characteristics of woven composite panel for change in excitation location (Reproduced by permission of Harold Kess, Purdue University, 2006)

Figure 5.79 Test for examining variation in vibration characteristics of woven composite panel due to change in boundary condition (Reproduced by permission of Harold Kess, Purdue University, 2006)

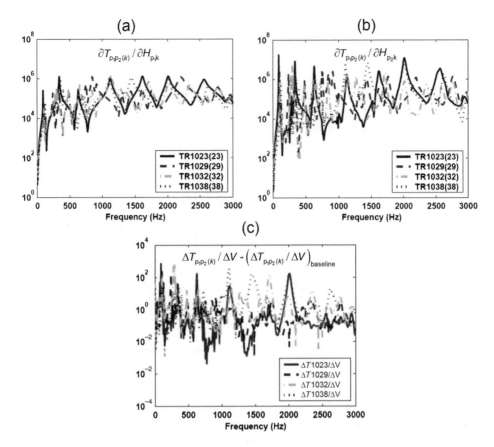

Figure 5.80 Sensitivity of $T_{p_1 p_2(k)}$ (a) $\partial T_{p_1 p_2(k)}/\partial H_{p_1 k}$, (b) $\partial T_{p_1 p_2(k)}/\partial H_{p_2 k}$ showing large sensitivity to FRF in denominator of transmissibility and (c) $\Delta T_{p_1 p_2(k)}/\Delta V$ for change in bolt preload, V (in in lb torque), in Figure 5.79 (Reproduced by permission of Society for Experimental Mechanics, Inc.)

also indicated ('TR1023(23)' for excitation location 23). Note that the sensitivity to the denominator FRF in the transmissibility (Figure 5.80(b)) is roughly an order of magnitude higher than the sensitivity of the transmissibility to the numerator FRF (Figure 5.80(a)). This result suggests that all sources of variability that change the FRF in the denominator will more greatly influence the transmissibility. Also, the sensitivities of all transmissibility pairs trend together; however, certain pairs are more sensitive at certain frequencies. This information can be used when selecting the sensor locations in practice if the frequency range of the operating excitation is known.

When a source of variation is introduced, the transmissibility changes. This change is recorded and then normalized by the variable that has been varied. For example, Figure 5.80(c) shows the change in transmissibility from the baseline, $\Delta T_{p_1 p_2(k)}/\Delta V$, in units of 1 per in-lb, minus the baseline change in transmissibility when no change is made in any variable, $(\Delta T_{p_1 p_2(k)}/\Delta V)_{\text{baseline}}$, for a change in the torque applied at the bolts which clamp both ends of the plate in Figure 5.79. In certain frequency ranges, the variation due to a boundary condition change is much larger for specific transmissibility functions, for example variation in transmissibility from 10 to 38 is much larger than for other sensor pathways in the frequency

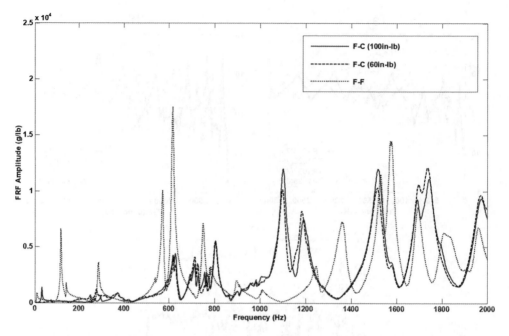

Figure 5.81 Changes in FRF magnitude $H_{10,10}(\omega)$ due to change in boundary condition from free-free (...) to free-clamped at 60 in-lb (- - -) and 100 in-lb (__) (Reproduced by permission of Harold Kess, Purdue University, 2006)

range surrounding 1500 Hz. The same is true near 750 Hz. These changes in the transmissibility function occur because of the measured changes in the FRFs used to calculate the transmissibility. One *driving-point*[12] FRF is shown in Figure 5.81 for three different sets of boundary conditions (free-free and clamped-clamped with 60 and 100 in-lb of torque applied to the bolts in Figure 5.79). Although the changes in the FRF are largest when moving from the free–free to either of the clamped–clamped boundary conditions, there are also significant changes when the torque used to preload the clamps on either end of the plate is varied. These variations due to clamping torque are encapsulated by the plot in Figure 5.80(c) for the transmissibility function.

This same procedure was followed to investigate all of the sources of variability listed in Table 5.6. In order to present all of the results, each of the $\Delta T_{p_1p_2(k)}/\Delta V$ spectra was first divided by the RMS value across the 0–3000 Hz frequency range, and then the baseline variability divided by its RMS value was subtracted at each frequency. Finally, this difference was divided by the baseline variability divided by its RMS value at each frequency. The results in Figure 5.82(a) for the transmissibility $T_{10,23(23)}$ and the results in Figure (b) for $T_{10,29(29)}$ were obtained for the various sources of variability. The change due to damage of different levels (three crack lengths) is also included for comparison. The baseline result, which is zero by design, is also indicated. The objective of using this variability map is to identify in which frequency range damage can be detected despite sources of variability in the measurement or computational algorithm.

---

[12]A *driving-point* FRF is one in which the input and output location and direction are the same.

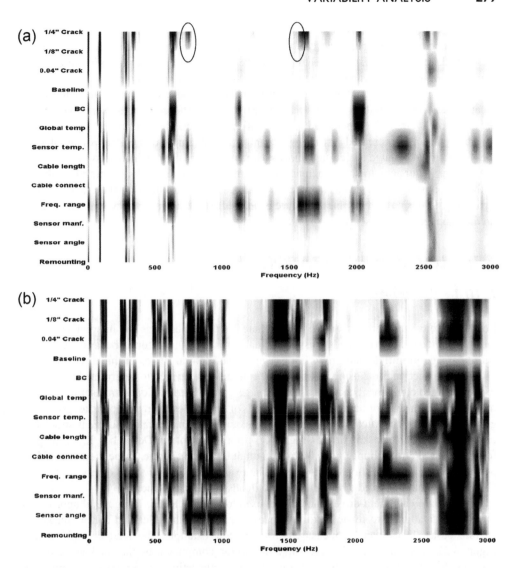

Figure 5.82 Normalized variability in measurement and computational algorithm for transmissibility function with (a) $T_{10,23(23)}$ and (b) $T_{10,29(29)}$ (Reproduced by permission of DEStech Publications, Inc.)

First, consider the result in Figure 5.82(a) for $T_{10,23(23)}$. The largest sources of variability appear to come from boundary condition changes, global temperature changes and the change in sensor frequency bandwidth (e.g. 3 versus 10 kHz). In most frequency ranges where the damage is indicated by changes in the transmissibility function, the sources of variability also produce changes. In some applications, the sensor frequency range, for instance, might not be expected to vary because a fixed set of sensors would be installed for health-monitoring purposes. In this instance, variability due to the sensor frequency range could be neglected. Note that there are two frequency ranges (circled) where the damage causes changes in the transmissibility that are least likely to be masked by changes due to sources of variation (around 700 and 1600 Hz).

Furthermore, if variation due to temperature changes to the sensor and the frequency range of the sensor can be neglected, then these two frequency ranges would provide features that correlate well with the presence of damage. Features could be extracted from these ranges by summing the magnitude of the transmissibility function as discussed in previous sections.

Second, consider the result in Figure 5.82(b) for $T_{10,29(29)}$ with a change in one of the sensor locations and the excitation location. It is clear from the figure that the sensitivity to the presence of a crack has increased significantly; however, the sensitivity to sources of variability has also increased just as significantly. This example demonstrates that although the sensitivity to damage detection (or loads estimation) can often be increased by repositioning sensors and/or actuators, this increase in sensitivity to damage (or loads) must be balanced with the robustness of a given damage (or load) feature to sources of static and dynamic variability. In this case, the damage feature is derived from transmissibility functions but this method applies to other sources of features as well.

5.10 LOADS IDENTIFICATION

5.10.1 Overview

Loads identification methods are used to estimate both external loads and internal loads from measurement data. Internal loads must be estimated for health monitoring because these loads cause damage to initiate and grow. For example, the internal forces in the clevis that were estimated in Section 5.8.2 using restoring force models are needed to determine how loading to the damaged component changes as the damage grows. External loads are useful for health monitoring because information about these loads often enables operators to make 'go' or 'no-go' decisions about a structural component. For example, if an airplane fuselage is struck by a cart as occurred on Alaska Airlines flight 536 prior to takeoff in December 2005, the peak force measured during impact may be high enough to warrant further inspection to avoid the potential for depressurization during flight. The example in Section 2.3.2 demonstrated how FRF models could be used to estimate impact loads such as this one.

Physics-based and data-driven models are used to identify loads. The general approach to loads identification is to relate unknown loads to response variables that can be measured. In rare cases, loads can be measured directly. For example, the load cell pictured in Figure 4.32(a) can measure the force imparted to the component by the stack actuator also shown in Figure 4.32(a). Usually, loads must be indirectly estimated using response measurements and a model of some kind. Consider the beam component of constant thickness h, density ρ, modulus E and cross-sectional area $A(x)$ shown in Figure 5.83. The beam responds with displacement $u(x, t)$ at position x and time t to the force distribution $f(x, t)$. A closeup of the beam is also shown where a crack of depth c is located at position x_c. The load can be estimated using the model for the beam given by

$$f(x,t) = \underbrace{\rho A(x) \frac{\partial^2 u(x,t)}{\partial t^2}}_{Mu} + \underbrace{\frac{\partial^2}{\partial x^2} \left(EI(x) \frac{\partial^2 u(x,t)}{\partial x^2} \right)}_{Ku}. \tag{5.64}$$

Ordinarily, the discretized model of the beam would be used to estimate the *external load* as denoted by the inertia and stiffness terms on the right-hand side of the equation. The

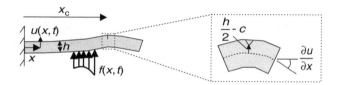

Figure 5.83   Beam component with damage subject to external forces

acceleration would be measured and then integrated twice to obtain the displacement for use in this differential equation model. Knowledge of this external load would be useful for making future predictions based on the operational loading that is anticipated. Measurements of the displacement could also be used to estimate the *internal loading* within the beam. For example, the stress at the location of the crack within the beam is found by calculating the curvature (see Section 5.7.1) from the measurement of $u(x,t)$ and then multiplying by the distance to the crack tip from the neutral axis:

$$\sigma(x_c, t) = -E\left(\frac{h}{2} - c\right) \frac{\partial^2 u(x_c, t)}{\partial x^2}. \tag{5.65}$$

Stress concentration factors would also need to be applied at the crack tip. Such factors for various geometry components can be found in many textbooks such as Grandt (2004).

5.10.2   Estimation Errors

One of the key issues in loads identification is that errors in the model and measurement lead to errors in the load estimates. To reduce these errors, *overdetermined measurements* are usually taken using multiple datasets and/or multiple sensors. Then these measurements are used to estimate loads such that the error is minimized in some sense. For example, suppose that a constant point force, P, is applied at the end of the beam in Figure 5.83 in the vertical direction. Assume that the beam can be modeled like a Bernoulli–Euler beam (i.e. planar sections remain planar after the load is applied and the beam length is much larger than its cross-sectional dimensions). If one static displacement variable, x, is measured at the end of the beam, then a model relating the unknown force and measured displacement of the undamaged beam is

$$x = \hat{x} + \eta = \frac{L^3}{3EI} P, \tag{5.66}$$

where η is uncorrelated noise on the response measurement, $\hat{x}$. If only one measurement is used to estimate the force based on this model, then there will be a bias error in the estimated force, $\hat{P}$:

$$P - \underbrace{\frac{3EI}{L^3} \eta}_{\text{Bias error}} = \frac{3EI}{L^3} \hat{x} = \hat{P}. \tag{5.67}$$

An overdetermined measurement approach is required to minimize the bias error. Recall from Section 5.6.2 that least-squares solution techniques are used to estimate FRFs that are contaminated by noise on the input and output data. Assume that N measurements of the static displacement are taken for the same applied force, P. These measurements could be taken one after the other. The set of equations obtained relating the applied force to the measured displacements is given below:

$$\left\{ \begin{array}{c} \hat{x}_1 \\ \hat{x}_2 \\ \vdots \\ \hat{x}_N \end{array} \right\} + \left\{ \begin{array}{c} \eta_1 \\ \eta_2 \\ \vdots \\ \eta_N \end{array} \right\} = \frac{L^3}{3EI} P \qquad \text{(5.68a, b)}$$

$$\{\hat{x}_n\} + \{\eta_n\} = \frac{L^3}{3EI} P.$$

The technique described in Section 5.6.2 is then used to estimate the solution with the minimum squared errors summed across all of the equations. The vector of displacement measurements is premultiplied on both sides of the equation to calculate this solution. Because the noise is uncorrelated with the displacement measurement, $\{\hat{x}_N\}^T \{\eta_N\} = 0$. This result leads to the following solution for the estimated load:

$$\hat{P} = \frac{3EI}{L^3} \frac{\sum\limits_{i=1}^{N} \hat{x}_i^2}{\sum\limits_{i=1}^{N} \hat{x}_i}, \qquad \text{(5.69)}$$

where the vector operations have been replaced with scalar summations. If η is a random variable that is uncorrelated with the force and response, then the expected value of $\hat{P}$ converges to P (the true force acting on the beam) as more measurements are taken. The same approach can be used to incorporate overdetermined response measurements at different locations on the beam and in different directions (see end of chapter problems).

5.10.3   Conditioning of Loads Identification Algorithms

The algorithm utilized for loads identification depends on the type of load being estimated and the measurements available for estimation. Algorithms are a combination of mathematical operations and constraints based on assumptions made about the load. As in all *inverse problems*, in which an unknown variable is estimated using known variables, loads estimation algorithms must be properly conditioned in order to provide unique and accurate results. The conditioning of a loads estimation algorithm depends on the nature of the load being estimated and the measurements used to estimate the load. For example, suppose that two point loads act on the beam in Figure 5.83. Now suppose that one displacement measurement is taken on the end of the beam. With this one measurement, the inverse problem is *underdetermined*, i.e., there are not enough equations to solve for the two unknown loads. Underdetermined inverse problems are said to be poorly conditioned because there is more than one possible solution. It would be necessary in this case to identify the slope of the beam by measuring the displacement at a second point on the beam in order to estimate the two unknown loads.

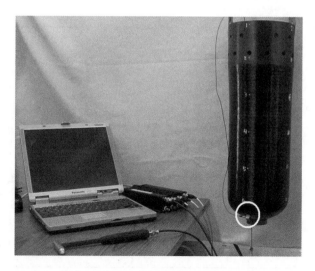

Figure 5.84   Experimental setup for impact loads identification in filament-wound cylinder using one triaxial accelerometer positioned at bottom

An alternative to acquiring more measurements to properly condition a loads estimation problem is *iteration*. Iterative methods in loads identification make use of *a priori* knowledge about the load being estimated and the component being tested. For example, consider the filament-wound cylinder shown in Figure 5.84. This type of component is used in fuel tanks and in missiles. A triaxial accelerometer (PCB 356A22) is shown attached to the nose of the cylinder. This sensor measures the acceleration in three orthogonal directions with a sensitivity of 100 mV/g and frequency bandwidth of 10 kHz. A modal impact hammer (PCB 086C03) with sensitivity 10 mV/lb is used to impact the canister to simulate operational loads. Impacts have been identified as the primary cause of damage leading to failure in these types of cylinders when they are pressurized. Twenty four locations are indicated on the cylinder.

The loads identification algorithm used is based on a data-driven model relating impact spectra at each of these twenty four locations, $\{F(\omega)\}_{24 \times 1}$, to the three response spectra measured by the accelerometer, $\{A(\omega)\}_{3 \times 1}$. An FRF model similar to the one discussed in Section 2.3.2 is used to relate the forces and responses. The training set of data for one cylinder is expressed as follows:

$$\{A(\omega)\}_{3 \times 1} = [H(\omega)]_{3 \times 24} \{F(\omega)\}_{24 \times 1} \tag{5.70}$$

With 24 possible impact forces and only three response measurements, this inverse problem is clearly underdetermined. There are only three equations but 24 unknown force spectra.

In order to use this model to estimate impact loads, the following algorithm was used to overcome the poor conditioning due to the low number of response measurements:

- The training set of FRFs in Equation (5.70) was acquired using the modal impact hammer at low impact force levels to avoid damage to the filament winding. These FRFs were stored on the laptop computer shown in Figure 5.84.

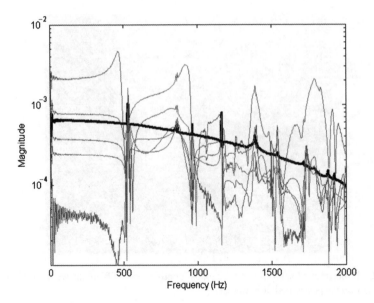

Figure 5.85   Estimated force spectra with force applied at one of the 24 locations on the cylinder; correct estimate (dark line) and incorrect estimate (gray lines)

- Response acceleration spectra in all three measurement directions were acquired during an impact at a given point on the canister. Both points in the training set of data and points between the original training points were tested.
- It was then assumed that a *single* impact occurred at *one* of the 24 points on the cylinder. These two assumptions help to condition the force estimation algorithm in the following way. First, by assuming that only one impact force, $F_n(\omega)$, occured at one of the 24 points marked on the cylinder, the inverse problem to estimate this one force was overdetermined. The mathematical operation to estimate this force only required three of the FRFs. The solution with minimum error in the least squares sense is found as follows:

$$\{A(\omega)\}_{3\times1} = \begin{Bmatrix} H_{1,n}(\omega) \\ H_{2,n}(\omega) \\ H_{3,n}(\omega) \end{Bmatrix} F_n(\omega) \tag{5.71a}$$

$$\hat{F}_n(\omega) = (\{H_{i,n}(\omega)\}^H \{H_{i,n}(\omega)\})^{-1} \{H_{i,n}(\omega)\}^H \{A(\omega)\} \tag{5.71b}$$

$$= \{H_{i,n}(\omega)\}^+ \{A(\omega)\} \tag{5.71c}$$

- The matrix $\{H_{i,n}(\omega)\}^+$ in Equation (5.71c) is called the *pseudoinverse* of $\{H_{i,n}(\omega)\}**$. Because this vector of FRFs is not a square matrix, it cannot be inverted in the traditional way. Theoretically speaking, there is no unique solution to the set of three equations in Equation (5.71a). Practically, however, the solution with minimum least squared error can be obtained using the pseudoinverse. The MATLAB® function 'pinv' can be used to calculate the pseudoinverse of a matrix. The pseudoinverse is calculated at each frequency and the force is estimated at each frequency using Equation (5.71c). This calculation is repeated for each n of the 24 possible locations on the cylinder.

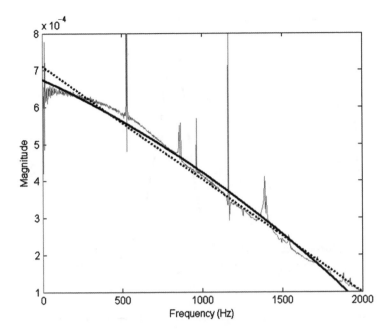

Figure 5.86 Estimated force spectra and (. . .) linear and (__) quadratic functions fit to the magnitude to test single impact constraint

- Examples of the estimated force magnitude solutions obtained using this technique are shown in Figure 5.85 for an impact at location 1 on the cylinder. There are 24 force estimates but only one of them can be correct because there is only one impact force. The correct estimate is shown with the dark line. The incorrect estimates are shown with gray lines. The next step in the algorithm is to distinguish between the correct and incorrect estimates. This step is carried out by enforcing a constraint on the solution using the second assumption made about the impact. Assuming that a single impact occurs, it is known that the force spectrum should be relatively flat in nature without severe roll-off. In order to test each of the force spectra against this constraint, two different polynominals, a straight line and quadratic function, were fitted through the force spectra. Then the differences between these curve fits and the estimated force magnitudes for each of the assumed impact locations were computed. The assumed impact location with the least deviation from the linear and quadratic functions was selected as the correct location. Figure 5.86 shows an example of these two polynomial functions fit to the force spectrum magnitude at one assumed impact location. The MATLAB® functions 'polyfit' and 'polyval' can be used to perform these fits.
- The final estimate of the impact force time history using this algorithm is shown in Figure 5.87. The agreement between the estimated (__) and measured (. . .) force time histories is good. The force was measured using the load cell on the modal impact hammer. Tests were also conducted to identify the percent accuracy obtained using this algorithm. Two thousand four hundred impacts were made on the training cylinder, and the exact location was identified nearly 98 % of the time as indicated in Table 5.7 in the first column.

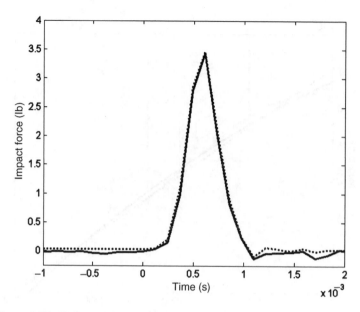

Figure 5.87 Estimated force time history (__) and measured time history (...)

The second column of the table indicates the percent of impact locations that were identified in the vicinity of the exact location. When the boundary conditions of the canister were removed and replaced, and the sensor was removed and replaced, the percentages change very little. Only 240 impacts were made in these two sets of tests resulting in artificially higher percentages of correctly identified impact locations. A final set of 3480 impacts were performed on different canisters using the same set of FRF training data from Equation (5.70). The percentage of correctly identified impact locations is listed in the table. Although there is a drop in the number of exact locations identified, 90 % of the impact locations were identified within the vicinity of the correct location.

This example has demonstrated that load estimation algorithms can be better conditioned by structuring them such that they are overdetermined and constrained in some way based on prior assumptions about the loads. Once loads are accurately identified in location and magnitude, this information can be used to construct FRFs and other types of input–output functions for use in damage identification. The next chapter discusses several examples of loads identification using physics-based and data-driven models.

Table 5.7 Test results for training cylinder in three configurations and other cylinders

	Exact percentage	Adjacent percentage
Same cylinder	97.8	97.9
Mounts replaced	100	100
Sensor replaced	95.8	100
Cross cylinder	67.9	90.1

5.11 DAMAGE IDENTIFICATION

Various types of damage mechanisms were described in Section 3. Damage mechanisms are sometimes referred to as *faults*. The damage identification process involves three basic steps: (1) detecting damage, (2) locating damage and (3) quantifying damage (Rytter, 1993). Each of these steps is described, and then an example where these three steps are applied to an S2 glass filament-wound pressure vessel is provided.

5.11.1 Damage Detection

In step 1, damage due to a permanent physical change in a structural material or component must be detected in the presence of various sources of environmental variability and measurement errors. Damage is sometimes local in nature as in the case of cracking making it more difficult to detect the damage in one area of a component using limited sensing capabilities. In other cases, damage is global as in the case of oxidation affecting larger areas of the component. Some forms of damage cause linear changes in the way components respond. Other forms of damage cause components to respond in a nonlinear manner. Both of these changes must be recognized to detect damage. Examples of linear and nonlinear damage mechanisms have been used in discussions throughout this chapter and Chapters 2 and 3.

In some applications, it is more convenient to detect damage when the structural component is out of service. Examples might include thermal tiles on a spacecraft and discs in a gas turbine engine. Thermal tiles are heated to temperatures so high that sensors may not survive if installed in regions exposed to severe heating. In other applications, damage may be more observable in measurement data when the component is under load. The rolling tire discussed in Section 3.1.4 is an example where the rotational loads in service tend to accentuate the presence of bead area cracks. Examples of these types of load interaction effects are given in case studies later in the book.

Issues in damage detection relating to hypothesis tests and thresholds were already addressed in Section 5.8.4. The threshold settings established for distinguishing between healthy and unhealthy component behavior in a damage detection algorithm determine the minimum size damage that can be detected. This threshold must take into account both the physical limitations for detecting small damage and the measurement limitations due to variability that can mask the presence of damage. Together these limitations determine the *probability of detection*. If β is the probability of false negative detection as described in Section 5.8.4, then $1-\beta$ is the probability of detection.

Thresholds for damage detection in data analysis algorithms are set using *a priori* information about the way healthy components should respond to operational or active excitations. In some damage detection methods, baseline measurements from the population of data produced by healthy components similar to the one being monitored are used to establish thresholds for damage detection. These methods are said to be *supervised* because comparisons are continuously being made between new data that is acquired and data previously acquired from one or more healthy components. Numerous supervised damage detection algorithms have been described in this chapter. For example, the statistical data analysis techniques in Section 5.4.1 used changes in various statistics from measured response time histories before and after damage was introduced to detect the presence of damage in the two-DOF system model. Section 5.4.2 also used supervised methods to detect damage by

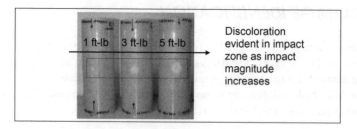

Figure 5.88   S2 glass filament wound composite cylinder with impact damage causing visible discoloration for 1, 3 and 5 ft·lb impacts

observing changes in discrete time model parameters and restoring force projections for the two-DOF system model. Supervised methods require that measurement data be acquired from the component in question before it becomes damaged; therefore, if a component is already damaged when health monitoring measurement hardware is installed, then the health-monitoring algorithms can only detect the growth of preexisting damage.

In other damage detection methods, thresholds of normal (or healthy) component behavior are set using physical assumptions about how a healthy component should respond. These methods are referred to as *unsupervised* because they use prior assumptions about healthy versus nonhealthy components as examples behavior to detect damage without requiring data acquired from healthy components as examples. For instance, if it is assumed that the two-DOF system examined in Section 5.5.1 should exhibit linear behavior if it is healthy before acquiring any measurement data from the system, then the appearance of harmonics in Figure 5.33(a) would indicate the presence of damage, which introduces nonlinear harmonics. Unsupervised methods are particularly useful in cases where health monitoring is to be performed on a component that has already aged and contains at least some damage. Damage can be detected using unsupervised methods despite the lack of healthy data to which new data can be compared.

As an example of damage detection, consider the S2 glass fiber epoxy filament wound composite cylinders shown in Figure 5.88. From left to right, the cylinders have been impacted with 1, 3 and 5 ft·lb energy impacts at the center. Discoloration due to these impacts is evident in the figure. Three different methods were used to detect, locate and quantify the impact damage. These methods are illustrated in Figure 5.89. Each method is selected to provide

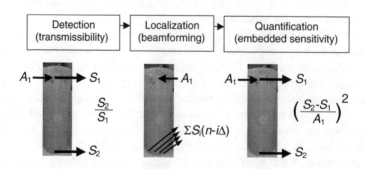

Figure 5.89   Damage identification approach to detect, locate and quantify damage in S2 glass filament wound composite cylinder

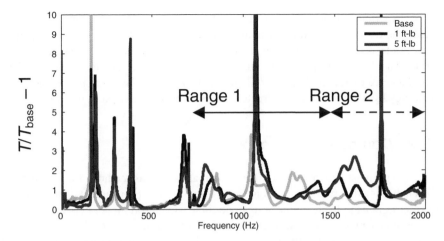

Figure 5.90 Normalized transmissibility difference for healthy cylinder, 1 ft·lb and 5 ft·lb impacted cylinders

certain information about the damage using the sensor and actuator measurements shown in the figure. Damage detection is discussed in this section. An actuator (modal impact hammer) was used to excite the cylinder in the location shown in the illustration beneath the damage detection block. Two sensors were used to measure the response normal to the cylinder in the locations shown. The transmissibility function was then calculated using the two corresponding FRF measurements (refer to Section 5.8.1). Recall that transmissibility can also be computed without an actuator.

In order to detect the damage, two healthy cylinders were tested in addition to the three impacted cylinders. The first cylinder was used to measure a baseline transmissibility function, which was then compared to the transmissibility functions for the second healthy cylinder and the three impacted cylinders. Two frequency ranges were used in the data analysis: 700 Hz–1.5 kHz and 1.5–2 kHz. Figure 5.90 shows the normalized difference, $T/T_{\text{base}} - 1$, between the transmissibility function for the baseline (*healthy*) cylinder and the second healthy cylinder and the 1 and 5 ft·lb impacted cylinders. The feature extracted from the transmissibility function was the cumulative sum of the magnitude of the normalized difference in transmissibility across a given frequency range. The values of these features are listed in Table 5.8 for the two frequency ranges. The values are provided for two of the damaged cylinders and the healthy cylinder. A maximum error of ±0.1 is indicated in the table

Table 5.8 Transmissibility damage indicators across two frequency ranges for S2 glass filament wound cylinders (Reproduced by permission of Society for Experimental Mechanics, Inc.)

Cylinder	Frequency range	
	700–1.5 kHz (±0.1 max. error)	1.5–2 kHz (±0.1 max. error)
Baseline comparison	0.71	0.58
1 ft lb impact	1.12	0.88
5 ft lb impact	1.22	1.50

based on experimental observations of the change in damage indicator when the test is repeated. The damage indicators for both frequency ranges detect the presence of the 1 and 5 ft lb impacts because the baseline damage indicator defines the threshold for damage detection. The 1.5–2 kHz frequency range more clearly quantifies the difference between the two damaged cylinders than the 700 Hz–1.5 kHz range. This example demonstrates the methodology by which damage in structural components is detected in general.

5.11. 2 Damage Localization

Spatial analysis must be performed to locate damage. Various methods for locating damage have been discussed in previous sections. For example, wave propagation data was processed using spectrograms to locate damage in a one-dimensional rod model in Section 5.5.4 and a two-dimensional plate model in Section 5.7.2. Propagating waves were also analyzed in Section 5.7.4 to identify the direction from which waves scattering from damage arrive in order to triangulate the location of damage. The curvature of modal vibration deflection shapes was also used in Section 5.7.1 to locate damage. All of these techniques utilize data from more than one spatial location to locate damage.

From a systems perspective, damage location is tantamount to identifying the component that is damaged. For example, a suspension system like the one shown in Figure 1.2(d) has many components. In order to locate damage in this suspension, the component that is damaged must be identified. The wheel hub could be cracked, the stabilizer bar could be cracked or the seals in the strut could be leaking. Similarly, the metallic panel shown in Figure 2.1 can become damaged in the panel or the struts that support the panel. The damage location algorithm in this example would need to identify if damage is located in one or both of these components. From a component perspective, damage is located by pinpointing the region of the material component that is damaged. For example, the rolling tire illustrated in Figure 5.61 is a single component that can become damaged in the bead area as shown in Figure 3.7(a), the sidewall or the tread. In order to locate damage in this example, data must be acquired and analyzed accordingly within each of these regions of the tire.

It is important to locate damage in applications where it is desirable to predict the growth of damage. The location of existing damage determines in large part the internal or external load that acts to increase the size and extent of the damage. For example, the stabilizer bar link shown in Figure 5.68(a) contains a notch in which a crack can grow until the bar fractures. In order to predict how quickly the crack grows, the location of the damage must be known; otherwise, the loading across the crack cannot be estimated. Similarly, the stabilizer strut in the unmanned air vehicle shown in Figure 3.16 can become cracked as indicated in the figure. In order to predict how quickly this crack will grow or if the strut will fail when turbulence is encountered by the aircraft, the location of damage along the strut must be identified so that the proper bending moment can be calculated.

Consider again the S2 glass composite cylinder described in Section 5.11.1. Damage was detected in that section using transmissibility functions measured across the impacted region of the cylinder. Transmissibility functions can also be used to approximately locate the damage. Figure 5.91 illustrates the two original sensor locations (1 and 2) in addition to three other locations at 3, 4 and 5 (behind the cylinder). Recall that transmissibility functions are more sensitive to local changes in stiffness; therefore, the transmissibility function across the path from sensor 1 to 2 should change more than the transmissibility functions across the

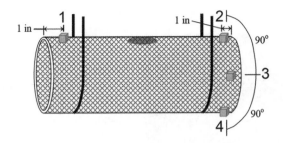

Figure 5.91 Cylinder with five acceleration sensors attached (one in the back of cylinder)

other pathways created by this group of sensors. Table 5.9 lists the damage indicators calculated using the method described in Section 5.11.1 for four of these pathways. Note that the highest damage indicator is obtained for path 1–2, which is where the impact damage is located. This approach provides an approximate location of the damage but it is not known where the impact damage is situated along the path between 1 and 2.

To obtain a more precise estimate of the damage location, the beamforming method illustrated in Figure 5.89 can be used. An actuator is shown exciting the cylinder with a pulse waveform like those described previously with a center frequency of 19.75 kHz. Then an array of sensors measures the wave front as it travels through the cylinder, interacting with any damage in the material along the way. A photograph of the cylinder with the actuator and sensors installed is shown in Figure 5.92(a). The sensors are installed in a T shape to provide enough data to localize the damage. A linear array of sensors would be insufficient because the sensors would be linearly dependent resulting in the detection of damage on both sides of the array. Damage was simulated using a local mass addition to change the impedance of the material at the location shown.

Beamforming calculations were performed with respect to the sensor marked 1 and the sensor marked 2 to identify the direction of waves that scatter from the attached mass through triangulation. Figure 5.89 illustrates that beamforming is achieved by delaying and then summing the signals measured at each of these sensor locations. The time delays introduced at each sensor location in the beamformer were calculated based on the group velocity of the propagating flexural wave at 19.75 kHz taking into account the curved nature of the cylinder. Figure 5.92(b) shows a planar image of the estimated location of the mass (x) and the actual location (∇). The actuator is shown with a solid circle and the sensors are shown with open

Table 5.9 Transmissibility damage indicators for S2 glass filament-wound cylinder after 5 ft ·lb impact across 1.5–2 kHz frequency ranges for four pathways (Reproduced by permission of Society for Experimental Mechanics, Inc.)

Path	Damage indicators (± 0.1 max. error)
1–2 (damage path)	1.50
1–3	0.80
1–4	0.92
1–5	1.11

(a)

(b)

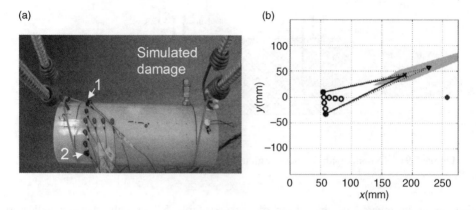

Figure 5.92   (a) Cylinder with actuator and sensors installed and (b) triangulated damage position using beamforming

circles. The estimated and actual locations of the mass are in relatively good agreement. This example demonstrates that damage can be located either approximately or precisely depending on the transducers and the data analysis algorithms that are used.

5.11.3 Damage Quantification

The damage quantification task provides the operator of a structural component with a sense of the operational risk associated with the damage. As discussed in Chapters 2 and 3, damage can be modeled as a change in material properties (density and modulus), geometric parameters (crack length and depth of corrosion), mechanical properties (stiffness, damping and mass) and in other ways. When quantifying damage, the objective is to estimate a physically meaningful parameter that characterizes the damage level. For instance, the transmissibility function damage indicators described in 5.11.1 provide an indication of the presence of damage, but the indicators by themselves do not provide a clue as to the extent of damage. A more physically meaningful indicator such as the reduction in stiffness due to fiber breakage caused by the impacts would provide a direct measure of the damage level. When quantifying damage, it is also usually necessary to first *classify the type of damage* so that the damage level can be assessed with a model.

Because damage is often nonlinear in nature, the presence of nonlinearity can be helpful in quantifying damage. For example, if the attachment bolt for the metallic panel shown in Figure 2.1 begins to loosen, then it will slide and rattle on the panel interface. By monitoring the change in nonlinear response, the level of damage can be quantified as shown in later chapters. The methods discussed in this chapter including HOS and restoring force projections in addition to other methods can be used to track changes in the nonlinear behavior of a component that becomes damaged.

As a specific example of damage quantification, consider again the S2 glass composite cylinder described in Section 5.11.1. Damage due to fiber breakage was detected in that section using transmissibility functions measured across the impacted region of the cylinder. It was observed in Table 5.8 that the transmissibility-based damage indicators for the 5 ft lb

Table 5.10 Estimated reduction in bending stiffness of cylinder wall for two impacts (Reproduced by permission of Society for Experimental Mechanics, Inc.)

Cylinder	Stiffness reduction (lb/in)
1 ft lb impact	60
5 ft lb impact	250

impacted cylinder were large relative to the damage indicators for the 1 ft lb impacted cylinder. This result provides a measure of the relative level of damage; however, in order to utilize this difference to quantify damage, both sets of these results are needed so that the 5 ft lb result can be compared to the 1 ft lb result. When sufficient test data is available on damaged components, this type of data-driven approach to quantifying damage in a relative sense is appropriate. If test data is unavailable or too expensive to acquire on damaged components, then damage must be quantified using a physics-based approach.

Damage in the S2 glass cylinders can be quantified more directly using physics-based modeling techniques. For example, the embedded sensitivity model described in Section 2.3.2.1 can be used to estimate the reduction in bending stiffness of the cylinder across the impacted area of the cylinder wall. Knowledge of this reduction in stiffness would be helpful in ascertaining the potential for failure when the cylinder is pressurized. Figure 5.89 illustrates that the embedded sensitivity method uses an actuation force applied and measured at location $A1$ along with two response measurements from sensors $S1$ and $S2$. This calculation produces the stiffness reduction estimates listed in Table 5.10 for the 1 and 5 ft lb impact energies. When compared to the original static stiffness of the cylinder wall at the midpoint, 880 lb/in, the 250 lb/in reduction for the 5 ft lb impact represents a 28 % loss in strength. This information is useful to an operator and component engineer because it provides a measure of the severity of damage and potential effect on the component performance. This example demonstrates that although relative indications of damage level can be obtained using data-driven methods, physics-based models provide a direct measure of the damage level.

As a second example of damage quantification, consider the aluminum beam shown in Figure 5.93(a). This beam is made from 6061-T6 aluminum and is meant to represent a helicopter rotor. The beam has length 27.5 in, width 1.5 in and thickness 0.125 in. One end is clamped to a shaker, which exerts a vertical force causing the beam to vibrate. The first two resonant frequencies of the beam were identified to be at 5.5 and 33 Hz. The second

(a)   (b)

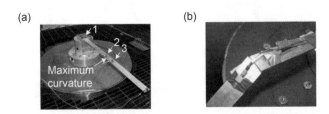

Figure 5.93 (a) Aluminum beam representing a rotor with three sensors to measure transmissibility and (b) beam after it has fractured (Reproduced by permission of SAGE Publications, Inc.)

resonant frequency exhibited its maximum curvature at the location indicated in the figure; therefore, symmetric notches with 1 mm on the top and bottom were milled into the beam at this location of maximum curvature to ensure that crack growth would occur in the notches. The location of maximum curvature is more susceptible to crack growth for the reasons discussed in Section 5.10.1 relating to the stress in a beam. Three acceleration sensors (PCB 333A32, 100 mV/g) were installed at locations 1–3 as shown in the figure where locations 2 and 3 lay on either side of notch. These sensors were used to measure the transmissibilities between the motion at the root of the beam and the motions on either side of the notch (Nataraju *et al.*, 2005).

In order to initiate and accelerate the growth of a crack in the notch of the beam, the excitation frequency and amplitude were selected to excite the beam's second resonance. As the crack in these experiments grew, the second natural frequency decreased; therefore, the shaker was driven at 32.5 Hz for 53 min, 30.6 Hz for 2 h and so on in order to continuously excite the second resonance, which exhibited the largest curvature across the crack. A background random vibration component was also used in the excitation spectrum in order to excite the beam throughout the frequency range from 0 to 500 Hz. The beam eventually failed as shown in Figure 5.93(b).

During the tests, the transmissibility function $T_{13}(\omega)$ was measured and is plotted in Figure 5.94(a) in the frequency range containing the fourth natural frequency. Note that the natural frequency decreases. This trend in natural frequency was used along with the finite element model(FEM) of this beam (refer to Section 2.2.8.4 for mass and stiffness matrices) to estimate the crack depth in the notch continuously throughout the test. Each time the

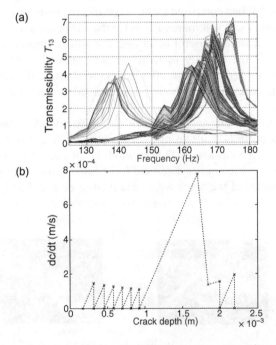

Figure 5.94 (a) Change in measured transmissibility between root and sensor across notch and (b) estimated rate of change in crack depth versus crack depth during test (Reproduced by permission of SAGE Publications, Inc.)

natural frequency decreased, the FEM was simulated iteratively to identify the crack depth that could cause the measured decrease in natural frequency. A symmetric crack was assumed in these finite element simulations. This approach produced a measure of the damage level (crack depth) at each point in the test. In order to use the estimated crack depth for quantification of damage and for making predictions, the rate of change in crack depth versus the crack depth was plotted. This plot is shown in Figure 5.94(b). The points shown on this plot at $dc/dt = 0$ are artifacts of the plotting process. Note that the rate of change in crack depth is nearly constant until the estimated crack depth of 1.6 mm is reached. At this crack depth, the beam undergoes rapid crack growth due to the effects of plastic deformation along the notch.

5.12   REGRESSION ANALYSIS FOR PROGNOSIS

Both the loading information and damage information generated from the previous two sections are useful for health monitoring. Individually, these pieces of information can be used to guide inspections, maintenance, logistics and operations. In order to make predictions about future performance, however, loading and damage information must be combined using other information relating to the operation of a component. *Regression models* are generally used to develop these relationships for use in prediction (Bevington and Robinson, 1992). Regression models for health monitoring use parametric models to relate one or more variables that affect component performance.

Two types of regression models are commonly used. One type of model describes the *growth of damage with some parameter(s)*. These models are useful for predicting the damage level at some future point in time. For example, the plot of *dc/dt* versus *c* in Figure 5.94(b) can serve as a regression model relating the rate of change in damage level to the damage level. Because of the large excursion in the plot for the crack length of 1.6 mm, an actual polynomial model of low order cannot be fit to this data. This model would be useful for predicting the future crack depth given the current estimated depth.

The second type of regression model describes how the *performance of a component is related to one or more parameters*. These models are useful for predicting if a component will perform at an acceptable level in the future. For example, a set of data was illustrated in Figure 1.4(b) for filament wound cylinders subjected to impacts. This data can be used to generate a regression model that is either linear or quadratic based on the method of least squares discussed in several earlier sections. The test data used to generate the regression model relates the pressure at which the cylinders burst when pressurized to the energy of impacts that act on those cylinders. The resultant model can be used for prediction. For instance, if an impact of 3 ft lb occurs, then the burst pressure can be estimated and compared to the performance requirement for a missile casing constructed with this cylinder.

Regression models can be identified using physics-based methods or data-driven methods. In general, physics-based methods can be used when more information is available about the specific damage mechanism and loading of interest. For example, the FEM utilized in the example above involving the aluminum beam to estimate the crack depth is the basis of a physics-based method to develop the regression model in Figure 5.94(b). Physical models are used to develop regression models in physics-based approaches. On the contrary, data-driven methods are based solely on the use of test data that relates damage or loading data to performance data. Data-driven models can be used in cases even when the underlying physical nature of the damage or performance criterion is uncertain. For example, the burst pressure versus impact load energy regression model in Figure 1.4(b) is based only on test data and no *a priori* physical model of the cylinder. Additional examples using both types of methods are given in the next two sections.

5.12.1 Physics-Based Methods

Consider the shock module clevis shown in Figure 3.2(a). This clevis has a notch in it; the notch causes a fatigue crack to initiate and grow under dynamic loading. A vehicle corner test rig was used to impose and measure these loads along the shock module. Accelerometers were used to measure the responses on either side of the clevis. At intermediate points within the durability test schedule, a sine sweep was used with 0.058 Hz/s from 0–35 Hz to measure the restoring forces and FRFs across the clevis. The areas inside the restoring force projections at 29.2 Hz, RF_{area}, were then calculated using methods described previously (see Section 5.4.2). The measured FRFs across the clevis were also used to estimate the reduction in stiffness, ΔK, across the clevis due to the growth of the crack. This reduction in stiffness was estimated using the embedded sensitivity modeling technique described in Section 2.3.2.1.

To develop a regression model relating the internal loading, damage level and rate of change in the damage level, the expression in Equation (3.14) was used with $m = -1$, $n = 1$ and $C = 0.015$. The resulting regression model is plotted in Figure 5.95. This model indicates what the rate of change in the damage level (e.g. loss in stiffness) is as a function of the number of loading cycles. The change in internal load is incorporated as indicated in Equation (3.14) to account for the change in loading with damage. This type of regression model is useful for predicting the rate of change in damage and future damage levels in future tests involving this shock module subjected to damage in the clevis.

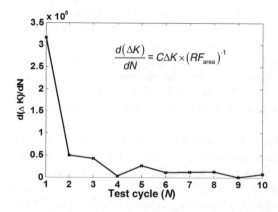

Figure 5.95 Regression model relating estimated change in stiffness to change in restoring force area for suspension module clevis with crack shown in Figure 3.2(a)

5.12.2 Data-Driven Methods

The S2 glass cylinder discussed in Sections 5.10 and 5.11 must withstand a certain burst pressure during operation. In order to characterize how the burst pressure changes as a function of impact loading, a regression law similar to the one shown in Figure 1.4(b) for the larger filament-wound canister can be identified experimentally. This model could then be used to predict the future performance of the cylinder if the impact energy can be estimated. Likewise, a regression model relating the burst pressure to the reduction in static stiffness of the cylinder wall that was estimated in Section 5.11.3 can also be developed. If the load is not estimated, then this regression model can be used to predict the future performance given only the estimated damage level (reduction in stiffness). In general, regression models relating any of the damage indicators described in this chapter to future performance can be developed in this manner. Two types of regression models will be described in case studies presented in Chapter 8.

5.13 COMBINING MEASUREMENT AND DATA ANALYSIS

The measurement and data analysis techniques described in this chapter and the previous chapters are combined with models of components and damage to develop health-monitoring systems. Figure 5.96 illustrates the health-monitoring process applied to the S2 glass filament wound composite cylinders analyzed in Sections 5.10–5.12. This procedure was also illustrated in Figure 1.6 for a gas turbine engine wire harness that is analyzed in case studies later in the book.

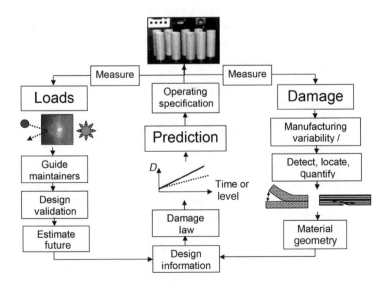

Figure 5.96 Health-monitoring process applied to S2 glass filament wound cylinders

5.14 SUMMARY

This chapter has reviewed data analysis techniques for health monitoring. The primary issues in filtering, state inference, temporal analysis, spectral analysis, time-frequency analysis, averaging, spatial analysis, loads identification, damage identification and regression modeling were discussed. Variability analysis in health monitoring was also described to ensure that loading and damage features can be interpreted properly in the midst of operational, environmental and computational variability. Appendix B provides a summary of references and a list of citations from the literature on data analysis. Some of the key issues are highlighted below.

- Models are *the means by which load and damage information are extracted from raw health-monitoring data.*
- A filter is *any weighting function used to emphasize certain parts of the data and de-emphasize other parts of the data.*
- Windows are *weighting functions multiplied by the time data to reduce the effects of leakage.*
- Frequency–domain filters are *weighting functions in amplitude and phase applied to either analog or digital data to emphasize certain frequency data and de-emphasize other data.*
- Spatial filters *weight data from different measurements according to some rule that emphasizes certain components of the data and de-emphasizes other components.*
- Phased arrays are *sets of omnidirectional sensors (or actuators) that function together as one directionally sensitive sensor (or actuator).* In phased sensor arrays, spatial filters are produced when *sensors receive data and then the data is weighted, and phase shifted to increase the sensitivity to waves approaching from specific directions.*
- *In many applications, the unmeasured variables have different dynamics than the measured variables.* In order to estimate unmeasured variables in these applications, state observers are required.
- The advantage of analyzing time history data directly is that *no transforms are required leading to fewer errors.* The disadvantage of analyzing time histories is that *loading and damage characteristics may not be as easily identified nor interpreted using raw time data.*
- Deterministic temporal analysis techniques utilize *time series models often used for prediction to analyze time history data.*
- Time–delay embeddings are sets of delayed measurements that relate one variable to time-delayed versions of that variable.
- The advantage of transformed data analysis is that broadband data can be analyzed more easily in addition to narrowband data. The disadvantage of transforming data is that errors are introduced by the transformation process.
- *DFT analysis is useful for distinguishing between frequency components that are sensitive to changes in the component and frequencies that are insensitive to changes.*
- The basic idea behind HOS is that interactions between various frequency components can be detected by analyzing the spectral content of the data along more than one frequency axis.
- By including both time and frequency as independent variables in the analysis, changes in frequency content can be detected as a function of time and vice versa.
- Wavelets are functions that can be stretched and shifted to represent a variety of signals. The wavelet kernel shortens the time interval for analyzing high-frequency signal components and lengthens the time interval for analyzing low frequency signal components.

- *Synchronous averaging* is used if there is a reference signal either from an actuator applied to actively sense the component response or from some other sort of trigger. *Asynchronous averaging* is used when there is no reference signal to which the averaging can be performed.
- *Locating damage is tantamount to detecting damage in different spatial pathways* from one measurement to another.
- *The process of locating some kind of source using distributed measurements is called triangulation.*
- When analyzing raw data, important parameters are extracted, called *features*, from which conclusions can be drawn about the nature of loading and damage in a component.
- *Multiple* features are sometimes needed to fully characterize the change in loading (or damage) in a component.
- The reason that *fewer features are preferred is because less measurement data is needed to identify loads and damage with confidence using fewer features.*
- Training data is used in health monitoring to establish the normal and abnormal condition for a component. *Thresholds* are then set up to differentiate between the normal and abnormal condition of the component.
- *Hypothesis testing* is the field of statistics concerned with choosing between two (or more) different conjectures about a population of features using a *test statistic. More samples result in more confidence in the sample mean and, therefore, the diagnosis. Less damage results in less confidence due to the variance in the sample mean.*
- Some sources of variability that are random in nature can be lumped together and dealt with using the statistical methods presented in the previous section. Other sources of variability should be analyzed one by one from a physical point of view.
- The process of decomposing the total variation of a given feature into two parts, *one due to the sensitivity of the data to the source of variation and the other due to the way in which the feature is calculated*, helps to identify and minimize computational variability.
- Loads identification methods are used to estimate both external loads and internal loads from measurement data.
- As in all *inverse problems*, in which an unknown variable is estimated using known variables, loads estimation algorithms must be properly conditioned in order to provide unique and accurate results.
- The damage identification process involves three basic steps: (1) detecting damage, (2) locating damage and (3) quantifying damage.
- When quantifying damage, it is also usually necessary to first *classify the type of damage* so that the damage level can be assessed with a model.
- In order to make predictions about future performance, loading and damage information must be combined using design information relating to the operation of a component.

REFERENCES

Allemang, R. (1999) 'Vibrations: analytical and experimental modal analysis', UC-SDRL.

Allen, D.W., Castillo, S., Cundy, A.L., Farrar, C.R. and McMurry, R.E. (2001) 'Damage detection in building joints by statistical analysis', *Proceedings of the 19th International Modal Analysis Conference*, 955–961.

Bendat, J.S. and Piersol, A.G. (1993) *Engineering Applications of Correlation and Spectral Analysis*, 2nd edition, John Wiley Sons, Ltd, New York.

Bevington, P.R. and Robinson, D.K. (1992) *Data Reduction and Error Analysis for the Physical Sciences*, 2nd edition, McGraw-Hill, Inc, New York.

Bishop, C. (1995) *Neural Networks for Pattern Recognition*, Oxford University Press, New York.

Brigham, E.O. (1988) *The Fast Fourier Transform and Its Applications*, Oppenheim, A. (Ed.), Prentice Hall, Englewood Cliffs, New Jersey.

Chui, C. K. (1992) *An Introduction to Wavelets*, Volume 1, Academic Press, San Diego, California.

Cohen, L. (1995) *Time-Frequency Analysis*, Prentice-Hall, Upper Saddle River, New Jersey.

Dietrich Jr., C.B. (2000) 'Adaptive arrays and diversity antenna configurations for handheld wireless communication terminals', Ph.D. dissertation, Virginia Tech.

Dougherty, E.R. (1990) *Probability and Statistics for the Engineering, Computing, and Physical Sciences*, Prentice Hall, Englewood Cliffs, New Jersey.

Farrar, C.R., Duffey, T.A., Doebling, S.W. and Nix, D.A. (1999) 'A statistical pattern recognition paradigm of vibration-based structural health monitoring', *2nd International Workshop on Structural Health Monitoring*, Stanford, California, 764–773.

Farrar, C.R., Doebling, S.W. and Nix, D.A. (1999) 'Vibration-based structural damage identification', The Royal Society, Physical Trans.: Math., Phys., and Eng. Sci., **359**(1778), 131–149.

Farrar, C., Fasel, T.R., Gregg, S.W., Johnston, T.J. and Sohn. H. (2002) 'Experimental modal analysis and damage detection in a simulated three story building', *Proceedings of the 20th International Modal Analysis Conference*, Paper no. 379.

Friedland, B. (1986) *Control System Design: An Introduction to State-Space Methods*, McGraw-Hill, Inc., New York.

Frost III, O.L. (1972) 'An algorithm for linearly constrained adaptive array processing', *Proc. IEEE*, **60**, 926–935.

Grandt, A. (2004) *Fundamentals of Structural Integrity*, John Wiley Sons, Ltd, New York.

Griffiths, L.J. and Jim, C.W. (1982) 'An alternative approach to linearly constrained adaptive beamforming', *IEEE Trans. AP*, **AP-30**, 27–34.

Hamming, R.W. (1989) *Digital Filter*, 3rd edition, Dover Publications, Mineola, New York.

Jiang, H., Adams, D. and Jata, K. (2006) 'Material damage modeling and detection in a homogeneous thin metallic sheet and sandwich panel using passive acoustic transmission', Struct Health Monit, v. 5, pp. 373–387.

Johnson, T. (2002) 'Analysis of dynamic transmissibility as a feature for structural damage detection', *Masters Thesis*, School of Mechanical Engineering, Purdue University, West Lafayette, Indiana.

Johnson, T.J. and Adams, D.E. (2006) 'Rolling tire diagnostic experiments for identifying incipient bead damage using time, frequency, and phase-plane analysis', (invited) Society of Automotive Engineering World Congress on Experiments in Automotive Engineering, Paper no. 2006-01-1621.

Kess, H. (2006) 'Investigation of operational and environmental variability effects on damage detection algorithms in heterogeneous (woven composite) plates', *Masters Thesis*, School of Mechanical Engineering, Purdue University, West Lafayette, Indiana.

Nataraju, M., Adams, D.E. and Rigas, E. (2005) 'Nonlinear dynamics simulation and observations of damage evolution in a cantilevered beam', *Int. J. Struct. Health Monit.*, **4**, 259–282.

Newland, D.E. (1993) *An Introduction to Random Vibrations, Spectral and Wavelet Analysis*, 3rd edition, Addison Wesley, Longman Limited, London, United Kingdom.

Nikias, C.L. and Petropulu, A.P. (1993) *Higher Order Spectra Analysis: A Nonlinear Signal Processing Framework*, Oppenheim, A. (Ed.), PTR Prentice Hall, Upper Saddle River, New Jersey.

Packard, N.H., Crutchfield, J.P., Farmer, J.D. and Shaw, R.S. (1980) 'Geometry from a time series', *Phys. Rev. Lett.*, **45**(9), 712–716.

Rytter, A. (1993) 'Vibration based inspection of civil engineering structures', *PhD Thesis*, University of Aalborg, Denmark.

Stark, J., Broomhead, D.S., Davis, M.E. and Huke, J. (1997) 'Takens embedding theorems for forced and stochastic systems', *Nonlinear Anal. Theory Method. App.* **30**, 5303–5314.

Strogatz, S.H. (1994) *Nonlinear Dynamics and Chaos*, Addison-Wesley, Reading, Massachusetts.

Sundararaman, S. (2003) Structural diagnostics through beamforming of phase arrays: characterizing damage in steel and composite plates, *Masters Thesis*, School of Mechanical Engineering, Purdue University, West Lafayette, Indiana.

Sundararaman, S. and Adams, D.E. (2002) 'Phased transducer arrays for structural diagnostics through beamforming', *Proceedings of American Society of Composites 17th Annual Technical Conference*, WA4, paper no. 177, ISBN: 0-8493-1501-8.

Sundararaman, S., Adams, D.E. and Rigas, E. (2005) 'Structural damage identification in homogeneous and heterogeneous structures using beamforming', *Int. J. Struct. Health Monit.*, **4**(2), 171–190.

Takens, F. (1981) 'Detecting strange attractors in turbulence', *Dynamical Systems and Turbulence- Lecture Notes in Mathematics*, Rand, D. A. and Young, L. S. (Eds), **898**, Springer-Verlag, Berlin.

Van Veen, B.D. and Buckley, K.M. (1988) 'Beamforming: a versatile approach to spatial filtering', *IEEE ASSP Mag.*, 1–24.

Virgin, L. (2000) *Introduction to Experimental Nonlinear Dynamics: A Case Study in Mechanical Vibration*, Cambridge University Press, New York.

Widrow, B. and Stearns, S. (1985) *Adaptive Signal Processing*, Prentice-Hall, Englewood Cliffs, New Jersey.

Wooh, S.C. and Shi, Y. (1999) 'Optimum beam steering of linear phased arrays', *Wave Motion*, **29**, 245–265.

Worden, K. and Tomlinson, G.R. (1990) 'The high frequency behavior of frequency response functions and its effect on their hilbert transforms', *Proc. Int. Modal Anal. Conf.*, **1**, 121–130.

PROBLEMS

(1) Use the three different types of windows shown in Figure 5.4 to compute and plot the magnitude of the frequency spectrum of a swept sinusoidal signal from 100 to 200 Hz with unity amplitude. Use a time step of 0.001 s and a sweep rate of 10 Hz/s. Comment on the differences obtained in terms of amplitude and frequency accuracy using the three windows.

(2) Use the simulation codes 'threedoftrans.m' and 'filterdemo.m' to calculate and plot the three modal responses obtained from the three-DOF system model shown in Figure 2.12. Assume a Gaussian excitation force is applied with unity variance and zero mean to DOF 1. Comment on the frequency content of these responses.

(3) Use the simulation code 'beamsteer.m' to animate the array gain factor for a beamformer used to spatially filter incoming propagating waves. Analyze the change in beamformer performance for different values of N_s (number of sensors; 2, 4 and 8) and d (spacing between sensors; $d = \lambda/2, 2\lambda, \lambda$. Which set of parameters provides the best overall performance?

(4) Use the simulation code 'threedoftrans.m' along with the code 'thermalobserver.m' to develop a dynamic observer for estimating the response x_2 and x_3 given a measurement of x_1 of the three-DOF system model shown in Figure 2.12. Plot these actual and estimated responses as a function of time when the three-DOF system model is excited by one-half of a sinusoid over 0.1 s.

(5) Examine the change in how quickly the thermal observer developed in Section 5.3 can sync up with the actual responses of the three-DOF system model when the entire observer gain matrix is decreased. What happens to the speed of the estimated response when only one element in the matrix is decreased? The simulation code 'thermalobserver.m' can be used to conduct this study.

(6) For the two-DOF system model shown in Figure 2.2, use the simulation code 'twodoftemporal.m' to calculate all of the statistical features listed in Table 5.2 for a 10 % decrease in the mass M_1 simulating corrosion damage. Calculate and plot each feature for five data sets where the excitation is random (given in 'twodoftemporal.m'). Which feature(s) could be used to detect a change in M_1?

(7) Consider the longitudinal rod illustrated in Figure 2.23. Use the simulation code 'rodtemporal.m' to calculate the differences between baseline and damaged responses at various DOFs along the rod (recall Figure 2.26) for a crack located at various locations along the rod. Use the excitation force already provided in the simulation code. Calculate the mean time and temporal variance. How could this information be used to locate the crack?

(8) Calculate and plot the auto-correlation and cross-correlation functions for the long-itudinal rod with the crack located at various positions along the rod. Also, vary the magnitude of the damage by increasing the percent decrease in modulus used to model the crack. Use the simulation code 'rodtemporal.m' to conduct the analysis. Comment on how the damage can be located and quantified using the information supplied by the auto and cross-correlation functions.

(9) Use the simulation code 'twodoftemporal.m' to examine the change in restoring forces obtained for the two-DOF system model in Figure 2.2 for increasing damage level. Use the 5 rad/s excitation force already provided in the code. Calculate the restoring force areas for DOFs 1 and 2 when the nonlinear parameter μ_2 is increased from 0 to 150 N/m^2 in steps of 25 N/m^2. Plot the change in restoring force areas as a function of μ_2. Which of the DOFs provides a better means for quantifying damage using restoring force areas?

(10) Repeat problem (9) using time–delay embeddings and return maps. Use the excitation force containing 2 and 8 rad/s provided in the code 'twodoftemporal.m'. Use a delay of 0.15 s to construct the return map for the displacement of DOF 2, x_2. Calculate and plot the cumulative Euclidean distance between the return maps before and after the non-linear damage is inserted as a function of μ_2.

(11) Use the simulation code 'twodofembed.m' to repeat the analysis leading up to Figure 5.28, except in this case plot the Poincare section corresponding to the return map by sampling the response x_1 once per period of oscillation, first, for the 2 rad/s component and then for the 8 rad/s component. Comment on the changes in the Poincare section for increases in the nonlinear parameter.

(12) Calculate and plot the DFT magnitudes for the two-DOF system model in Figure 2.2 when different springs within the model are decreased by 10 % to simulate damage as was done in Figure 5.32. Which damage locations lead to the most significant changes in the DFT? Which frequency ranges change the most? Use the code 'twodofdft.m' to conduct this analysis along with the forcing function defined in the code.

(13) Consider the two-DOF system model in Figure 2.2. Use the simulation code 'two-dofdft.m' along with the forcing function defined in the code to calculate and plot the magnitude of the DFT of x_1 when a friction nonlinearity is introduced at DOF 2 as described in Section 2.2.7. Increase the parameter μ_1 in the MATLAB® code from 0 to 0.05 N to simulate increases in friction at a bolted attachment. Describe a method for quantifying this change in damage level using the nonlinear nature of the response.

(14) Calculate, plot and explain the spectrogram of the signal $y(t) = \sin(2\pi(1 + 0.02 * t)t) + \sin(2\pi(10 - 0.05 * t)t)$. The code 'timefreqanalysis.m' can be used to solve this problem.

(15) Move the simulated crack to element 4 and then element 14 in the longitudinal FEM of a rod found in the code 'timefreqanalysis.m.' Then generate the difference spectrograms corresponding to before and after damage is introduced for all of the nodes as was done in Figure 5.40. Explain how damage can be located using these spectrograms.

(16) Suppose simulated damage is introduced first at element 4 and then at element 14 in the longitudinal FEM of a rod found in the code 'timefreqanalysis.m.' Generate the difference spectrograms needed to locate the damage. Recall that difference spectro-grams correspond to before and after damage is introduced in the rod (see Figure 5.40) (Hint:Use a different baseline to locate each damage location).

(17) Use the simulation code 'frfestimation.m' to estimate the FRFs for the two-DOF system in Figure 2.2(a) using the H_2 estimator corresponding to those shown in Figure 5.48

estimated using the H_1 estimator. Comment on the differences. Now add noise to the force time history and recalculate the results.

(18) Move the simulated stiffness damage in the plate model (Figure 4.6(a)) found in 'ninedofspatial.m' to different locations, calculate the modal deflections and plot the difference between damaged and undamaged deflections. Explain how the damage can be located using this approach. Simulate damage using a mass change instead of a stiffness change and repeat the steps above to locate the damage.

(19) Repeat the steps used to generate Figure 5.54 and Figure 5.55 by implementing the simulation code 'ninedofspatial.m'; however, add 5 % Gaussian random noise to the deflection shapes before computing the spatial derivatives and plotting them. Comment on the errors obtained in these results relative to those in Figure 5.54 and Figure 5.55.

(20) Use the simulation code 'platefemsim.m' to generate the wave propagation response of the two-dimensional FEM of an aluminum plate with damage located at various locations. Then process the nodal difference data before and after the simulated damage is introduced using spectrograms ('timefreqanalysis.m') to locate the damage as was done in Figure 5.40 for the longitudinal rod containing simulated damage.

(21) Use the discrete time domain model demonstrated in Section 5.4.2 with the code 'twodoftemporal.m' to generate model coefficients relating the excitation force at DOF 1 to the displacement response measured at DOF 2 in the two-DOF system for use as features in damage detection. Then introduce damage in the form of a 10 % mass reduction at DOF 2. Examine the changes in model coefficients with increases in the percent reduction in mass.

(22) Repeat the example in Section 5.8.4 for the nine-DOF system model; however, simulate damage by reducing the spring connecting nodes 4 and 5 by 10 % instead of with a mass change at DOF 4. Use the same excitation force as in the code 'ninedoffeature.m.' What is the hypothesis test to detect damage in this case? What is the threshold for detection? What are the probabilities of false positive and negative detection? Decrease the change in stiffness to 1 %. How do the results of the hypothesis test change?

(23) Use the code 'ninedofdata.m' to conduct a computational variability analysis for the nine-DOF system model in Figure 4.6. In which frequency ranges will the transmissibility function between DOFs 1 and 9 exhibit the largest change in magnitude as the two corresponding FRFs vary? What sources of environmental and measurement variability might cause these FRFs to vary? Compare the sensitivity of this transmissibility pair to the transmissibility pair containing DOFs 1 and 4. Explain the significance of the difference between these two pairs of measurements with respect to variability.

(24) Calculate the variability formulae given in Equation (5.63) for transmissibility functions using the embedded sensitivity function instead (recall Section 2.3.2.1).

(25) Formulate the least squares estimation problem (Equation (5.69)) needed to estimate the point load P applied to the beam shown in Figure 5.83 given one additional response data point along the beam. Use strength of materials formulas for the beam to relate the displacement at the second location to the tip displacement.

6
Case Studies: Loads Identification

This chapter and the next two chapters give brief descriptions of case studies in loads identification, damage identification, and damage and performance prediction. The techniques introduced in Chapters 2–5 are the basis for these case studies. A mix of data-driven and physics-based approaches is discussed. Some of these case studies involve transient loads (impacts), whereas others involve steady-state loads (vibration and thermal).

6.1 METALLIC THERMAL PROTECTION SYSTEM PANEL

The metallic thermal protection system (TPS) panel shown in Figure 2.1 is used to demonstrate two different types of loads identification techniques. This panel is used as a heat shield on aerospace vehicles. The first technique for estimating loads uses a data-driven model consisting of frequency response functions (FRFs). The second technique uses a Newtonian model of the panel. Both techniques use the measured acceleration response of the panel as it undergoes an impulsive load to estimate that load. This type of load can be caused by a micrometeorite, dropped tool, hail storm or other sources of debris impact.

6.1.1 Data-Driven

Heat shield panels like the one shown in Figure 2.1 can reach temperatures as high as 1650 °F and acoustic pressures up to 180 dB during launch and re-entry. This type of mechanically attached TPS panel is envisioned to replace the current bonded tiles used on the Space Shuttle Orbiter, which replaced ablative tiles used on the predecessors of the Space Shuttle (e.g. Apollo capsule). The ease with which these mechanically attached panels can be replaced is the primary advantage of this type of panel. The panel in Figure 2.1 is constructed from two-face sheets and a honeycomb core all manufactured from Inconel® 617 (Hundhausen, 2004). The standoff attachments are constructed from MA 754. These nickel metal alloys can withstand extremely high temperatures.

The corner of the panel that is instrumented with a triaxial accelerometer (1 V/g) is shown in Figure 6.1(a). The accelerometer is shown attached to the topside of the panel; however, the

---

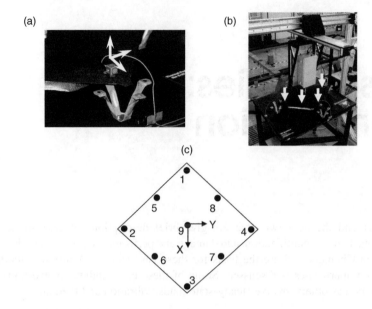

Figure 6.1 (a) Corner of TPS panel with triaxial accelerometer, (b) panel undergoing low-velocity impacts from automatic impactor and (c) nine impact locations (Reproduced by permission of Jason Hundhausen, LANL, 2004)

accelerometer would be attached to the bottom of the panel in practice to avoid high temperatures and debris. The panel is installed on a heavy steel table in (b).

The table is meant to simulate the airframe to which these panels are attached. An automatic impact machine is used to excite the panel in one location with an impulsive load in the vertical direction. Each impact has the same amount of energy associated with it. A load cell measures the force input, and four triaxial accelerometers installed on the corners of the panel measure the acceleration response of the panel as a whole. Nine impact locations (shown in (c)) are struck five times each to estimate the FRFs between the measured impact forces and responses using the H_1 FRF estimation technique discussed in the previous chapter.

This set of FRFs is used to construct the training model used for estimating impulsive loads that act on the panel. The model relating the measured response acceleration spectra, $\{A\}$, to the measured force spectra, $\{F\}$, is of the following form:

$$\left\{ \begin{array}{c} A_{1x} \\ A_{1y} \\ A_{1z} \\ A_{2x} \\ \vdots \\ A_{4z} \end{array} \right\}_{12\times1} = \left[\begin{array}{cccccc} H_{1x,1z} & H_{1x,2z} & H_{1x,3z} & H_{1x,4z} & \cdots & H_{1x,9z} \\ H_{1y,1z} & H_{1y,2z} & H_{1y,3z} & H_{1y,4z} & \cdots & H_{1y,9z} \\ H_{1z,1z} & H_{1z,2z} & H_{1z,3z} & H_{1z,4z} & \cdots & H_{1z,9z} \\ H_{2x,1z} & H_{2x,2z} & H_{2x,3z} & H_{2x,4z} & \cdots & H_{2x,9z} \\ \vdots & \vdots & \vdots & \vdots & \vdots & \vdots \\ H_{4z,1z} & H_{4z,2z} & H_{4z,3z} & H_{4z,4z} & \cdots & H_{4z,9z} \end{array} \right]_{12\times9} \left\{ \begin{array}{c} F_{1z} \\ F_{2z} \\ F_{3z} \\ F_{4z} \\ \vdots \\ F_{9z} \end{array} \right\}_{9\times1} \quad (6.1a)$$

$$\{A\} = [H]\{F\} \tag{6.1b}$$

The various subscripts indicate the directions from which and to which the FRFs are calculated. The frequency ω is not included in the equation to simplify the notation.

Table 6.1 Results of test impacts using data-driven FRF model for loads estimation (Reproduced by permission of Jason Hundhausen, LANL, 2004)

Location	Actual impact force (lb$_f$)	Estimated impact force (lb$_f$)	Error (%)
1	23.4	23.2	−0.72
2	10.2	10.6	4.02
3	14.1	14.1	−0.23
4	14.8	14.8	−0.09
5	23.5	23.5	0.35
6	10.0	10.0	−0.42
7	9.8	9.8	0.22
8	8.3	8.5	2.04
9	15.2	14.9	−2.01

After calculating the training FRF model in Equation (6.1), it was used to identify new impacts that were made at one of the nine locations shown in Figure 6.1(c). The impacts were located and quantified by solving the overdetermined set of equations in Equation (6.1). There are 12 equations but only nine unknown forcing entries assuming only one impact occurs in the vertical direction. After transforming the measured acceleration responses into the frequency domain using the discrete Fourier transform (DFT), the pseudoinverse of the rectangular FRF matrix was calculated as described following Equation (5.71) to produce an estimate of the force vector: $\{\hat{F}\} = [H]^+\{A\}$. The pseudoinverse produces a solution that minimizes the sum of the squared errors across all 12 equations in Equation (6.1). The vector of the estimated force spectra was then transformed back into the time domain using the inverse DFT. Examples of the estimated maximum forces compared to the actual measured maximum forces at each of the nine locations are listed in Table 6.1. The percentage error indicates that the force amplitude estimates were relatively accurate.

Figure 6.2(a) illustrates these results in graphical form. The panel location versus the estimated maximum force is shown on the plot when each of the nine locations was impacted. The results indicate that the maximum force levels occur when the panel is impacted at

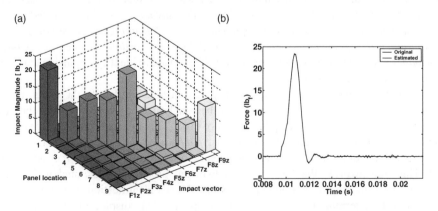

Figure 6.2 (a) Graphical illustration of results from impact tests after developing training model and (b) example of estimated and measured forces (Reproduced by permission of Jason Hundhausen, LANL, 2004)

locations 1, 5 and 9. Because the force levels are highest at these locations and all the impacts possess the same amount of energy, it can be concluded that the panel displaces the least at locations 1, 5 and 9. This conclusion can be drawn because the integral of force and displacement must be constant across equal energy impacts. The displacements vary locally in the panel due to small differences in how the face sheets are attached to the honeycomb core as well as to the differences in modal deflections at the nine locations. By measuring the acceleration response, the maximum force upon impact can be estimated and provided to maintainers who determine whether to inspect certain panels or replace them altogether. Figure 6.2(b) shows an example of the estimated impact force–time history obtained after solving Equation (6.1) and transforming into the time domain. The estimated and measured force–time histories are nearly equivalent with the exception of some error in the estimated force due to measurement noise.

Issues to consider beyond those discussed here include the effects of larger amplitude impacts than those used to construct the training model. Also, the ability to locate and quantify impacts that act on locations other than those used in the training model would be important. This issue was partially addressed in Section 5.10.3 for the filament-wound cylinders. Although 12 measurements (four triaxial accelerometers) were used to estimate the impact forces in this example, the forces could be estimated with as few as nine measurements. If glancing impacts occur causing relatively large components of the force in the x and y directions, then this approach would also need to be modified. Finally, nonimpulsive loads applied to more than one location simultaneously could also be estimated using the data-driven FRF model.

6.1.2 Physics-Based

The data-driven approach described above utilized 12 measurements to estimate nine forces. Fewer measurements generally result in less accurate loading estimates when data-driven techniques are used. Also, data-driven techniques usually lose accuracy when there is variability from one panel to another because the training data is only valid for the panel(s) on which the training data is acquired. There are exceptions to this limitation as demonstrated in Section 5.10.3 for the filament-wound cylinder with 7 in. diameter. Physics-based techniques use first-principle models to estimate loads. *The advantage of physics-based models is that they usually generalize better to components for which training data is not acquired. The disadvantage is that assumptions made in the modeling process limit the accuracy of physics-based models.* A Newtonian model (see Section 2.2) is used to identify impulsive loads on the Inconel[®] panel in this section.

For the panel shown in Figure 2.1, two steps are needed to identify the loads. First, the acceleration response data from the corners of the panel must be used to estimate the translational and rotational motions of the panel center of mass. This step involves the *kinematics*. It is assumed here that the panel can be treated like a *rigid body* because the struts are more flexible than the panel. Second, these motions must then be used to estimate the forces acting on the panel. This step involves the *kinetics*. This two-step process is illustrated in Figure 6.3. In general, this is the approach used in physics-based loads identification. Measured responses are first used to determine the motion of the component (e.g. displacement of beam shown in Figure 5.83). Then these motions are used to determine the forces that produce the motion (e.g. force distribution acting on beam shown in Figure 5.83).

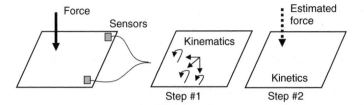

Figure 6.3 Two-step process for identifying impulsive load on panel

To estimate the translational and rotational motions with respect to the center of mass of the metallic panel, the motions are measured at a point p on the panel. Four triaxial accelerometers like the one shown in Figure 6.1(a) measure the response in the x, y and z directions. The motions at these points p are related to the motions with respect to the center of mass in the following way if it is assumed that the rotations are small:

$$
\left\{\begin{array}{c} \ddot{x} \\ \ddot{y} \\ \ddot{z} \end{array}\right\}^{p}_{3 \times 1} = \begin{bmatrix} 1 & 0 & 0 & 0 & z_{p,cm} & -y_{p,cm} \\ 0 & 1 & 0 & -z_{p,cm} & 0 & x_{p,cm} \\ 0 & 0 & 1 & y_{p,cm} & -x_{p,cm} & 0 \end{bmatrix}^{p}_{3 \times 6} \left\{\begin{array}{c} \ddot{X} \\ \ddot{Y} \\ \ddot{Z} \\ \ddot{\Theta}_x \\ \ddot{\Theta}_y \\ \ddot{\Theta}_z \end{array}\right\}^{cm}_{6 \times 1}
\tag{6.2a}
$$

$$
\{\ddot{x}\}^{p} = [T_R]^{p}\{\ddot{R}\}^{cm}
\tag{6.2b}
$$

where the $x_{p,cm}$, $y_{p,cm}$ and $z_{p,cm}$ coordinates in the transformation matrix $[T_R]$ define the distance of p relative to the panel center of mass, cm. Equation (6.2) is a result of Chasle's theorem for a rigid body; that is, all motions of a rigid body are the sum of a rigid body translation of cm and rotation relative to cm. By making triaxial measurements at two or more such points like p, a least-squares solution can be obtained for the center of mass motions, $\{\ddot{R}\}^{cm}$. The triaxial motions at four points were measured in this case.

To estimate the forces applied to the panel, the Newton–Euler equations of motion for the panel were used. Assuming the panel behaves like a rigid body, the equations of motion can be expressed as follows:

$$
\left\{\begin{array}{c} F_x \\ F_y \\ F_z \\ M_x \\ M_y \\ M_z \end{array}\right\}^{cm}_{6 \times 1} = \begin{bmatrix} m & 0 & 0 & 0 & 0 & 0 \\ 0 & m & 0 & 0 & 0 & 0 \\ 0 & 0 & m & 0 & 0 & 0 \\ 0 & 0 & 0 & I_{xx} & I_{xy} & I_{xz} \\ 0 & 0 & 0 & I_{yx} & I_{yy} & I_{yz} \\ 0 & 0 & 0 & I_{zx} & I_{zy} & I_{zz} \end{bmatrix}_{6 \times 6} \left\{\begin{array}{c} \ddot{X} \\ \ddot{Y} \\ \ddot{Z} \\ \ddot{\Theta}_x \\ \ddot{\Theta}_y \\ \ddot{\Theta}_z \end{array}\right\}^{cm}_{6 \times 1}
\tag{6.3a}
$$

$$
\{F\}^{cm} = [M]\{\ddot{R}\}^{cm}
\tag{6.3b}
$$

The inertial parameters for the panel were computed using a solid model: $m = 3.15$ lb, $I_{xx} = 77.1$ lb·in.^{2}, $I_{yy} = 89.5$ lb·in.^{2}, $I_{zz} = 165.6$ lb·in.^{2} and $I_{xy} = I_{yz} = I_{xz} = 0$. When modeling the panel, the inertia of the insulation beneath the panel was neglected. The sequence of operations is given below to estimate the external forces applied to the panel:

- Forces were applied to the panel, and the triaxial motions at point p were measured. These motions were used to estimate the motion of the panel with respect to cm:

$$\{\ddot{R}\}^{cm} = ([T_R]^P)^+ \{\ddot{x}\}^p \tag{6.4}$$

- The estimated motions of the panel with respect to cm were then used to estimate the equivalent forces and moments acting through cm:

$$\{F\}^{cm} = [M]\{\ddot{R}\}^{cm} \tag{6.5}$$

- The estimated forces and moments acting through cm were then transformed into equivalent applied forces acting at five points on the panel; four triaxial forces acting through the standoff attachments (see Figure 6.1(a)) and one triaxial force acting at the point of impact. There were 17 unknowns including the 12 forces through the standoffs, three forces at the impact point and the x and y coordinates of the impact point (note the z coordinate was zero). To reduce the number of unknowns, the triaxial displacements at the four corners of the panel were estimated using the acceleration measurements. Then these displacements were multiplied by the stiffnesses of the standoffs in the three directions to directly calculate the forces due to the standoffs. This approach left only five unknowns with six equations of the form similar to Equation (6.4) where forces were substituted for motions. The unknown forces and location were then obtained using an overdetermined least-squares solution.

A comparison between the measured and estimated maximum applied forces obtained using this physics-based modeling approach is given in Table 6.2. The errors are considerably larger using the physics-based approach than for the data-driven FRF modeling approach. The primary sources of these errors are (1) the assumption that the panel behaves like a rigid body,

Table 6.2 Results of physics-based model for loads estimation (Reproduced by permission of Jason Hundhausen, LANL, 2004)

Location	Actual impact force (lb$_f$)	Estimated impact force (lb$_f$)	Error (%)
1	21.1	19.3	−8.40
2	33.0	17.3	−47.58
3	33.9	37.5	14.01
4	35.0	16.8	−51.92
5	43.9	16.6	−62.13
6	34.3	23.9	−30.20
7	43.8	34.6	−21.00
8	35.0	21.6	−38.31
9	39.3	22.5	−42.74

(2) errors in the stiffness estimates for the panel attachments and (3) damping in the attachments was neglected. These sources of errors led to the force being underpredicted in most cases. Also, the location of the impact was only identified correctly in five out of the nine impact locations. Despite these errors in the physics-based approach to loads identification, this approach is often preferable to the data-driven approach for the reasons mentioned above

6.2 GAS TURBINE ENGINE WIRE HARNESS AND CONNECTOR

The wire harness connector panel shown in Figure 1.6 is used in gas turbine engines to deliver power to components and control their performance. As described in Section 1.4, the connectors that screw into these panels can become damaged due to hard landings or inadvertent loads that occur when the engine is serviced. Corrosion inside the connectors (see Figure 3.11(a)) can also prevent the harnesses from fully screwing into the panel and can cause weakening of the connector pins. Information about the loads acting on the various connectors that screw into these panels would be helpful to service engineers when they are called upon to diagnose faults (Stites *et al.*, 2006).

In order to estimate the loads that act on the connectors individually, consider the simplified wire harness connector panel shown in Figure 6.4(a). The panel has three connector bases with three harnesses connected to those bases. Figure 6.4(b) shows a miniature triaxial accelerometer (10 mV/g) installed on the rear of the panel. This sensor was used to measure the response of the panel in order to develop a data-driven training model for loads identification. The back panel is an advantageous location for the sensor because the panel's vibration provides information relating to all of the wire harness connectors.

A medium modal impact hammer (10 mV/lb) was used to excite the panel in the six locations and directions shown in Figure 6.4(a). Five averages were used to estimate the H_1 FRFs relating these input forces, $\{F\}$, to the three measurements made by the triaxial accelerometer, $\{A\}$. The FRF model constructed from these FRFs is as follows:

$$\{A\}_{3\times 1} = [H]_{3\times 6}\{F\}_{6\times 1} \tag{6.6}$$

(a)

(b)

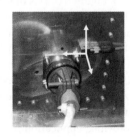

Figure 6.4 (a) Wire harness connector panel and (b) backside of panel showing single triaxial accelerometer (Reproduced by permission of DEStech Publications, Inc.)

where the frequency variable, ω, is removed for simplicity. The six forces in this equation were estimated in an iterative manner to compensate for the underdetermined nature of Equation (6.6). There are six unknowns but only three equations for calculating those unknowns. The following procedure was used for estimating the impulsive forces:

- For a given loading case (e.g. striking one of the three connectors in the vertical direction), the three acceleration–time histories were measured when the connector was struck. The force–time history was also measured using a load cell for comparison with the estimated values. The time history data was transformed into the frequency domain using the DFT. The pseudoinverse of $[H]$ was then calculated so $\{\hat{F}\}$ could be estimated using the expression $\{\hat{F}\} = [H]^{+}\{A\}$.
- The entries in $\{\hat{F}\}$ were then analyzed to determine the most likely location of the applied load. For example, the maximum force and its location among the six estimated forces were identified. The magnitudes of the combined estimated forces on each connector were also determined. Also, the magnitudes of the estimated forces acting on connectors 1 and 2 were compared with those of connector 3 using a statistical model based on the training data because the panel was asymmetric. Similar statistical comparisons were used to analyze the other two possible connector groupings. This process led to a conclusion as to which connector the impulsive force actually struck.
- After identifying the connector to which the force was applied, the data-driven FRF model was rewritten in reduced form taking into account this information:

$$\{A\}_{3\times1} = [H']_{3\times2}\{F'\}_{2\times1} \tag{6.7}$$

- This set of equations is overdetermined with only two unknowns in three equations in contrast to the original set of equations in Equation (6.6). The pseudoinverse (least-squares) solution for $\{\hat{F}'\}$ was then calculated.
- Smoothing (low-pass) filters were then applied in the frequency domain to reduce the variance in the estimated force spectra. The estimated spectra were compared to the measured spectrum to determine the accuracy of the estimate.
- Finally, the estimated force spectra were transformed back into the time domain to calculate the peak force and compare the estimated and measured force–time histories.

Figure 6.5 shows the results of the data-driven loads identification approach for the wire harness connector panel subject to a vertical (y) impulsive load on the first connector in the left of Figure 6.4(a). Figure 6.5(a) is the frequency domain comparison and (b) is the time domain comparison between the actual (measured) and estimated forcing function. Figure 6.5(a) shows that the algorithm correctly identifies the direction of the loading because the estimate of the x direction force is nearly zero. The spectral magnitude of the y direction force (vertical) also matches the measured force spectrum reasonably well. It is evident that the algorithm for calculating the estimated forces is poorly conditioned in the frequency range below 50 Hz and above 600 Hz due to numerical noise in the estimates in those ranges. Figure 6.5(b) shows that the estimated force–time histories are also relatively accurate.

As in the previous application involving the metallic panel, the data-driven model for the wire harness connector panel would need to be evaluated for forces much larger in amplitude than the training force levels to determine the effectiveness of the model. Also, more connectors would need to be considered to ensure that the technique scales to the realistic

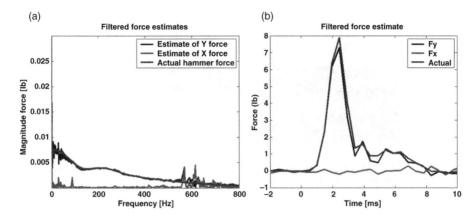

Figure 6.5 Actual force compared to estimated force in the (a) frequency domain and (b) time domain

panel shown in Figure 1.6. Additional tests were also performed by removing and then reattaching the harnesses onto the connector bases and removing and reattaching the accelerometer to determine the effects on the trained FRF model. It was determined that the changes due to these sources of variability have minimal effect on the accuracy of identifying loads if the connectors are fully tightened.

6.3 FUSELAGE RIVET PROCESS MONITORING

Riveting is a dynamic process similar to cutting, grinding and polishing. Rivets are inserted in a vulnerable region of a structural component near holes where stress concentrations are found. Damage mechanisms like stress corrosion, exfoliation corrosion and fatigue cracking are a function of the quality of 2000/7000 series aluminum alloys around rivet holes. Poor transverse mechanical properties can result from riveting leading to the initiation of these damage mechanisms. In addition, all repairs of aircraft fuselage are carried out manually by riveting teams. Human error caused by the skewed delivery of a rivet, insufficient preload supplied by the rivetor, etc. can cause variability in rivet quality. The objective of loads identification in riveting is to extract information from the rivet and part dynamic response that can be used to assess the quality of the rivet process.

A US4R pneumatic rivet gun is shown in Figure 6.6(a). The gun is operated by pressing the trigger causing high air pressure to propel a piston down the barrel. When the piston strikes the rivet head, the rivet is hammered into place. Common errors that occur during riveting include the following: (a) not maintaining a straight gun on the rivet head causing 'eyebrows' to form on the aircraft skin; (b) removing the bucking bar too early in the process resulting in damage to the rivet ('smiley') and (c) countersinking of the aircraft skin with too great of a depth causing damage to the aircraft skin.

The schematic in Figure 6.6(b) shows the other components of the riveting process. A bucking bar is held in place by a second operator while the rivet gun operator drives the rivet through the part. Shock accelerometers are shown attached to the butt of the gun barrel and were used because of their low sensitivity (0.1 mV/g) and ability to measure high accelerations indicative of the riveting process. Size 6 rivets with universal head were used.

(a)   (b)

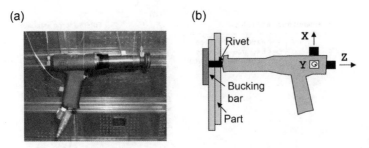

Figure 6.6   (a) Pneumatic rivet gun and (b) schematic of tool-part components with shock accelerometer to measure response of gun (Johnson *et al.*, 2006, AIAA)

The fuselage section shown in Figure 6.7(a), which is based on a Boeing 737 with 0.050 in.
thick 2024T3 aluminum and 7075T6 stringers/frames, was used in the experiments described
here. As riveting was performed, the three accelerometers on the gun were used to acquire
operational time history response data (see Figure 6.7(b)). A 10 kHz sample rate was used
along with a 500 Hz analog low-pass filter. The riveters were asked to perform five operations
of good quality and also several poor quality operations. The errors introduced in these poor
quality operations were a $10°$ skewed delivery (gun at $10°$ relative to the normal at the part
surface), skewed bucking bar at $10°$ with respect to the tangential direction at the part surface,
bucking bar removal part of the way through the operation, and extended riveting causing
damage to form on the rivet heads producing 'smiley' patterns.

The acceleration-time histories were transformed into the frequency domain using the DFT.
The results of good- and poor-quality rivet processes are shown in Figure 6.8. Figure 6.8(a)
corresponds to five good-quality rivets, whereas (b)–(e) correspond to rivet operations with
errors. The two boxed regions shown in the plots highlight two frequency ranges from 65 to
75 Hz and from 88 to 112 Hz that encapsulate the peaks in the good quality rivet gun response
spectra.

The acceleration spectra, A_j, were then used to calculate features for use in distinguishing
between good-quality and bad-quality rivets. The magnitudes of the three directional response
spectra were summed over a given frequency range, p, from $n_{p,min}$ to $n_{p,max}$ (where n denotes
the discrete frequency counter in the DFT) to produce these features:

$$F_p = \sum_{n_{p,min}}^{n_{p,max}} \sum_{j=1}^{3} \left\| A_j \left(\frac{n}{BS\Delta t} \right) \right\| \tag{6.8}$$

(a)   (b)

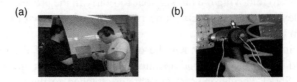

Figure 6.7   (a) Manual rivet gun process on test fuselage and (b) rivet gun instrumented with shock
accelerometers (Johnson *et al.*, 2006, AIAA)

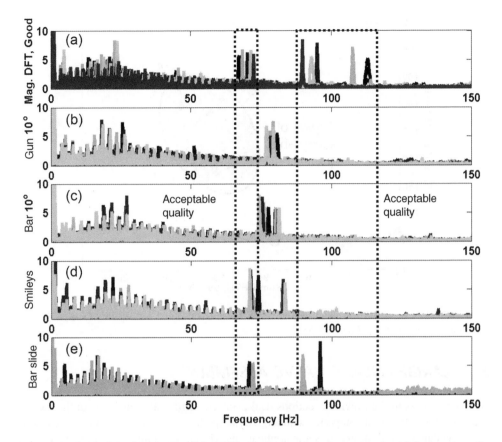

Figure 6.8 Sum of the magnitudes of the DFTs for the three accelerometers for (a) five good-quality operations, (b) four operations with 10° skewed delivery, (c) four operations with bucking bar oriented at 10° with respect to part, (d) four extended rivet operations producing 'smiley' pattern and (e) two operations with slippage of bucking bar (Johnson *et al.*, 2006, AIAA)

A two-dimensional feature vector was formed using the two frequency ranges mentioned above that are evident in Figure 6.8. A plot of this two-dimensional feature for each of the rivet operations is shown in Figure 6.9. The features for all except one of the poor-quality rivet operations fall outside the elliptical shape shown in the figure.

Johnson *et al.* (2006) developed an analytical nonlinear dynamic model that interprets these results from a physical point of view. A Simulink® model was used to show that changes in preload due to the errors during riveting are primarily responsible for the changes in the feature vector between good- and poor-quality rivet operations. By performing experiments with different rivet gun operators and bucking bar handlers, the experiments discussed here observed some of the variability that would be anticipated in practice. Other issues to address in practice would include the effects of changes in the rivet gun (mass), the type of rivet and the component undergoing riveting (stiffness, damping, etc.). A rigorous statistical model for hypothesis testing would also need to be developed to implement this type of rivet quality assurance technique in the field.

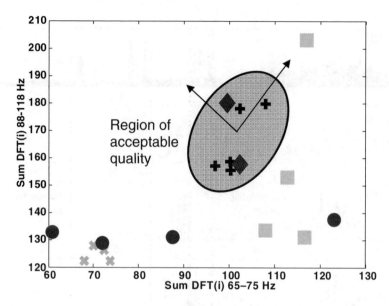

Figure 6.9 Plot of two-dimensional feature vector for good- (+) and poor-quality (●,◆,×,■) rivet operations (Johnson *et al.*, 2006, AIAA)

6.4 LARGE ENGINE VALVE ASSEMBLY

The case studies in the previous sections deal primarily with transient loads in the form of impacts. Many structural components undergo dynamic loads that are cyclic in nature leading to fatigue damage and failure. In some applications, vibration loading acts in the presence of high temperatures. These combined loads can result in combined failure modes of a component. For example, the engine valve shown in Figure 1.9 experiences both vibrational and temperature loads. This valve is used in large engines to shut off the supply of air–fuel mixture sent to the cylinder for compression. In the event of an engine fire, the valve is designed to close. In order to determine which health-monitoring measurements should be made and how to process the data to identify damage in this valve, the various types of loading must be well understood.

The internal components of the valve are shown in Figure 6.10(a). The circular component shown in the picture is used for manually rotating the valve. The component in the bottom

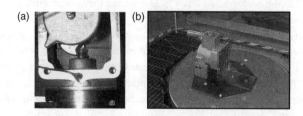

Figure 6.10 (a) Photograph of internal valve components (Adams, 2005) and (b) valve undergoing vibration testing (Reproduced by permission of John Wiley & Sons, Inc.,)

portion of the picture is the solenoid used to operate the valve. When the solenoid is activated with a specified DC voltage and current, the solenoid retracts the vertical plunger downward causing the triangular-shaped hammer to close the valve. Valve closure occurs because the valve axis is spring loaded.

If the valve becomes jammed during operation, it may fail to close. The bushing damage shown in Figure 1.9 caused the valve to jam in this manner. Vibration loading must be estimated to understand the source of this bushing damage. Also, if the solenoid pull force is not large enough to overcome resistance in the mechanism due to friction and other sources, then the valve can also fail to close. The effects of changes in temperature and actuation voltage in the electromechanical solenoid must be measured to understand these limitations in the solenoid.

In order to simulate field failures of this valve, it was tested using the electromagnetic shaker shown in Figure 6.10(b). The shaker applies a vertical acceleration to the valve along the valve axis. The fixture shown in the figure was designed to simulate a rigid attachment to the engine intake line. All of the resonances of the fixture were higher than the highest frequency excitation applied to the valve. The fixture and valve weighed approximately 40 lb in total. The Thermotron DS-4001 shaker is capable of imposing 1000 lb forces with up to 1 in. of displacement. The shaker was programmed to excite the valve with the operating spectrum shown in Figure 6.11. This spectrum was measured along the valve axis on an operating engine. A 24 VDC power supply was used to energize the valve frequently throughout the testing. After energizing the solenoid, it was noted whether the valve closed or failed to close and for what reason.

There are multiple frequencies in the operating spectrum (Figure 6.11). Each of these frequencies played a role in damaging the valve preventing it from closing:

- The low frequency 30 Hz component at a harmonic of the engine rotational speed caused the valve to respond with relatively large amplitude displacements. These displacements caused the valve to impact the bushing.
- The 120 Hz component due to a higher harmonic of the engine rotational speed caused the valve to rapidly impact the bushing. These high-frequency impacts when combined with the 30 Hz response caused the bushing to erode.
- The broadband frequency components above 166 Hz were due to a turbocharger positioned near the valve on some engine models of this type. This broadband response resulted in high

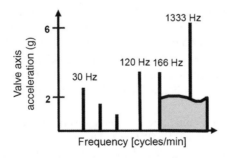

Figure 6.11   Vibration spectrum showing multifrequency nature of mechanical loading acting along the valve axis (Reproduced by permission of John Wiley & Sons, Inc.,)

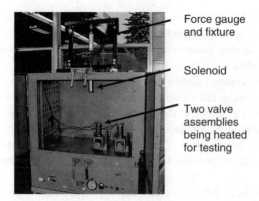

Figure 6.12   Thermal testing setup for valve assembly and solenoid

frequency rubbing between the plunger of the solenoid and hammer of the valve. It was observed in testing that this rubbing could eventually cause sticking of the valve.

In summary, the combination of these frequency components in the vibration loading along the valve axis caused two potential types of mechanical faults in the valve mechanism: (a) jamming of the valve due to erosion in the powdered metal bushing and (b) sticking due to material transfer (cold welding) between the solenoid plunger and hammer. As the valve is exposed to these vibration loads for increasing amounts of time, the valve is more likely to exhibit one or both of these faults.

As mentioned above, the thermal loads acting on the valve also played a role in preventing it from closing as designed. In order to identify the effects of thermal loading on the valve, the setup in Figure 6.12 was used. A solenoid from one of the valves was mounted on the top of an Environmetrics chamber capable of temperature and humidity control. A force gauge and fixture were used to measure the pull force of the solenoid at different operating temperatures. The solenoid skin temperature was measured using a thermocouple inserted between the solenoid and the ceiling of the test chamber. Steady-state conditions were ensured by applying the temperature to the solenoid for 5 min. The effects of sustained temperature exposure on the performance of the solenoid were also examined by heating the valve and then testing it on the shaker (see Figure 6.10(b)).

Figure 6.13 shows the change in solenoid pull force with the increasing solenoid skin temperature from 20 to 180 °C for three different energizing voltages (28, 24 and 20 VDC). The solenoid was rated for a 12.5 lb pull force when used with a 24 VDC power supply. It is evident from the data that a thermal load higher than 35 °C causes the solenoid pull force to drop below 12.5 lb. At the operating temperature of 80–100 °C, the pull force drops to below 6 lb. If the energizing voltage is below 24 VDC, the solenoid provides an even lower pull force.

In summary, increases in temperature and decreases in solenoid energizing voltage cause significant decrease in the solenoid pull force. These drops in pull force coupled with the mechanical sticking mechanisms described above resulted in the failure of the valve to close. The combination of vibration and thermal loads that has been identified in this case study using experimental techniques results in combined failure mechanisms.

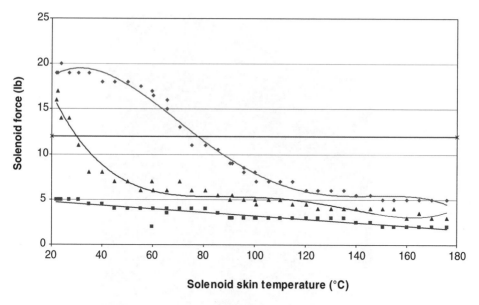

Figure 6.13   Solenoid pull force versus temperature (°C) for (◆) 28 V, (Δ) 24 V and (■) 20 V applied DC voltage

6.5   SUSPENSION WITH LOOSENING BOLT

Section 5.8.2 described an example of loads identification in a suspension system stabilizer bar link with a notch from which a fatigue crack grew. These loads were internal in nature and were largely a function of the interactions between the components surrounding the stabilizer bar. As a second example of internal loads identification, consider the bolt shown in Figure 3.1(a) from an Isuzu Impulse. This bolt connects the steering knuckle to the control arm through a ball joint. As indicated in Chapter 3, bolted fasteners are the primary cause of suspension recalls in the automotive industry. This case study of loads identification represents an important class of problems.

The hydraulic shaker apparatus shown in Figure 6.14 was used to force the left front tire vertically with a 0–15 Hz sine sweep excitation at a rate of 0.025 Hz/s. A sampling frequency of 600 Hz was used to acquire the data after low-pass filtering with a 100 Hz cutoff frequency. High sensitivity triaxial accelerometers (1 V/g) like the one shown in Figure 3.1(a) were used to acquire the response in several locations including at the top strut mount and the steering knuckle-control arm connection. The interest in this case study is to identify how changes in preload through the ball joint due to bolt damage change the internal forces within the suspension. This information is useful in determining how susceptible the suspension design is to bolt damage and how the load path changes when this damage occurs. To simulate damage to the bolt, the bolt torque was reduced from an initial value of 400 in · lb to 250, 100 a significant decrease in the and finally 0 in · lb by completely removing the bolt.

Velocity and displacement restoring forces were generated using the techniques described in Section 2.3.1.2 to characterize the change in internal load within the strut. Figure 6.15 shows the damping internal force within the strut at four different frequencies. The damping

Figure 6.14 Hydraulic shaker apparatus with Isuzu Impulse (Haroon *et al.*, 2005, AMSE)

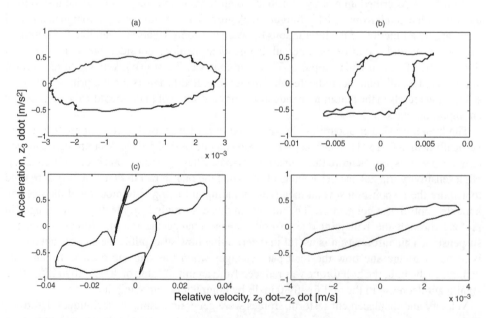

Figure 6.15 Frequency characteristic of damping internal force in the strut at (a) 4.05 Hz, (b) 4.17 Hz, (c) 6.67 Hz and (d) 12.5 Hz in the sweep data

force initially exhibits a central hysteresis loop in (a). As the frequency increases in (b), the force assumes the shape of a Coulomb friction curve (refer to Section 2.2.7). Then the shape changes to a piecewise-linear characteristic in (c), which is expected in a strut. For low velocities across the strut, the damping assumes one value in (c); for high velocities across the strut, the damping assumes a different value. Finally, at the highest frequency in (d) the damping internal force becomes linear in nature with some hysteresis. This information is useful because it indicates for what frequencies the strut dissipates the largest amount of response energy (4.05 and 6.67 Hz). At these frequencies, it is anticipated that the effects of damage in the suspension would be the greatest.

Because it was anticipated that the frequency range from 4 to 6 Hz would exhibit the most significant changes in damping restoring force due to damage in the bolt, the restoring forces were generated for the four different damage conditions in Figure 6.16 for the frequency range from 4.1 to 4.7 Hz. A bolted joint derives most of its stiffness from the stiffness of the components, which are clamped together by the bolt preload. Bolts extend as they are tightened to a point, but then the clamped components compress to provide the desired stiffness across the joint. As the bolt loosens in this case study at the steering knuckle, the stiffness across the ball joint decreases leading to smaller relative displacements and higher stiffness experienced by the strut, which is the component mounted above the ball joint. This increase in stiffness can be observed in the restoring forces plotted in Figure 6.17. The slope of the elliptical stiffness restoring force increases as the preload across the bolt decreases. These higher stiffness forces are accompanied by less dissipation (narrower hysteresis loop). This decrease in damping can also be observed in Figure 6.16. All of the damping restoring force curves in that figure lose their tails when the preload decreases.

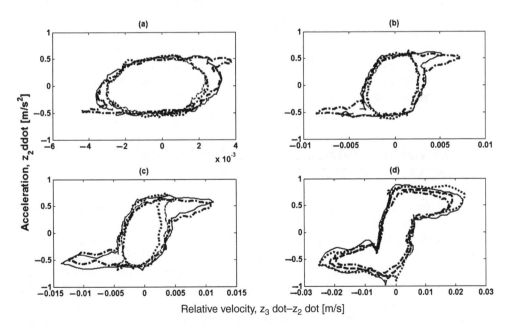

Figure 6.16 Change in frequency characteristic of damping internal force in the strut with damage at (a) 4.09 Hz, (b) 4.17 Hz, (c) 4.26 Hz and (d) 4.67 Hz in the sweep data for 400 in · lb (_), 250 in · lb (. . .), 100 in · lb (– – –) and no bolt (---) conditions

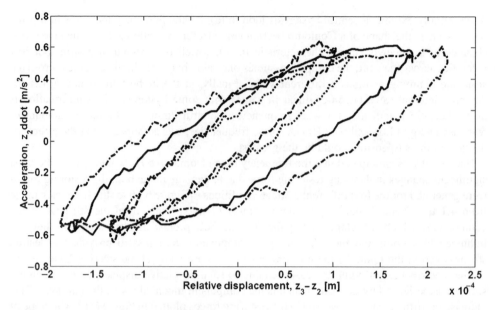

Figure 6.17 Change in stiffness internal force in the strut with damage at 4.17 Hz for 400 in·lb (__), 250 in·lb (...), 100 in·lb (– – –) and no bolt (-··-) conditions

When the bolt is completely removed, the stiffness drops again because the components are no longer clamped together to provide stiffness; consequently, the strut must deform by a larger amount to absorb the same amount of force as evident in Figure 6.17. This drop in stiffness then leads to more dissipation as shown in Figure 6.16. By computing the areas within these restoring force curves, these area values can be used to quantify changes in stiffness and damping restoring forces as was done in Section 5.4.2. This example demonstrates that both stiffness and damping forces vary as a function of damage in a system of components due to interactions among the components.

6.6 SANDWICH PANEL UNDERGOING COMBINED THERMO-ACOUSTIC LOADING

The case study in Section 6.4 demonstrated that combined loading in the form of temperature and vibration can result in more rapid failure of mechanical components. Each load contributes to different modes of component degradation that together lead to failure. This section considers the effects of combined loading in structural panels subject to high temperature and acoustic loads. Modest temperature changes up to 300 °F and acoustic levels up to 140 dB are considered. Aircraft and rotorcraft components in the hot gas path as well as satellite rocket motor casings experience loads much more severe than these in addition to vibration loading.

Figure 6.18 shows a picture of a combined loading environment for exposing structural components to thermal, acoustic and vibration loads simultaneously. The frame structure is used to support the vibration platform, which is 4 ft × 4 ft. Two shafts run through bearings on the base of this table at a fixed offset to provide a 10 g maximum acceleration at either 100 Hz

Figure 6.18 Combined vibration, acoustic and thermal loading environment (Reproduced by permission of Elsevier)

frequency or 1 in. amplitude oscillation of the table. The two motors used to drive these shafts are linked through a feedback control system. A sandwich panel is shown attached through struts to the table for testing. This panel is constructed from gamma TiAl with two face sheets and a honeycomb core. Four steel posts are used to support the panel. Two small quartz lamps are also shown and are used to heat the panel to temperatures as high as 300 °F. With quartz lamps, the proximity to the specimen determines the maximum achievable temperature. Much higher temperatures are possible using these types of lamps. A flexible tube entering at the top of the apparatus is also shown in the figure. This tube is attached to an acoustic air driver, which produces sound pressure pulsations at whatever frequencies and amplitudes the user requires. Sound pressure levels up to 140 dB are possible using this particular driver.

The focus of this discussion is on the effects of combined temperature and acoustic loads. In order to understand the effects of this combined loading, the free vibration response of the panel was measured at several temperatures in the range from 75 to 300 °F. As shown in Figure 6.19(a), a medium impact hammer was used to excite the panel at 25 points in the vertical direction. This array of points was chosen so that an acoustic sound pressure field could be simulated acting on the top panel surface. Four accelerometers were attached to the bottom of the panel to measure the response in four locations (see Figure 6.19(b) for one location). Actuators were also installed as shown in Figure 6.19(b) although they were not utilized in this particular case study. An insulation blanket was used to prevent high temperatures behind the panel. The actuator and sensor types that were used were shown in

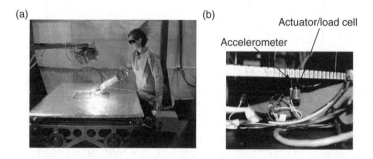

Figure 6.19 (a) Testing of panel to determine change in free vibration response with temperature and (b) backside of panel (Reproduced by permission of Elsevier)

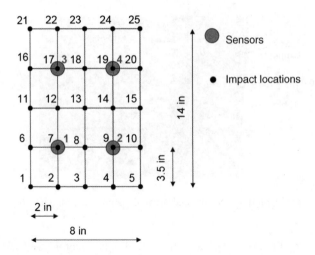

Figure 6.20 Panel measurement DOFs (Reproduced by permission of Elsevier)

Figure 4.33. The measurement DOFs are illustrated in Figure 6.20. Twenty-five input locations and four response locations are highlighted in the figure.

After estimating the FRFs between the 25 input and four response channels, a uniform
acoustic pressure pulsation amplitude of 120 Pa (135 dB) across the entire frequency range
was applied analytically to the panel. In practice, the acoustic loading would actually roll off
as a function of frequency; however, this pressure field represents a worst case scenario. Also,
the pressure would be higher in some locations on the panel and lower at other locations, and
the phase of each oscillating pressure component would be different at different locations. As
mentioned above, these effects were ignored because it was assumed that the pressure was
uniform to simulate a worst case scenario.

The FRF matrix was multiplied by the equivalent oscillating force amplitude, f_o in lb, at
each of the 25 input locations assuming a 120 Pa acoustic pulsation (1 Pa = 0.02 lb/ft^2). The
pressure was distributed evenly across the 25 input locations of the 1.5 ft^2 panel. The equation
for calculating the four response accelerations is given by

$$
\begin{Bmatrix} A_1 \\ A_2 \\ A_3 \\ A_4 \end{Bmatrix}_{4\times1} = \begin{bmatrix} H_{11}(T) & H_{12}(T) & \cdots & H_{1,25}(T) \\ H_{21}(T) & H_{22}(T) & \cdots & H_{2,25}(T) \\ H_{31}(T) & H_{32}(T) & \cdots & H_{3,25}(T) \\ H_{41}(T) & H_{42}(T) & \cdots & H_{4,25}(T) \end{bmatrix}_{4\times25} \begin{Bmatrix} f_o \\ f_o \\ \vdots \\ f_o \end{Bmatrix}_{25\times1}
\tag{6.9}
$$

The frequency variable is not shown in this equation to simplify the notation. Note that each
FRF is a function of temperature.

The predicted responses at the first sensor location in g's of acceleration as a function of
frequency at two different temperatures (75 and 300 °F) are shown in Figure 6.21. Note that
the response at the first two resonant frequencies increases somewhat when the frequency
increases. This increase in response at lower frequencies would cause a decrease in the fatigue
life of the component. Only the data up to 4000 Hz could be analyzed in this case because the
modal impact hammer could only impose a 4000 Hz bandwidth excitation. The most

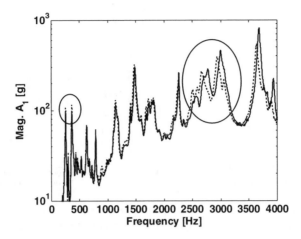

Figure 6.21   Data showing change in vibration response assuming uniform 120 Pa (135 dB) acoustic pressure applied at 75 °F (_) and 300 °F (...) surface temperatures (Reproduced by permission of Elsevier)

significant changes in the response occur above 2500 Hz. In general, the resonant frequency values decrease above 2500 Hz. This result suggests that the panel is more susceptible to acoustic loading when exposed to higher temperatures because acoustic excitations are generally band limited to 3 or 4 kHz. Consequently, as the resonant frequencies decrease, the panel resonant responses are more likely to become excited by the acoustic loading.

The measurements in this case study were limited to 300 °F because at this temperature of the top surface, the bottom surface temperature was 200 °F, the limit of the sensors being used. In order to predict how acoustic loads would affect the response of a panel heated to 2000 °F, the following data-driven modeling techniques were used:

- The temperature FRF data at 75, 115 and 300 °F were plotted for a given response DOF (e.g. A_1) and given input location (e.g. point 15 in Figure 6.20). These FRFs exhibited changes in amplitude and frequency with increases in temperature.
- The changes in FRF amplitude with increases in temperature were recorded at one frequency. It was found that these changes occurred throughout the entire frequency range of interest from 0–4000 Hz.
- These changes in FRF amplitude with temperature were plotted as shown in Figure 6.22(a). Three data points are shown (__,o). These data points were used to develop a quadratic model indicated with a dotted line (...,x). This model was then used to extrapolate to a temperature of 2000 °F. The amplitude ratio at this temperature was estimated to be 1.47.
- The same procedure was then applied to the frequencies in the FRF as a function of temperature. Figure 6.22(b) shows the three data points (_,o) in addition to the quadratic model (...,x) used to extrapolate to 2000 °F. The frequency scale factor used to model the reduction in resonant frequencies when the temperature increased to 2000 °F was estimated to be 0.9903. Note that the natural frequencies of a flat plate are multiples of E; therefore, this constant frequency scale factor is consistent with the anticipated theoretical shift in frequency.
- These two scale factors were then applied to the amplitude and frequency of the response, A_1, calculated using Equation (6.9). The resulting prediction of the panel response at a

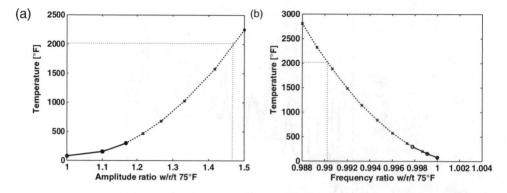

Figure 6.22 Data (_,o) and curve fits (...,x) of (a) temperature versus amplitude ratio and (b) temperature versus frequency ratio with respect to 75 °F (Reproduced by permission of Elsevier)

temperature of 2000 °F to a 120 Pa distributed, uniform acoustic pressure pulsation is shown in Figure 6.23. Note that there is a significant increase in response amplitude above 1000 Hz due to the temperature increase relative to the 300 °F temperature condition.

It was assumed in this case study that the panel material does not undergo any qualitative changes in behavior up to 2000 °F. If the material undergoes any such changes in its microstructure, then the model used to extrapolate is invalid. The material used in the posts for supporting the panel was steel. If this material experiences any changes in microstructure, then the extrapolation model will also fail to predict the high temperature response behavior of the mechanically attached panel.

This example demonstrates that combined loads can affect the health of structural components. Temperature increases affect the acoustic response of components by increasing the response amplitude and decreasing the resonant frequencies. Shifts in dynamic response due to temperature changes can also result in false indications of damage if the temperature change is not considered. If data is not available at higher temperatures in order to predict the effects

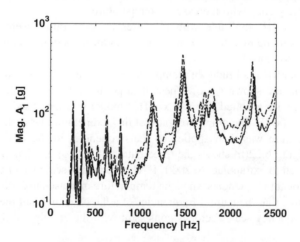

Figure 6.23 Data showing change in vibration response at 75 °F (_), 300 °F (...) and 2000 °F (– – –) surface temperatures (Reproduced by permission of Elsevier)

of combined loads, then data at lower temperatures can be used to extrapolate to higher temperature conditions.

6.7 SUMMARY

This chapter has demonstrated loads identification methods in several case studies involving a metallic sandwich panel, gas turbine engine wire harness and connector panel, engine valve assembly and suspension system. Physics-based and data-driven techniques have been applied to locate and quantify impact loads. Transient and steady-state loads have been studied. Combined loading from mechanical and thermal sources has also been examined, and failure modes due to combined loads have been described in one case study. Changes in stiffness and damping forces as a function of frequency and damage level have also been identified. Some of the key issues are highlighted below:

- Data-driven models are not as limited by the assumptions made when modeling how components respond to loads. Fewer measurements generally result in less accurate loading estimates when data-driven techniques are used. Data-driven techniques usually lose accuracy when there is variability from one component to another because the training data is only valid for the component(s) on which the training data is acquired.
- *The advantage of physics-based models is that they usually generalize better to components for which training data is not acquired. The disadvantage is that assumptions made in the modeling process limit the accuracy of physics-based models.*
- In physics-based load identification, measured responses are first used to determine the motion of the component (e.g. displacement of the beam shown in Figure 5.83). Then these motions are used to determine the forces that produced the motion (e.g. force distribution acting on the beam shown in Figure 5.83).
- When estimating loads, it is important to use *overdetermined* as opposed to *underdetermined* sets of equations so that least-squares solution techniques can be used.
- If the location of a load can be identified first, then inverse problems for loads identification can be more readily solved.
- Manufacturing processes like riveting, polishing and grinding can introduce damage mechanisms. Loads identification methods can be used during these processes to identify the susceptibility to future damage initiation.

REFERENCES

Adams, D.E. (2005) 'Prognosis applications and examples', Chapter 18 in *Damage Prognosis*, Inman, D. and Farrar, C. (Eds), John Wiley & Sons, Ltd, New York.

Ellmer, C., White, J. and Adams, D.E. (2006) 'Design and operation of a vibration-acoustic-thermal apparatus for identifying variations in free and forced response of healthy and damaged sandwich panels due to combined loading', *Proceedings of the 9th International Symposium on Multiscale and Functionally Graded Materials*, Honolulu, HI.

Haroon, M., Adams, D.E. and Luk, Y.W. (2005) 'A technique for estimating linear parameters using nonlinear restoring force extraction in the absence of an input measurement', Am J Vibration and Acoustics, **127**, 483–492.

Hundhausen, R. (2004) *Mechanical Loads Identification and Diagnostics for a Thermal Protection System Panel in a Semi-Realistic Thermo-Acoustic Operating Environment*, Masters Thesis, School of Mechanical Engineering, Purdue University, West Lafayette, Indiana.

Johnson, T.J., Manning, R., Adams, D.E., Sterkenburg, R. and Jata, K. (2006) 'Vibration-based diagnostics of tool-part interactions during riveting on an aluminum aircraft fuselage', J Aircraft, **43**(3), 779–786.

Stites, N., Adams, D.E., Sterkenburg, R. and Ryan, T. (2006) 'Loads and damage identification in gas turbine engine wire harnesses and connectors', *Proceedings of the European Workshop on Structural Health Monitoring*, Granada, Spain, 996–1003.

PROBLEMS

(1) Use the nine-DOF system model pictured in Figure 4.6 and simulated in 'ninedofdata.m' and 'ninedofenviron.m' to develop an FRF training model for transient loads identification. First, establish the baseline model by generating the FRF matrix for the nine input DOFs in the model. Second, apply a transient input force to the model and obtain responses at all nine DOFs. Then estimate the transient force after applying 3 % Gaussian noise to the response data. Change the width of the transient force and examine changes in the accuracy of the estimated force. Next, reduce the number of responses to DOFs 1, 3, 7 and 9 (corners of the model). How can this limited dataset be used to estimate the transient force?

(2) Repeat problem (1) but apply an excitation with three frequency components at 10, 50 and 100 Hz instead of a transient input force.

(3) Use the nine-DOF system model pictured in Figure 4.6 and simulated in 'ninedofdata.m' and 'ninedofenviron.m' to apply a similar physics-based methodology to the one described for the metallic panel in Section 6.1.2. First, assume the panel behaves like a rigid body. Then apply a transient force input to the panel, and use the response data from the four corners of the panel to estimate the motions with respect to the center of mass. Apply Newton's law to obtain the forces and moments with respect to the center of mass. Assuming the stiffness of the panel standoffs are known, estimate the external forces acting on the panel (location and magnitude).

(4) Repeat problem (3) but account for the panel flexibility in the model using a modal expansion to describe the responses at the corners.

(5) For the valve studied in Section 6.4, assume that 5 lb is required to close the valve without any mechanical damage. Then assume the rate at which the mechanism jams can be expressed in terms of jamming force, F_j. Assume the jamming force is a function of the time exposed to the spectrum shown in Figure 6.11 and is given by $F_j = 1 - e^{-at}$, where a is a small constant. Develop a performance model as a function of time and temperature that determines whether or not the valve will close for a 24 VDC energizing voltage to the solenoid.

(6) Use the simulation code 'platefemsim.m' to generate the response of the plate model to a transient vertical force (see Figure 2.29) at node 50. Use the displacement responses at nodes in the four corners of the plate (nodes 1, 11, 111 and 121) to locate the applied load.

(7) If a transient vertical force is applied to the two-dimensional square plate in Figure 5.64(a) at node 50, develop and implement a data-driven FRF model for use in identifying this force given the input-output behavior at nine different positions distributed across the panel. Use the code 'platefemsim.m' to develop the FRF model, simulate the transient response and estimate the impulsive force. Attempt to estimate the force using six positions. Comment on the errors obtained when using smaller numbers of DOFs in the data-driven model.

7

Case Studies: Damage Identification

Numerous case studies on damage identification are described in this chapter. Several of the components analyzed in these case studies were examined in the previous chapter on loads identification. Vibration-based methods are discussed first. The premise of vibration-based methods for damage identification is that damage is accentuated in measurement data when components vibrate. Both linear and nonlinear data analysis techniques are applied. Then wave propagation methods are discussed. As propagating waves travel, they encounter damage, which attenuates and scatters waves. Measurement data can be used to detect and locate damage by analyzing these disrupted waveforms. Examples of damage identification in the presence of operational loading are described in the last section of this chapter.

7.1 VIBRATION-BASED METHODS

7.1.1 Metallic Thermal Protection System Panel

7.1.1.1 Passive Method

Damage in the panel shown in Figures 2.1 and 6.1 is detected and located using passive operating vibration data in this section. The primary damage mechanism of interest in panels of this nature is bolt damage due to either self-loosening or cracking. In order to simulate authentic operational vibration responses, the panel was acoustically excited using a 12 in. loudspeaker with 1 kHz bandwidth Gaussian random noise. A 130 dB sound pressure level was created at the surface of the panel. The test setup is shown in Figure 7.1(a) and (b). A 240 Watt infrared heat lamp capable of 200 °F heating was also used to simulate moderate temperature fluctuations experienced by thermal heat shields in acreage thermal protection systems installed on the underside of aerospace vehicles. The objectives of this case study were to detect and locate damage using passive vibration data in a semi-realistic thermo-acoustic combined loading environment.

The first set of tests was performed at room temperature. Four 1 V/g accelerometers were attached with epoxy to the bottom of the panel in the locations shown in Figure 7.1(c). Each of the bolts was initially torqued to 75 in · lb ±5 in · lb for four baseline tests. Bolts were removed

Health Monitoring of Structural Materials and Components: Methods with Applications   D. Adams
© 2007 John Wiley & Sons, Ltd

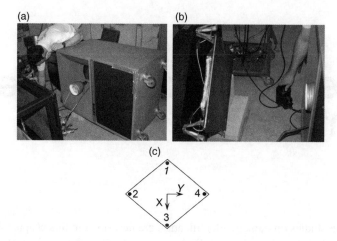

Figure 7.1   (a) Loudspeaker and infrared heat lamp used to excite metallic panel shown in (b) and (c) sensor locations on rear of panel (Reproduced by permission of SPIE)

and then replaced before conducting each test. The transmissibility functions were then estimated between the various pairs of sensors indicated in Figure 7.1(c) (i.e. 1–2, 1–3, 1–4, 2–3, 2–4 3–4). To estimate the transmissibility functions, a blocksize of 8192 was used with 878 averages and 50 % overlap. A Hanning window was applied to each block of data to reduce the effects of leakage. The transmissibility functions estimated in test 1 were stored and used to calculate the following transmissibility damage indices for the other three baseline tests:

$$DI_{i,j} = \frac{1}{N} \left\| \sum_{f=0}^{800} 1 - \frac{T_{i,j}^B(f)}{T_{i,j}^D(f)} \right\|, \tag{7.1}$$

where B denotes the test 1 transmissibility and D denotes the tests 2, 3 and 4 transmissibilities. Figure 7.2 shows the absolute values of the damage indices corresponding to tests 2, 3 and 4

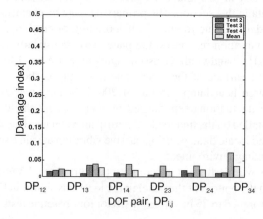

Figure 7.2   Absolute value of transmissibility damage index for three baseline tests relative to test 1 as a function of sensor pair where DP_{ij} denotes the index for sensor pair (i,j) (Reproduced by permission of SPIE)

Table 7.1 Different damage cases tested with four standoff bolts (Reproduced by permission of SPIE)

Case	Standoff 1	Standoff 2	Standoff 3	Standoff 4
1	TIGHT	TIGHT	TIGHT	TIGHT
2	15% PRELOAD LOSS	TIGHT	TIGHT	TIGHT
3	TIGHT	15% PRELOAD LOSS	TIGHT	TIGHT
4	TIGHT	TIGHT	15% PRELOAD LOSS	TIGHT
5	TIGHT	TIGHT	TIGHT	15% PRELOAD LOSS
6	35% PRELOAD LOSS	TIGHT	TIGHT	TIGHT
7	TIGHT	35% PRELOAD LOSS	TIGHT	TIGHT
8	TIGHT	TIGHT	35% PRELOAD LOSS	TIGHT
9	TIGHT	TIGHT	TIGHT	35% PRELOAD LOSS
10	15% PRELOAD LOSS	35% PRELOAD LOSS	TIGHT	TIGHT
11	35% PRELOAD LOSS	TIGHT	15% PRELOAD LOSS	TIGHT

*Preload Torque: 75 in-lb

for the six sensor pairs. The mean damage indices across these three tests are also shown. Because test 3 produced damage indices close to the mean, the damage indices for test 3 were used as the baseline with which all subsequent indices were compared to detect and locate damage. Note that many more baseline (healthy) tests would need to be conducted in order to estimate an accurate mean and variance for use in statistical hypothesis testing. Therefore, this case study will draw conclusions about damage based on qualitative comparisons of the damage indices.

The various damage experiments that were conducted are listed in Table 7.1. The results for damage cases 7 and 8 are discussed here. Figure 7.3 shows the damage indices for a 35 % reduction in torque in the standoff attachment bolt 3 (see Figure 7.1(c)). The largest changes occur for sensor pairs 1–3, 2–3 and 3–4. Each of these sensor pairs includes measurement DOF 3, which is the location of the loose bolt. It can be concluded from these results that transmissibility has detected and located damage in standoff bolt 3. Similarly, Figure 7.4 shows the damage indices for a 35 % reduction in torque in the standoff attachment bolt 2. In this damage case, sensor pairs 1–2, 2–3 and 2–4 exhibit the largest damage indices relative to those of the baseline test. Because the loose bolt is located at measurement DOF 2, it can again be concluded that damage has been detected and located using the transmissibility damage indices.

When Figure 7.4 is compared to Figure 7.3, it is evident that the damage indices for the loose bolt at standoff 2 are much larger than those for the loose bolt at standoff 3. The bolts utilized in each standoff were nominally identical. The difference in damage indices can be explained by

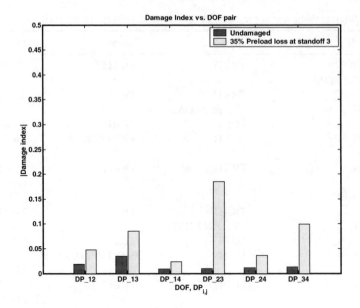

Figure 7.3   Absolute value of transmissibility damage index for 35 % loss in preload at standoff 3 bolt relative to baseline test (Reproduced by permission of SPIE)

noting that though a torque wrench was used to estimate the reduction in torque (35% reduction from 75 in · lb), each of the bolt interfaces at standoffs 2 and 3 is slightly different in contact area and friction. These differences led to more slippage during loosening of the bolt at standoff 2 than the bolt at standoff 3 and, hence, more reduction in stiffness at standoff 2.

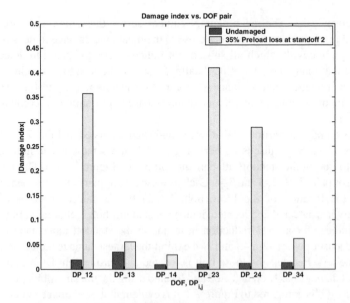

Figure 7.4   Absolute value of transmissibility damage index for 35 % loss in preload at standoff 2 bolt relative to baseline test (Reproduced by permission of SPIE)

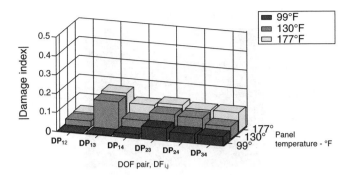

Figure 7.5 Absolute value of transmissibility damage index for heated panel at three temperatures with respect to data at 66 °F indicating large change in transmissibility associated with standoff bolt 3 (Reproduced by permission of SPIE)

The effects of temperature on these results were also assessed. A type T (copper-constantan) thermocouple was positioned at the center of the panel to measure the temperature at that location due to the heat lamp. All of the bolts were torqued to their healthy values (75 in·lb ±5 in·lb). A test at 66 °F was conducted first and used to calculate the baseline set of transmissibility functions for use in the damage index calculation in Equation (7.1). The panel temperature was then increased to 99, 130 and 177 °F, and the damage indices were recalculated. The results of these tests are shown in Figure 7.5. As the temperature increases, the damage indices for sensor pairs 1–3 and 2–3 (to a lesser extent) increase more than the other indices. In particular, the damage index for sensor pair 1–2 is small compared to all of the other indices. This finding suggests that the bolt at standoff 3 may be torqued to a different level than the other bolts. When the surface temperature of the panel increases, the panel expands into the standoff attachments. This increase in the in-plane loading to the attachments causes larger forces and moments across the bolted attachments. Because the bolt at standoff 3 is torqued to a different value, the increase in temperature affects this standoff differently than the other standoffs. It can be concluded that operational temperature loading of this panel accentuates differences in preload from bolt to bolt. At room temperature, this difference cannot be observed in the damage indices (see Figure 7.2). This example demonstrates that health monitoring is often preferable to offline inspection due to the presence of operational loads in online measurements that accentuate the presence of damage.

7.1.1.2 Active Methods

Recall from Chapter 4 that active measurements are taken using an auxiliary actuator to impose a measurable excitation and response. Also, recall from Chapter 5 that active sensing along with a measurement of the input force is required for damage quantification. Two different methods for active sensing and damage identification are demonstrated in the case studies below. The first method uses offline inspection with a modal impact hammer for actuation. Embedded sensitivity functions (see Section 2.3.2.1) are used to process this first set of data. Damage is classified and then quantified using this method. The second method uses online inspection with integrated piezo stack actuators and load cells. Virtual force models are used to process this second set of data (see Section 2.3.2.2). Both case studies apply these methods to metallic sandwich panels.

Automatic impact hammer

Sensor 1

Sensor 2

Sensor 3

Sensor 4

Figure 7.6 Metallic panel tested with automatic impact hammer with four sensors attached to identify mass and stiffness changes (Yang *et al.*, 2005)

Offline Measurements

The Inconel® metallic panel examined in this case study was discussed in Section 7.1.1.1 and is shown instrumented for testing in Figure 7.6. Four 100 mV/g accelerometers are attached to the panel using super glue. Sensor pair 1–2 spans the diagonal path of the panel, and sensor pair 2–3 spans the path through the attachment bolt at a standoff. An automatic hammer is used to excite the panel in the center, and a load cell measures the excitation force. The baseline set of FRFs between the force load cell and four accelerometers were acquired before simulating damage to the panel.

The first material damage simulated in this test was mass damage due to oxidation or corrosion. Strips of paper tape 15 cm and 45 cm in length with low bending stiffness were adhered to the panel in the orientations and positions shown in Figure 7.7. A 4.5 cm long wire was used in a few tests to increase the mass change. The total mass addition (meant to simulate mass reduction) ranged from 1 to 5 g. Strips running in two directions were considered to assess the assumption made regarding lumped parameters using the embedded sensitivity method. A 1 Hz frequency resolution was used for data acquisition with 1024 frequency points. The magnitudes of the measured FRFs for damage cases (a–d) are shown in Figure 7.8 along with the baseline case where no tape was attached. There is no obvious change in the FRFs when viewed in this way.

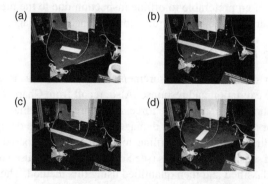

(a)   (b)

(c)   (d)

Figure 7.7 Metallic panel with paper tape strips attached to simulate mass change in four configurations: (a) 15-cm tape transverse, (b) 45-cm tape transverse, (c) 45-cm paper tape with 4.5-cm wire and (d) 15-cm paper tape longitudinal (Yang *et al.*, 2005)

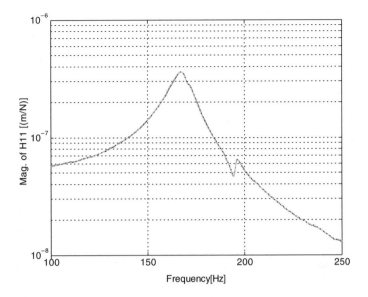

Figure 7.8 $H_{11}(\omega)$ for four damage cases in addition to baseline case shown in previous figure with no obvious differences between the various cases (Yang *et al.*, 2005)

The first step in quantifying damage was to classify it as either a mass, damping, or stiffness change. For example, if the damage is due to creep, then the modulus changes (stiffness). On the contrary, if the damage is due to oxidation, then the density on the surface of the panel changes (mass). Figure 7.9 shows the difference in FRF $H_{11}(\omega)$ for damage case (a) (_) compared to the embedded sensitivity functions with respect to the stiffness (– – –), damping

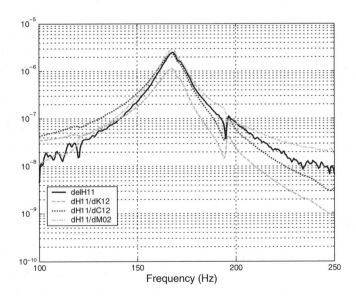

Figure 7.9 Change in $H_{11}(\omega)$ for damage case (a) along with sensitivity to stiffness, damping and mass across sensor path 1–2

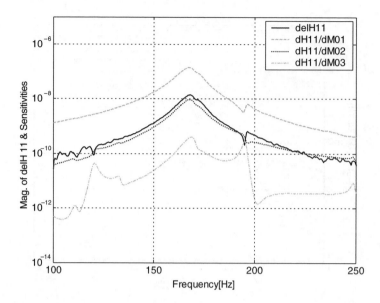

Figure 7.10   Comparison of $\Delta H_{11}(\omega)$ for damage case (a) with sensitivities with respect to sensor 1 and 2 measurement DOFs showing good match (Reproduced by permission of DEStech Publications, Inc.)

(...) and mass (-.-) spanning sensor path 1–2. The sensitivity with respect to stiffness changes in path 1–2 was computed using the formula:

$$\frac{\partial H_{11}}{\partial K_{12}} = -(H_{11} - H_{12})^2 \quad \text{with} \quad H_{12} = H_{21}. \tag{7.2}$$

The damping and mass sensitivities were calculated by multiplying with factors of $j\omega$ (refer to Section 2.3.2.1). These comparisons clearly show that the sensitivity to a mass change at sensor location 2 is a better match to $\Delta H_{11}(\omega)$ than the other sensitivity functions. It can be concluded that the damage is due to a mass change.

The next step in quantifying the damage was to locate it. Figure 7.10 shows the change in $H_{11}(\omega)$ compared with the sensitivity functions with respect to mass at three different locations (sensors 1, 2 and 3 in Figure 7.6). It is clear from this figure that the sensitivity functions with respect to masses at sensor locations 1 and 2 both match the change in FRF relatively well. It can be concluded that damage is located at one or both of these locations. This ambiguity is due to the fact that the mass change actually lies between sensors 1 and 2. Because the sensitivity with respect to mass changes at sensor location 2 matched the change in FRF very well, it was assumed that the mass change occurred at sensor location 2.

Having classified and located the damage, the mass damage could then be quantified using the finite-difference inversion technique demonstrated in Section 2.3.2.1. Figure 7.11 shows the estimated change in mass at sensor location 2 for the four damage cases in Figure 7.7. Due to measurement variability, the only range over which these estimated changes in mass were flat was 140–180 Hz; therefore, this range was used to estimate the mean change in mass by averaging the values shown in the figure. The results comparing estimated changes in mass to actual added masses after unit conversions are listed in Table 7.2. In damage cases (a) and (b)

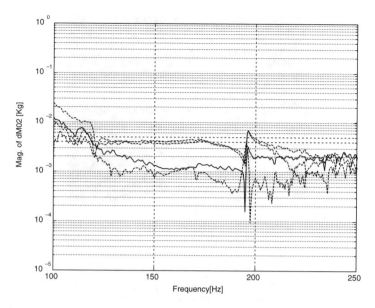

Figure 7.11 Estimate of ΔM_2 for four damage cases (a) (_), (b) (– – –), (c) (-.-) and (d) (...) (Reproduced by permission of DEStech Publications, Inc.)

for which strips of tape were attached transverse to the line connecting sensor locations 1 and 2 (see Figure 7.7), the estimation errors were 5 % or less. Given the small change in the mass introduced, these errors were considered small. In damage case (d) for which the strip of tape was attached along the line connecting sensor locations 1 and 2, the error was much greater (25.3 %). This greater error for damage case (d) relative to cases (a) and (b) was due to the violation of the lumped parameter assumption used in the formulas listed in Section 2.3.2.1. The mass is more lumped for damage cases (a) and (b) than for case (d).

The same methodology that was applied above to estimate mass changes due to damage was also used to estimate stiffness changes due to loosening of the bolt in the standoff attachment shown in Figure 7.12. To simulate damage, the bolt was loosened in 1/16 turns (22.5°). Sensor pair 2–3 spanned the path across the loosened bolt; therefore, this sensor pair was used in the embedded sensitivity analysis. The same steps described above were used to classify this damage mechanism as a stiffness change. The change in stiffness was located using the graph shown in Figure 7.13. This figure indicates that the stiffness change occurred between sensor locations 2 and 3. To quantify the damage due to loosening, the ratio of the change in $H_{11}(\omega)$ to the sensitivity with respect to K_{23} was calculated and plotted as shown in

Table 7.2 Comparison of estimated (average over 140–180 Hz) and actual added masses simulating damage on face sheet (Reproduced by permission of DEStech Publications, Inc.)

Damage case	Actual mass (g)	Estimated mass (g)	Error (%)
(a)	1.33	1.31	1.5
(b)	3.99	3.79	5.01
(c)	4.88	4.18	14.3
(d)	1.33	0.99	25.3

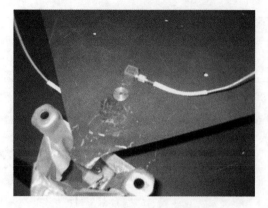

Figure 7.12 A close-up of panel corner shows sensor 2 and sensor 3 (beneath panel), and standoff attachment bolt loosened to simulate damage (Reproduced by permission of DEStech Publications, Inc.)

Figure 7.14 at the various stages of loosening. The mean values of these estimates were then calculated over the frequency range from 140 to 180 Hz. Table 7.3 shows the resultant estimated reduction in stiffness across the bolted interface due to bolt loosening. Because the actual stiffness could not be measured, these values could not be verified; however, it is interesting to note that the change in stiffness is a nonlinear function of bolt rotation. Olson *et al.* (2006) also implement an active sensing approach to estimate loss of torque in mechanically attached panels.

Online Measurements

The active sensing method discussed in the previous section made use of an offline actuation method that is appropriate only when the component is not operating. An integrated active

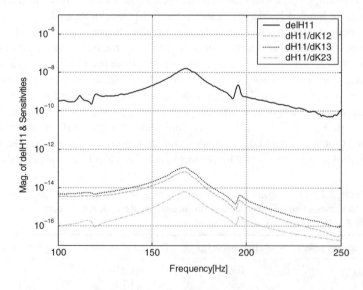

Figure 7.13 Comparison of $\Delta H_{11}(\omega)$ for loosened bolt at 22.5° with sensitivities with respect to the sensor 1–2, 1–3 and 2–3 measurement paths showing good pattern match with path 2–3 (Reproduced by permission of DEStech Publications, Inc.)

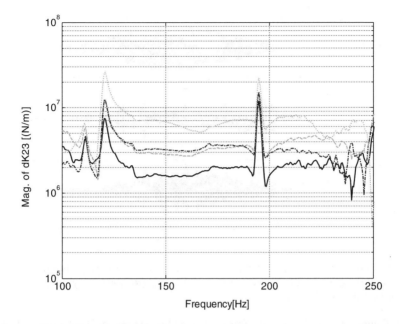

Figure 7.14 Estimate of magnitude of ΔK_{23} for four damage cases (_) 22.5°, (– – –) 45°, (-.-) 67.5° and (...) 90° loosening of standoff bolt showing increase in magnitude of stiffness change (Reproduced by permission of DEStech Publications, Inc.)

measurement technique is discussed in this section for online use. This technique enables tuning of the excitation force to provide more desirable data for damage identification. The actuator setup pictured in Figure 4.33 was used in this case study. A P-010.10 transverse actuator (1000 V, 589 kHz) was attached to a PCB 209C01 (2.2 V/lb) load cell. The four actuators were driven with up to ±10V. Four PCB 352C65 accelerometers were used to measure the response adjacent to the actuator. In order to ensure that the entire 16 bits of resolution in the ADC were used, autoranging was automatically performed prior to each set of measurements. The complete set of actuators and sensors is shown in Figure 4.2(b).

A sample sweep signal used to drive one of the actuators is shown in Figure 7.15(a). This signal is a 7.5 V Hanning windowed swept sine from 18 kHz down to 2 kHz. The signal was swept from high to low frequency in order to avoid any modulation of lower frequency responses with higher frequency responses. A 1.28 s sweep length was used. The measured force in the load cell beneath the actuator is shown in Figure 7.15(b), and the measured acceleration from the accelerometer is

Table 7.3 Estimated (average over 140–180 Hz) stiffness changes due to loosening bolt of standoff attachment to panel (Reproduced by permission of DEStech Publications, Inc.)

Angle (degree)	Stiffness change (N/m)
22.5	1.71E+06
45.0	2.97E+06
67.5	3.32E+06
90.0	6.26E+06

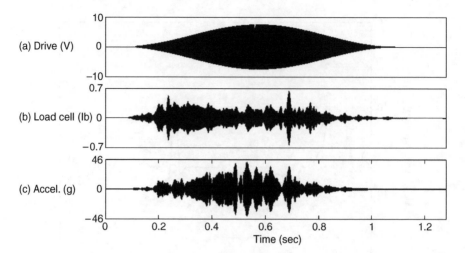

Figure 7.15   (a) Drive signal, (b) force measured using load cell showing distortion in sweep and (c) accelerometer response positioned next to actuator-load cell (Reproduced by permission of Jonathan White, Purdue University, 2006)

shown in Figure 7.15(c). Note that the force delivered by the actuator is not a sweep. Instead, the force time history exhibits dynamics that are characteristics of the actuator and adhesive bond frequency response. The disadvantage of this type of actuation force is that it depends on the adhesive bond used as well as the actuator characteristics. Therefore, there can be a significant variability in measurements made using multiple actuators.

In order to obtain a more uniform swept excitation force, the FRF between the drive signal and the load cell measurement was calculated using this data. Then this FRF was inverted and multiplied by the desired swept excitation force to produce the modified (tuned) drive signal shown in Figure 7.16(a) followed by the modified load cell and accelerometer measurements

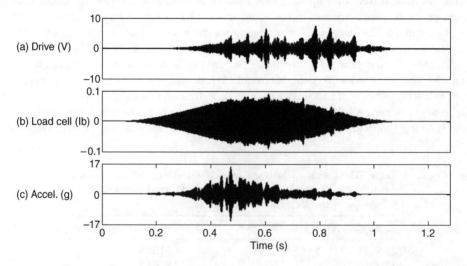

Figure 7.16   (a) Tuned drive signal using a load cell, (b) force measured using load cell showing improvement in swept shape and (c) accelerometer response positioned next to actuator-load cell (Reproduced by permission of Jonathan White, Purdue University, 2006)

(a)   (b)

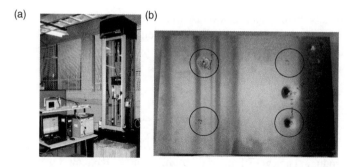

Figure 7.17   (a) Instron 9250 apparatus for imposing impacts with VXI data acquisition system and (b) impacts at several locations at different levels (Reproduced by permission of Jonathan White, Purdue University, 2006)

in Figure 7.16(b) and Figure 7.16(c). Although the drive signal deviates significantly from a windowed sweep, the force measured at the load cell is a good approximation to a swept excitation. This improved excitation force leads to more repeatable and uniform measurements in the measurement system used in this work.

An aluminum honeycomb sandwich panel was used in this damage identification case study. Two face sheets were attached to a honeycomb core using an adhesive. To impose damage on these panels, the Instron 9250 high velocity drop tower pictured in Figure 7.17(a) was used. The panel is visible beneath the impact apparatus. Impact energies of 1, 4, 8 and 16 ft lb were used to produce the increasing amounts of plastic deformation and penetration damage shown in Figure 7.17(b). The circles in the figure indicate where the actuator and sensor pairs were attached to the backside of the panel shown. These locations allowed for FRFs to be measured at a distributed array of measurement DOFs. The challenge in this sample was to diagnose damage using measurements on the opposite side of the panel. As the impacts were being made, the impact forces were measured using the instrumented tup on the tip of the impact head. The force measurements are shown in Figure 7.18.

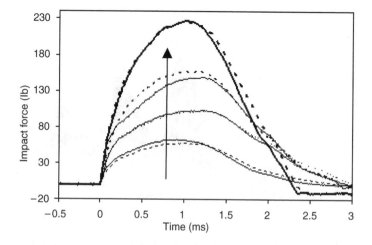

Figure 7.18   Impact force data acquired for 1, 4, 8 and 16 ft lb energy impacts in increasing order (Reproduced by permission of Jonathan White, Purdue University, 2006)

Before any impacts were made, 25 sweeps were used to drive the healthy panel so that 25 averages could be taken to estimate the FRFs between each of the four load cells and all four of the accelerometers. As mentioned above, the drive signal was tuned for each actuator and the ranges were automatically set to reduce ADC errors. The resultant FRF model was expressed as follows:

$$\{X(\omega)\}_{4\times1} = [H(\omega)]_{4\times4}\{F(\omega)\}_{4\times1}. \tag{7.3}$$

Then each of the damage cases shown in Figure 7.17(b) was imposed, and the FRF matrix was recalculated producing a new FRF matrix, $[H'(\omega)]$. A new FRF matrix was calculated for each successive increase in damage level (i.e., the baseline FRF matrix for a given damage case was taken to be the previous FRF matrix prior to imposing that damage case). Equations (2.70) and (2.71) were then applied to estimate the virtual forces due to that damage case over the frequency range from 2 to 18 kHz. Recall that by associating virtual forces with damage, the assumption is being made that changes in material and geometrical properties of the component produce changes to the internal force field within the component. These virtual forces were then reduced in dimension to one number by averaging (a) over the virtual forces produced by the four different actuators and (b) over the 2 – 18 kHz frequency range. Recall from Equation (2.70) that only one column of $[H'(\omega)]$ corresponding to one actuation DOF is needed to estimate the virtual force due to damage; therefore, results are available across the four actuators.

The results of this averaging process are shown in Figure 7.19. The damage site is indicated on the damage axis; each damage case displays four increasingly more severe impact damages. The average virtual force in units of lb is shown on the vertical axis. Note that the virtual force has engineering units because the actuation force was measured using a load cell. The location axis indicates which DOF exhibits the largest estimated virtual force. If the damage is located at DOF A, then the virtual force at location A should be the largest in theory. Consider the results for impact damage introduced on the top face sheet above the actuator

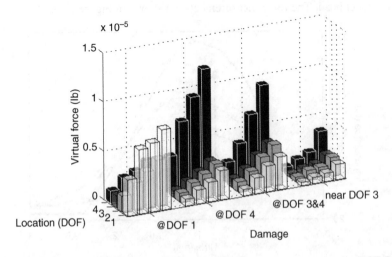

Figure 7.19 Average virtual force calculated for impact damage at measurement DOF 1, DOF 4, between DOFs 3 and 4, and near DOF 3 for increasing damage (Reproduced by permission of Jonathan White, Purdue University, 2006)

attached at DOF 1. This data is indicated with white, light gray, dark gray and black shaded bars at the left end of Figure 7.19. First, it is evident that the virtual force is largest for DOF 1 (white bars), at which the damage was introduced. It is also evident that the virtual force due to damage increases with increases in the impact energy. It appears that for impacts in this location, the step from 1 to 4 ft lb is much larger than that of the increases in impact energy above 4 ft lb.

The results obtained when impact damage was introduced above the actuator attached to DOF 4 are similar to those in the previous case. However, there is an even larger difference in virtual force estimates at the location of damage (DOF 4) relative to the other locations. Also, there is a more gradual change in virtual force level as the damage level increases. One reason for the difference between the results for damage located at DOF 1 versus DOF 4 could be local variations in panel construction. For example, if the impact occurs directly above a stiff area of the honeycomb core, then the resulting plastic deformation will be less severe. Another reason that the virtual force varies in level for the two damage cases is that the actuators and mounting conditions are different for each actuator. In other words, the bond may be somewhat better for the actuator at DOF 4 than the actuator at DOF 1 leading to higher virtual forces for damage at DOF 4. The results obtained for impacts made above the actuator attached to DOF 3 are the least conclusive. The quality of the actuator bond, local variations in the panel and potential damage to the transducers and their attachment interfaces due to high impact forces could explain these differences in the results for damage located above DOF 3.

Thermal damage in the form of local heating was also introduced on the panel. An acetylene torch was used to heat the panel as shown in Figure 7.20 for different periods of time to simulate increasingly larger damage levels (5, 10, 15 and 20 s). Heating caused the face sheet to debond from the honeycomb core to which the sheet was glued (see Figure 3.9(a)). The same measurement system and approach were used to estimate the virtual forces due to damage. The estimated virtual forces due to thermal damage are shown in Figure 7.21. As with the impact damage results, there was again an increasing trend in the virtual force estimates for each damage site as the damage level increased. The thermal damage results were more conclusive regarding the perceived damage location. In each case, damage was properly located. For example, the dark gray bars in Figure 7.21 at the right of the figure are highest among the four DOFs indicating that damage was located at DOF 3. Also, the black and dark gray bars are of equal magnitude when the damage was introduced between DOFs 3 and 4.

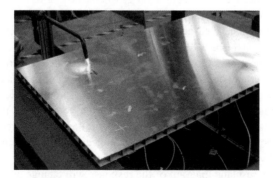

Figure 7.20 Panel undergoing local heating to cause debonding between face sheet and core of aluminum sandwich panel (Reproduced by permission of Jonathan White, Purdue University, 2006)

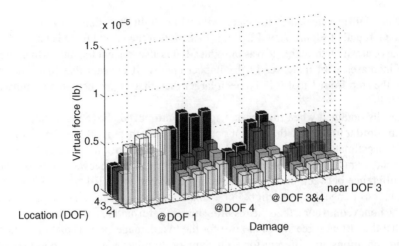

Figure 7.21   Average virtual force calculated for thermal damage at measurement DOFs 1, 4, between DOFs 3 and 4, and near DOF 3 for increasing damage (Reproduced by permission of Jonathan White, Purdue University, 2006)

7.1.2   Gas Turbine Engine Wire Harness and Connector

The reliability of electrical distribution systems is critical to aircraft safety. Wiring issues are difficult to diagnose, leading to unnecessary grounding of aircraft and component removal. Maintenance inspections of aircraft grounded because one or more components exhibit faults often result in 'No Fault Found' incidences in which the cause of problems is actually the wiring and wire connectors of those components. These 'No Fault Found' instances deplete spare part stockpiles, increase maintenance time, dissatisfy customers and reduce profits. Health monitoring is needed to manage the wiring and electrical components of an aircraft effectively.

A method for loads identification in a gas turbine engine wire harness and connector panel was presented in Section 6.2. A complementary method for damage identification is discussed in this section. The method uses passive response data to detect loose or damaged connectors due to corrosion and other sources of damage. Phase-plane analysis is used for sinusoidal type data due to engine rotational forces (refer to Section 5.4.2) whereas transmissibility analysis is used for broadband response data due to aerodynamic type response components (refer to Section 5.8.1).

First, consider phase-plane data analysis for detecting loose connectors. A single triaxial accelerometer mounted on the back of the panel as described in Section 6.2 was used to measure the panel response. The test setup shown in Figure 7.22 was used to impose a sinusoidal excitation force in the vertical direction on the connector panel. A function generator was used to drive the shaker with a constant amplitude force. A 51.2 kHz sampling frequency was used in the Agilent VXI data acquisition system shown in the figure to measure the acceleration response. The Z direction is along the axis of the connector whereas the X and Y directions are in the plane of the panel. The response was measured with the connectors fully tightened and then with one connector loosened by 0.5 turns, 2 turns, etc. Based on the discussion in Section 5.4.2, it is anticipated that the phase-plane diagram of the response will change as the stiffness and mass properties of the connector change. The acceleration in the vertical direction was integrated numerically once to obtain the velocity and then twice to obtain the displacement response.

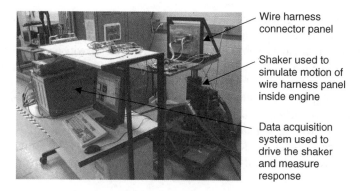

Wire harness
connector panel

Shaker used to
simulate motion of
wire harness panel
inside engine

Data acquisition
system used to
drive the shaker
and measure
response

Figure 7.22   Test setup for simulating vibration of wire harness connector panel

When integrating, trends were removed by subtracting the mean and slope from the data before integrating and then high pass filtering the data after integrating.

The phase-plane diagrams at 45 Hz for the fully tightened (_), 0.5 turn loose (– – –) and 2 turn loose (-.-) connectors are shown in Figure 7.23. The velocity and displacement have been normalized to suppress changes in this ellipse from test to test. Although the phase-plane diagram appears to shift and change in size, it is not obvious that the ellipse is changing. To quantify these changes, the connector was loosened by 0.5, 1.0, 1.5 and 2.0 turns. A 40 s sweep excitation was then used to drive the shaker from 10 to 100 Hz. It was desirable to determine which frequency would provide the clearest indication of loosening. At each stage of loosening the area inside the ellipse was estimated numerically at each frequency using the methods described in Section 5.8.2. The differences in area inside the phase-plane ellipse generated using the vertical direction acceleration measurement on the panel for loose connectors and the fully tightened connector are shown in Figure 7.24(a) in the 10–100 Hz frequency range for the four levels of loosening in addition to the fully tightened condition. Note that the

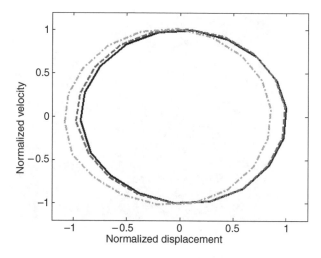

Figure 7.23   Phase plane maps for the wire harness system at 45 Hz: tight (–-), 0.5 turns loose (– – –) and 2 turns loose (-.-) (Reproduced by permission of DEStech Publications, Inc.)

difference for the fully tightened connector is zero. Lighter colors indicate a larger area, and darker colors indicate a smaller area. The increases and decreases in the phase-plane area correspond to a decrease in the connector bending stiffness of the wire harness. These changes in stiffness cause the major and minor axis of the phase-plane diagram to change. The most significant changes in area occur for the 0.5 and 2 turn loose conditions. The results at 40, 70, 75 and 80 Hz exhibit the largest changes suggesting that these frequencies would be optimal for detecting loose connectors.

For a comparison with these phase-plane results, the FRF between the shaker displacement input and the panel vertical acceleration response was calculated. The differences between these FRF magnitudes and the fully tightened FRF magnitude are shown in Figure 7.24(b) for

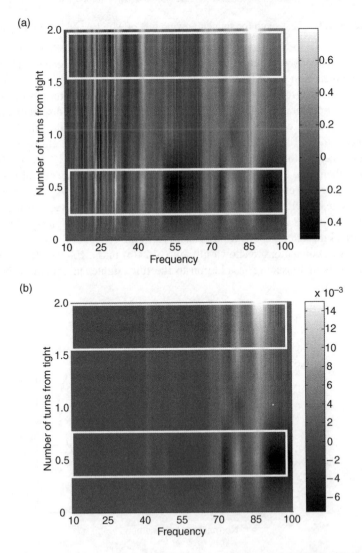

Figure 7.24 (a) Area of phase-plane diagram for loose connector and (b) FRFs of panel in vertical direction for loose connectors (Reproduced by permission of DEStech Publications, Inc.)

the various loosened conditions from 10 to 100 Hz. For the most part, this set of results agrees with the phase-plane results in Figure 7.24(a). The 0.5 and 2 turn loose conditions exhibit the largest changes in FRF magnitude as do the phase-plane area calculations. Also, the same frequency ranges exhibit the largest changes in FRF magnitude. It would be desirable in practice to use this FRF measurement instead of a phase-plane diagram because the FRF measurement incorporates knowledge of the input amplitude and phase. However, the input is rarely known so it is fortunate that the phase-plane diagram provides the same conclusions as the FRF regarding the loose connector.

This example demonstrates that passive measurements made using a single sensor can be used to detect and quantify (on a relative scale) damage in structural components. The disadvantage of phase-plane analysis is that the data must exhibit one dominant frequency in order to effectively use this approach to integrate into the phase-plane diagram. Also, the amplitude of the sinusoidal measurement should remain approximately constant; otherwise, the area of the phase-plane diagram may change as the amplitude of response changes. In an operating engine, these assumptions might be appropriate for long segments of time. For example, a measured acceleration response PSD is shown in Figure 4.4 for an operating gas turbine engine on a test stand. It is clear that there is strong sinusoidal response at certain frequencies (e.g. 40 Hz).

The triaxial accelerometer used to monitor severe loads and loose connectors in the gas turbine engine wire harness connector panel measures three acceleration responses in three orthogonal directions. With these three measurements, it is possible to calculate three transmissibility functions. These transmissibility functions should indicate a loose connector, which will influence the response in certain directions more than the response in other directions. After collecting sine sweep data for 15 trials using the setup in Figure 7.22, the transmissibility function, T_{YZ}, in the vertical direction (Y) with respect to the connector axial direction (Z) was calculated by averaging the time histories, calculating the DFT and then taking the ratio at every frequency in the sweep. This function is plotted in Figure 7.25 for the

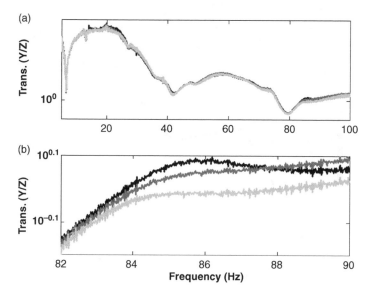

Figure 7.25   (a) T_{YZ} for tight (--), 0.5 (– – –) and 2 turns loose (-.-) from 10 to 100 Hz and (b) from 82 to 90 Hz (Reproduced by permission of DEStech Publications, Inc.)

fully tightened, 0.5 turn and 2 turn loosened connectors. T_{YZ} is shown from 10 to 100 Hz in Figure 7.25(a) and from 82 to 90 Hz in Figure 7.25(b). There is a clear trend in the transmissibility as the connector is increasingly loosened. It is also clear that there is little variation in the transmissibility due to other factors because only the changes due to loosening are evident.

In order to quantify the change in transmissibility, a damage index, DI, is computed. The index relates a potentially damaged (comparison) transmissibility, $T_{YZ(dam)}$, to the baseline, undamaged transmissibility, $T_{YZ(und)}$, for N_f frequencies. A dot product involving the real and imaginary parts of both the transmissibility functions is taken at each frequency in order to incorporate information about the relative amplitude and phase between the responses at two Cartesian acceleration measurements:

$$DI = \left(\frac{1}{\displaystyle\sum_{k=1}^{N_f} |T_{YZ(und)}(\omega_k)|^2} \sum_{k=1}^{N_f} |T_{YZ(und)}(\omega_k)| \, \| T_{YZ(dam)}(\omega_k)| \right.$$

$$\left. \cos(\sphericalangle T_{YZ(dam)}(\omega_k) - \sphericalangle T_{YZ(und)}(\omega_k)) \right) - 1. \tag{7.4}$$

The $\sphericalangle$ symbol denotes the angle of the corresponding complex number with respect to the real axis. The frequency range used in this equation was 10–100 Hz. The results of the calculation in Equation (7.4) are plotted in Figure 7.26 for all three transmissibility functions, T_{YZ}, T_{XZ} and T_{XY}. For the different levels of connector loosening, the damage indices for the three

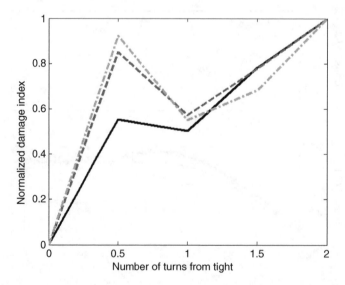

Figure 7.26 Normalized damage index for three transmissibility combinations: T_{XY} (--), T_{XZ} (– – –) and T_{YZ} (-.-) (Reproduced by permission of DEStech Publications, Inc.)

transmissibility functions indicate approximately the same level of damage. The drop in damage index at 0.5 and 1.0 turn is due to the stiffening observed across the connector at these damage levels. The T_{XY} damage index provides the most desirable result because it exhibits a nearly monotonic increase as a function of damage level. T_{XY} is not as sensitive to this stiffening effect because it relates responses in the plane of the panel rather than normal to the panel.

7.1.3 Suspension System

The example in Section 6.5 dealt with a loosened bolt in an automotive suspension system. Changes in load due to bolt loosening were detected in that section using restoring force projections. An alternative method for characterizing the nonlinear nature of this type of joint damage is presented in this section. The basic approach is to identify nonlinearity due to rubbing (friction) and rattling (impacts) that take place as the suspension bolt loosens. The advantage of this approach is that damage can be detected without a baseline dataset if nonlinearity across the bolted joint indicates that it is damaged. A discrete frequency model that has been applied by Adams and Farrar (2002) on damage detection in joints is used to analyze response data from the suspension.

The discrete frequency model that is applied to relate two response measurements, $Y(m\Delta f)$ and $X(m\Delta f)$, at discrete frequency $m\Delta f$ is of the following form:

$$Y(m\Delta f) = T(m\Delta f) \times X(m\Delta f) + A_{-1}(m\Delta f)Y((m-1)\Delta f) + A_1(m\Delta f)Y((m+1)\Delta f), \quad (7.5)$$

where $T(m\Delta f)$ accounts for the linear transmissibility between the two responses and $A_{-1}(m\Delta f)$ and $A_1(m\Delta f)$ are used to model nonlinear correlations between the responses at one frequency below and one frequency above the excitation frequency of interest, $m\Delta f$. Analogous to the discrete time models used in Section 2.3.1.3, $T(m\Delta f)$ is called the exogenous coefficient and $A_{-1}(m\Delta f)$ and $A_1(m\Delta f)$ are called autoregressive coefficients. If there are no nonlinearities acting on or within a component, then the two response measurements are linearly correlated and $A_{-1}(m\Delta f) = 0 = A_1(m\Delta f)$. If nonlinearities occur due to damage introduced into the component, then these coefficients are not zero. The magnitudes of the coefficients determine the level of nonlinearity. The three coefficients in Equation (7.5) are estimated using the method of least squares that was discussed in Section 5.6.2. $T(m\Delta f)$, $A_{-1}(m\Delta f)$ and $A_1(m\Delta f)$ are treated as unknowns in the equation. Then multiple averages are used to estimate these coefficients as a function of frequency.

Consider the test setup shown in Figure 6.14 with the vehicle tire forced vertically by the electro-hydraulic shaker, which is driven with a broadband random Gaussian excitation signal. The wheel pan motions used to force the tire were in the amplitude range from 0.5 to 6.0 mm RMS displacement. Figure 7.27 shows the estimated damage feature, $1 - |A_{jd}/A_{jun}|$, relating the damaged and undamaged $A_1(m\Delta f)$ coefficients for the vertical motion data from measurement DOFs x_2 (top of the strut) and x_3 (steering knuckle), with x_3 taken as x and x_2 taken as y in Equation (7.5). Any change in the A coefficients from zero causes this damage feature to deviate from 1 indicating a change in nonlinearity. Figure 7.27 shows that there is a clear change in the A coefficients in the 40–55 Hz range where the 100 in·lb torque case exhibits more nonlinear correlation than the other cases. The 250 in·lb torque case does not introduce any significant changes in the nonlinear correlations, and the undamaged case and the case with no bolt exhibit similar behavior with no nonlinear correlations evident.

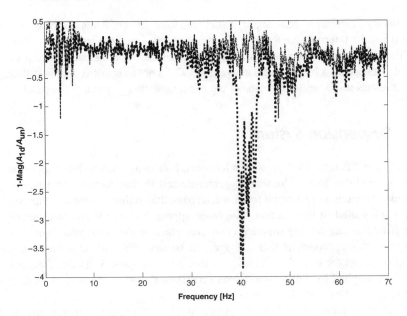

Figure 7.27 $1 - |A_{jd}/A_{jun}|$ for first-order linear model (vertical direction); 250 in·lb torque (– – –), 100 in·lb torque ($\cdots$) and no bolt (–.–)

This result can be explained by considering the nature of a bolted joint. When the bolt is fully tightened, rubbing and rattling do not occur so the nonlinear correlation between the responses measured across the joint is zero. As the bolt loosens, rubbing and rattling can occur leading to the nonlinear correlation observed for the 100 in·lb torque condition on the bolt in Figure 7.27. When the bolt is loosened completely and removed, the preload across the joint becomes zero and cannot enable friction and rattling, leading to zero nonlinear correlation of the response measurements across the joint as in the fully tightened case. This example demonstrates that nonlinear data analysis techniques can be used to characterize damage. In the case of a bolted joint, the level of damage can be estimated directly by considering the amount of nonlinear correlation detected between two response measurements taken across the joint.

7.2 WAVE PROPAGATION-BASED METHODS

Wave propagation case studies in damage identification are described in the following sections. In some of the case studies, wave propagation measurements are used to detect damage and in other cases damage is located and quantified as well. All of the case studies use what is referred to as the 'pitch-catch' approach to active sensing. In this approach, an actuator transmits a waveform and a sensor (or sensors) measures the waveform in the specimen at some distance from the actuator. An alternative approach, the 'pitch-echo' method, transmits and receives waveforms from the same location on the specimen. In this approach, it is common to use multiplexers to actuate and sense from the same transducer. All of the following case studies also use externally attached sensors as opposed to embedded sensors, which are advantageous in some applications.

7.2.1   Wheel Spindle

Two wave propagation methods for detecting cracks in the wheel end spindle, which is pictured in Figure 1.5(b), are described here. Wave propagation is used for detection because the wheel assembly is heavily damped leading to few, if any, standing waves of vibration when the assembly is excited using actuators. The spindle that can become cracked is shown outside of the wheel assembly in Figure 7.28(a). The portion to the left of this picture is the flange of the spindle (inboard side) and the portion to the right is the spline (outboard side). In both the crack identification methods described in this section, elastic waves propagate from right to left through the spindle. The approximate location of the crack is denoted in Figure 7.28(a) with an arrow. The spindle cannot be removed from the assembly before testing the spindle for cracks.

In the first crack identification method, modal impacts were made on the hub of the wheel in the longitudinal direction to produce broadband elastic wave propagation to approximately 5 kHz through the entire wheel end including the spindle. The wheel (tire) was removed from the vehicle before testing the assembly. The wheel hub and modal impact location on a rib of the hub cover are pictured in Figure 7.28(c). A medium modal hammer PCB 086C03 (10 mV/lb) was used to apply the impact excitation. The response data on the flange of the spindle in the radial direction was then acquired. The PCB 333B50 (1 V/g) accelerometer that was used to sense the response is shown in Figure 7.28(b). Paint was removed from the flange surface before attaching the sensor to ensure adequate wave transmission and to lessen the variability due to differences in paint coverage from wheel to wheel. Superglue was used to attach the sensor shown in Figure 7.28(b); however, an alternative method of attaching the sensor to reduce variability in the measurements is described below.

A one-dimensional finite element model was used to simulate the spindle's response to a half-sine wave impulsive excitation in order to anticipate the effects of cracks on the measured data. A seven-element model with varying cross-sectional area and length and the properties of steel was developed (refer to Section 2.2.8.2 for the approach). An impulsive excitation that spanned 0–30 kHz was used although the excitation supplied by an actual impulsive force is band limited to 5 kHz. The modal frequencies of the spindle model were compared to modal impact test results performed on the spindle shown in Figure 7.28(a) to match the first resonant frequency, which occurs at 4.2 kHz. A match was achieved between the model and the data at this first resonant frequency by modifying the stiffness boundary conditions on the spindle finite element model.

To simulate a crack, the stiffness matrix corresponding to the fourth element of the model was modified. This element corresponded to the section of the spindle that becomes cracked.

(a)   (b)   (c)

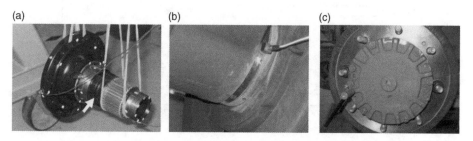

Figure 7.28   (a) Spindle, (b) exposed metallic surface of spindle after paint has been removed and (c) wheel hub with modal hammer shown impacting the spindle on a rib (Reproduced by permission of SPIE)

(a)   (b)

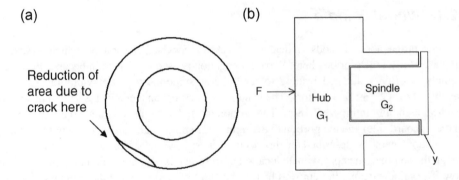

Figure 7.29 (a) Schematic of crack that is modeled in the fourth segment of the spindle and (b) schematic of spindle in wheel hub assembly (Reproduced by permission of SPIE)

As discussed in Section 3.1.2, a crack in a cylindrical component like the spindle reduces the cross-sectional area at that location as illustrated in Figure 7.29(a). This change is modeled by a reduction in the stiffness term for the fourth element.

Figure 7.30 shows the FRFs between an impulsive excitation force applied at the spline (hub of the wheel assembly) and the displacement response at the flange of the spindle. The solid line corresponds to the baseline response with the initial stiffness values for the healthy spindle. The dashed lines correspond to cracked spindles with increasing crack depth from right to left. In the experimental data acquisition approach using modal impacts, the hammer excites wave

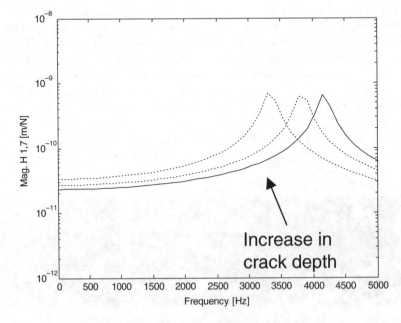

Figure 7.30 FRFs from spline to flange of spindle finite element model for increasing crack depth and / or length (Reproduced by permission of SPIE)

propagation response out to 3 kHz; therefore, the frequency range of interest is 0–3 kHz. The first resonant peak of the spindle model decreases as the crack depth and length increase. This decrease in the first resonant frequency causes as increase in the response FRF below 3 kHz.

When interpreting the spindle finite element modeling results, it is important to note that measurements are acquired on the spindle when it is embedded in the wheel. The spindle and the surrounding parts of the assembly are illustrated in Figure 7.29(b). Assume that the FRF of the surrounding parts of the wheel assembly is denoted by G_1 and that the FRF of the spindle is denoted by G_2. Both of these FRFs are functions of frequency. The response at the back of the spindle on the flange is the product of these two FRFs along with the excitation force applied to the hub, $Y = G_2G_1F$. A crack in the spindle changes G_2, but does not change G_1. It can be concluded that the increase in the FRF magnitude of the spindle, G_2, that is illustrated in Figure 7.30 leads to an increase in the overall response of the wheel assembly, Y, assuming the variations in G_1 are relatively small.

The data acquisition system used to acquire experimental modal impact data consists of a Panasonic Toughbook with a National Instruments 4 channel USB 2.0 device sampling at 50 kHz. The software for the system was developed in MATLAB®. A modal impact was made on the rib of the hub as shown in Figure 7.28(c) because this location was one of the easiest ones for an operator to consistently strike to produce repeatable measurements. In order to determine the optimum position at which to strike the hub, data was collected for impacts on all of the ribs around the face of the hub. The resulting FRF magnitudes from 0 to 3 kHz are plotted in a circular plot for four assemblies (two healthy and two with cracked spindles) in Figure 7.31. The hub illustration in the figure indicates how the impact locations correspond to the FRF magnitude circle plots. The magnitude is denoted with color on a grayscale (white colors correspond to higher amplitudes and dark colors correspond to lower amplitudes). Note that the FRF magnitudes for the two assemblies containing cracked spindles are larger than for the two assemblies containing healthy spindles. This finding is consistent with the finite

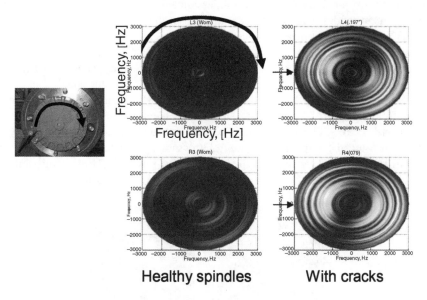

Figure 7.31 A plot of FRF magnitudes for four different assemblies (two healthy and two with cracks) as a function of impact rib location along with the circumference of the wheel (Reproduced by permission of SPIE)

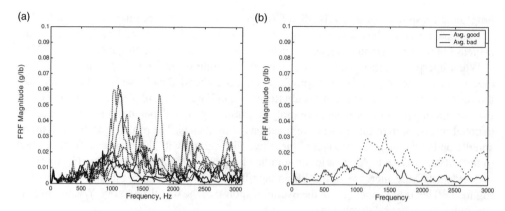

Figure 7.32   (a) FRFs of assemblies with healthy (_) and cracked (...) spindles and (b) mean FRFs for healthy (_) and cracked (...) spindles (Reproduced by permission of SPIE)

element analysis results described above using Figure 7.30. These experimental results indicate that impacts should be made on the rearmost ribs (indicated with arrows in the two cracked FRF plots) to highlight the presence of cracks.

Instead of using superglue to attach the accelerometer to the spindle flange, a vinyl strap was used to mechanically preload the accelerometer to the spindle. Figure 4.13(a) demonstrated that this strap provided more repeatable measurements than superglue. Five averages were taken using five modal impacts on the rear hub location in order to estimate the FRF between the hub impact force and the flange response measurements. Figure 7.32(a) shows the FRF magnitude results of modal impact tests performed on different assemblies mounted on vehicles. The dotted lines correspond to FRFs for assemblies containing cracked spindles, and the solid lines correspond to FRFs for assemblies containing healthy spindles. As predicted in the finite element model, cracks cause an increase in the FRF magnitudes from 0–3 kHz. Figure 7.32(b) shows the mean FRF magnitude at each frequency for the assemblies containing healthy (_) and cracked (...) spindles. The mean FRF for the cracked spindles is higher than the mean for the healthy spindles over the frequency range from 1 to 3 kHz.

In the second wave propagation method used to detect cracks in the spindle, an automatic piezo-stack actuator (P-841.20) was utilized to excite the wheel hub with a swept sine signal extending from 10 Hz to 12 kHz at the base of the flange in the radial direction. Three accelerometers (PCB 333B50) were used beneath the actuator and on either side of the actuator to measure the responses. Figure 7.33(a) shows the actuator and sensor package and (b) shows the strapped package installed on the wheel assembly. The transmissibility ratio of the various sensor pairs was then used to detect cracking. This method of calculating

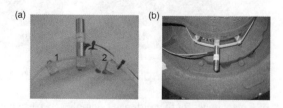

Figure 7.33   (a) Actuator with sensors on either side, and (b) strap installed on wheel

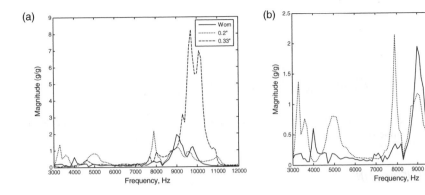

Figure 7.34 (a) Results from automatic actuator test using transmissibility from sensor 1 to sensor 2 for worn (_), 0.2 in. cracked (. . .) and 0.33 in. cracked (– – –) spindles and (b) close up of worn and 0.2 in. cracked spindle results (Reproduced by permission of SPIE)

transmissibilities using active sensing provided more repeatable data beyond the first resonant frequency of the raw spindle than the modal impact data. Because the actuator and sensors are attached to the flange, the variability due to the other parts of the assembly was reduced.

Consider the ratio of the response measurements taken at sensor 2 and sensor 1 denoted in Figure 7.33(a). Both the sensors are used in order to normalize the measurement and reduce variability from assembly to assembly. Figure 7.34(a) shows the results for three assemblies in the 3–12 kHz frequency range with worn (_), 0.2 in. deep cracked (. . .) and 0.33 in. deep cracked (– – –) spindles. The 0.33 in. cracked spindle in its assembly exhibits a larger response than the other two spindles throughout the frequency range. The results for the assemblies with worn and 0.2 in. deep cracked spindles are shown in Figure 7.34(b). It is not obvious from this figure that the assembly containing a cracked spindle has a higher response than the assembly containing a worn spindle. The trapezoidal method of numerical integration was used to calculate the areas under each of the transmissibility curves. The results of the integration are listed in Table 7.4. These results indicate that the cracked assemblies exhibit overall higher response transmissibilities in the range from 3 to 12 kHz.

In this section, frequency response analysis demonstrated that spindle cracks produced a filter, which amplified the elastic response of the surrounding components of the wheel assembly. Modal impacts were effective at detecting cracks below the first resonant frequency of the spindle; however, this approach required that the transmission FRF through the spindle be measured resulting in variability due to the other assembly components. Automatic actuators were used to excite the assembly above the first resonant frequency of the spindle and lessen the variability by making all measurements on the flange.

Table 7.4   Results of integration of transmissibility measurements for three assemblies (Reproduced by permission of SPIE)

Assembly	Integral (g/g)
Worn	25.6
0.2 in	38.8
0.33 in	90.7

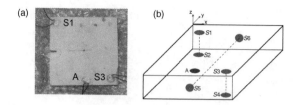

Figure 7.35 (a) Ceramic tile with actuator (A) and two sensors (S1 and S3) to detect notch damage and (b) schematic of ceramic tile with six piezoelectric disk sensors and one disk actuator

7.2.2 Ceramic Tile

The spindle discussed in the previous section was treated as a one-dimensional component. A plate-like two-dimensional specimen is discussed in this section. The specimen is a solid piece of ceramic tile, which is used in police protective bulletproof vests. The tile is shown in Figure 7.35(a). A notch is shown in the tile in the leftmost edge. This notch was used to simulate damage due to impacts causing cracking. The dimensions of the tile are 9.9 cm × 9.9 cm × 1.9 cm. A single PI 402815 transducer was used as a transverse (tapping) actuator and six of these discs were used as transverse sensors. The attachment locations are shown in the schematic in Figure 7.35(b). Two of these sensors and the actuator are visible in Figure 7.35(a). The transducers were attached using superglue. The tile is shown sitting on a soft boundary condition (bubble wrap) to ensure that the response data characterizes the tile rather than the boundary condition. The bubble wrap also helps to damp out the response of the tile.

A 33220A 20 MHz Agilent function generator was used to drive the actuator with a 10-V peak-to-peak signal. A 5-cycle pulse consisting of a sinusoid modulated with a Hanning window was used to drive the actuator with a 10 kHz width narrowband signal centered at two different frequencies: 100 kHz and 200 kHz. These frequency pulses excited a variety of transverse displacement propagating elastic waves in the specimen. A Tektronix TDS5054B-NV-T digital oscilloscope was used to acquire the response data. 64 time domain averages were taken at 71 ms intervals to minimize the effects of noise. Data was collected twice for the baseline condition and each of the three damage conditions to identify the effects of measurement variability. The three damage conditions are illustrated in Figure 7.36. Different depths of the notch were used to simulate different levels of cracking of the tile. The response data was analyzed using DFT and spectrogram analysis techniques (refer to Sections 5.5.1 and 5.5.4).

Two baseline response spectra for sensor 1 obtained using a 100 kHz pulse are plotted in Figure 7.37(a). The 100 kHz center frequency is evident as is the 10 kHz bandwidth. The units

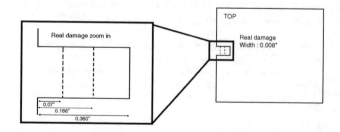

Figure 7.36 Illustration of tile with different depths of notch damage

of the vertical axis are volts scaled by the block size of the time history. The two spectra are nearly identical indicating that measurement variability is negligible assuming the transducers are not removed and replaced. The resonant peaks correspond to traveling and standing modes of response of the tile. Figure 7.37(b) shows a comparison between one of the baseline measurements and a damage measurement with the deepest notch. There is a clear difference between these two measurements. The resonant peaks have shifted, and the amplitude of response of the tile has increased throughout the frequency range of the pulse.

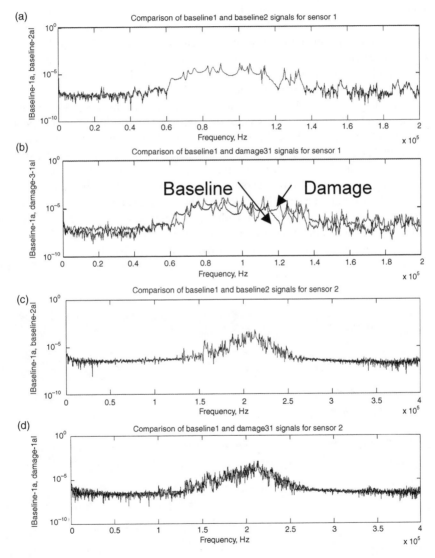

Figure 7.37   (a) Two baseline 100 kHz pulse spectral responses of ceramic tile with no damage and (b) response for deepest notch compared to baseline response for sensor 1, (c) two baseline 200 kHz pulse spectral responses of ceramic tile with no damage and (d) response for deepest notch compared to baseline response for sensor 2

The 200 kHz pulse results are shown in Figure 7.37(c) and (d). These results are similar to those for the 100 kHz pulse. There is a negligible variability between the baseline spectra in (c), but significant differences between the baseline and damage measurements in (d). A damage detection feature could be extracted from this spectral data by calculating the area under the magnitude curve, the sum of the squared differences between the two signals or by another suitable feature extraction algorithm.

A different way of processing the response data was also applied using spectrogram analysis. Figure 7.38 shows the spectrogram contour plots for the 100 kHz pulse response data. In Figure 7.38(a), each of the three plots is the magnitude (in color) as a function of time (in Hertz along the vertical axis) and frequency (in hertz along the vertical axis). In general, brighter colors denote larger amplitude responses. The top subplot in Figure 7.38(a) is for the baseline measurement, the middle subplot is for the deepest notch damage measurement and the bottom subplot is the difference between the top two plots. The six plots (a–f) correspond to the six sensors shown previously in Figure 7.38(b).

As in the DFT spectra plots, it is again evident that the damage measurement in the middle subplot exhibits a higher amplitude response than the baseline (healthy) measurement in the top subplot of Figure 7.38(a). It is also interesting to note that sensors 4, 5 and 6 are not as sensitive to the damage as sensors 1, 2 and 3. Sensor 3 is the most sensitive because it displays the highest magnitude difference. It is also interesting to note that in all six of the different plots, the highest magnitude differences occur in the initial portions of the time history at the left end of the horizontal time axis. This result can be explained by referring back to Section 2.2.2 where the damping factor associated with higher frequency modal vibration frequencies was shown to increase as a function of frequency. In the initial portion of the time history, the higher frequency response components exhibit larger differences in the baseline and damage measurements because the wavelength of those frequency components is smaller. Toward the

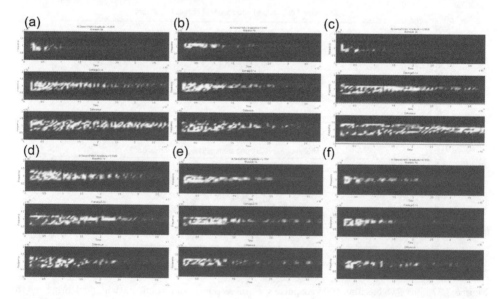

Figure 7.38 100 kHz pulse of ceramic tile showing measurements for the baseline case, deepest notch in the tile (damage) and the difference spectrogram for (a) sensor 1, (b) sensor 2, (c) sensor 3, (d) sensor 4, (e) sensor 5 and (f) sensor 6

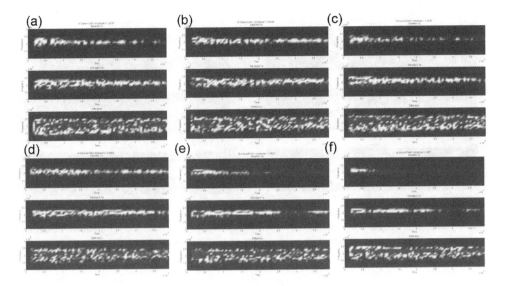

Figure 7.39   200 kHz pulse of ceramic tile showing measurements for the baseline case, deepest notch in the tile (damage) and the difference spectrogram for (a) sensor 1 (b) sensor 2 (c) sensor 3 (d) sensor 4 (e) sensor 5 and (f) sensor 6

end of the time history, the lower frequencies dominate the response because of attenuation and those low frequency components exhibit less sensitivity to the damage.

The 200 kHz results are similar to those for the 100 kHz pulse and are shown in Figure 7.39. The primary difference in the 200 kHz results is that all six sensors exhibit a high sensitivity to the deepest notch damage. The 200 kHz results are more sensitive to the notch damage because the higher frequency waves possess smaller wavelengths in general leading to more interaction with the localized notch damage. The notch damage was also quantified using the response spectrograms for the 200 kHz pulse. Figure 7.40 shows the spectrograms for the 200 kHz pulse at sensor 1 for the shallowest notch damage (top), medium notch damage (middle) and difference between these two measurements in part (a). Spectrograms comparing the deep notch to the shallow notch are shown in part (b). The difference spectrogram for the deep notch damage is larger in amplitude than the spectrogram of the medium notch damage. This result suggests that the spectrogram can be used to quantify the depth of the notch. A suitable feature for damage detection and quantification can be extracted from the spectrogram by applying the techniques discussed in Chapter 5.

7.2.3 Gamma Titanium Aluminide Sheet

In many applications, structural components undergo changes in their geometrical characteristics as a result of operational loading. For example, thin sections in the fuselage of an aerostructure can buckle into different equilibrium states with different radii of curvature. Changes in curvature cause changes in wave speed (refer to Equation (2.98)) as well as changes in vibration natural frequencies (refer to Equation (3.7)). Face sheets of sandwich panels that are heated to high temperatures and cooled rapidly can also undergo buckling as described in

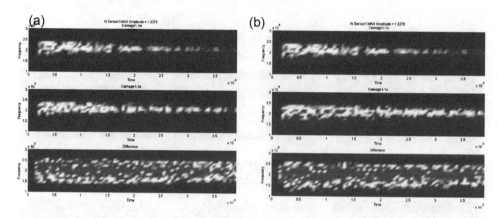

Figure 7.40   200 kHz pulse of ceramic tile showing (a) shallowest notch damage, middle notch damage and difference spectrogram response for sensor 1 and (b) shallowest notch damage, deepest notch damage, and difference spectrogram response

Section 3.1.5. When detecting damage by comparing the response of a component before and after damage occurs, it is important to take into consideration these differences to avoid false positive or negative indications of damage.

A gamma titanium aluminide (TiAl) face sheet is tested in this section. This sheet is shown in Figure 7.41. The sheet is 600 mm × 400 mm and 1 mm thick with an elastic modulus of 150 GPa and a density of 4250 kg/m^3. There is a slight curvature in the sheet due to residual stresses introduced during manufacturing. If a small load is applied on the sheet, it can buckle into one of two positions with convex or concave curvature. The goal of this section is to demonstrate the effects of this change in curvature on damage detection using wave propagation active sensing data.

An APC850 circular piezoelectric transducer with a resonant frequency of 1 MHz and a 10 mm diameter was attached with beeswax to the sheet. This transducer was used as an actuator to transmit propagating waves through the plate. The schematic in Figure 7.42 shows the location of this actuator (denoted by Actuator-1). An Agilent 33220A arbitrary signal generator was used

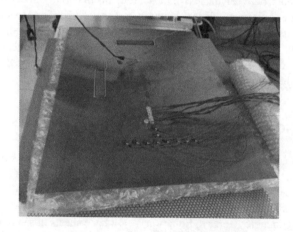

Figure 7.41   Photo of gamma TiAl sheet with actuators and sensor arrays used in tests

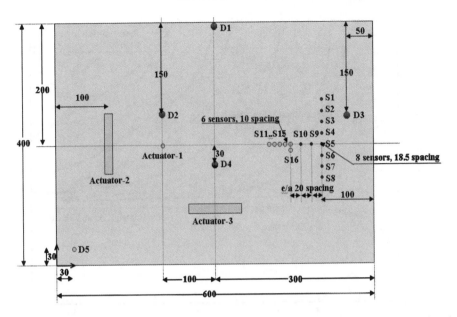

Figure 7.42 Schematic of gamma TiAl sheet showing three actuator locations and 16 sensor locations along with five simulated mass damage locations (dimensions in mm)

to generate a sine sweep signal between 100 and 200 kHz. Figure 7.43(a) shows the 10 V peak-to-peak time domain voltage signal used to drive the actuator, and Figure 7.43(b) shows the spectrum of this signal. One hundred sixty averages were taken to minimize the effects of measurement noise. Response data was acquired using a four-channel Tektronix TDS3014B Digital Oscilloscope sampling at 2.5 MHz. Six APC850 piezoelectric sensors were used to measure the response in the locations denoted by S11–S16 in the schematic. The sensors were spaced 1 cm apart. Other actuator and sensor positions were also examined, but those results are not discussed here. Damage due to oxidation (mass change) was simulated by attaching a small mass in five different locations that are denoted by D1–D4 in the schematic (Figure 7.42).

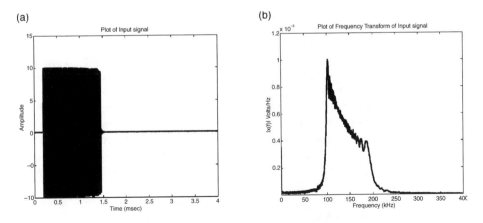

Figure 7.43 (a) Voltage sweep signal from 100 to 200 kHz and (b) spectrum of signal

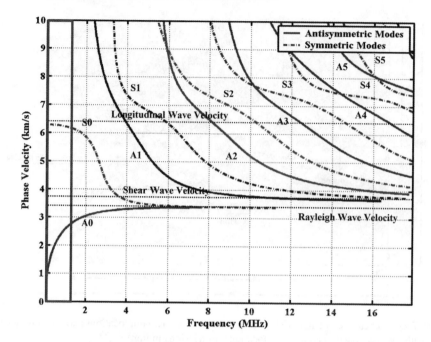

Figure 7.44 Solutions of Rayleigh-Lamb wave equation showing symmetric and antisymmetric mode solutions and frequency range of operation in experiments

In order to determine what types of propagating waves the actuator transmits into the sheet, the Rayleigh-Lamb wave equation was solved and the results were plotted as shown in Figure 7.44. These solution curves indicate the waves that can exist in the gamma TiAl sheet at any particular frequency. The curves also indicate the phase velocity of the individual traveling waves. Because a 100–200 kHz sweep signal was used to excite the sheet, the area enclosed by the rectangle in the lower frequency range is the region of this graph that is relevant in this section. A transverse-type piezoelectric actuator was used to excite the sheet; therefore, the antisymmetric Lamb wave denoted with an A in the figure is the primary wave that was measured in the responses of the sheet. The symmetric Lamb wave denoted with an S can be excited as well in the sheet by utilizing a pinching-type actuator such as the long actuators shown in both Figures 7.41 and 7.42. Each of these types of waves is sensitive to different types of damage mechanisms based on the respective wave displacement mode shape. Refer to Table 2.3 for examples of the wave displacement mode shapes associated with symmetric and antisymmetric propagating waves. The phase velocity of the antisymmetric wave can be approximated by the flexural wave velocity given in Equation (5.43).

The sheet was initially placed in the concave position. The APC actuator was then driven with the 100–200 kHz swept excitation. The discrete wavelet transform (DWT) was then calculated for each of the six sensor signals using Newland's harmonic wavelet (refer to Section 5.5.4). The magnitudes of the responses in wavelet levels 7–10 corresponding to the 100–200 kHz frequency range were then calculated, and the mean value was estimated over the length of each time history. The results are shown in Figure 7.45 for the following 16 cases:

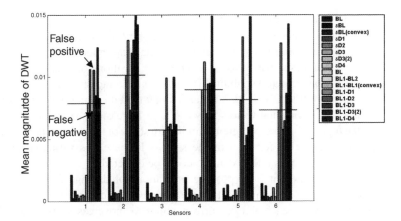

Figure 7.45 Magnitude of Fourier series coefficients as a function of sensor number, 1–6, for 16 cases involving changes with respect to the baseline wave propagation response of the sheet in its concave position

- BL – baseline concave;
- δBL – difference between two concave baseline datasets taken back to back;
- δBL(convex) – difference between two convex baseline datasets taken back to back;
- δD1 – difference between two concave damage datasets with mass placed at location D1 taken back to back;
- δD2 – difference between two concave damage datasets with mass placed at location D2 taken back to back;
- δD3 – difference between two concave damage datasets with mass placed at location D3 taken back to back;
- δD3(2) – difference between two concave damage datasets with mass placed at location D3 taken back to back (repeat test);
- δD4 – difference between two concave damage datasets with mass placed at location D4 taken back to back;
- BL – baseline concave (repeat of first case for reference in figure);
- BL1-BL2 – difference between baseline concave datasets separated by 1 h;
- BL1-BL1(convex) – difference between baseline concave dataset and baseline convex dataset;
- BL1-D1 – difference between baseline concave dataset and damage dataset with mass placed at location D1;
- BL1-D2 – difference between baseline concave dataset and damage dataset with mass placed at location D2;
- BL1-D3 – difference between baseline concave dataset and damage dataset with mass placed at location D3;
- BL1-D3(2) – difference between baseline concave dataset and damage dataset with mass placed at location D3(2) (repeat test);
- BL1-D4 – difference between baseline concave dataset and damage dataset with mass placed at location D4.

The results for the δ tests indicate that for the most part the variability in measurement magnitude for the baseline and damage datasets is relatively small compared to the magnitude of the original baseline dataset acquired in the concave position. The variation in the baseline measurement magnitude acquired in the convex position (third bar from the left at each sensor location) is relatively large compared to the variations in the measurements taken in the concave positions. This result suggests that different equilibrium positions of a component lead to different amounts of measurement variability. The difference in measurements taken 1 h apart in the baseline concave position, BL1-BL2, is much larger than the variations in measurements taken back to back. Environmental factors play a role in increasing the variation in baseline datasets as does the flipping between different equilibrium states of the sheet. A horizontal line is drawn at this magnitude value for each of the sensors because this variation represents a threshold beyond which changes due to damage could be detected.

All six of the sensors exhibit false positive indications of damage for the cases when baseline measurements are compared between the convex and concave positions (BL1-BL1(convex)). The bars for each sensor corresponding to those cases are larger than the baseline variability in the concave baseline datasets. The measurements at sensors 1 and 2 indicate that damage cases D2, D3 and D4 are detected but not D1. The bars corresponding to those damage cases are also larger than the baseline variability. False negative indications of damage are evident in the results at sensor 1. For certain sensors, very few of the damage cases are detected beyond the threshold of baseline variability. For example, the results for sensor 5 indicate that damage is present in only one of the two tests that were conducted with damage located at D3. The second test with a mass placed at D3 exhibits a larger response magnitude than the baseline variability, but the magnitude for the first test with simulated mass damage located at D3 does not. This type of variability analysis is essential in damage identification especially when a component can occupy more than one equilibrium position.

7.2.4 Aluminum Plate

Damage in an aluminum plate is located in this section using the spatial filtering technique called beamforming discussed in Section 5.2.3. The plate that was tested is shown in Figure 7.46. The plate is made from 6061 aluminum and has dimensions 560 mm × 304 mm × 6.5 mm. A PIC stack actuator (P.010.00P) was used to generate a group of circularly diverging flexural waves. It is not possible to generate a single wave at one frequency experimentally. The actuator was attached to the left edge in the middle of the plate in Figure 7.46. Transverse responses were sensed with an array of nine PCB 352C22 accelerometers that are shown attached to the right side of the plate. A four-channel Tektronix TDS5054B-NVT digital oscilloscope (1 GHz/s sample rate) and a 4 to 1 × 4 switch box were used to acquire the response data. An Agilent 33220A waveform generator was used to generate a 2.5V peak-to-peak 5 cycle Hanning windowed sinusoidal pulse centered at 20 kHz with a 2 kHz bandwidth to drive the actuator. Two hundred fifty six synchronous averages were taken at a burst rate of 71 ms. Damage was induced using a hammer and chisel to produce scores in the surface of the plate at certain locations.

An example of the family of acceleration time histories measured on the plate for a 20 kHz pulse transmitted by the actuator is shown in Figure 7.47. The actuator signal is shown along with the nine sensor signals. Note that the times of arrival for the nine sensors are different. The wave arrives at sensor 5 first because it is located in the center. The wave then arrives at sensors 4 and 6, which display similar patterns at least initially due to their symmetric placement with respect to sensor 5. Similar observations can be made regarding the other

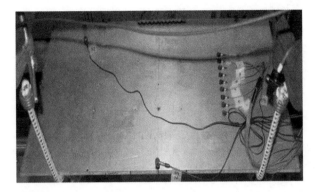

Figure 7.46   Aluminum plate with linear array of accelerometers (right), and actuator (left)

symmetrically placed sensors. The effects of reflections in the sensor data are evident beyond the initial arrival of the waveform. This data suggests that waves transmitted from the actuator have not had time to develop into plane waves. Therefore, data should be processed assuming the approaching wave is circularly diverging.

Suppose that sensors 4, 3, 2 and 1 are delayed in time by the same amount as sensors 6, 7, 8 and 9, respectively. The delays could be chosen so that all of the delayed waves arrive at the same time as the wave at sensor 5. If the resulting set of delayed signals were added, the combined signal would be a maximum because the actuator is directly located across from sensor 5. Instead of locating the source of the transmitted waves, a similar approach could be taken to locate the

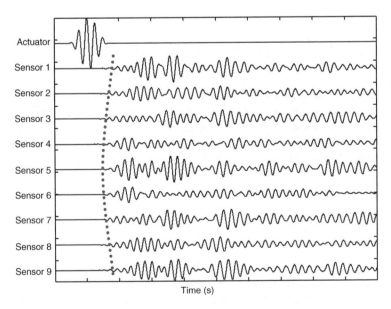

Figure 7.47   Actuator drive signal and nine sensor responses showing different times of arrival of propagating wave

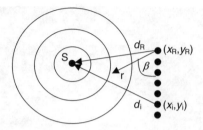

Figure 7.48 Circularly diverging waveform transmitted from actuator and arriving at sensor array from which the location of the scatter due to damage source can be located.

source of scattered waves due to discontinuities attributed to damage in the plate. In general, the problem of damage location using phased sensor arrays is as illustrated in Figure 7.48. Only seven of the nine sensors (sensors 2-8) shown in Figure 7.46 are drawn in Figure 7.48 because those were the only ones used in the remainder of this section for beamforming.

In the figure, a source, S, of waves in a component is shown transmitting circularly diverging waves. An array of sensors is also shown measuring the component response at locations, (x_i, y_i), in the plane of the component. The group velocity of the propagating waves, c, and the distance of the source from each of the sensors determine the sensor time delays of arrival. In this case study, the propagating wave velocity is the group velocity of the 20 kHz pulse given by Equation (5.44). The time delay is calculated with respect to a reference sensor located at (x_R, y_R). The time delay, τ_i, determines the time it takes for a source signal, which could be due to damage, to reach the ith sensor relative to the time it takes to reach the reference sensor. This time delay is calculated as follows for each of the N_s sensors,

$$\tau_i = \frac{d_i - d_R}{C} = \frac{1}{C}\left(\sqrt{\begin{matrix}(x_i - x_R - r\cos\beta)^2 + \\ (y_i - y_R - r\sin\beta)^2\end{matrix}} - r\right); i = 1, \ldots, N_s \qquad (7.6)$$

where (r, β) are the scan distance and look angle measured with respect to the reference sensor in the array. All of the location coordinates are known based on the actuator and sensor placements. As discussed in Section 5.2.3, the sensors should be spaced approximately half a wavelength apart to enable beamforming. The array gain factor is calculated by delaying each sensor signal, multiplying by a weight, w_i, and then averaging the resulting delayed signals:

$$y(t) = \frac{1}{N_s}\sum_{i=1}^{N_s} w_i x_i(t - \tau_i). \qquad (7.7)$$

The weights in this case study were chosen to be the calibration factors of the accelerometers.

In order to locate damage, data was acquired before and after the aluminum plate was damaged. Damage was imposed by scoring the plate in certain locations. The location of the damage was then estimated by scanning at every (r, β) combination. The peak response in the array gain factors in the difference signals, $y_{baseline} - y_{damage}$, served to locate the source of scattering due to damage. Damage can also be triangulated as described in Section 5.7.4 using two (or more) scanning angles, β_1 and β_2, with respect to two (or more) reference sensors. If the distance between the reference sensors is d_{12}, then the location of the source is given by:

$$x_s = \frac{d_{12} \sin \beta_1 \cos \beta_1}{\sin(\beta_2 - \beta_1)} \quad \text{and} \quad y_s = \frac{d_{12} \sin^2 \beta_1}{\sin(\beta_2 - \beta_1)}. \tag{7.8}$$

The damage identification results obtained using beamforming of difference signals taken before and after damage was imposed are shown in Figure 7.49 for three different damage cases. In each of these plots, the actuator location is indicated with a box and the damage location is indicated with a circle. The figures were generated by calculating the array gain factors (Equation (7.7)) for each pair of (r, β). Then the magnitude of the discrete Fourier transform of $y(t)$ for those values of (r, β) was plotted. Darker gray colors correspond to higher magnitude array gain factors. Darker colors should also indicate the damage locations. Figure 7.49(a) shows a result for damage located in the upper left quadrant of the plate. The bottom sensor in the array was used as the reference sensor in this case. The dark wedge in the upper left quadrant of the plate indicates that the beamforming process has identified this area as the scattering source due to damage. The highest magnitude point in this wedge could be selected as the most likely damage location; however, in practice it is usually best to report a zone suspected of damage.

Figure 7.49(b) shows the array gain factor magnitude for damage located in both the lower middle quadrant of the plate and the upper left quadrant. The difference between the baseline (healthy) signal and the signal measured after introducing damage in both of these locations was used in the beamforming process. The top sensor in the array was used as the reference sensor in this case. The dark wedges in the upper left quadrant of the plate and the lower middle quadrant of the plate indicate that the beamforming process has identified these areas as the scattering sources due to damage.

Two additional damage sites were introduced in the lower left quadrant of the plate and the middle right quadrant of the plate, and then beamforming was performed again. In this case, the signal measured after the first two damage cases were imposed was used as the baseline signal in an attempt to remove the effects of the first two damage cases. In practice, this choice of a baseline signal would be common because as damage accumulates, it would be desirable to reset the baseline signal in order to isolate new damage locations. The beamforming results in Figure 7.49(c) indicate that although the new damage locations have been approximated reasonably well by the dark wedges in the lower left quadrant and middle right quadrant of the plate, the remnants of the previous damage cases are visible as well. This example demonstrates that components with many geometrical features (holes, fillets, etc.) and/or multi-site damage can lead to distorted damage location results.

7.3 DAMAGE IDENTIFICATION UNDER LOAD

Health monitoring is ideally applied online to structural components as they are operating. Loads acting on a component tend to change the way damage is perceived in health monitoring measurements. The effects of loading on damage detection are examined in this section in two case studies.

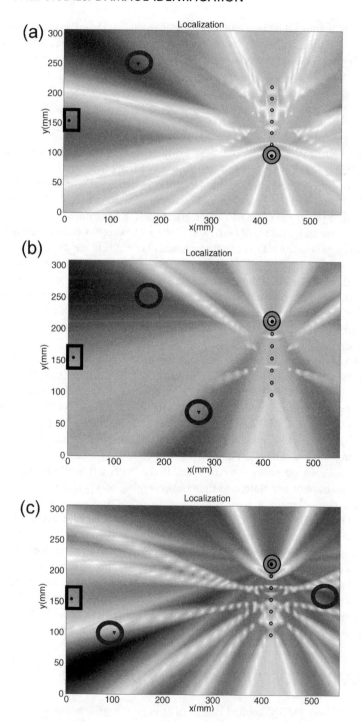

Figure 7.49 Plots of the array gain factor magnitudes for a 20-kHz pulse with (a) damage in upper left quadrant of plate, (b) damage in both upper left quadrant and bottom middle quadrant of plate and (c) damage in the bottom left quadrant and middle right quadrant of the plate (Sundararaman *et al.*, 2006, American Institute of Physics)

7.3.1 *Metallic Panel Under Thermo-Acoustic Loading*

Impact and thermal damage was detected, located and quantified in the metallic sandwich panel case study discussed in Section 7.1.1.2.2. Active sensing was used to excite and measure vibration responses in the panel in that section. A sandwich panel was also tested in the case study discussed in Section 6.6 to determine the effects of combined temperature and acoustic loading on the panel forced vibration response. In this section, damage is detected and located in a metallic sandwich panel that is simultaneously loaded thermally and acoustically.

Figure 6.18 showed the experimental setup that is used in this case study to expose a gamma TiAl sandwich panel to 300 °F temperatures and 110 dB acoustic sound pressures. The panel was supported by standoff attachments as shown in Figure 7.50. The measurement system that is used in this case study to identify damage was also shown previously in Figure 4.2(b) and described in Section 7.1.1.2.2. A transducer array consisting of four piezoelectric stack actuators bonded to four load cells along with four accelerometers attached adjacent to these actuators was installed on the backside of the panel using thermal epoxy. The actuators were used to excite panel vibrations in the frequency range from 4 to 18 kHz. Because the sweep is relatively fast, a decreasing swept sinusoidal waveform was used to drive the actuators to avoid modulation of harmonics from lower frequency signals with higher frequency responses. The load cells and accelerometers were used to measure the input forces and responses, which were then used as inputs to the virtual force damage identification algorithm. Two different damage mechanisms were introduced into the panel: simulated oxidation using a local mass addition and simulated damage in the standoff attachment by loosening the bolt.

Both of these damage mechanisms were introduced into the panel in the vicinity of response DOF 3 (see Figure 7.50). Measurements were conducted at two different temperatures: 75 and 300 °F. Virtual forces at DOF 3 were estimated for both temperatures and damage mechanisms by transmitting 4–18 kHz sweeps into the panel from each of the four actuators to estimate the corresponding FRF matrix. 10 averages were taken to calculate the FRFs. Then a damage indicator was calculated by averaging the virtual forces at DOF 3 for each of the four different excitation DOFs in the frequency range from 4 to 8.6 kHz. The results of this averaging process are shown in Figure 7.51. The simulated mass damage results are indicated by (_) and (– – –) line types at the 75 and 300 °F conditions. The results for the loosened bolt are indicated by the (. . .) and (-.-) line types at 75 and 300 °F. These results indicate that simulated mass and loosened bolt damage are both correctly identified by the virtual force algorithm because the highest virtual force is at response DOF 3 regardless of temperature. The results also show that the simulated mass damage produces a larger virtual force than the loosened bolt damage no matter what the temperature is.

Figure 7.50   Gamma TiAl sandwich panel supported by mechanical attachments

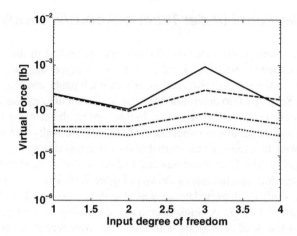

Figure 7.51   Average virtual force at response DOF 3 across the four excitation DOFs in the 4–8.6 kHz frequency range with simulated mass damage at response DOF 3 at 75 (_) and 300 °F (– – –), and bolt loosening in corner closest to DOF 3 at 75 (...) and 300 °F (-.-) (Reproduced by permission of Elsevier, 2006)

There are also differences in the virtual force damage indicators for the various excitation DOFs. The damage indicators using excitation DOFs farthest from the damaged area do not vary significantly with temperature. This result is especially clear in the data for the simulated mass addition where the virtual forces were estimated using excitation DOFs 1, 2 and 4. It is also evident that for the excitation DOF located adjacent to the damage, temperature changes affect the perceived damage levels in the two damage conditions differently. In the case of simulated mass damage, an increase in temperature causes a decrease in the estimated virtual force calculated at DOF 3. This decrease in virtual force for the simulated mass damage is due to the combination of (a) softening of the superglue adhesive used to attach the mass and (b) softening of the standoffs resulting in less response from the mass. In the case of simulated attachment damage by loosening the bolt, an increase in temperature causes an increase in the virtual force calculated at DOF 3. The virtual force increases for the loosened bolt when the temperature increases because the temperature gradient across the panel increases in this case causing the bolt to bend.

The effects of acoustic loading on damage identification was also examined by applying 110 dB of sound pressure loading to the panel as it was heated in the simulated mass damage experiments. A loudspeaker was played with a 0.1 Hz–7 kHz frequency random signal in the reverberant chamber shown in Figure 6.18. The virtual force results of these experiments are plotted in Figure 7.52. The estimated virtual forces for the 75 and 300 °F temperatures are the same as in Figure 7.51. The virtual force damage indicators that were calculated when the panel was radiated acoustically are denoted with a (x) symbol. At both temperatures, the acoustic loading causes an increase in the magnitude of the virtual forces at the damage location, DOF 3. As in the previous case, higher temperatures lead to larger virtual forces at DOF 3 even in the presence of acoustic loads. The acoustic loading also causes increases in the virtual force damage indicators. These increases highlight the presence of damage in the active sensor measurements due to the mass addition. The virtual force increases when the sound pressure is applied because the distributed force due to the sound pressure increases the effective force of the attached mass near DOF 3 meant to simulate oxidation damage.

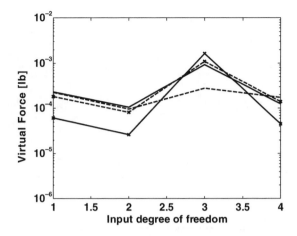

Figure 7.52   Average virtual force at response DOF 3 across the four excitation DOFs in the 4–8.6 kHz frequency range with simulated mass damage at response DOF 3 at 75 (_) and 300 °F (– – –) where (x) denotes data with 110 dB of acoustic loading (Reproduced by permission of Elsevier, 2006)

This case study demonstrates that operational loading (thermal, acoustic, etc.) affects the damage identification process. When active vibration and wave propagation sensing is used to identify damage, operational loads can highlight or suppress indications of damage in measurement data.

7.3.2 *Aluminum Plate Under Vibration Loading*

Damage in an aluminum plate due to surface defects was located using beamforming in Section 7.2.4. In this section, simulated mass damage due to oxidation or corrosion is detected in the plate while it undergoes vibration loading. This loading simulates the operational vibration that such a plate would experience if it were a component in a fuel tank, fuselage or other type of transportation system. The previous section demonstrated that thermal and acoustic loads cause changes in the apparent damage level of components when health monitoring measurements are used to identify damage. The vibration loading considered in this section also influences the damage indicators used to detect damage. Wave propagation with active sensing is again used in this section for damage detection.

The 6061 Al plate that was studied in Section 7.2.4 is shown in Figure 7.53(a) with the actuator (Acuator-1) and sensor (B1-B12) locations that were used in this case study. A PIC stack actuator and PCB 352C22 accelerometer sensors were again used in making measurements. The locations (D1-D20) at which a small mass was placed to simulate damage due to corrosion or oxidation are illustrated in Figure 7.53(b). Figure 7.53(c) shows where the MB Dynamics 50 lb electrodynamic shaker was attached through a stinger to excite the plate with a background (ambient) vibration input. The shaker imposes a vertical excitation force on the plate. The same data acquisition equipment that was used in Section 7.2.4 was used in this case study.

The detection of localized damage in positions D1–D20 was investigated with and without ambient vibration at 110–110 Hz was chosen as an excitation frequency because the first

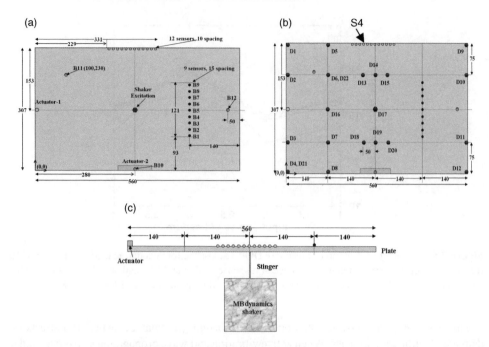

Figure 7.53   (a) Plate configuration with actuator, A, and sensors B1-B12, (b) simulated mass damage locations D1-D20 and (c) shaker to provide vibration

bending mode of vibration of the plate was identified at this frequency. The vibration levels of the plate were measured during these tests to be approximately 2–10 g's depending on sensor location. Several baseline datasets were taken before simulating damage to the plate. BL1 and BL2 denote datasets that were acquired back to back. BLn denotes a dataset that was taken after completing all of the simulated damage experiments. Two hundred fifty six averages were taken of the measured responses to a 5–20 kHz sweep from the PIC actuator in conjunction with the 110 Hz shaker input at the first bending mode of the plate. The average DFT magnitude of the voltage signal from each sensor was calculated. Figures 7.54(a) and (b) show the results of these experiments.

The difference DFT magnitude, which serves as the damage feature comparing two datasets, is shown on the vertical axis. The pair of datasets that are compared using the DFT magnitude is denoted on the horizontal axis. There are three different lines indicating differences in how the ambient vibration entered the experiment. The ($\square$) symbol denotes the cases for which no ambient vibration was introduced. The ($\circ$) symbol denotes the cases for which ambient vibration was introduced only in the second dataset. The ($\diamond$) symbol corresponds to the cases for which ambient vibration was introduced in both datasets. The first two comparisons on the horizontal axis correspond to baseline dataset comparisons. When no ambient vibration is introduced in either dataset ($\square$) or vibration is introduced in both datasets ($\diamond$), the two back-to-back baseline datasets are similar producing a small difference magnitude. However, when ambient vibration is introduced in the second dataset but not in the first dataset, the difference between the DFT magnitudes of the datasets is much larger. The threshold for damage detection is set at this value.

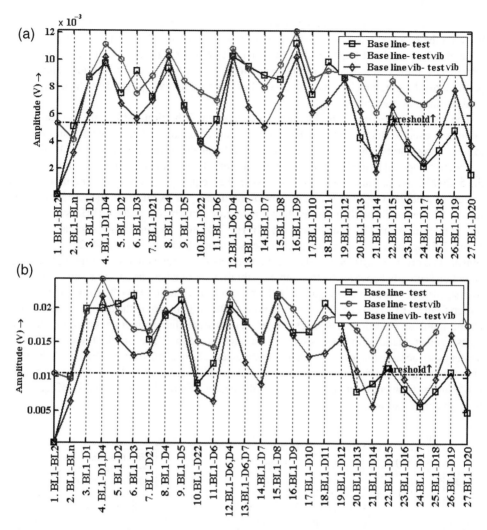

Figure 7.54 Difference DFT magnitude results for damage detection in a 6061 Al plate with excitation from a shaker at 110 Hz along with a 5–20 kHz sweep input measured at (a) sensor B9 and (b) sensor B11 with locations indicated in Figure 7.53 for (□) no ambient vibration, (○) ambient vibration in second dataset, and (◇) ambient vibration in both datasets

In Figure 7.54(a), the majority of damage cases are detected with and without vibration. Also, the results for BL1-BLn do not indicate a false positive indication of damage. The general trend is that the results for no ambient vibration on either dataset (□) are lower than the results for ambient vibration on the damage dataset but not on the baseline set (○) and higher than the results for ambient vibration on both datasets (◇). In fact, there are zero false negative indications of damage for the (□) case but several false negatives for the other two cases. This trend suggests that ambient vibrations are beneficial to damage detection if the vibrations are not present in the baseline dataset but present in the comparison dataset. Vibration increases the effective change in impedance due to the localized mass addition.

The results also indicate that damage locations closer to antinodes exhibit larger differences between the baseline and damage datasets. Datasets with multiple damage locations such as D6 and D4 also exhibit larger changes in DFT difference magnitude. There are only subtle differences in the results shown in Figure 7.54(b) for a different sensor location. The shaker point of attachment, which is at an antinode, has more influence on the results than the sensor location in these two cases.

The results in Figure 7.54 illustrate that ambient vibrations affect the measurement indicators used to detect damage. However, the only benefit to damage detection observed in that figure corresponded to cases in which ambient vibration was present in the datasets that were compared to the baseline (healthy) dataset. The benefits were not very significant for the 5–20 kHz active sensor measurement. When the frequency for active sensing increase, the effects of ambient vibration become more pronounced. For example, Figure 7.55 shows the results for sensor S4 with location shown in Figure 7.53(b) for a 20–80 kHz swept active sensor measurement with and without the 110 Hz shaker excitation. Trends similar to those in Figure 7.54 are observed in this figure. The sensor S4 is a high frequency PI 402815 piezo disk sensor. The primary difference between the two sets of results is that in Figure 7.55 damage detection is greatly benefited by the ambient vibration input when the baseline and comparison (damaged) datasets are both acquired in the presence of vibration ($\diamond$). The higher frequency sweep from 20 to 80 kHz results in a more noticeable shift upward in the damage indicator due to the ambient vibration. These examples demonstrate that ambient vibration influences measurement features that are used for damage detection. The ambient vibration has a tendency to increase the localized change in mechanical impedance due to damage. Vibrations also tend to increase damage detection thresholds. Measurement features that are extracted using higher frequency active sensing are influenced in a more beneficial and significant way by ambient vibrations.

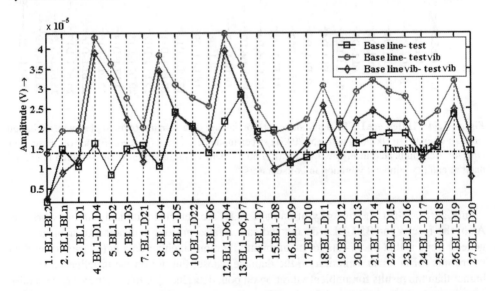

Figure 7.55 Difference DFT magnitude results for damage detection in a 6061 Al plate with excitation from a shaker at 110 Hz along with a 20–80 kHz sweep input for sensor S4 (type PI 402815) coincident with B3 with locations indicated in Figure 7.53(c) for ($\square$) no ambient vibration, (o) ambient vibration in second dataset and ($\diamond$) ambient vibration in both datasets

7.4 SUMMARY

This chapter has demonstrated vibration-based and wave-propagation-based methods for damage identification. Passive measurement data due to vibration and acoustic excitations in the operational frequency range of a component has been acquired and processed to detect and locate damage. Active measurement data above the operating frequency range has also been used to detect smaller defects and locate them more precisely. Damage has been detected using frequency domain measurements, phase-plane diagrams and wavelet analysis. Damage has been located using inverse frequency response function methods and beamforming techniques. Damage has been classified and quantified using experimental sensitivity functions. The following key points were made in the case studies:

- Loss of preload in bolted connections due to bolt loosening, cracking or warping is a common type of damage mechanism. *Higher temperature gradients* across a mechanically attached component lead to more pronounced changes in measurements due to bolt loosening.
- Active methods are required for quantifying damage in an absolute manner because an *input measurement is needed*. Damage must be classified (e.g., mass, damping, stiffness) before quantifying the level of damage.
- Offline measurements can be used to identify damage, for example, using modal impact testing as in Section 7.1.1.2.1. Such measurements are not as prone to variability due to operational excitations.
- Online measurements are preferred in some health monitoring applications. Section 7.1.1.2.2 demonstrated one type of active measurement technique in which the excitation force was measured using an automatic actuator with an integrated load cell.
- Difference techniques in health monitoring help to identify relatively small changes in measurements between baseline and damage datasets that are not apparent.
- Sensitivity methods can be used to quantify damage by processing raw measurements using a physics-based model. Changes due to damage should be relatively small in order to avoid bias errors in estimated damage levels.
- Actuators are dynamic devices that exhibit frequency response limitations. Actuators cannot be driven with signals that are faster than the bandwidth of the actuator and attachment mechanism; otherwise, the actuator dynamics will dominate the measurement data. Feedback and feedforward control methods can be used to tune drive signals to suppress the actuator dynamics.
- Phase-plane diagrams can be used to detect and quantify damage on a relative scale if measurement data contains one dominant frequency or at most a few dominant frequencies. Transmissibility functions can be used to detect and quantify damage if measurement data is broadband in nature.
- Triaxial and other multidirectional sensors are useful in health monitoring because multiple responses at a point can be measured and processed to produce an array of information. Damage often produces *different effects in different measurement degrees of freedom*; therefore, measurements in multiple directions are more sensitive to general damage cases.
- Nonlinear correlations in response measurements taken across interfaces due to rubbing or impacting can be characterized to identify damage. These correlations can be characterized by determining how the response at one frequency is correlated with the excitation at that frequency and the responses at neighboring frequencies.

- If it can be assumed that a healthy component is linear in nature, then nonlinear features in measurement data can be used to *directly quantify damage* in a component without requiring a history of the nonlinear damage feature.
- In components that are heavily damped, wave propagation methods of damage identification are preferred because standing waves of vibration do not become established due to the high damping levels.
- If damage must be detected in embedded components, then the sources of variability due to the surrounding components (boundary conditions) must be minimized. Multiple measurements are sometimes needed to detect damage in these applications in order to avoid false positive indications of damage.
- Wave propagation methods can be applied using measurement data below the resonant frequencies of a component or within the resonant frequency range of the component. Data in the resonant frequency range is advantageous because the component response is amplified in this range.
- The actuator and sensor locations for wave propagation must be chosen carefully to increase sensitivity to damage mechanisms of interest and decrease the sensitivity to sources of measurement and component variability.
- Higher frequency waveforms are generally more sensitive to small defects because higher frequency waves often have shorter wavelengths, which interfere more with small geometry defects.
- When components are loaded in operation, components can undergo changes in their geometrical characteristics. These changes lead to changes in the way waves propagate through components (e.g., wave speeds increase in curved components).
- In many wave propagation applications, it cannot be assumed that plane waves are propagating because defects often act as point sources that scatter waves that circularly diverge as they propagate away from the defects.
- Multiple damage mechanisms lead to more spatial background noise when attempting to locate damage using phased sensor arrays.
- Operational loads affect the measurement features used to identify damage. When active sensing techniques are used to measure propagating waves while components simultaneously undergo operational acoustic, thermal and vibration loads, damage can become either less or more apparent in measured data.

REFERENCES

Olson, S., DeSimio, M., and M. Derriso, 'Fastener Damage Estimation in a Square Aluminum Plate,' *Structural Health Monitoring*, June 2006, Vol. 5, pp. 173–183.

Hundhausen, R.J, Adams, D.E. and Derriso, M. (2005) 'Identification of damage in a standoff metallic thermal protection system panel subjected to combined thermo-acoustic excitation', *Proceedings of the SPIE Conference: Health Monitoring and Smart Nondestructive Evaluation*, Vol. 5768, pp. 145–156.

Hundhausen, R. (2004) 'Mechanical loads identification and diagnostics for a thermal protection system panel in a semi-realistic thermo-acoustic operating environment', *Masters Thesis*, School of Mechanical Engineering, Purdue University, West Lafayette, Indiana.

Yang, C., Adams, D.E., Derriso, M. and Grant, G. (2005) 'Damage quantification of simulated fastener and oxidation damage in an inconel mechanically attached thermal protection system panel using experimental sensitivity functions', *Proceedings of the International Workshop on Structural Health Monitoring*, Stanford, CA, pp. 1676–1683.

White, J. (2006) 'Impact and thermal damage identification in metallic honeycomb thermal protection system panels using active distributed sensing with the method of virtual forces', *Masters Thesis*, School of Mechanical Engineering, Purdue University, West Lafayette, Indiana.

Stites, N., Adams, D.E., Sterkenburg, R. and Ryan, T. (2006) 'Loads and damage identification in gas turbine engine wire harnesses and connectors', *Proceedings of the 3rd European Workshop on Structural Health Monitoring*, Granada, Spain, pp. 996–1003.

Ackers, S., Kess, H., White, J., Johnson, T., Evans, R., Adams, D.E. and Brown, P. (2006) 'Crack detection in a wheel spindle using modal impacts and wave propagation', *Proceedinjgs of the SPIE Conference on Nondestructive Evaluation and Smart Structures and Materials*, Vol. 6177. Paper #: 61770B.

Sundararaman, S., White, J., Adams, D.E. and Jata, K. (2006) 'Application of wave propagation and vibration-based structural health monitoring techniques to friction stir weld plate and sandwich honeycomb panel', *Proceedings of the 2006 Quantitative NDE Conference, Seattle, WA*.

White, J., Ellmer, C. and Adams, D.E. (2006) 'Health monitoring for reliability testing of metallic sandwich panels using integrated active sensing with dual actuator-sensor pairs and the method of virtual forces to identify thermal and impact damage', *Proceedings of the 9th International Symposium on Multiscale and Functionally Graded Materials*, Honolulu, Hawaii.

PROBLEMS

(1) Consider the nine-DOF system model pictured in Figure 4.6 and simulated in 'nine-dofdata.m' and 'ninedofenviron.m.' Estimate the transmissibility functions between DOFs 1, 3, 7 and 9 for 10 % and 50 % reductions in the stiffness of the attachment connected to DOF 3. Estimate the damage index given in Equation (7.1) for each of these transmissibility functions using the entire frequency range for which the FRFs are calculated. Plot the damage indices for each transmissibility function. Can you locate the reduction in stiffness using these damage indices? Cut the frequency range in half and re-estimate the damage indices. Comment on the differences in the ability to locate the stiffness reductions.

(2) Repeat problem (1) using the one-dimensional finite element model of the cracked rod pictured in Figure 2.24(b). Calculate the transmissibility functions between nodes 1 and 5, 5 and 9, 9 and 13, and 13 and 17 using the corresponding FRFs that can be calculated using the simulation code 'rod1dwave.m.' Insert cracks in element 3, 7, 11 and 15 using a 50 % reduction in cross-sectional area to simulate the cracks. Comment on the ability to locate damage in the rod using transmissibility measurements.

(3) Repeat problem (1); however, add 5 % Gaussian random noise to the FRFs that are used to estimate the transmissibility functions. Use the hypothesis testing methods discussed in Section 5.8.4 to select a threshold for damage detection based on the features calculated from Equation (7.1). Assume that a 2 % probability of false positive indication of damage is desired. What is the percentage of false negative indications of damage for the 10 % reductions in stiffness?

(4) Consider the nine-DOF system model pictured in Figure 4.6 and simulated in 'nine-dofdata.m' and 'ninedofenviron.m.' Estimate the embedded sensitivity functions of H_{11} with respect to stiffness changes between DOFs 3 and 6. Reduce the stiffness between these two DOFs by 2%, 5%, and 10% and then estimate the reduction in stiffness using the embedded sensitivity technique. Comment on the accuracy of the results obtained for the different levels of damage.

(5) Repeat problem (4); however, consider reductions in mass at DOF 6 due to damage instead of the stiffness reductions considered in problem (4).

(6) Use the simulation code 'platefemsim.m' to generate the transverse displacement FRFs of the plate model. Calculate the embedded sensitivity functions of H_{11} with respect to stiffness changes between nodes 26 and 27. Then reduce the flexural stiffness of element

24 by 2%, 5%, and 10% and estimate the reduction in stiffness across this element using the transverse FRFs. Consider paths in both the x and y directions in the plane of the plate. Comment on the results.

(7) Consider the simplified panel model shown in Figure 2.10(b). Assume the single DOF model shown in Figure 2.10(a) represents an actuator used to excite the panel. Use the simulation code 'twodofwsensor.m' to calculate and plot the transient response of the panel to sinusoidal, Hanning-windowed pulses of varying bandwidths (e.g. 0.5–1 Hz, 2–3 Hz, 7–8 Hz) applied to the actuator mass, M_s. Comment on the nature of the responses as the center frequency of the pulse increases. Use Equation (2.19) to plot the spectrum of the excitation force applied to the panel through the actuator attachment and explain the sensitivity of the panel response to the excitation bandwidth.

(8) Repeat problem (1) using the method of virtual forces to model and identify stiffness damage in the nine-DOF model. Assume that four actuator/sensor pairs are used at DOFs 1, 3, 7 and 9 to calculate the FRFs required.

(9) Repeat problem (8); however, move the location of the simulated stiffness damage in the nine-DOF model from the support stiffness at DOF 3 to the stiffnesses connecting DOFs 2 and 3, DOFs 3 and 5, and DOFs 3 and 6. Examine each of these stiffness elements independently. Comment on the effects on the virtual force when reductions in the different stiffness elements are considered.

(10) Consider the three-DOF system model shown in Figure 2.12. Use the simulation code 'threedoffrfload.m' to generate the steady-state phase-plane diagrams relating x_1 and its derivative at 1, 2.2, 4, and 6 rad/s. Generate the phase-plane diagrams before and after the springs K_1 and K_3 are reduced one at a time by 5 %, 10 %, and 25 %. Using the techniques discussed in Section 5.4.2, explain the changes observed in these phase-plane diagrams. Calculate the area inside the phase-plane diagrams and plot this area as a function of the damage level at each oscillation frequency.

(11) Consider the one-dimensional finite element model of the cracked rod pictured in Figure 2.24(b). Assume that an excitation force is applied at the left end at node 1 and the displacement response is measured at the right end at node 17. By using the simulation code 'rod1dwave.m,' insert simulated cracks in elements 3, 7, 11, and 15 with a 50 % reduction in area and then calculate the FRF between the input force and displacement response. Plot these FRFs in addition to the FRF for the baseline model. Comment on the changes observed due to the location of the crack. Could these differences be used to locate the damage?

(12) Repeat problem (11); however, use the spectrograms for the displacement responses at nodes 9 and 17 for damage detection. Apply a swept excitation force from 40 to 240 kHz with a 1e6 N force to node 1 using the code 'rod1dwave.m.' Use the 'timefreqanalysis.m' code to generate the spectrograms. Comment on the sensitivity of the changes in the spectrogram at the two different nodes due to the simulated cracks in various elements along the rod.

(13) Use the simulation code 'platefemsim.m' to generate the transverse displacement responses of the plate model at nodes 111–121. Then reduce the flexural stiffness of element 24 by 2 %, 5 % and 10 % and calculate the responses. Use the difference between the two sets of responses to locate the damage using beamforming per Equation (7.7). Move the measured displacements to nodes 56–66 (down the center of the plate). Comment on the differences observed using these two different sets of measurements.

8

Case Studies: Damage and Performance Prediction (Prognosis)

The two previous chapters demonstrated methods for identifying loads and damage in structural components. This chapter demonstrates how loading and damage information can be used together with models to predict the future performance of damaged components. The first case study uses a physics-based model in conjunction with damage estimates to predict the burst strength of a pressure vessel. The second case study uses a generalized phenomenological damage model for predicting the fatigue life of a component in an automotive suspension system. Although these two case studies are focused on specific components, they demonstrate the fundamental approaches used for damage and performance prediction.

8.1 S2 GLASS CYLINDER (PERFORMANCE PREDICTION)

The S2 glass woven composite cylinder discussed in Chapter 5.11 is used in this section to demonstrate a methodology for prognosis of cylinder burst strength. Such cylinders are used in a variety of pressure vessels including missile casings. When a cylinder is impacted, fibers break causing the cylinder to lose stiffness in the cylinder wall. This loss of stiffness results in a decrease in the pressure at which the cylinder will burst due to catastrophic fiber breakage in the hoop direction when pressurized. The procedure used in this section to predict this loss in burst strength after an impact occurs is given below (Kess *et al.*, 2006):

Step 1: Estimate the loss in bending stiffness of the cylinder wall in the vicinity of the impact zone using the embedded sensitivity approach.

This step was carried out in Section 5.11.3. One sensor was positioned on each side of the impact zone, and an excitation force was applied adjacent to one of the sensors using a modal impact hammer. The FRFs between the two response measurements and the force measurement were then computed. Note that the excitation force must be measured in engineering units in order to calculate the FRFs. After estimating the FRFs, the formula in Equation (2.66a) was applied to estimate the loss in stiffness. The results of these estimates are given in Table 5.23. A 10 ft lb energy impact resulted in a 250 lb/in loss in stiffness, and 1 ft lb impact

Health Monitoring of Structural Materials and Components: Methods with Applications D. Adams
© 2007 John Wiley & Sons, Ltd

Table 8.1 Composite layup thicknesses and orientations from inner to outer diameters

Thickness (in)	Material orientation (°)	
0.008	90	Tube OD
0.008	90	
0.008	30	
0.008	−30	
0.008	90	
0.008	90	
0.008	30	
0.008	−30	Tube ID

resulted in a 60 lb/in loss in stiffness. As expected, more fibers break for the 10 ft lb impact leading to a larger drop in stiffness. For the 10 ft lb impact, it was estimated that the cylinder wall stiffness, which was approximately 880 lb/in in the center and 528 lb/in at the quarter point of a healthy cylinder, decreased by approximately 28 %.

Step 2: Remove as many layers from an FEM of the cylinder in the vicinity of the impact zone as required to match the estimated loss in stiffness.

The FEM of the cylinder was constructed in order to link the estimated loss in stiffness to the reduction in strength of the cylinder. The estimated loss in stiffness is in engineering units of lb/in in this case study; therefore, this estimate of the damage level can be applied directly to the numerical FEM for use in prediction (Step 3). Shell elements with composite material properties were used to construct the FEM. Table 8.1 lists the properties of the composite material layup. Table 8.2 lists the ply properties for the fiberglass used to wind these cylinders. In order to improve the approximation of transverse shear stresses in the cylinder, reduced integration elements were selected for use in the model.

In order to update the FEM to reflect the loss in bending stiffness due to a given impact level, the static stiffness was estimated in the following way. First, the model was constrained along the bottom against vertical motion, along one side to restrain lateral motion and on one end to restrain axial motion. These boundary conditions were used in order to focus attention on the cylinder wall and estimate the static stiffness. Second, in three distinct load cases, a vertical force was applied at three locations along the length of the cylinder: middle point of the cylinder, quarter point and at the end point of the cylinder. The displacements at the load location were calculated for use in estimating the tube bending stiffness at that location. The ratios of the displacements to forces were then calculated to estimate the equivalent stiffness at

Table 8.2 S2 fiberglass ply properties

E_{11}	5.95E6 psi
E_{22}	1.3E6 psi
v_{22}	0.28
G_{12}	0.86E6 psi
G_{13}	0.48E6 psi
G_{23}	0.48E6 psi

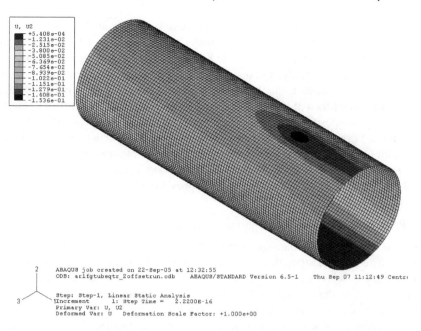

Figure 8.1 FEM of cylinder for load applied at quarter point in order to estimate wall-bending stiffness for use in modeling impact damage

those points. This information was needed because the available damage estimate was expressed in terms of this stiffness parameter as opposed to some other damage indicators (e.g. the number of fibers that has failed due to the impact and area damaged due to impact).

The FEM with deformation contours for one static loading case with the force applied at the quarter point is shown in Figure 8.1. Darker colors correspond to larger displacements. The estimated baseline stiffness values at the three load locations in these simulations are listed in Table 8.3. As expected, the wall-bending stiffness is highest at the center point of the cylinder. The simulated stiffness parameters were slightly higher than the experimentally estimated values. Therefore, it was anticipated that the numerical model would provide an estimate of burst strength that would be greater than the actual burst strength.

Third, the FEM was used to simulate the damaging effects of the impact in the form of stiffness reductions. Outer layers across the entire cylinder were removed until the stiffness reduction that was estimated from the FEM matched the experimentally estimated stiffness reduction. Although removing layers across the entire cylinder does not represent the

Table 8.3 Finite element baseline stiffness estimates

Load location	Stiffness (lb/in)
Mid point	1179
Quarter point	1078
End point	583

localized nature of the fiber damage due to impacts, a worst case analysis was thought to be appropriate. The most critical area of the cylinder would be the first to fail under pressure loading so this approach is reasonable. Table 8.3 shows that the wall-bending stiffness of the cylinder is highest at the middle and quarter points and lowest at the end point. In order to bound the stiffness reduction using the FEM, layers were removed at the quarter point in one set of simulations and then at the end point in a second set of simulations. It is known that fiberglass epoxy composites are stronger in tension than in compression. At peak deflection during an impact event, the outer hoop layers go into compression, which can be devastating for the health of the cylinder, whereas the inner layers go into tension. To approximate the stiffness reduction due to impact damage, the outer hoop layers were removed. It was also known that the failure mode of interest in these tubes would be the hoop failure because the tubes were open ended.

Step 3: Predict the loss in burst strength using the updated FEM.

After modeling the impact damage as an equivalent loss in cylinder wall-bending stiffness, the approximate residual strength could then be estimated by loading the remaining unda-maged hoop layers with an internal pressure load. This pressure load was meant to simulate pressurization when the missile is launched. The undamaged fiberglass tube contained approximately 0.032 in of hoop fibers after the impact damage broke the outer fiber layers. Failure was assumed to occur when the fibers were stressed beyond their maximum normal stress in the hoop direction.

Table 8.4 summarizes the finite element stiffness reduction and damage approximation results. Residual strengths were calculated for both the quarter point and the end point in order to identify lower and upper bounds on the actual residual strength reduction. A 5 ft lb impact energy that produces a 250 lb/in stiffness loss in the cylinder wall resulted in an 18 % loss in residual strength assuming damage occurs at the quarter point or a 35 % loss in residual strength assuming damage occurs at the end point.

In order to validate these results for the S2 glass cylinders, the estimated residual strength results were compared to measured residual strength data acquired for various other types of cylinder materials. These measurements were made by bursting cylinders impacted at different energy levels. The residual strength corresponded to burst strength in these experiments. Figure 8.2 shows these measurement results for carbon epoxy (♦), braided Kevlar overwrap (■), filament wound Kevlar overwrap (Δ), braided S glass overwrap (●) and filament wound S glass overwrap (*). Carbon epoxy has the least resistance to impacts, whereas braided Kevlar has the most resistance to impacts. The figure also shows the upper and lower bounds for the estimated residual strength reduction for the S2 fiberglass-woven cylinders using the FEM in conjunction with the embedded sensitivity function estimates of

Table 8.4 Finite element approximation of stiffness reduction and residual strength

Location	Impact energy (ft lb)	Stiffness loss (lb/in)	Thickness removed (in)	Residual strength (%)
Quarter point	1	60	0.0013	0.96
Quarter point	5	250	0.0057	0.82
End point	1	60	0.0025	0.92
End point	5	250	0.0100	0.65

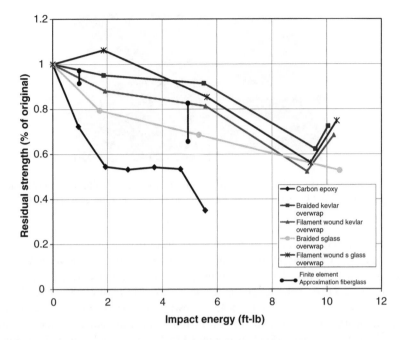

Figure 8.2 Experimentally measured percentage of residual strength versus impact energy for various types of cylinder materials superimposed with upper and lower bounds of finite element prediction for S2 glass cylinders

the reductions in stiffness due to impacts at 1 ft lb and 5 ft lb energy levels (vertical lines in figure). Note that the predicted residual strengths of the cylinders at these two impact energy levels fall in the range of measured residual strengths for the other material layups signaling that the results are reasonable. These values would consequently serve as the prognosis, or prediction, of the residual cylinder burst strengths. The estimates could be improved with further testing of the burst strengths of the S2 glass cylinders to ascertain whether the upper or lower bound for the residual strength is more accurate for this type of material.

8.2 STABILITY BAR LINKAGE (DAMAGE GROWTH MODELING)

The residual strength prediction method described in Section 8.1 was based on a physics-based FEM of the component. In a single component, a physics-based method is often feasible. In a system of components, physics-based models are time consuming and usually difficult to develop because of the complicated interactions among the components. This section discusses an empirical (phenomenological) modeling approach for predicting the rate of fatigue damage accumulation and component failure in a suspension system. The model is based on the approach described in Sections 3.2.2 and 5.12.2. Recall that both loading and damage indicators are needed to develop a regression model that can predict fatigue failures because internal loads redistribute as damage accumulates.

Figure 8.3   (a and b) Fixture pieces for gripping link and (c) positioning of fixture inside fatigue machine grips (Reproduced by permission of Muhammad Haroon, Purdue University, 2006)

Numerous tests were conducted on the sway bar links from an Isuzu automobile (Haroon, 2007). An MTS® 810 Material Test System was used to apply dynamic tension–tension fatigue loads to the links in order to initiate and propagate a fatigue crack. The fixture used to hold the link in the grips of the fatigue machine is shown in Figure 8.3(a) and (b). The link positioned within the fatigue machine is also shown in (c). Two single axis accelerometers (1000 V/g) were attached to the two ends of the link to measure axial acceleration along the link. An initial tensile load of 7000 N was applied to the link in order to avoid rocking of the sway bar link inside the fixture. Then the loading was cycled at 5 Hz to simulate a typical durability test profile used in the automotive industry. At certain increments during the test, sine sweep and random inputs were used to estimate the internal loading and damage indicators so that these indicators could be tracked over the course of the test. The test parameters used in these experiments are listed in Table 8.5. Restoring forces were used to characterize the loading (Section 2.3.1.2), whereas transmissibility functions were used to characterize the damage (Section 2.2.6).

The link was subjected to 2500 cycles of a 0.15 mm amplitude load at a frequency of 5 Hz. Then sine sweep and random excitations were used to measure the restoring forces and transmissibility functions. An additional 2500 cycles were then applied, and this test sequence was repeated until the link failed. The weakest part of the sway bar link was thought to be the welded regions at both ends where the main rod attaches to the bushing housing (see arrow in Figure 8.3(c)). The cyclic loading caused a circumferential crack to initiate at the lower weld location at the side of the link where the hydraulic actuator applied the loading. Figure 8.4 shows the velocity and displacement restoring forces for the link. Figure 8.5 shows the transmissibility measurement across the link.

After 5000 cycles of loading (two blocks of the durability test), a circumferential crack was visually detected in the lower weld. The corresponding changes in the restoring

Table 8.5   Parameters for sine sweep and random inputs for link tests (Reproduced by permission of Muhammad Haroon, Purdue University, 2006)

Input type	Amplitude (mm)	Frequency content (Hz)	Sampling frequency (Hz)	Length (s)
Sine sweep	0.05	0–15	2000	100
Random	0.05 RMS	0–30	2000	100

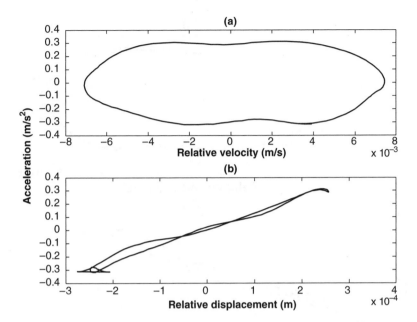

Figure 8.4   (a) Velocity and (b) displacement restoring force in the sway bar link under a tensile load of 7000 N at 14.5 Hz excitation frequency (Reproduced by permission of Muhammad Haroon, Purdue University, 2006)

forces and the transmissibility measurements are shown in Figures 8.6 and 8.7. Figure 8.6(a) shows that the area within the damping restoring force increases. Figure 8.7 shows that the transmissibility between 5 Hz and 15 Hz decreases. In order to track the change in load, the entrained area inside the damping restoring force was estimated (refer to

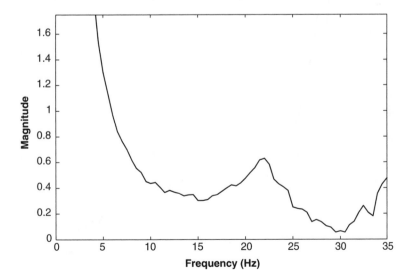

Figure 8.5   Transmissibility measurement across the sway bar link under a 7000 N tensile load (Reproduced by permission of Muhammad Haroon, Purdue University, 2006)

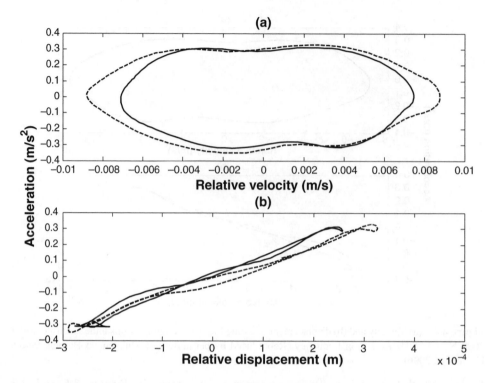

Figure 8.6 Change in (a) velocity and (b) displacement restoring forces when a circumferential crack initiates in link undamaged (_) and crack (– – –) (Reproduced by permission of Muhammad Haroon, Purdue University, 2006)

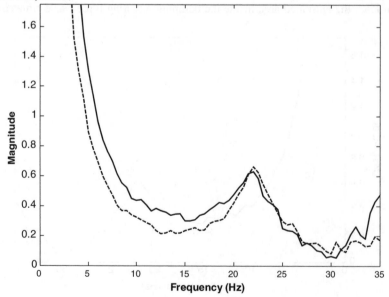

Figure 8.7 Change in transmissibility measurement when crack initiates in link undamaged (_) and initial crack (– – –) (Reproduced by permission of Muhammad Haroon, Purdue University, 2006)

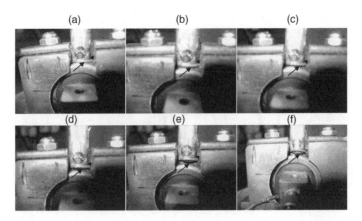

Figure 8.8　((a) and (b)) Appearance and ((c)–(f)) growth of circumferential crack to failure in sway bar link lower weld location under tension–tension fatigue loading (Reproduced by permission of Muhammad Haroon, Purdue University, 2006)

Section 5.4.2). To track the change in damage, the following transmissibility-based damage index, ΔT_k, was used:

$$\Delta T_k = \frac{\sum\limits_{i=a}^{b} |\mathrm{Re}(Ln(T_k(\omega_i))) - Re(Ln(T_{k-1}(\omega_i)))|}{N_f} \tag{8.1}$$

Where $T_k(\omega_i)$ is the transmissibility measurement for measurement number k at frequency ω_i, $T_{k-1}(\omega_i)$ is the transmissibility for measurement k-1, $Ln(\cdot)$ denotes the natural logarithm, i is the index that determines the range of frequencies (ω_a to ω_b) over which the change in transmissibility is summed and N_f is the total number of frequencies in the summation.

The cyclic loading tests were continued until the link failed at the lower weld location as anticipated. The loading and damage indicators were continuously estimated throughout

Figure 8.9　Fatigue failure of sway bar link under constant amplitude cyclic loading (Reproduced by permission of Muhammad Haroon, Purdue University, 2006)

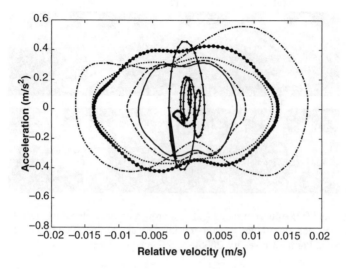

Figure 8.10 Change in velocity restoring force with initiation and growth of circumferential crack in sway bar link under tension–tension fatigue loading with undamaged (_), initial crack (– – –), progression 1 (· · · ·), progression 2 (-.-), just prior to failure (-+-) and just after failure (-o-) (Reproduced by permission of Muhammad Haroon, Purdue University, 2006)

this test to failure. Figure 8.8 shows the progression of crack growth leading to failure. Figure 8.9 shows the fracture surfaces of the two elements on either side of the failed region.

Figures 8.10 and 8.11 show the progressive changes in the velocity restoring force at 14.5 Hz and the transmissibility measurement across the link as the crack grows. The area of the velocity restoring force increases as the crack grows and seems to indicate the approaching failure. Just prior to failure, the restoring force becomes nonsimply connected (Figure 8.10 (-+-)). The source of these intersections in the restoring force projection is seen in the acceleration responses at the lower end of the link as shown in Figure 8.12. It is clear from the acceleration responses that increasingly higher frequencies appear in these responses as the crack grows. These higher frequencies appear because the crack introduces nonlinearity. Table 8.6 lists the areas of the velocity restoring forces as the crack progresses in addition to the transmissibility-based damage index in Equation (8.1) for the frequency range from 20 to 30 Hz.

A data-driven model for prognosis was then developed using the differences from test cycle to test cycle of the indicators in Table 8.6. The empirical regression model relates the rate of change of the damage indicator to the current value of the damage indicator and the current value of the internal loading indicator (Section 3.2.2). This law is expressed as follows:

$$\frac{\mathrm{d}(\Delta T)}{\mathrm{d}N} = C|\Delta T|^{m} \times |\Delta(RF_{\mathrm{area}})|^{n} \qquad (8.2)$$

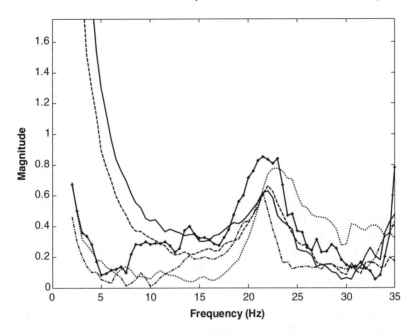

Figure 8.11 Change in transmissibility measurement with initiation and growth of circumferential crack in sway bar link under tension–tension fatigue loading with undamaged (_), initial crack (– – –), progression 1 (····), progression 2 (-.-) and just prior to failure (-+-) (Reproduced by permission of Muhammad Haroon, Purdue University, 2006)

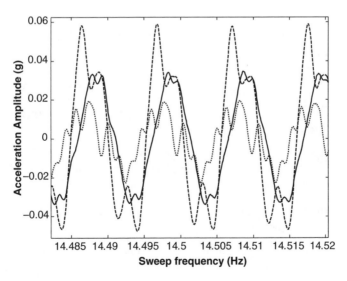

Figure 8.12 Increase in the frequency content of the link response as the fatigue crack grows toward failure, undamaged link (——), partially grown crack (– – –), just before failure (···) (Reproduced by permission of Muhammad Haroon, Purdue University, 2006)

Table 8.6 Changes in velocity restoring force area and transmissibility damage index with initiation and growth of a circumferential crack in the sway bar link (Reproduced by permission of Muhammad Haroon, Purdue University, 2006)

	Undamaged	Undamaged (comparison)	Crack visible 5000 cycles	Progression 10 000 cycles	Progression 12 500 cycles	Before failure 15 000 cycles
RF_{area} (mm^3/s^2)	.0074	.0073	.0088	.0149	.0246	.0028
ΔT_k	–	0.1568	0.1971	0.6267	0.9602	0.7472

Δ (test-to-test difference) quantities are used instead of the single-test values to emphasize changes from one cycle to another in the model. Absolute values are used for the difference in the damage and restoring force indicators in order to reduce the effects of small amounts of variability in the estimated values. This empirical model is intended to describe the rate of change of the damage indicator as a function of the current loading and the damage level as in the case of Paris' equation for a component (Equation (3.11)). As shown in Figure 8.13, the constant values $m = 0.5$, $n = 0.1$ and $C = 0.9947$ produce the same rate of change of the damage indicator (darker line) as the actual change in the damage indicator, ΔT_k (lighter line). The empirical model follows the same trend as the damage indicator; hence, the model could be used to predict the growth of damage by predicting the growth of the associated damage indicator.

In order to demonstrate that this type of regression model can be used to predict the growth of damage in components (i.e. the rate of change of transmissibility damage indicator, ΔT_k), multiple tests were run for the same material properties, geometries, loading and crack mechanisms in the sway bar link. If the crack growth model that is found in one of these

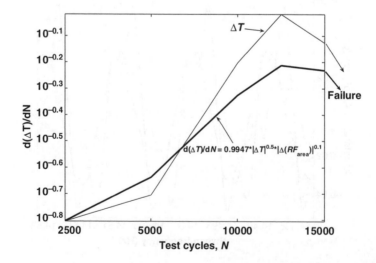

Figure 8.13 Empirical regression model relating estimated change in transmissibility measurement to the change in restoring force area for circumferential crack in sway bar link with actual (lighter line) and modeled (darker line) behavior (Reproduced by permission of Muhammad Haroon, Purdue University, 2006)

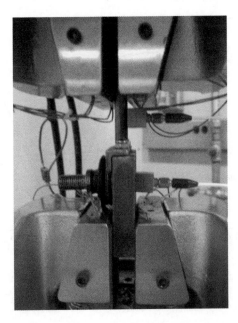

Figure 8.14   Fatigue testing of upper weld of sway bar link (Reproduced by permission of Muhammad Haroon, Purdue University, 2006)

tests can be used to describe the rate of change of the damage indicators in the other tests, then such an empirical model is useful for damage prognosis.

After failing the lower welds in three sway bar links, each of the links was flipped and inserted in the fatigue test machine in order to test the upper weld locations (Figure 8.14). One accelerometer was placed on the lower end and a second accelerometer was placed on the central rod. Three sway bar links were tested by subjecting them to an initial tensile load of 5500 N, and 2500 durability cycles of a 0.10 mm amplitude load at a frequency of 5 Hz were applied to the links. As in the previous test on the lower weld of the link, 5–15 Hz sine sweep and random tests were conducted in between the durability cycles in order to estimate the restoring force and transmissibility damage indicators.

These two indicators of loading and damage were then used to develop the empirical regression model for predicting the rate of change of transmissibility with damage. The constants used in the model were $m = 1.1$, $n = 0.02$ and $C = 1.15$. As in the case of the lower weld, the empirical model (darker line) follows the same trend as the measured damage indicator (lighter line) as shown in Figure 8.15. Note the large drop in the rate of change of damage growth around 10 000 test cycles. This behavior indicates that the load has lessened in this portion of the test due to a redistribution of loading as the crack grows.

Identical tests on two other links were repeated, and the damage growth model developed for the first link was used to predict the change in the damage indicator. Figures 8.16 and 8.17 show the results of these predictions relative to the measured transmissibility damage indicators for these two additional links. These results indicate that for the same component, material, boundary conditions, loading and damage mechanism, an empirical relationship between the load and damage can be used to predict the growth of damage to failure. The small

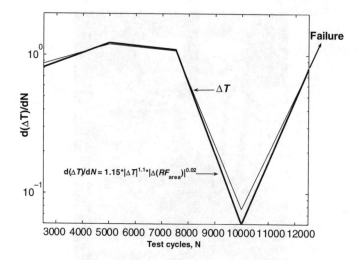

Figure 8.15 Empirical regression model relating estimated change in transmissibility measurement to the change in restoring force area for circumferential crack in sway bar link with actual (lighter line) and modeled (darker line) behavior (Reproduced by permission of Muhammad Haroon, Purdue University, 2006)

offsets in the prediction can be attributed to the variability in how the links are inserted in the hydraulic test system.

This type of approach could be applied to other components and damage mechanisms in the suspension system such as in shock damping loss, jounce bumper degradation, loosening bolts, cracked clevis, etc. This library of models could then be used in future durability tests to detect and predict the growth of damage. This case study has demonstrated that damage growth in components can be predicted at the system level without requiring specific

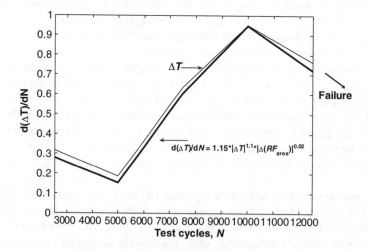

Figure 8.16 Empirical regression model for second test link with actual (lighter line) and modeled (darker line) behavior (Reproduced by permission of Muhammad Haroon, Purdue University, 2006)

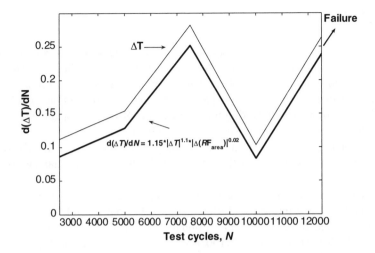

Figure 8.17 Empirical regression model for third test link with actual (lighter line) and modeled (darker line) behavior (Reproduced by permission of Muhammad Haroon, Purdue University, 2006)

information about the damage (crack length, etc.) as long as the boundary conditions, loading and damage mechanism are repeatable from test to test.

8.3 SUMMARY

This chapter has demonstrated methods for predicting the future performance and growth of damage in structural components. The following key points were made in the case studies:

- FEM (numerical) can be used to predict the future performance of pressure vessels, which have been impacted at various impact energies. The estimated damage levels are used to update the baseline FEM, and a regression model relating the impact energy to the residual strength after impact can be developed for prediction purposes.
- Empirical regression laws for predicting the rate of damage accumulation in components with loading can be developed using system level measurements in conjunction with loading and damage indicators. In this approach, a Paris like law is curve fit to the measured data for a particular component, damage mechanism, etc. for use in predicting the future damage level in similar components.

REFERENCES

Haroon, M. (2007) 'A Methodology for Mechanical Diagnostics and Prognostics to Assess Durability of Ground Vehicle Suspension Systems', *Doctoral Thesis*, School of Mechanical Engineering, Purdue University, West Lafayette, Indiana.

Kess, H., Sundararaman, S., Shah, C., Adams, D.E., Walsh, S., Pergantis, C. and Triplett, M. (2006) 'Development of a Sensor-Based Structural Integrity Measurement Technique for Potential Application to Missile Casings', *Proceedings of the Army Science Conference*. Proceedings are published online at http://www.asc2006.com/ by TMC Design Group.

PROBLEMS

(1) Consider the nine-DOF system model pictured in Figure 4.105 and simulated in 'nine-dofdata.m' and 'ninedofenviron.m.' Assume that the stiffness between DOFs 4 and 5 decreases at the rate $dK/dt = 0.001t^2$, where t is time in seconds. Enforce this rate of change in stiffness in the model and generate an empirical regression model relating the rate of change in K to the restoring force between DOFs 4 and 5 and the transmissibility across DOFs 4 and 5 similar to the law in Equation (8.2).

(2) Repeat problem (1) after modifying the boundary condition stiffness slightly by 5 % in the nine-DOF model. Comment on the differences in the regression models for these two problems.

(3) Consider the FEM for a beam developed in Section 2.2.8.4. An impulsive force with amplitude F acts on the left end of the beam. Assume the relationship between the magnitude of the force, F, and the residual strength, R, is $R = 1$ for $F < 100$ N and $R = 1-0.3183*atan(F/100)$ for $F \geq 100$. $R = 1$ corresponds to the healthy condition of the beam. For what force value does the residual strength drop to 50 % of the full strength of the beam?

(4) Consider again the setup described in problem (3). If the variability in the estimated force amplitude is ± 10 %, what is the variability in the estimated residual strength of the beam? Answer the question again when 20 % and 80 % drops in residual strength are of interest.

(5) Consider the damaged strut shown in Figure 3.16. What is the relationship between the applied force P and the maximum stress at the crack location? What changes would the displacement restoring force at the end of the strut exhibit as a function of crack depth? What changes would the transmissibility function between the end and the crack location on the strut exhibit as the crack grows?

(6) Consider the two-dimensional plate model developed in Section 5.7.4. Assume that the flexural stiffness, D, decreases in element 55 at the rate $dK/dt = 0.001t$, where t is time in seconds. Use the simulation code 'platefemsim.m' to enforce this rate of change in stiffness in the model. Generate an empirical regression model as expressed in Equation (8.2) relating the rate of change in the transmissibility measured, first, between nodes 1 and 121 and, second, between nodes 60 and 61 to the restoring force measured across those same nodes. Now develop the model using D in place of the transmissibility. Comment on the differences in the models.

Appendix A

Descriptions of the MATLAB® simulation codes and Simulink® models utilized in this book are given in this appendix. These descriptions are not meant to replace manuals supplied by Mathworks™. If at any point the reader needs more information about a certain command, `help command_name` can be typed on the MATLAB® workspace to get more details. Some of these commands may change as new versions of the software are issued by MATLAB®. A few of these commands like `ss` require the Control Systems toolbox in addition to the standard version of MATLAB®. The MATLAB® computer programs, which are described in this Appendix, can be downloaded from the companion website for this book at http://www.wiley.com/go/adams_health.

A.1 MATLAB® SIMULATION CODES

In the following sections, command lines are explained for the simulation codes. The first simulation code is described in detail. Then in the subsequent codes, only new commands are described as they arise.

A.1.1 'twodofeig.m'

This code computes the eigenvalues and eigenvectors for a linear two-DOF system model and then identifies the modal frequencies and modal deflection shapes. Changes in the two natural frequencies for changes in two stiffness parameters are also plotted. The code is listed below. The following commands are used:

- The workspace is cleared and all open figures are closed.

```
clear all;
close all;
```

- The mass and stiffness parameters of the model are defined (note the semicolon ';' is used to suppress outputs to the command line when the simulation runs). Also note a % sign is used to insert comments into the code.

```
% Define parameters
m1=1;
m2=1;
```

```
k1=10;
k2=20;
k3=10;
```

- The mass, stiffness and proportional viscous damping matrices are defined along with the system matrix, [A]. When creating matrices, a semicolon is used at the end of a row to start a new row of a matrix.

```
% Generate matrices
M=[m1 0;0 m2];
K=[k1+k2 −k2; −k2 k2+k3];
C=.01*M+.01*K;
A=[zeros(2) eye(2);−inv(M)*K −inv(M)*C];
```

- The commands zeros and eye are used to generate matrices of zeros and the identity matrix (with ones on the diagonal) of the specified size.
- The command inv is used to invert square matrices.
- The command eig is used to compute the eigenvalues and eigenvectors. Two output variables are specified: the diagonal matrix of eigenvalues d and the matrix of eigenvectors v.

```
% Solve for eigenvalues (d) and eigenvectors (v)
[v,d]=eig(A);
```

- The first and the third eigenvalues and eigenvectors are used to calculate the modal frequencies and modal deflection shapes as described in the text by scaling the vectors so that the largest element has magnitude 1.

```
% Find natural frequencies and deflection shapes
d(3,3)
d(1,1)
v(1:2,3)/v(1,3)
v(1:2,1)/v(1,1)
```

- A pause command is inserted for user to proceed on with the next portion of the simulation code.

```
% Continues on with variation in eigen-properties as K1 and
% K2 change
disp('Press any button to continue');
pause;
```

- A vector of stiffness values k_1 is created starting at 10 and ending at 5 in steps of -1 (10, 9, 8, ..., 5).

```
% Calculate and plot change in natural frequency with
% stiffnesses K1
k1=10:−1:5;
```

- This vector is then used to calculate the stiffness and damping matrices, system matrix and eigenvalues so that the damped natural frequency can be recorded by taking the imaginary parts of the eigenvalues. A `for` loop is used to loop through these different values of k_1.

```
for ii=1:max(size(k1)),
  K=[k1(ii)+k2 -k2; -k2 k2+k3];
  C=.01*M+.01*K;
  A=[zeros(2) eye(2);-inv(M)*K -inv(M)*C];
  [v,d]=eig(A);
  wd(:,ii)=abs(imag([d(3,3) d(1,1)]'));
end
```

- The command `abs` finds the magnitude of a real or complex number, and the command `imag` finds the imaginary part of a complex number.
- An apostrophe following a vector or matrix definition as in `[d(3,3) d(1,1)]`' is used to calculate the Hermitian transpose (complex conjugate transpose) of the vector or matrix.
- Then the two damped natural frequencies are plotted against the values of k_1; the notation `wd(1,:)` is used to access the entire first row of a matrix.
- The command `figure(1)` is used to generate a plotting window.

```
figure (1);
subplot(211);
plot(k1,wd(1,:),'k-x',k1,wd(2,:),'k--x');
xlabel('K_1 [N/m]');
ylabel('Natural Freqs. [rad/s]');
ylim([2 8]);
```

- The command `subplot(211)` is used to access the first row of a two-row plotting window.
- The command `plot(k1,wd(1,:),'k-x',...)` is used to plot the first natural frequency versus the stiffness values in the color of black with a solid line and x symbols at the plotting points.
- The commands `xlabel` and `ylabel` are used to attach labels to the abscissa and ordinate axes.
- The command `ylim([2 8])` is used to set the ordinate scale from 2 to 8 units.
- The same steps are then carried out to examine changes in the damped natural frequencies for changes in the other stiffness variable, `k2`.

A.1.2 'twodoffree.m'

This code computes and plots the free response of a linear two-DOF system model to a set of initial conditions. Many of the commands used are the same as in the previous code. The following new commands are used:

- The input matrix `[B]` is created as described in the text.

```
B=[0 0 1 0]';
```

- The initial conditions on displacement and velocity for each degree of freedom are then defined. Recall that an initial condition on velocity is essentially equivalent to an impulsive (or sudden) excitation.

```
% Set initial conditions
% Use an initial condition on the velocity to simulate
% impact
x10=0;
xd10=1;
x20=0;
xd20=0;
```

- The command ss is used to create a state-variable model in MATLAB®. The A and B matrices are inputs to this command. The eye command is used to create an identity matrix that selects the two displacement states as the output from the simulation. The zeros command is used to create a two by two zero matrix. The zeros(2,1) command is used to define the D matrix in the state variable form.

```
% Generate system in MATLAB
sys=ss(A,B,[eye(2) zeros(2)],zeros(2,1));
```

- The command initial is used to simulate the free response of a linear state-variable model to initial conditions. The input arguments are the state-variable structure sys (in MATLAB® terminology) and the initial conditions. The command help initial can be used to get more information about the command. The output variables on the left-hand side of the command line in the square brackets are x and t, the displacement and velocity response variables.

```
% Simulate response
[x,t]=initial(sys,[x10 x20 xd10 xd20]');
```

- The plotting command is the same as in the previous code. The two displacement responses are accessed from the x matrix, which contains as many rows as there are time points in the time vector t. x contains as many columns as there are states (4 in this example).

```
figure(1);
plot(t,x(:,1),'-k',t,x(:,2),'--k');
xlabel('Time [sec]');
ylabel('x_1 and x_2 [m]');
legend('x_1','x_2');
xlim([0 10]);
```

A.1.3 'twodoffrf.m'

This code computes and plots the frequency response functions of a linear two-DOF system model. Many of the commands used are the same as in the previous code. The following new commands are used:

- The command w=(0:.05:20)' is used to generate a frequency vectcor from 0 to 20 rad/s in steps of 0.05 rad/s.

```
% Construct frequency vector in rad/sec
w=(0:.05:20)';
```

- The impedance matrix of the system is inverted at each frequency w(ii) using a for loop as described in the text. A three-dimensional FRF matrix is produced with rows corresponding to response DOFs, columns corresponding to excitation DOFs and the third dimension corresponding to frequency. The symbol ^2 is used to square a number.

```
% Invert the impedance matrix at each frequency
for ii=1:max(size(w)),
   H(1:2,1:2,ii)=inv(-w(ii)^2*M+j*w(ii)*C+K);
end
```

- The subplot command is used to access the first row (..1) of four rows of plots (4..). The plot command plots the magnitude of the H_{11} FRF versus the frequency in Hz. The squeeze command is used to convert the three-dimensional FRF for a given row and column into a vector.

```
figure(1);
subplot(411);
plot(w/2/pi,abs(squeeze(H(1,1,:))),'k-');
xlabel('Frequency [Hz]');
ylabel('Mag. H_{11} [m/N]');
xlim([0 3]);
ylim([0 1.5]);
```

- The angle command is used to calculate the argument (phase angle) of a complex number. The unwrap command is used to calculate a phase angle value outside the range $+180°$ and $-180°$. The conversion factor 180/pi is included to plot in degrees instead of radians, which is the default angle unit.

```
subplot(412);
plot(w/2/pi,180/pi*unwrap(angle(squeeze(H(1,1,:)))),'k-');
xlabel('Frequency [Hz]');
ylabel('Phs. H_{11} [deg]');
xlim([0 3]);
ylim([-400 0]);
```

- The subsequent lines of code are used to repeat the previous steps for different values of stiffness and mass parameters to simulate changes in the FRF magnitude for different damage cases.

A.1.4 'twodofwsensor.m'

This code computes and plots the FRFs for a two-DOF linear system with a small sensor attached. The algorithm uses the impedance modeling technique described in the text. All of the commands utilized are the same as in the previous simulation codes. See those codes for more details.

A.1.5 'threedoftrans.m'

This code computes and plots the transmissibility functions for a three-DOF linear system. Most of the commands utilized are the same as in the previous simulation codes. New commands utilized are as follows:

- The two FRF vectors are divided using the command ./, which divides one entry of the first vector by one entry in the second vector (element wise division). This division produces the transmissibility function as described in the text.

```
figure(1);
subplot(311);
plot(w/2/pi,abs(squeeze(H(2,2,:))./squeeze(H(1,2,:)))),'k-
');
xlabel('Frequency [Hz]');
ylabel('Mag. T_{21} [m/m]');
xlim([0 3]);
ylim([0 20]);
```

A.1.6 'twdofnlfrf.m.'

This code simulates the response of a two-DOF system with nonlinear Coulomb friction and quadratic stiffness components. Most of the commands utilized are the same as in the previous simulation codes. New commands utilized are as follows:

- In order to simulate the dynamic response of a nonlinear system, time-domain simulation parameters must be defined. These parameters are used to compute FRFs by averaging data across many different blocks of data. The parameter `ovlap` is the overlap used when estimating the FRFs in broadband random data. The time step for the simulation is `dt`. The frequency vector `fv` is generated using a frequency resolution equal to the inverse of period of a data block, $1/t(BS)$.

```
% Create time simulation parameters and excitation force
Nt=2^17;     % Number of time points
BS=1024*10;    % Size of data blocks
navg=30;     % Number of averages
ovlap=(Nt/BS-navg)/(1-navg);    % Data block parameters
Fs=256; dt=1/Fs;    % Sampling frequency and time step
t=(0:dt:((Nt-1)*dt))';    % Time vector
fv=(0:1/t(BS):(BS-1)/t(BS))';    % Frequency vector
```

- The command `randn(Nt,1)` generates a random signal with zero mean and a normal distribution.

```
% Input
f1=1*randn(Nt,1);
```

- The simulation model `two_dof_model_nl.mdl` is executed using the `sim` command.

```
% Simulate linear model
sim('two_dof_model_nl');
```

- The command `rfrf` is used to compute FRFs, using block averaging of data for `navg` averages and `ovlap` overlap. The code that defines this command is also included on the CD accompanying the book. The input variable is given as the first argument (`f1`) followed by the response variable (`x1`). The output variables are the FRF (real and imaginary parts) and the ordinary coherence function.

```
% Calculate FRFs
[frf_lin1,coh_lin1]=rfrf(f1,x1,navg,ovlap);
[frf_lin2,coh_lin2]=rfrf(f1,x2,navg,ovlap);
```

- The command `fv(1:BS/2)` provides the frequency vector to plot the FRF magnitude up to the maximum frequency in the simulation. This frequency is half of the sampling frequency.

```
% semilogy linear FRFs
figure(1);
subplot(211);
semilogy(fv(1:BS/2),abs(frf_lin1(1:BS/2)),'k-'); hold on;
xlabel('Frequency [Hz]');
ylabel('Mag. H_{11} [m/N]');
xlim([0 3]);
ylim([1e-3 1e0]);
```

- The commands assigning parameter values to `mu1` and `mu2` are used to introduce Coulomb friction and quadratic stiffness nonlinearities in the model.

```
% Coulomb friction in joint 2
mu1=.05;
```

A.1.7 'rod1dwave.m'

This code computes and plots the transmissibility functions for a three-DOF linear system. Most of the commands utilized are the same as in the previous simulation codes. New commands utilized are as follows:

A.1.8 'rod1dtranswave.m'

This code assembles the elemental matrices for a rod to simulate its one-dimensional long-itudinal wave response. Most of the commands utilized are the same as in the previous simulation codes. New commands utilized are as follows:

- Parameters such as the number of elements Ne, length of the rod L, length of each element Le and stiffness of the boundary condition supports Ks are defined. All units are in SI (metric) unless otherwise specified (e.g. E=70e9 GPa)

```
% Parameters
Ne=17;
A=(2.54e−2)^2;
E=70e9;
L=0.1;     % Length of component
Le=L/Ne;     % Length of each element
V=A*L;
rho=2700*V/L;     % Density per unit length
Ks=0.01*A*E/Le;
```

- The two by two mass and stiffness element matrices are generated as described in the text.

```
% Generate matrices
m=[rho*Le/3 rho*Le/6;rho*Le/6 rho*Le/3];
k=[A*E/Le −A*E/Le; −A*E/Le A*E/Le];
```

- The global mass and stiffness matrices are then assembled by combining elemental matrices along the diagonal of the global matrices.

```
% Generate total mass and stiffness matrices
M=zeros(Ne+1);
K=zeros(Ne+1);
for ii=1:Ne,
  M(ii:ii+1,ii:ii+1)=M(ii:ii+1,ii:ii+1)+m;
  K(ii:ii+1,ii:ii+1)=K(ii:ii+1,ii:ii+1)+k;
end
```

- Boundary conditions are applied on either ends of the rod to simulate bushing stiffnesses as described in the text. The stiffness on the right end of the rod is attached at node Ne+1 because there are Ne+1 nodes for Ne elements.

```
% Apply the boundary conditions
K(1,1)=K(1,1)+Ks;
K(Ne+1,Ne+1)=K(Ne+1,Ne+1)+Ks;
```

- The time-domain simulation parameters are set, such as the time step dt. The maximum time and response are also set for use in plotting the responses. The forcing function applied on the left end of the rode is created using a sinusoidal pulse consisting of a sinusoidal signal with three periods multiplied by a `hanning` window to construct a pulse-like signal that starts and ends gradually. Then the command `lsim` is used to simulate the response of this system to the forcing function.

```
% Simulate response
Nt=1024;
dt=2e−7;
xmax=.00004;
ymax=5e−7;
t=(0:dt:(Nt−1)*dt)';
f1=zeros(Nt,1);
f1(1:100)=1e6*hanning(100).*sin(linspace(0,6*pi,100))';
[x,t]=lsim(sys,f1,t);
```

- The force frequency spectrum is plotted using the `fft` (fast Fourier transform) command to calculate the Discrete Fourier series coefficients as described in the text. The output from this command must be divided by half the number of time points (`Nt/2`) in order to produce the correct scaling for the coefficients.

```
% Plot force
plot((0:1/t(Nt):(Nt−1)/t(Nt)),abs(fft(f1)/Nt/2),'k−');
ylabel('F_{1}(\omega) [N]')
xlabel('Frequency [Hz]');
xlim([0 500e3]);
```

- A crack is modeled using a 10 % reduction in stiffness of the eighth element. This reduction in stiffness could correspond to a reduction in modulus or a reduction in cross-sectional area.

```
% Insert the crack
crackind=8;
K(crackind:crackind+1,crackind:crackind+1)=K(crackind:crack
  ind+1,crackind:crackind+1)−0.1*k;
```

A.1.9 'twodofdpe.m'

This code computes the mass, damping and stiffness coefficients for a two-DOF system model using simulated time data. Most of the commands utilized are the same as in the previous simulation codes. New commands utilized are as follows:

- The forcing function used in this code is given a uniform distribution using the command `rand` instead of a normal distribution. The function creates a vector of random numbers `Nt` rows long and 1 column wide.

```
f1=rand(Nt,1);
```

- The derivatives of the simulated velocity time histories are differentiated using the `diff` command. Then the result is divided by `dt` to scale to acceleration units (i.e. $a = dv/dt$).

```
% Compute derivatives
a1=diff(x(:,3))/dt;
a2=diff(x(:,4))/dt;
```

- Navg averages are used to estimate the parameters in the two-DOF model. For each average ii, two matrices are assembled as discussed in the text. The matrix D1 contains the acceleration, velocity and displacement simulated response data. The matrix D2 contains the force time history data. Each of these matrices contains two rows corresponding to the two equations of motion in the two-DOF model. Then the pseudoinverse command pinv is used to estimate the parameters such that the sum of the squared errors across all Navg equations is minimized (least squares).

```
% Estimate parameter matrices
Navg=1000;
for ii=1:Navg,
  D1(1+2*(ii-1):2+2*(ii-1),1:8)=[a1(ii) 0 x(ii,3)
    x(ii,3)-x(ii,4) 0 x(ii,1) x(ii,1)-x(ii,2) 0; 0
    a2(ii) 0 x(ii,4)-x(ii,3) x(ii,4) 0 x(ii,2)-x(ii,1)
    x(ii,2)];
  D2(1+2*(ii-1):2+2*(ii-1),1)=[f1(ii); 0];
end
P=pinv(D1)*D2;
```

A.1.10 'twodofrf.m'

This code simulates the response of the two-DOF nonlinear system with Coulomb friction and quadratic stiffness. Then the restoring forces are plotted. All of the commands utilized are the same as in the previous simulation codes.

A.1.11 'twodofdtm.m'

This code computes and plots the transmissibility functions for a three-DOF linear system. Most of the commands utilized are the same as in the previous simulation codes. New commands utilized are as follows:

- The forcing function is created using a summation of sinusoids at 1, 2 and 3 rad/s.

```
f1=sin(2*pi*t)+sin(2*pi*2*t)+sin(2*pi*3*t);
```

- Before estimating the parameters in the discrete time model, each of the columns of the simulation data matrix is divided by its standard deviation using the command std to help alleviate numerical problems when the pseudoinverse command is applied.

```
% Normalize columns by standard deviation
stdnorm=std(D1);
```

```
D1(:,1)=D1(:,1)/stdnorm(1);
D1(:,2)=D1(:,2)/stdnorm(2);
D1(:,3)=D1(:,3)/stdnorm(3);
D1(:,4)=D1(:,4)/stdnorm(4);
D1(:,5)=D1(:,5)/stdnorm(5);
D1(:,6)=D1(:,6)/stdnorm(6);
D1(:,7)=D1(:,7)/stdnorm(7);
P=pinv(D1)*D2
```

- The response is simulated using the discrete time equation after setting the first few time points in the simulated response equal to those from the original simulation data.

```
% Simulate response
xls(1:st+Navg-1)=0;
xls(1:st)=x(1:st,1);

for ii=1:Nsim,
  xls(st+ii)=P(1)*xls(st+ii-1)+P(2)*xls(st+ii-2)
  +P(3)*xls(st+ii-3)+P(4)*xls(st+ii-4)
  +P(5)*f1(st+ii)+P(6)*f1(st+ii-1)+P(7)*f1(st+ii-2);
end
```

A.1.12 'threedoffrfload.m'

This code computes and uses the FRF matrix for a three-DOF system model to estimate forces applied to the system from the simulated response. Most of the commands utilized are the same as in the previous simulation codes. See those codes for more details. New commands utilized are as follows:

- After calculating the FRF matrix using matrix inversion as in previous codes, the upper part (negative frequency content) of the FRF is constructed. This part of the FRF is needed when dealing with discrete time series that last only for a certain period of time and is sampled.

```
% Invert the impedance matrix at each frequency
for ii=1:max(size(w)),
  H(1:3,1:3,ii)=inv(-w(ii)^2*M+j*w(ii)*C+K);
end
% Generate negative frequency content of frequency response
% functions
for ii=1:max(size(w))-2,
  H(1:3,1:3,ii+max(size(w)))=conj(H(1:3,1:3,max(size(w))-
  ii));
end
```

- The response is calculated in the frequency domain first using the standard formula for FRF models. The forcing function is applied to only DOF 2 so the other elements of the forcing vector are zero.

```
% Calculate response in frequency domain
for ii=1:max(size(H)),
  X(1:3,ii)=H(1:3,1:3,ii)*[0; F2(ii); 0];
end
```

- The responses are then transformed into the time domain using the inverse fast Fourier transform command `ifft`.

```
% Calculate response in time domain
x(:,1)=real(ifft(X(1,:)))'*max(size(t))/2;
x(:,2)=real(ifft(X(2,:)))'*max(size(t))/2;
x(:,3)=real(ifft(X(3,:)))'*max(size(t))/2;
```

A.1.13 'onedofesens.m'

This code computes and plots the embedded sensitivity functions and parameter estimates for a single DOF linear system. Most of the commands utilized are the same as in the previous simulation codes. New commands utilized are as follows:

- The sensitivity function to changes in stiffness is calculated as described in the text. Each element of the FRF is squared using the `.^2` command.

```
% Compute sensitivity function
SHK=-H.^2;
```

- The sensitivity function is used along with the difference between the FRF before and after changes are made to the system parameters to estimate those changes in the parameters. Refer to the text for the relevant formulas.

```
subplot(212);
plot(w/2/pi,abs(squeeze(H4-H)./SHK),'k-
  ',w/2/pi,abs(squeeze(H5
  H)./SHK),'k:',w/2/pi,abs(squeeze(H6-H)./SHK),'k-.');
xlabel('Frequency [Hz]');
ylabel('Mag. \Delta K [N/m]');
xlim([0 1]);
ylim([0 1.5]);
```

A.1.14 'threedofesense.m'

This code applies the embedded sensitivity modeling technique to the three-DOF system model to estimate changes in parameters using FRF simulation data. All of the commands utilized are the same as in the previous simulation codes.

A.1.15 'threedofvforce.m'

This code calculates virtual forces for the three-DOF system model after changes in its parameters are made. All of the commands utilized are the same as in the previous simulation codes.

A.1.16 'threedofmodal.m'

This code extracts modal frequency and modal vectors from FRF data. Most of the commands utilized are the same as in the previous simulation codes. New commands utilized are as follows:

- The frequency values where resonances are found are defined in `fr`. The imaginary parts of the FRFs between DOF 2 and the other DOFs at these frequency values are then obtained. Note that the number of the index in the vector is obtained from the frequency value by dividing by the frequency resolution.

```
% Extract mode shapes using peak-pick method
fr=2*pi*[.39. 68. 96]; % Damped natural freqs. in rad/s
imag(H(:,2,round(fr(1)/w(2))))/max(abs(imag(H(:,2,round(fr(
    1)/w(2))))))
imag(H(:,2,round(fr(2)/w(2))))/max(abs(imag(H(:,2,round(fr(
    2)/w(2))))))
imag(H(:,2,round(fr(3)/w(2))))/max(abs(imag(H(:,2,round(fr(
    3)/w(2))))))
```

- The global least squares method is used to estimate the modal frequency and modal vector using FRF data near the first mode of the three-DOF system model. Refer to the text for the relevant formulas.

```
% Mode #1
for ii=1:6,
    D1(1+3*(ii-1):3+3*(ii-1),1:4)=[squeeze(H(1:3,2,fri(1)-
        4+ii)) eye(3)];
    D2(1+3*(ii-1):3+3*(ii-1),1)=[j*w(fri(1)-
        4+ii)*squeeze(H(1:3,2,fri(1)-4+ii))];
end
p=pinv(D1)*D2;
p(1)
p(2:4)/max(abs(p(2:4)))
```

A.1.17 'bbandpdf.m'

This code computes and plots the autospectral density and uses this information to find mean square values and other statistical parameters for random forcing time histories. All of the commands utilized are the same as in the previous simulation codes.

A.1.18   'panelsimprestemp.m'

This code simulates the response of a panel model to thermo-mechanical loading including the effects of changes in modulus with temperature and randomness of the sound pressures acting on the panel during an aerospace launch and reentry cycle. All of the commands utilized are the same as in the previous simulation codes. This file loads parameter data for the model from the file 'panelmatrices.mat'.

A.1.19   'ninedofdata.m'

This code computes and plots the FRFs of a nine-DOF system model with and without simulated damage. The code also calculates and plots the time response of the nine-DOF system model to sinusoidal and transient forcing functions. Most of the commands utilized are the same as in the previous simulation codes.

- The following command is used to reduce the mass at DOF seven by 10 % in the model to simulate corrosion damage.

```
M=diag([m m m m m m m-0.10*m m m]); % Simulate damage
```

A.1.20   'ninedofenviron.m'

This code computes and plots the FRFs of a nine-DOF system model with and without changes in mass and stiffness due to environmental factors such as humidity and temperature. Most of the commands utilized are the same as in the previous simulation codes.

- The following command is used to increase the mass at all nine DOFs to simulate moisture absorption.

```
M=diag([m+0.1*m m+0.1*m m+0.1*m m+0.1*m m+0.1*m m+0.1*m m+0.1*m
m+0.1*m m+0.1*m]); % Simulate change in mass
```

A.1.21   'attachtrans.m'

This code computes and plots the transmission FRF for a generic sensor attachment. The frequency is normalized so that the natural frequency of the attachment is 1 rad/s. All of the commands utilized are the same as in the previous simulation codes.

A.1.22   'twoDshapesensor.m'

This code generates two-dimensional vibration mode shapes for a square plate with simply supported boundary conditions. Many of the commands utilized are the same as in the previous simulation codes.

- The `for` loops below are used to generate a two-dimensional sinusoidally varying deflection shape with spatial period 1 length unit.

```
for ii=1:max(size(x)),
  for jj=1:max(size(y)),
    S(ii,jj)=sin(2*pi/1*x(ii))*sin(2*pi/1*y(jj));
  end
end
```

- The `mesh` command is used to plot the deflection shape S generated above. The `view` command is used to view the three-dimensional mesh plot from above. The `colormap` command is used to change the color of the plot to grayscale in this case.

```
mesh(x,y,S);
xlabel('x');
ylabel('y');
view(0,0);
colormap('gray');
```

A.1.23 'piezoaccelfrf.m'

This code calculates and plots the mechanical FRF of a piezoelectric accelerometer element relating the input acceleration to the strain across the piezo element for three different values of the damping ratio. The parameters are normalized so that the resonant frequency of the element is 1 rad/s. All of the commands utilized are the same as in the previous simulation codes.

A.1.24 'piezosensorfrf.m'

This code calculates and plots the electrical FRF relating the piezoelectric low-output impedance voltage to the measured voltage in the data acquisition system for three different values of the data acquisition input impedance. All of the commands utilized are the same as in the previous simulation codes.

A.1.25 'ninedofselect.m'

This code calculates the modal frequencies and deflection shapes for a nine-DOF system model. The modal deflection is then plotted for the fifth mode. Then the observability test matrices for different sensor configurations are formed, and the singular values are calculated and plotted in order to determine the rank of those matrices. Many of the commands utilized are the same as in the previous simulation codes.

- The undamped natural frequencies in Hertz are calculated using the `eig` command with the stiffness and damping matrices. Then the indices in the frequency vector f are calculated by dividing each natural frequency by the frequency resolution and rounding to the nearest integer using the `round` command.

```
fn=imag(sqrt(eig(-inv(M)*K))/2/pi);
ind=round(fn/f(2));
```

- The eigenvalues and eigenvectors are computed. Then the vector of eigenvalues, `diag(d)`, is sorted in increasing order using the `sort` command, and the indices used for sorting are stored in `Is`. These indices are used to sort the eigenvectors, `v`, from which the modal deflection shapes, `shps`, are extracted from the first nine elements of every other eigenvector corresponding to the positive natural frequencies.

```
[v,d]=eig(A);
[d,Is]=sort(abs(diag(d)));
v=v(:,Is);
shps=imag([v(1:9,1)/max(abs(v(1:9,1)))
v(1:9,3)/max(abs(v(1:9,3)))   v(1:9,5)/max(abs(v(1:9,5)))
v(1:9,7)/max(abs(v(1:9,7)))   v(1:9,9)/max(abs(v(1:9,9)))
v(1:9,11)/max(abs(v(1:9,11)))   v(1:9,13)/max(abs(v(1:9,13)))
v(1:9,15)/max(abs(v(1:9,15)))
v(1:9,17)/max(abs(v(1:9,17)))]);
```

- The modal deflection shape for mode 5, md=5, is plotted in three dimensions using the `plot3` command. The x and y axes correspond to the nodal point locations of the nine-DOF system model. The static shape is plotted with '.' symbol and the deflection shape is plotted with 'o' symbol.

```
figure(2);
md=5; % Mode number to consider
plot3([0 1 2]',[0 0 0]',shps(1:3,md),'k-o'); hold on;
plot3([0 1 2]',[1 1 1]',shps(4:6,md),'k-o');
plot3([0 1 2]',[2 2 2]',shps(7:9,md),'k-o');
plot3([0 0 0]',[0 1 2]',shps([1 4 7],md),'k-o');
plot3([1 1 1]',[0 1 2]',shps([2 5 8],md),'k-o');
plot3([2 2 2]',[0 1 2]',shps([3 6 9],md),'k-o');
plot3([0 1 2]',[0 0 0]',0*shps(1:3,md),'k:'); hold on;
plot3([0 1 2]',[1 1 1]',0*shps(4:6,md),'k:');
plot3([0 1 2]',[2 2 2]',0*shps(7:9,md),'k:');
plot3([0 0 0]',[0 1 2]',0*shps([1 4 7],md),'k:');
plot3([1 1 1]',[0 1 2]',0*shps([2 5 8],md),'k:');
plot3([2 2 2]',[0 1 2]',0*shps([3 6 9],md),'k:');
axis([0 2 0 2 -3 3]);
xlabel('X');
ylabel('Y');
zlabel('Modal deflection');
```

- The output matrices, `Ct`, corresponding to various sensor configurations are formed. Then the associated observability test matrices are formed. For example, the first line below defines an output matrix with only one sensor measurement at node 1.

```
Ct=[1 zeros(1,17)];
N1=[Ct'  A'*Ct'  (A')^2*Ct'  (A')^3*Ct'  (A')^4*Ct'  (A')^5*Ct'
(A')^6*Ct'   (A')^7*Ct'   (A')^8*Ct'   (A')^9*Ct'   (A')^10*Ct'
(A')^11*Ct'  (A')^12*Ct'  (A')^13*Ct'  (A')^14*Ct'  (A')^15*Ct'
(A')^16*Ct' (A')^17*Ct'];
...
Ct=eye(18);
N9=[Ct'  A'*Ct'  (A')^2*Ct'  (A')^3*Ct'  (A')^4*Ct'  (A')^5*Ct'
(A')^6*Ct'   (A')^7*Ct'   (A')^8*Ct'   (A')^9*Ct'   (A')^10*Ct'
(A')^11*Ct'  (A')^12*Ct'  (A')^13*Ct'  (A')^14*Ct'  (A')^15*Ct'
(A')^16*Ct' (A')^17*Ct'];
```

- The 18 singular values of the observability test matrices are plotted in log space after normalizing by the maximum element in the test matrix using the `norm` command.

```
figure(3);
semilogy(1:18,svd(N1)/norm(N1),'k.-',1:18,svd(N2)/
norm(N2),'k-o',1:18,svd(N3)/norm(N3),'k-x',1:18,svd(N4)/
norm(N4),'k-+',1:18,svd(N5)/norm(N5),'k-s',1:18,svd(N9)/
norm(N9),'k-d')
```

A.1.26 'adcerrors.m'

This code plots digital signals with DC and 60-Hz components to examine the effects of analog-to-digital conversion errors. Many of the commands utilized are the same as in the previous simulation codes.

- A sinusoidal signal is biased by a 60-Hz noise component. Then the resulting signal is quantized into discrete amplitude bins using the Simulink® model `quantizeinput`. The quantization interval is set to $10/2^4$ for 4 bits of resolution.

```
y3=y+y1;
u=y3;
qinterval=10/2^4;
sim('quantizeinput');
```

A.1.27 'aliasdemo.m'

This code demonstrates the effects of undersampling of analog signals resulting in aliasing. Many of the commands utilized are the same as in the previous simulation codes.

- The signal y is sampled at a different rate using the indexing command, e.g. `y(1+20:50:Nt)`.

```
subplot(413);
plot(t,y,'k',t(1:50:Nt),y(1:50:Nt),'k:*',t(1+20:50:Nt),
y(1+20:50:Nt),'k:+');
```

A.1.28 'filterdemo.m'

This code demonstrates some simple time, frequency and spatial filtering techniques. Time data is normalized, and then FIR and IIR filters are applied to develop bandpass filters. A modal filter for a two-DOF system is then applied. Many of the commands utilized are the same as in the previous simulation codes.

- The signal y is interpolated using the `interp` command.

```
yli=interp(y1,2);
```

- The signal y is decimated using the `decimate` command.

```
yld=decimate(y1,2);
```

- The modal deflection shapes for the two-DOF system model are extracted and normalized to unity maximum deflection for use in spatial filtering.

```
[v,d]=eig(A);
shp2=real(v(1:2,3)/max(v(1:2,3)));
shp1=real(v(1:2,1)/max(v(1:2,1)));
```

- The transient responses of the two-DOF system model are spatially filtering using the modal deflection shapes by taking the dot product of the response vector with the modal deflection shape vectors.

```
plot(t,shp1(1)*x(:,1)/sqrt(sum(shp1.^2))+shp1(2)*x(:,2)/
sqrt(sum(shp2.^2)),'k',t,shp2(1)*x(:,1)/
sqrt(sum(shp1.^2))+shp2(2)*x(:,2)/sqrt(sum(shp2.^2)),'k--');
```

- Coefficients in the vectors `Num1` and `Den1` for a 500–1000 Hz bandpass filter (FIR) are loaded, and then a discrete time model defined by these coefficients is defined in MATLAB® using the `filt` command. The `bode` command is then used to generate the magnitude and phase in the frequency domain corresponding to this discrete time FIR model.

```
load bpassfilt
sys=filt(Num1,Den1);
w=(0:.01:pi)';
[m,p]=bode(sys,w);
```

A.1.29 'windowdemo.m'

This code demonstrates various time-domain windows for reducing errors in estimating signal frequencies and amplitudes in digitally sampled data. Many of the commands utilized are the same as in the previous simulation codes.

- A frequency vector extending from negative half the sampling frequency to positive half the sampling frequency is generated.

```
f=(-m/t(m)/2:1/t(m):m/t(m)/2-1/t(m))';
```

- A normalized frequency vector from -0.5 to 0.5 is defined. This vector is then used to generate a P301 flattop window using coefficients that can be found in books on signal processing.

```
ftv=linspace(-0.5,0.5,BS)';
tmp2=0.9994484+2*(0.955728* cos(2*pi*1*ftv)+0.539289*cos(2*pi
*2*ftv)+0.091581*cos(2*pi*3*ftv));
```

- The magnitude of the DFT is plotted using the `fftshift` command to arrange the `fft` from the negative frequency components to the positive frequency components. Note that the plot is scaled such that only the positive frequency components are visible.

```
semilogy(f,fftshift(abs(Y1f)),'kx-');
```

A.1.30 'beamsteer.m'

This code animates the beamforming pattern obtained for spatially filtering data acquired from multiple sensors in a straight line. The user is prompted for the number of sensors and the spacing between the sensors. Many of the commands utilized are the same as in the previous simulation codes.

- Certain weights for sensors not to be included in the beamformer are zeroed out. The variable `throw_out_sens` should be set to 1 to throw out sensors listed in the vector `bc`.

```
bc=[2,4,6];     % Sensors to be thrown out
throw_out_sens=1;     % Throw out sensors
if throw_out_sens==1
  w(bc,:)=zeros(length(bc),size(w,2));
end;
```

- The `input` command is used to get input from the user and store it in the variable `tempp`.

```
tempp=input('Sensor Spacing: lambda/? ');     % Get spacing
```

- The `for` loop is used to cycle through steering angles from 0 to 135° and then back to 45° and then up to 90° .

```
for m=[0:1:135 135:-1:45 45:1:90],     % Loop through various
    desired steering angles
```

- The command `polar` is used to plot polar plots.

```
polar(theta,gain,'b-');
```

A.1.31 'thermalobserver.m'

This code develops a state observer to estimate the temperatures in a panel undergoing heating with only one temperature state available for measurement. All of the commands utilized are the same as in the previous simulation codes.

A.1.32 'twodoftemporal.m'

This code generates time history responses from the two-DOF system model and analyzes the data using time-domain techniques including statistical analysis and deterministic analysis (restoring forces, discrete time models and return maps). All of the commands utilized are the same as in the previous simulation codes.

- A random time series is generated and added to the forcing function. In each loop of the code, this random number generator is reset to ensure a truly random experiment. After simulating the response of the system model, the statistical parameters for each response DOF displacement are calculated. The `var` command is used to calculate the variance, and the `mean` command is used to calculate the mean value.

```
% Reset the state of random number generator
randn('state',sum(100*clock));
% Input
f1=1*sin(2*t)+1*sin(5*t)+1*sin(8*t)+randn(Nt,1);
% Simulate linear model
sim('two_dof_model_nl');
% Calculate statistical parameters
mu1=mean(x1);
mu2=mean(x2);
rms1=sqrt(mean(x1.^2));
rms2=sqrt(mean(x2.^2));
var1=var(x1);
var2=var(x2);
g1=Nt*sum((x1-mu1).^3)/sum((x1-mu1).^2)^(3/2);
g2=Nt*sum((x2-mu2).^3)/sum((x2-mu2).^2)^(3/2);
rmsd=sqrt(mean((x1-x2).^2));
R12=mean(x1.*x2);
C12=mean((x1-mu1).*(x2-mu2));
```

- Restoring forces are plotted for data points 20000 through 20330.

```
Spt=20000; Ept=20330;
figure(3);
subplot(221);
plot(vb2(Spt:Ept)-vb1(Spt:Ept),ab1(Spt:Ept),'-k',
vd2(Spt:Ept)-vd1(Spt:Ept),ad1(Spt:Ept),':k'); hold on;
```

- The index for the minimum relative displacement is found as is the index for the maximum relative displacement. Then these values are used to integrate under the acceleration using the `trapz` command.

```
% Compute areas of baseline and damaged datasets
[ymin,inmin]=min(xb2(Spt:Ept)−xb1(Spt:Ept));
[ymax,inmax]=max(xb2(Spt:Ept)−xb1(Spt:Ept));
[ymin2,inmin2]=min(xb2(Spt−1+inmax:Ept)−xb1(Spt−
    1+inmax:Ept));
delt1=ymax−ymin;
no1=abs(inmax−inmin);

intdown=trapz(ab2(Spt−1+inmin:Spt−1+inmax));
intup=trapz(ab2(Spt−1+inmax−1:Spt−1+inmax−1+inmin2));
inttotb=abs(intdown)+abs(intup);
inttotb=inttotb*delt1/no1;
```

- The parameter matrices for a discrete time model are constructed.

```
% Estimate parameter matrices
Navg=21;
st=20000;
for ii=1:Navg,
  D1(1+(ii−1)*1,1:7)=[x(st+ii−1,1) x(st+ii−2,1) x(st+ii−3,1)
x(st+ii−4,1) f1(st+ii) f1(st+ii−1) f1(st+ii−2)];
  D2(1+(ii−1)*1,1)=[x(st+ii,1)];
end
```

- The parameters are estimated after dividing each column by its standard deviation to improve the numerical conditioning of the matrices. Then the parameters are estimated using the pseudoinverse command, `pinv`.

```
stdnorm=std(D1);
D1(:,1)=D1(:,1)/stdnorm(1);
D1(:,2)=D1(:,2)/stdnorm(2);
D1(:,3)=D1(:,3)/stdnorm(3);
D1(:,4)=D1(:,4)/stdnorm(4);
D1(:,5)=D1(:,5)/stdnorm(5);
D1(:,6)=D1(:,6)/stdnorm(6);
D1(:,7)=D1(:,7)/stdnorm(7);
P=pinv(D1)*D2;
```

A.1.33 'rodtemporal.m'

This code simulates the response of a rod to a longitudinal excitation force at the left end with and without simulated damage along the rod. Statistical analysis is then performed on the data. All of the commands utilized are the same as in the previous simulation codes.

A.1.34 'twodofembed.m'

This code generates return maps for the two-DOF system model and then calculates damage indicators using these maps before and after simulated damage is introduced in the model. Many of the commands utilized are the same as in the previous simulation codes.

- The cumulative Euclidean distance is computed by summing up the distance between each point on the undamaged and damaged return maps.

```
del=40;
ind1(1)=sum(sqrt((x1(Stp:Enp)-x1b(Stp:Enp)).^2+(x1(Stp-
del:Enp-del)-x1b(Stp-del:Enp-del)).^2+(x1(Stp-2*del:Enp-
2*del)-x1b(Stp-2*del:Enp-2*del)).^2));
```

A.1.35 'twodofdft.m'

This code calculates and plots the DFT for a two-DOF system model with and without simulated damage introduced. All of the commands utilized are the same as in the previous simulation codes.

A.1.36 'twodofhos.m'

This code computes and plots the higher order spectrum (bispectrum) for a two-DOF system model with nonlinear damage introduced. The function `hospec` is the algorithm for calculating the bispectrum and trispectrum. All of the commands utilized are the same as in the previous simulation codes.

A.1.37 'timefreqanalysis.m'

This code analyzes the time-frequency content of propagating waves and nonlinear time series from the longitudinal finite element model and two-DOF nonlinear system model. Many of the commands utilized are the same as in the previous simulation codes.

- The spectrogram is computed using the command `specgram`. Then the spectrogram is plotted using the `surf` command. Then the shading of the surface is modified to interpolate the colormap, and the view is changed to inspect the surface from above.

```
[S,fs,T]=specgram(y,BS2,Fs,wind,BS2/2);
surf(T,fs(1:fmax),abs(S(1:fmax,:))/(BS2/2));
shading('interp');
view(0,90);
```

A.1.38 'averagingdemos.m'

This code demonstrates various aspects of averaging including cyclic averaging and asynchronous averaging using time and spectral analysis. All of the commands utilized are the same as in the previous simulation codes.

A.1.39 'frfestimation.m'

This code demonstrates methods in FRF estimation in the presence of noise on the excitation and response time histories. The function `rfrf` is used to calculate the FRF and coherence function given the input and output time histories, percent overlap and number of averages. This function calculates the H_1 FRF, but can be easily modified to calculate the H_2 FRF. All of the commands utilized are the same as in the previous simulation codes.

A.1.40 'ninedofspatial.m'

This code demonstrates how data from a nine-DOF system model can be spatially analyzed to identify damage using modal deflection shapes and gradients and curvatures of those shapes. All of the commands utilized are the same as in the previous simulation codes.

A.1.41 'platefemsim.m'

This code simulates the response of a plate with free–free boundary conditions using a finite element model with four-node plate elements. The number of elements can be modified by the user as can the material parameters. The displacement responses are also plotted assuming 100 elements at 121 nodes in the vertical direction. The function `plate_matrix_assembly` is used to construct the mass and stiffness matrices. The elemental matrices are stored in the function PLATEELEM.

- The modulus of any element can be modified to simulate damage as in the following line of code for element 55, where the modulus has been reduced by 50%.

```
E(55)=E(55)/2;
```

A.1.42 'ninedoffeature.m'

This code calculates a damage feature using the FRFs for the nine-DOF system model for various damage cases. The projection of the calculated feature vector for each damage case is then taken with a given feature vector to attempt to locate the damage. Many of the commands utilized are the same as in the previous simulation codes.

- The feature vector calculated in this code is the sum of the FRF magnitude over five frequency ranges in 500 Hz increments from 0 to 2500 Hz.

```
Hf(:,1)=[sum(abs(H(1,1,1:500)))
sum(abs(H(1,1,1+500:500*2)))
sum(abs(H(1,1,1+500*2:500*3)))
sum(abs(H(1,1,1+500*3:500*4)))
sum(abs(H(1,1,1+500*4:500*5)))]';
```

A.2 SIMULINK® MODELS AND BLOCK DIAGRAMS

A.2.1 'two_dof_model_nl.mdl'

This block diagram model is used to demonstrate the effects of nonlinear friction and stiffness in models of structural components. The model is shown below and has the following attributes:

- The time vector and forcing function are applied at the block to the far left.
- The summing junction shown on the left (circle) represents the first equation of motion for M_1.
- The summing junction shown on the right (circle) represents the second equation of motion for M_2.
- The integrators in the forward paths proceeding from these summing junctions are used to obtain the dynamic response of the DOFs for use in the feedback paths containing damping and stiffness parameters as well as coupling paths that connect the two equations of motion.
- The blocks on the top of the diagram represent the coupling damping and stiffness terms between the two equations of motion.
- Many other types of nonlinear functions other than those used in this model can be implemented in Simulink®.
- The response displacements are measured using the two blocks labeled x1 and x2.

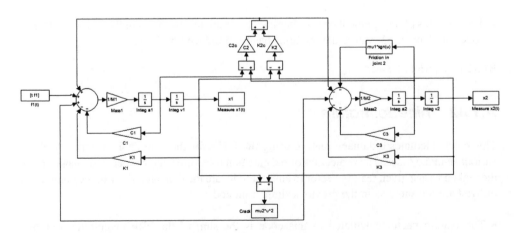

Figure A.1   Block diagram for nonlinear two degree of freedom system model

A.2.2 'quantizeinput.mdl'

This block diagram model is used to demonstrate the effects of quantization of the amplitude values of an analog signal. The model is shown below. The time and signal to be quantized are fed to the quantizer block, which then outputs the quantized signal to the workspace.

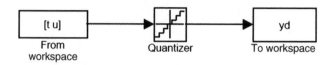

Figure A.2   Quantization block diagram

Appendix B

B.1 JOURNALS AND CONFERENCES DEALING WITH HEALTH MONITORING

Tables B.1 and B.2 in this section provide lists of technical journals and conferences that highlight developments in health monitoring.

The tables cited in this Appendix are also given on the companion website for this book at http://www.wiley.com/go/adams_health. These tables will be updated as necessary to provide up to date information.

B.2 SENSORS

In Tables B.3–B.10, different types of displacement, velocity, acceleration, strain, force, temperature and pressure sensors are summarized.

The references cited in this Appendix are also given on the companion website for this book at http://www.wiley.com/go/adams_health. These references will be updated as necessary to provide up to date information.

B.3 REFERENCES ON DATA ANALYSIS FROM THE LITERATURE

In Tables B.11–B.18, references from the literature on a wide range of data analysis topics in health monitoring are summarized and cited.

Health Monitoring of Structural Materials and Components: Methods with Applications   D. Adams
© 2007 John Wiley & Sons, Ltd

Table B.1 Technical Journals in health monitoring

Journal name	Publisher
AIAA journal	American institute of aeronautics and astronautics
Experimental mechanics	Society of experimental mechanics
International journal of analytical and experimental modal analysis	CSA illumina
International journal of engineering science	CSA illumina
International journal of fatigue	Elsevier science
International journal of fracture	Springer
Journal of applied mechanics	American society of mechanical engineers
Journal of dynamic systems, measurement and control	American society of mechanical engineers
Journal of engineering mechanics	American society of civil engineers
Journal of intelligent material systems and structures	Sage publishers
Journal of pressure vessel technology	American society of mechanical engineers
Journal of sound and vibration	Academic press
Journal of structural engineering	American society of civil engineers
Journal of vibration and acoustics	American society of mechanical engineers
Mechanical systems and signal processing	Academic press
NDT&E international	Elsevier science
Physical review letters	American physical society
Sensors actuators	CSA illumina
Smart materials and structures	Institute of physics
Structural health monitoring: An international journal	Sage publishers
The journal of the acoustical society of america	Acoustical society of america
The shock and vibration digest	Sage publishers

Table B.2 Technical conferences in health monitoring

Conference name
International modal analysis conference
European workshop on structural health monitoring
International workshop on structural health monitoring
The international society for optical engineering (SPIE)
International mechanical engineering congress
Asia-pacific conference on systems integrity and maintenance (ACSIM)
IEEE aerospace conference
International conference on adaptive structures and technologies
AIAA/ASME/ASCE/AHS/ASC structures, structural dynamics and materials conference
International conference on adaptive structures
IEEE conference on antennas and propagation
International conference on damage assessment of structures
International design engineering technical conference
Society for the advancement of material and process engineering conference
Integrated systems health management conference
Health and usage monitoring conference
Machinery failure prevention technology annual meeting
Materials science and technology conference
Quantitative NDE conference
AIAA/ASME/ASCE/ASC Structures, structural dynamics & materials conference

Table B.3 Displacement sensors

Sensor Type:	Advantages	Disadvantages	Applications	Example
Inductive	1. Low susceptibility to noise and interference 2. Accurate at high temperatures while being unsusceptible to environmental conditions	1. Accurate for small distance (1 mm–150 mm) 2. Surface conditions affect high resolution measurements	1. Crack detection in turbine blades 2. Corrosion thinning measurements on aircraft skins	
Capacitive	1. Can be used on conductive and non-conductive materials 2. Wide bandwidth and high resolutions	1. Sensitive to environmental parameters 2. Susceptible to electrostatic charge due to friction	1. Measure aircraft engine door cowling gaps 2. Monitor aircraft cargo door alignment.	
Gyroscope	1. Capable of operating under hostile environments 2. Have a high signal to noise ratio and low power consumption	1. Mechanical gyros accumulate drift between actual and sensed values over time 2. Provide only relative information	1. Measuring angular displacement of aircraft wings due to turbulence. 2. Satellite position monitoring and control	

(*Continued*)

Table B.3 Displacement sensors (*Continued*)

Sensor Type:	Advantages	Disadvantages	Applications	Example
Magnetic	1. Are more stable in noisy environments 2. Capable of achieving low temperature sensitivity.	1. Are susceptible to external magnetic interference. 2. Can only be used for ferromagnetic materials.	1. Monitor crackshaft for ignition timing and misfire 2. Monitoring weld health in welded steel armor plates.	
Optical	1. Absence of loading effects on the structure. 2. Insensitivity to stray magnetic field or electrostatic interference	1. Not suitable to be bent at steep angles due to refraction of light 2. Fibres are delicate and can be easily damaged.	1. Monitoring hull deflection of a composite patrol boat 2. Measure displacement of composite bridge decks due to automotive loading.	
Ultrasonic	1. Resistant to external disturbances such as vibration, ambient noise and EM radiation. 2. Capable of detecting small defects at large distances	1. Sensors have a "dead" region directly below them where damage cannot be detected. 2. Time consuming and requires higher level of user skill	1. Study of wear, chipping/ breakageand temperature in tooling parts 2. Examining bolts or rivets in aircraft wings	
Acoustic emission	1. Is less sensitive to material surface roughness or geometry 2. High sensitivities allow for crack formation detection.	1. Susceptible or extraneous noise 2. Sensors must be mounted to surface, resulting in possible mass loading issues	1. Monitor seal and blade-tip rubbing in turbo machinery 2. Damage assessment in a steel-concrete composite bridge deck	

Table B.4 Velocity sensors

Sensor Type:	Advantages	Disadvantages	Applications	Example
Magnetic induction	1. High sensitivity and excellent noise immunity 2. Lower output, for use in high speed applications	1. Susceptible to electromagnetic field interference 2. Must be mounted perpendicular to plane of motion	1. Measuring gear speed in an automotive gearbox 2. Measure rotational speeds in gas turbine engines	
Optical	1. High accuracy and high reliability 2. Not affected by surface roughness or color	1. Difficult to measure parts in hard to reach areas 2. Requires powered light source	1. Monitoring vibrations of automotive tires 2. Monitor molten plastic flow in injection molding process	
Piezoelectric	1. Larger frequency range then magnetic sensors 2. Reduces signal noise for high frequency measurements.	1. Response is nonlinear for low frequencies (<10Hz) 2. Requires sensor to be mounted to structure.	1. Measure vibrations in cavitating pumps. 2. Monitor seals in paper handling machines.	

Table B.5 Acceleration sensors

Sensor Type:	Advantages	Disadvantages	Applications	Example
Capacitive	1. Higher sensitivities than piezoresistive accelerometer. 2. Measures static acceleration	1. Must compensate for drift and interfernce affects. 2. Low resolution and fragile	1. Measure aircraft wing flutter response 2. Measure hard disk drive acceleration due to writing process	
MEMS	1. Small, lightweight, high g acceleration. 2. Lower cost then other accelerometers	1. Performance/Specifications can degrad over time 2. Expensive to repair due to their small size.	1. Used for automotive airbag development measurements 2. Monitor laptop computer vibration and stop harddrive processes to prevent damage	
Piezoelectric	1. Wide dynamic range, low output noise 2. Can produce high output voltage	1. Low bandwidth, not suited for low frequency testing. 2. Requires sensor to be mounted to structure resulting in possible mass loading affects	1. Measuring vibration response in an exhaust system 2. Measuring acceleration response of TPS panel impact.	
Piezoresistive	1. Not adversely affected by electromagnetic fields 2. Measures static acceleration	1. Limited resolution due to resistive noise 2. Primarily for low to mid frequency applications	1. Measure accelerations of ejection seats 2. Measure crash test dummy acceleration due to collisions	

Table B.6 Strain sensors

Sensor Type:	Advantages	Disadvantages	Applications	Example
Piezoresistive	1. Capable of recognizing static forces	1. Requires sensor to be mounted to structure	1. Measure strains in gas turbine fan blades	
	2. Simplicity of mounting to the surface	2. Susceptible to external sources of noise and temperature	2. Measure helicopter blade deflections	
Optical	1. Not susceptible to electromagnetic interference	1. Requires fiber optic cable to be run to each sensor	1. Strain monitoring of civil structures, for instance bridges, dams, buildings, pipelines	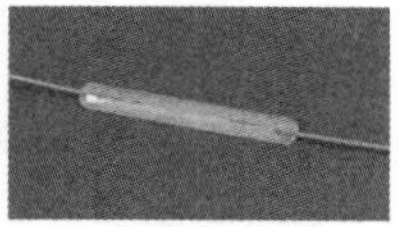
	2. Multiplexing capability	2. Requires a power source	2. Monitoring ship hul strains	

Table B.7 Force sensors

Sensor Type:	Advantages	Disadvantages	Applications	Example
Piezoresistive	1. High stiffness allows direct insertion in machine structures	1. More expensive then other types	1. Recording impact forces in military applications	
	2. High natural frequencies, ideal for quick transient forces	2. The output can be nonlinear	2. Measuring wave forces on off-shore oil platforms	
Optical	1. Multiplexing capability	1. Requires fiber optic cable to be run to each sensor	1. Monitoring traffic loads over the span of a bridge	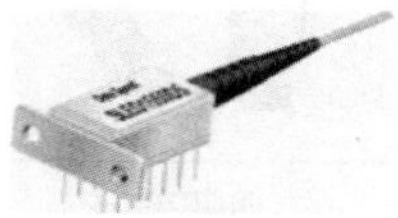
	2. Ideal for high temperature applications	2. Requires a power source	2. Measure clamping force of a car window closing	

Table B.8 Temperature sensors

Sensor Type:	Advantages	Disadvantages	Applications	Example
Acoustic	1. Capable of operating in cryogenic temperature range 2. Immune to high levels of radiation	1. Susceptible to external sources of noise 2. Sensors must be mounted to surface	1. Measuring temperature inside catalytic converters 2. Temperature measurement for feedback control of engine combustion	
Optical	1. Negligible electromagnetic interference affects. 2. Small and flexible for easy installation	1. fiber optic cables are delicate and limit maximum temperature 2. Slow data processing	1. Measuring temperature of electric generators 2. Temperature monitoring in semi-conductor manufacturing	
Thermoresistive	1. Typically cheaper than other sensors 2. Easy implementation due to small size	1. Resistance vs. temperature is nonlinear causing limited temperature range 2. Limited operating temperature range	1. Measuring automotive engine oil and coolant temperatures 2. Measure inside air temperature in HVAC systems	
Thermoelectric	1. Offer higher temperature range then thermoresistive sensors 2. Are cheapest of all temperature sensors	1. Have upper temperarure limit of 3100F 2. Measured temperatures drift over time	1. Engine and turbine exhaust gas monitoring 2. Heat treating and metals processing temperatures	

Table B.9 Pressure sensors

Sensor Type:	Advantages	Disadvantages	Applications	Example
Piezoresistive	1. Measure both static and dynamic pressures 2. Reliable under varying environment conditions	1. Increase in pressure might lead to transducer becoming nonlinear 2. Can produce significant electrical noise	1. Measure engine combustion chamber pressures 2. Measure jet engine pressure at inlet and outlet of each component	

Table B.10 Piezoelectric actuators

Configuration:	Sensing Direction	Advantages	Disadvantages	Applications	
Stack	Transverse	1. Ideal for static and low frequency applications. 2. Capable of applying tension and compression loads	1. Low electroc-mechanical coupling 2. Requires strong bonds to ensure high fidelity. 3. Stability problems for large displacement	1. Fine tuning of laser equipment 2. Alignment of fiber optics. 3. Control injection valves in the automotive industry	
	Shear	1. Extremely reliable ($>10^9$ cycles). 2. High resonant frequencies.	1. Needs to be pre-loaded to avoid un-poling resulting in lowered operation frequencies.	1. Atomic force microscopy. 2. Active vibration cancellation.	
Tube	Transverse	1. Capable of measuring displacements along all three axes. 2. Sub-nanometer resolution	1. Small Displacement 2. Relative to stack actuators, small force	1. Hard drive read/write head testing. 2. Needle valve actuation.	
Ring	Transverse	1. Available with clear aperatures for transmitted-light applications. 2. High resolution for static/dynamic applications.	1. More delicate than other configurations due to the center bore 2. Low foce	1. Image positioning. 2. Micropositioning.	
Disk	Transverse	1. Provide a relatively large travel range for their size. 2. Fast response w/sub-nanometer resolution	1. Low force	1. Knife edge control in extrusion tools. 2. Tuning of circular boring, drilling process.	
Bimorph (PVDF)	Transverse/ shear	1. Low operating voltage. 2. Excellent resistance to humidity.	1. Low frequency operation. 2. Low resolution (unsuitable for precision). 3. Low force and slow response	1. Position control of pneumatic values. 2. Measuring accelerations of flexible structures.	

Table B.11   References on methods for loads identification

Reference	Summary
Stevens, K.K. (1987) 'Force identification problems—an overview.'	Conference: Overview of indirect force estimation for linear systems.
Chae, *et al.* (1999) 'A feasibility study in indirect identification of transmission forces through rubber bushing in vehicle suspension system by using vibration signals measured on links.'	Journal: Relates the transmission force to the deformation of rubber bushings through an appropriate model.
Decker, M. and Savaidis, G. (2002) 'Measurement and analysis of wheel loads for design and fatigue evaluation of vehicle chassis components'	Journal: Discussed the interactions of wheel forces and moments, forces acting in a suspension and the stress response of an axle casing.
O'Connor, C. and Chan, T.H.T. (1988) 'Dynamic wheel loads from bridge strains'	Journal: Modeled the bridge deck as lumped masses interconnected by mass-less elastic beams and estimated loading of bridge due to wheels.
Chan, T.H.T., Law, S.S., Yung, T.H. and Yuan, X.R. (1999), 'An interpretive method for moving force identification.'	Journal: Modeled the bridge deck using Bernoulli-Euler beams and estimated loading of bridge due to wheels.
Zhu, X.Q. and Law, S.S. (2000) 'Identification of vehicle axle loads from bridge responses.'	Journal: Modeled the bridge deck as orthotropic plates and estimated loading of bridge due to wheels.
Wang, M.L. and Kreitinger, T.J. (1994) 'Identification of force from response data of a nonlinear system.'	Journal: Presented the sum of weighted acceleration technique (SWAT) to estimate the input force.
Giergil, J. and Uhl, T. (1989) 'Identification of the input excitation forces in mechanical structures.'	Journal: Presented an iterative formula for calculation of excitation forces in mechanical structures based on properties of the Toeplitz matrix.
Haas, D.J., Milano and Flitter, L. (1995) 'Prediction of helicopter component loads using neural networks.'	Journal: Used a neural network approach to relate rotor system component loads to flight data recorded using a flight recorder.
Giasante, *et al.* (1983) 'Determination of in-flight helicopter loads.'	Journal: Identified the external vibratory forces acting on a helicopter in flight using a calibration matrix.
Li, J. (1988) 'Application of mutual energy theorem for determining unknown force sources.'	Conference: Identified spectrum of loads based on vibration velocity response measurements.
Zion, L. (1994) 'Predicting fatigue loads using regression diagnostics.'	Conference: Presented an approach based on a regression model relating loads and flight data in a helicopter.
Uhl, T. and Pieczara, J. (2003) 'Identification of operational loading forces for mechanical structures.'	Journal: Based on the difference between measured and simulated system responses, genetic algorithm estimates loads.
Starkey, J.M. and Merrill, G.L. (1989) 'On the ill-conditioned nature of indirect force-measurement techniques.'	Journal: Investigated the ill-conditioned nature of the inverse problem and found that the condition of the FRF matrix is a good indicator of errors.
Bartlett, F.D., Jr. and Flannelly W.G. (1979) 'Model verification of force determination for measuring vibratory loads.'	Journal: Found that the pseudo-inverse method of force estimation worked well for identifying vibrations forces on the rotary hub of a helicopter model.

Table B.11   *(Continued)*

Reference	Summary
Hundhausen, R.J., Adams, D.E., Derriso, M., Kukuchek, P. and Alloway, R. (2005) 'Transient loads identification for a standoff metallic thermal protection system panel.'	Conference: Used two methods for identifying transient loads on standoff metallic panels: (1) rigid body approach and (2) inverse FRF approach.
Turco, E. (2005) 'A strategy to identify exciting forces acting on structures.'	Journal: Explores the use of the Tikhonov regularization technique to reduce ill-conditioning effects of frequency domain equations for pin-jointed trusses.
Kammer, D.C. (1996) 'Input force reconstruction using a time domain technique.'	Journal: Convolves the measured response and an inverse system of Markov parameters to estimate input forces on a structure in the time domain.
Jacquelin, E., Bennani, A. and Hamelin, P. (2003) 'Force reconstruction: analysis and regularization of a deconvolution problem.'	Journal: Applies Tikhonov and trunctation regularization techniques to the indirect force estimation problem and chooses the regularization parameters.
Fabunmi, J.A. (1986) 'Effects of structural modes on vibratory force determination by the pseudoinverse technique.'	Journal: Studied the implication of using the least-squares method of force identification without considering the modes and mode shapes.
Carne, T.G., Mayes, R.L. and Bateman, V.I. (1994) 'Force reconstruction using the sum of weighted acceleration technique—Max-Flat procedure.'	Conference: Used FRF data to determine appropriate scalar weights to use in the Sum of Weighted Acceleration Technique for force reconstruction.
Mayes, R.L. (1994) 'Measurement of lateral launch loads on re-entry vehicles using SWAT.'	Conference: Uses the SWAT method to reconstruct forces acting on a structure, but uses the free decay time histories to calculate the weights.
Liu, Y. and Shepard, S., Jr. (2005) 'Dynamic force identification nased on enhanced least squares and total least-squares schemes in the frequency domain.'	Journal: Utilizes and compares the least-square method of indirect force estimation without regularization and with truncated SVD and regularization.

Chae, C.K., Bae, B.K., Kim, K.J., Park, J.H. and Choe, N.C. (1999) 'A feasibility study in indirect identification of transmission forces through rubber bushing in vehicle suspension system by using vibration signals measured on links', *Vehicle Syst. Dyn.*, **33** (5), 327–349.

Decker, M. and Savaidis, G. (2002) 'Measurement and analysis of wheel loads for design and fatigue evaluation of vehicle chassis components', *Fatig. Fract. Engin. Mat. Struct.*, **25**(12), 1103.

O'Connor, C. and Chan, T.H.T. (1998) 'Dynamic wheel loads from bridge strains' *J. Struct. Div.* ASCE, **114**(8), 1703–1723.

Chan, T.H.T., Law, S.S., Yung, T.H. and Yuan, X.R. (1999) 'An interpretive method for moving force identification', *J. Sound Vib.*, **219**(3), 503–524.

Zhu, X.Q. and Law, S.S. (2000) 'Identification of vehicle axle loads from bridge responses', *J. Sound Vib.*, **236**(4), 705–724.

Wang, M.L. and Kreitinger, T.J. (1994) 'Identification of force from response data of a nonlinear system', *Soil Dynam. Earthquake Engin.*, **13**, 267–280.

Giergil, J. and Uhl, T. (1989) 'Identification of the input excitation forces in mechanical structures', *Arch. Transport*, **1**(1).

Haas, D.J., Milano and Flitter, L.(1995) 'Prediction of helicopter component loads using neural networks', *J. Am. Helicopter Soc.*, (1), 72–82.

Giasante, N., Jones, R. and Calapodas, N.J. (1983) 'Determination of in-flight helicopter loads', *J. Am. Helicopter Soc.*, **27**, 58–64.

Li, J. (1988) 'Application of mutual energy theorem for determining unknown force sources', *Proceedings of Internoise 88*, Avignion.

Zion, L. (1994) 'Predicting fatigue loads using regression diagnostics', *Proceedings of the American Helicopter Society 50 Annual Forum*, Washington D.C.

Uhl, T. and Pieczara, J. (2003) 'Identification of operational loading forces for mechanical structures', *Arch. Transport*, **16**(2).

Stevens, K.K. (1987) 'Force identification problems—an overview', *Proceedings of SEM Spring Conference on Experimental Mechanics*, pp. 838–844.

Starkey, J.M. and Merrill, G.L. (1989) 'On the ill-conditioned nature of indirect force-measurement techniques', *J. Modal Anal.*, 103–108.

Bartlett, F.D., Jr. and Flannelly W.G. (1979) 'Model verification of force determination for measuring vibratory loads', *J. Am. Helicopter Soc.*, **24**, 10–18.

Hundhausen, R.J., Adams, D.E., Derriso, M., Kukuchek, P. and Alloway, R. (2005) 'Transient loads identification for a standoff metallic thermal protection system panel', *Proceedings of the IMAC-XXIII: A Conference & Exposition on Structural Dynamics*, No. 394.

Turco, E. (2005) 'A strategy to identify exciting forces acting on structures', *Int. J. Num. Method Engin.g*, 64, 1483–1508.

Kammer, D.C. (1996) 'Input force reconstruction using a time domain technique', *Am. Inst. Aeronaut. Astronaut.*, Inc., pp. 21–30.

Jacquelin, E., Bennani, A. and Hamelin, P. (2003) 'Force reconstruction: analysis and regularization of a deconvolution problem', *J. Sound Vib.*, **265**, 81–107.

Fabunmi, J.A. (1986) 'Effects of structural modes on vibratory force determination by the pseudoinverse technique', *Am. Inst. Aeronaut. Astronaut.*, Inc., **24**(3), 504–509.

Carne, T.G., Mayes, R.L. and Bateman, V.I. (1994) 'Force reconstruction using the sum of weighted acceleration technique–Max-Flat procedure', *Proceedings of 12th International Modal Analysis Conference*, pp. 1054–1062.

Mayes, R.L. (1994) 'Measurement of lateral launch loads on re-entry vehicles using SWAT', *Proceedings of 12th International Modal Analysis Conference*, pp. 1063–1068.

Liu, Y. and Shepard, S., Jr. (1995) 'Dynamic force identification based on enhanced least squares and total least-squares schemes in the frequency domain', *J. Sound Vib.*, **282**, 37–60.

Table B.12   References on vibration-based damage identification methods

Reference	Summary
Doebling, *et al.* (1996) 'Damage identification and health monitoring of structural and mechanical systems from changes in their vibration characteristics: a literature review.'	Report: Comprehensive survey of vibrations-based techniques for damage detection, location and characterization.
Hoon, *et al.* (2001) 'A review of structural health monitoring literature: 1996–2001.'	Report: An update to the work by Doebling *et al.* (1996) that outlines feature extraction and damage quantification methods among other issues.
Afolabi, D. (1987) 'An anti-resonance technique for detecting structural damage.'	Conference: Showed how data around anti-resonances is much more sensitive to structural damage compared to the resonances.
Zhang, *et al.* (1999) 'Structural health monitoring using transmittance functions.'	Journal: Showed that transmissibility functions are reliable detection features to locate perturbations in experiments on a composite beam.
Johnson, T.J. and Adams, D.E. (2002) 'Transmissibility as a differential indicator of structural damage.'	Journal: Developed a transmissibility-based detection feature that was able to detect and locate damage.
Wang, W. and Zhang, A. (1987) 'Sensitivity analysis in fault vibration diagnosis of structures.'	Conference: Determined that certain frequency ranges in FRFs, including those near anti-resonances, are sensitive to changes in structural parameters.
Trendafilova, I., *et al.* (1998) 'Damage localization in structures. A pattern recognition perspective.'	Conference: Presented a pattern recognition approach for damage localization in structures.
Sohn, H. and Farrar, C.F. (2001) 'Damage diagnosis using time series analysis of vibration signals.'	Journal: Used standard deviation of residual errors from a combination of AR and ARX models as a damage-sensitive feature to locate damage.
Nair, *et al.* (2003) 'Application of time series analysis in structural damage evaluation.'	Conference: Previous algorithm is modified to increase the effectiveness in identifying small damage patterns by using normalized relative accelerations.
Adams, D.E. and Farrar, C.R. (2002) Classifying linear and non-linear structural damage using frequency domain ARX models.'	Journal: Used frequency domain autoregressive models to develop linear and nonlinear damage features in a three-story building frame.
Johnson, *et al.* (2005) 'Embedded sensitivity functions for characterizing structural damage.'	Journal: Presented the use of algebraic combinations of measured FRF data to estimate perturbations in mass, damping, or stiffness due to damage.
Adams, D.E. (2002) 'Nonlinear damage models for diagnosis and prognosis in structural dynamic systems.'	Conference: Demonstrated that model reduction near bifurcations caused by structural damage is a useful way to identify damage features.
Farrar, *et al.* (1999) 'A statistical pattern recognition paradigm of vibration-based structural health monitoring.'	Conference: Discussed the process of vibration-based structural health monitoring as a statistical pattern recognition problem.
Corbin, *et al.* (2000) 'Locating damage regions using wavelet approach.'	Conference: Detected damage using wavelet decomposition of acceleration response data.

(Continued)

Table B.12 (*Continued*)

Reference	Summary
Moyo, P. and Brownjohn, J.M.W. (2002) 'Detection of anomalous structural behavior using wavelet analysis.'	Journal: Used wavelet analysis to detect anomalies using strain data from a bridge but does not distinguish damage from other sources of variability.
Sun, Z. and Chang, C.C. (2002) 'Structural damage assessment based on wavelet packet transform.'	Journal: Developed a damage assessment method using the wavelet packet transform to produce inputs to neural network models.
Hou, *et al.* (2000) Application wavelet-based approach for structural damage detection.'	Journal: Showed that damage can be detected by decomposing response data using wavelets with the potential to locate damage as well.
Haroon, M. and Adams, D.E. (2005) 'Active and event-driven passive mechanical fault identification in ground vehicle suspension systems.'	Conference: Presented active and passive data interrogation methodologies for damage identification based on the frequency bandwidth of signals.
Haroon, M. and Adams, D.E. (2006) 'Nonlinear fault identification methods for ground vehicle suspension systems.'	Conference: Discussed nonlinear damage identification methods which track nonlinear changes accompanying damage using response acceleration data.
Worden, *et al.* (2003) 'Experimental validation of structural health monitoring methodology I: novelty detection on a laboratory structure.'	Journal: Presented experimental verification of the novelty detection method for damage identification based on transmissibility functions.
Manson, *et al.* (2003) 'Experimental validation of structural health monitoring methodology II: novelty detection on an aircraft wing.'	Journal: Applied the previously discussed outlier analysis based novelty detection algorithm on a realistic structure, the wing of a Gnat aircraft.
Monaco, E., Calandra, G. and Lecce, L. (2000) 'Experimental activities on damage detection using magnetorestricitve actuators and statistical analysis.'	Conference: Used averages of differences between healthy and damaged structure FRFs as damage detection features.
Natke, H.G. and Cempel, C. (1997) 'Model-aided diagnosis based on symptoms.'	Conference: Used changes in natural frequencies and mode shapes in a finite element model of a cable-stayed steel bridge to detect damage.
Garcia, *et al.* (1998) 'Comparison of the damage detection results utilizing an ARMA model and a FRF model to extract modal parameters.'	Conference: Time domain ARMA model and FRF modal extraction techniques are compared and ARMA model out performs modal parameters.
Garcia, G. and Osegueda, R. (1999) 'Damage detection using ARMA model coefficients.'	Conference: Parameters of time domain ARMA model are used for damage detection; location was possible with ambiguity for multiple damage sites.
Sohn, H. and Farrar, C.R. (2000) 'Statistical process control and projection techniques for structural health monitoring.'	Conference: Combined statistical process control with projection techniques, such as principal component analysis, for damage detection.

Table B.12   (*Continued*)

Reference	Summary
Bodeux, J.B. and Golinval, J.C. (2000) 'ARMAV model technique for system identification and damage detection.'	Conference: Demonstrated the use of time-domain Auto-Regressive Moving-Average Vector (ARMAV) models for detecting damage.
Heyns, P.S. (1997) 'Structural damage assessment using response-only measurements.'	Conference: Used a Multivariate Auto-Regressive Vector (ARV) model based approach to detect and locate damage in a cantilever beam.
Tsyfansky, S.L. and Beresnevich, V.I. (1997) 'Vibrodiagnosis of fatigue cracks in geometrically nonlinear beams.'	Conference: Attempted to detect and quantify fatigue cracks in a beam by analyzing the nonlinear harmonics in the Fourier spectrum of the response.
Masri, *et al.* (2000) 'Application of neural networks fort detection of changes in nonlinear systems.'	Journal: Presented a neural network technique for health monitoring using vibration measurements; prediction error was used for detecting damage.
Feng, M. and Bahng, E. (1999) 'Damage assessment of bridges with jacketed rc columns using vibration test.'	Conference: Proposed a jacketed column monitoring method that combines vibration testing, neural network and finite element techniques.
Worden, K. and Fieller, N.R.J. (1999) 'Damage detection using outlier analysis.'	Journal: Studied outlier analysis for damage detection with a Mahalanobis distance based on measured transmissibility functions as damage feature.
Salawu, O.S. (1997) 'Detection of structural damage through changes in frequency: a review.'	Journal: Reviewed methods for detecting damage using natural frequencies and discussed relationships between frequency changes and structural damage.
Farrar, C.R. (1997) 'Variability of modal parameters on the alamosa canyon bridge.	Conference: Showed that the sensitivity of frequency shifts to damage is low but shifts these exhibit less statistical variation from random error.
Doebling, *et al.* (1997) 'Effects of measurements statistics on the detection of damage in the alamosa canyon bridge.'	
Cawley, P. and Adams, R.D. (1979) 'Location of defects in structures from measurements of natural frequencies.'	Journal: Detected damage in composite materials using ratios between frequency shifts for two different modes.
Pandey, *et al.* (1991) 'Damage detection from changes in curvature mode shapes.'	Journal: Showed that absolute changes in mode shape curvature can be a good indicators of damage.
Pandey, A.K. and Biswas, M. (1994) 'Damage detection in structures using changes in flexibility.'	Journal: Presented a damage detection and location method based on changes in the measured flexibility matrix using lowest frequency vibration modes.
Pandey, A.K. and Biswas, M. (1995) 'Damage diagnosis of truss structures by estimation of flexibility change.'	

(*Continued*)

Table B.12   (*Continued*)

Reference	Summary
Lim, T.W. (1991) 'Structural damage detection using modal test data.'	Journal: Used the unity check methods for damage detection by defining a least-squares problem for the elemental stiffness changes in a truss.
Banks, H.T., Inman, D.J., Leo, D.J., Want,Y. (1996) 'An experimentally validated damage detection theory in smart structures.'	Journal: Developed a damage detection theory based on the derivative of frequency with respect to either stiffness or mass.
Doebling, S.W. (1996) 'Minimum-rank optimal update of elemental stiffness parameters for structural damage identification.'	Journal: Developed an optimal minimum-rank update of stiffness parameters for damage identification.
Escobar, J.A., Sosa, J.J., Gomez, R. (2005) 'Structural damage detection using the transformation matrix.'	Journal: Used transformation matrix in two- and three-dimensional analytical building models to detect damage.
Fritzen, C.P., Jennewein, D., Kiefer, T. (1998) 'Damage detection based on model updating methods.'	Journal: Applied a sensitivity approach that used both time and frequency to localize damage in a finite element beam model.
Hajela, P. and Soeiro, F.J. (1989) 'Structural damage detection based on static and modal analysis.'	Journal: Eigenmodes and static displacements were used to detect changes in stiffness.
Hwang, H.Y., Kim, C. (2004) 'Damage detection using a few frequency response measurements.'	Journal: Modeled damage using changes in the component stiffness matrix and treated the damage detection problem as a minimization problem.
Lew J.S. (1995) 'Using transfer function parameter changes for damage detection of structures.'	Journal: Found that changes in environmental factors contribute less significantly to the structural natural frequencies than actual damage.
Kaouk, M., Zimmerman, D.C. (1994) 'Structural damage assessment using a generalized minimum rank perturbation theory.'	Journal: Addressed unsymmetric impedance matrices with singular value decomposition to acquire a damage vector.
Samuel, P.D., Pines, D.J. (2004) 'A review of vibration-based techniques for helicopter transmission diagnostics.'	Journal: Points out progress in the area of vibration-based fault detection.
Sheinman, I. (1996) 'Damage detection and updating of stiffness and mass matrices using mode data.'	Journal: Damage was detected using minimal static and dynamic measurements through a closed form algorithm.
Tsuei, Y.G., Yee, E.K.L. (1989) 'A method for modifying dynamic properties of undamped mechanical systems.'	Journal: Modified mass and stiffness matrices by adding small changes in mass and stiffness to the forcing function of the unmodified structure.
Zimmerman, D.C., Kaouk, M. (2005) 'Model correlation and system health monitoring using frequency domain measurements.'	Journal: Addressed unsymmetric impedance matrices with singular value decomposition to acquire a damage vector.

Doebling, S.W., Farrar, C.R., Prime, M.B. and Shevitz. D.W. (1996) 'Damage identification and health monitoring of structural and mechanical systems from changes in their vibration characteristics: a literature review', *Los Alamos National Laboratory report*, LA-13070-MS.

Sohan, H., Farrar, C.R., Hemez, F.M., Shunk, D.D., Stinemates, D.W. and Nadler, B.R. (2003) 'A review of structural health monitoring literature: 1996–2001', *Los Alamos National Laboratory report*, LA-13976-MS.

Afolabi, D. (1987) 'An anti-resonance technique for detecting structural damag', *Proceedings of the 5th International Modal Analysis Conference*, pp. 491–495.

Zhang, H., Schulz, M.J., Naser, A., Ferguson, F. and Pai, P.F. (1999) 'Structural health monitoring using transmittance functions', *Mech. Sys. Signal Process.*, **13**(5), 765–787.

Johnson, T.J. and Adams, D.E. (2002) 'Transmissibility as a differential indicator of structural damage', *ASME J. Vib. Acoust.*, **124**(4), 634–641.

Wang, W. and Zhang, A. (1987) 'Sensitivity analysis in fault vibration diagnosis of structures', *Proceedings of the 5th International Modal Analysis Conference*, pp. 496–501.

Trendafilova, I., Heylen, W., Sas, P. (1998) 'Damage localization in structures. A pattern recognition perspective', *ISMA*, **23**, 99–106.

Sohn, H. and Farrar, C.F. (2001) 'Damage diagnosis using time series analysis of vibration signals', *Smart Mat. Struct.*, **10**, 446–451.

Nair, K.K., Kiremidjian, A.S., Lei, Y., Lynch, J.P. and Law, K.H. (2003) 'Application of time series analysis in structural damage evaluation', *Proceedings of the International Conference on Structural Health Monitoring*, Tokyo, Japan.

Adams, D.E. and Farrar, C.R. (2002) 'Classifying linear and non-linear structural damage using frequency domain arx models', *Struct. Health Monitor.*, **1**(2), 185–201.

Johnson, T.J., Yang, C., Adams, D.E. and Ciray, S. (2005) 'Embedded sensitivity functions for characterizing structural damage', *Smart Mat. Struct.*, **14**, 155–169.

Adams, D.E. (2002) 'Nonlinear damage models for diagnosis and prognosis in structural dynamic systems', *SPIE*, **4733**.

Farrar, C.R., Duffey, T.A., Doebling, S.W. and Nix, D.A. (1999) 'A Statistical pattern recognition paradigm of vibration-based structural health monitoring', *2nd International Workshop on Structural Health Monitoring*, Stanford, CA, pp. 764–773.

Corbin, M., Hera, A. and Hou, Z. (2000) 'Locating damage regions using wavelet approach', *Proceedings of the 14th Engineering Mechanics Conference (EM2000)*, Austin, Texas.

Moyo, P. and Brownjohn, J.M.W.(2002) 'Detection of anomalous structural behavior using wavelet analysis', *Mech. Sys. Signal Process.*, **16**(2–3), 429–445.

Sun, Z. and Chang, C.C. (2002) 'Structural damage assessment based on wavelet packet transform', *J. Struct. Engin.*, **128**(10), 1354–1361.

Hou, *et al.* (2000) 'Application wavelet-based approach for structural damage detection', *J. Engin. Mech.*, **126**(7), 677–683.

Haroon, M. and Adams, D.E. (2005) 'Active and event-driven passive mechanical fault identification in ground vehicle suspension systems', *Proceedings of IMECE: ASME International Mechanical Engineering Congress and Exposition*, Orlando, FL, paper no. 80582.

Haroon, M. and Adams, D.E. (2006) 'Nonlinear fault identification methods for ground vehicle suspension systems', *IMAC-XXIV*, St. Louis, MO, paper no. 44.

Worden, K., Manson, G. and Allman, D. (2003) 'Experimental validation of structural health monitoring methodology i: novelty detection on a laboratory structure', *J. Sound Vib.*, **259**, 323–343.

Manson, G., Worden, K. and Allman, D. (2003) 'Experimental validation of structural health monitoring methodology ii: novelty detection on an aircraft wing', *J. Sound Vib.*, **259**, 343–363.

Monaco, E., Calandra, G. and Lecce, L. (2000) 'Experimental activities on damage detection using magnetorestritcive actuators and statistical analysis', *Smart Structures and Materials 2000: Smart Structures and Integrated Systems, Proceedings of SPIE*, Vol. 3985, pp. 186–196.

Natke, H.G. and Cempel, C. (1997) 'Model-aided diagnosis based on symptoms', *Structural Damage Assessment Using Advanced Signal Processing Procedures, Proceedings of DAMAS '97*, Univ. of Sheffield, UK, pp. 363–375.

Garcia, G., Osegueda, R. and Meza, D. (1998) 'Comparison of the damage detection results utilizing an arma model and a frf model to extract modal parameters', *Smart Systems for Bridges, Structures, and Highways, Proceedings of SPIE*, Vol. 3325, pp. 244–252.

Garcia, G. and Osegueda, R. (1999) 'Damage detection using ARMA model coefficients', *Smart Systems for Bridges, Structures, and Highways, Proceedings of SPIE*, Vol. 3671, pp. 289–296.

Sohn, H. and Farrar, C.R. (2000) 'Statistical process control and projection techniques for structural health monitoring', *European COST F3 Conference on System Identification and Structural Health Monitoring*, Madrid, Spain, pp. 105–114.

Bodeux, J.B. and Golinval, J.C. (2000) 'ARMAV model technique for system identification and damage detection', *European COST F3 Conference on System Identification and Structural Health Monitoring*, Madrid, Spain, pp. 303–312.

Heyns, P.S. (1997) 'Structural damage assessment using response-only measurements', *Structural Damage Assessment Using Advanced Signal Processing Procedures, Proceeding of DAMAS'97*, Univ. of Sheffield, UK, pp. 213–223.

Tsyfansky, S.L. and Beresnevich, V.I. (1997) 'Vibrodiagnosis of fatigue cracks in geometrically nonlinear beams', *Structural Damage Assessment Using Advanced Signal Processing Procedures, Proceeding of DAMAS '97*, Univ. of Sheffield, UK, pp. 299–311.

Masri, S.F., Smyth, A.W., Chassiakos, A.G., Caughey, T.K. and Hunter, N.F. (2000) 'Application of neural networks fort detection of changes in nonlinear systems', *J. Engin. Mech.*, July, 666–676.

Feng, M. and Bahng, E. (1999) 'Damage assessment of bridges with jacketed rc columns using vibration test', *Smart Structures and Materials 1999: Smart Systems for Bridges, Structures, and Highways, Proc. of SPIE*, Vol. 3671, pp. 316–327.

Worden, K. and Fieller, N.R.J. (1999) 'Damage detection using outlier analysis', *J. Sound Vib.*, **229**(3), 647–667.

Salawu, O.S. (1997) 'Detection of structural damage through changes in frequency: a review', *Engin. Struct.*, **19**(9), 718–723.

Farrar, C.R., Doebling, S.W., Cornwell, P.J. and Straser, E.G. (1997) 'Variability of modal parameters on the alamosa canyon bridge', *Proceedings of the 15th International Modal Analysis Conference*, Orlando, FL, pp. 257–263.

Doebling, S.W., Farrar, C.R. and Goodman, E.S. (1997) 'Effects of measurements statistics on the detection of damage in the alamosa canyon bridge', *Proceedings of the 15th International Modal Analysis Conference*, Orlando, FL, pp. 919–929.

Cawley, P. and Adams, R.D. (1979) 'Location of defects in structures from measurements of natural frequencies', *J. Strain Engin. Design*, **14**(2), 49–57.

Pandey, A.K., Biswas, M. and Samman, M.M. (1991) 'Damage detection from changes in curvature mode shapes', *J. Sound Vib.*, **145**(2), 321–332.

Pandey, A.K. and Biswas, M. (1994) 'Damage detection in structures using changes in flexibility', *J. Sound Vib.*, **169** (1), 3–17.

Pandey, A.K. and Biswas, M. (1995) 'Damage diagnosis of truss structures by estimation of flexibility change', *Modal Analysis Int. J. Analytic. Experiment. Modal Anal.*, **10**(2), 104–117.

Lim, T.W. (1991) 'Structural damage detection using modal test data', *AIAA J.*, **29**(12), 2271–2274.

Lew, J.-S. (1995) 'Using transfer function parameter changes for damage detection of structures', *AIAA J.*, **33**(11), 2189–2193.

Banks, H.T., Inman, D.J., Leo, D.J., Want, Y. (1996) 'An experimentally validated damage detection theory in smart structures', *Journal of Sound and Vibration*, **191**(5), 2615–2621.

Doebling, S. W. (1996) 'Minimum-rank optimal update of elemental stiffness parameters for structural damage identification', *AIAA J.*, **34**(12), 2615–2621.

Escobar, J.A., Sosa, J.J., Gomez, R. (2005) 'Structural damage detection using the transformation matrix', *Comput. Struct.*, **83**, 357–368.

Fritzen, C.P., Jennewein, D., Kiefer, T. (1998) 'Damage detection based on model updating methods,", *Mech. Sys. Signal Process.* **12**(1), 163–186.

Hajela, P. and Soeiro, F.J. (1989) 'Structural damage detection based on static and modal analysis', *AIAA J.*, **28**(6), 1110–1115.

Hwang, H.Y., Kim, C. (2004) 'Damage detection using a few frequency response measurements', *J. Sound Vib.*, **270**, 1–14.

Lew, J. S. (1995) 'Using transfer function parameter changes for damage detection of structures', *AIAA J.*, **33**(11), 2189–2193.

Kaouk, M., Zimmerman, D. C. (1994) 'Structural damage assessment using a generalized minimum rank perturbation theory', *AIAA J.*, **32**(4), 836–842.

Samuel, P.D., Pines, D.J. (2004) 'A review of vibration-based techniques for helicopter transmission diagnostics', *J. Sound Vib.*, **282**, 475–508.

Sheinman, I. (1996) 'Damage detection and updating of stiffness and mass matrices using mode data', *Comput. Struct.*, **59**(1), 149–156.

Tsuei, Y.G., Yee, E.K.L. (1989) 'A method for modifying dynamic properties of undamped mechanical systems', *Dynamic Sys. Measurement Cont.*, **111**, 403–408.

Zimmerman, D.C., Kaouk, M. (2005) 'Model correlation and system health monitoring using frequency domain measurements', *Struct. Health Monitor.* **4**(3), 213–215.

Table B.13   References on wave propagation for damage identification

Reference	Summary
Doebling, *et al.* (1996) 'Damage Identification and health monitoring of structural and mechanical systems from changes in their vibration characteristics: a literature review.'	Report: Includes a review of literature on damage identification using propagating elastic waves.
Sohn, *et al.* (2001) 'A review of structural health monitoring literature: 1996–2001.'	Report: Includes a review of literature on damage identification using propagating elastic waves.
Kessler (2002) 'Piezoelectric-based in-situ damage detection of composite materials for structural health monitoring systems.'	Thesis: Damage identification using guided waves on an Al plate and composite cylinder. Literature review of guided waves.
Wilcox, *et al.* (1999) 'Mode selection and transduction for structural monitoring using Lamb waves.'	Conference: Developed mode selection and transduction rules for monitoring structures using Lamb waves.
Bar-Cohen, *et al.* (1998) 'Composite material defects characterization using leaky Lamb wave dispersion data.'	Conference: Monitored the changes in dispersion characteristics of a leaky Lamb wave to characterize porosity in a composite plate.
Grisso (2004) 'Considerations of the impedance method, wave propagation, and wireless systems for structural health monitoring.'	Thesis: Studied temperature influences on wave propagation. Presented a method to quantify damage using the impedance method.
Lakshmanan and Pines (1997) 'Modeling damage in rotorcraft flexbeams using wave mechanics.'	Journal: Used and developed a wave propagation method to identify delaminations and transverse cracks in Gr/Ep composite rotorcraft.
Pines (1997) 'The use of wave propagation models for structural damage identification.'	Conference: Identified damage in beams using wave propagation by modeling damage as a local change in dispersion; local and global defects.
Prosser, *et al.* (1995) 'Advanced, waveform based acoustic emission detection of matrix cracking in composites.'	Journal: Used acoustic emission to identify cracking of thin composite specimens; also outlined the difficulties associated with acoustic emission.
Wevers (1997) 'Listening to the sound of materials: acoustic emission for the analysis of material behavior.'	Journal: Outlined the advantages of acoustic emission techniques over other NDE methods for identifying damage in a loaded composite component.
Shah, *et al.* (2000) 'New directions in concrete health monitoring technology.'	Journal: Used stress waves (0–100 ï kHz) and found that changes in signal amplitude across a crack were sensitive to crack.
Adamou and Craster (2004) 'Spectral methods for modeling guided waves in elastic media.'	Journal: Spectral method for dispersion curve generation of inhomogeneous, curved, multi-layered and materially damped structures.
Alleyne and Cawley (1992a) 'The interaction of Lamb waves with defects.'	Journal: Numerical and experimental study of defect identification using Lamb waves and two-dimensional fast Fourier transforms.
Alleyne and Cawley (1992b) 'Optimization of Lamb wave inspection techniques.'	Journal: Tests conducted on a butt-welded steel plate using A1 mode Lamb wave.
Beard (2002) 'Guided wave inspection of embedded cylindrical structures.'	Thesis: Detailed literature review and numerical development of guided wave inspection of curved plates and cylindrical structures.

(Continued)

Table B.13   (*Continued*)

Reference	Summary
Banerjee, *et al.* (2003) 'Lamb wave propagation and scattering in layered composite plates.'	Conference: Lamb waves for crack identification in composite plates.
Bar Cohen (2000) 'Emerging NDE technologies and challenges at the beginning of the 3rd millennium – part I.'	Journal: Traditional NDE techniques (ultrasonics, radiography, shearography) and associated challenges are reviewed.
Mustafa, *et al.* (1997) 'Imaging of disbond in adhesive joints with Lamb waves.'	Online Journal: Detect and image disbonds in the tear-strap by using angle wedge transducers to excite select Lamb modes.
Chahbaz, *et al.* (1996) 'Corrosion detection in aircraft structures using guided Lamb waves.'	Online Journal: Demonstrated the use of Lamb waves to detect corrosion damage in an aluminum fuselage panel.
Fromme (2001) 'Defect detection in plates using guided waves.'	Thesis: Studied and compared scatter patterns of the antisymmetric Lamb wave mode using both experimental and analytical results.
Giurgiutiu (2003) 'Lamb wave generation with piezoelectric wafer active sensors for structural health monitoring.'	Conference: Used piezoelectric sensors for detecting damage in an aluminum plate.
Lamb (1917) 'On waves in an elastic plate.'	Journal: The first work dealing with guided wave propagation in thin elastic specimens.
Lord-Rayleigh (1889) 'On the free vibrations of an infinite plate of homogeneous isotropic matter.'	Journal: The first work dealing with wave propagation in a semi-infinite solid.
Lowe (1995) 'Matrix techniques for modeling ultrasonic waves in multilayered media.'	Journal: Literature review of work involving guided wave dispersion curve generation.
Pavlakovic, *et al.* (1997) 'Disperse: a general purpose program for creating dispersion curves.'	Conference: Outlines the software developed by researchers at Imperial College for generating guided wave dispersion curves and mode shapes.
Pavlakovic (1998) 'Leaky guided ultrasonic waves in NDT.'	Thesis: Provided design rules for generating Lamb waves; also carried out defect identification studies in plates and shells.
Pavlakovic and Lowe (1999) 'A general purpose approach to calculating the longitudinal and flexural modes of multi-layered, embedded, transversely isotropic cylinders.'	Conference: Outlined dispersion curve (longitudinal and flexural modes) characterization in a composite cylinder.
Purekar and Pines (2002) 'A phased sensor/actuator array for detecting damage in 2-D structures'	Conference: Outlined phased arrays for damage identification in 2-d structures; testing was carried out on aluminum beam and plate specimens.
Purekar and Pines (2005) 'Damage detection in plate structures using Lamb waves with directional filtering sensor arrays.'	Conference: Use of a directional filtering algorithm for defect localization in structures.
Raghavan and Cessnik (2005) 'Piezoelectric-actuator excited-wavefield solutions for guided-wave structural health monitoring.'	Conference: Analytical development of arbitrary shaped piezoelectric actuator to excite Ao and So mode Lamb waves from 3-D elasticity.

Table B.13   (*Continued*)

Reference	Summary
Rose (1999) 'Ultrasonic waves in solid media.'	Book: A detailed outline of structural wave propagation with specific emphasis on free and forced guided waves for NDE applications.
Schmerr Jr. (1998) 'Fundamentals of ultrasonic nondestructive evaluation: a modeling approach.'	Book: A mathematical approach to ultrasonic nondestructive evaluation using transfer functions including traditional ultrasonic testing methods.
Sohn, *et al.* (2004) 'Multi-Scale structural health monitoring for composite structures.'	Conference: Used Lamb waves to identify areas of delamination by implementing the ideas of time reversal acoustics.
Sundararaman (2003) 'Structural diagnostics through beamforming of phased arrays: characterizing damage in steel and composite plates.'	Thesis: Outlined a phased array directional filtering algorithm for damage localization in steel and woven composite structures.
Tucker (2001) 'Ultrasonic waves in wood-based composite panels.'	Thesis: Includes a literature review of the use of ultrasonics in NDE. Demonstrated defect identification in wood analytically and experimentally.
Viktorov I.A. (1967) 'Rayleigh and Lamb waves: physical theory and applications.'	Book: Includes models for the generation of Lamb and Rayleigh waves using ultrasonic transducers.
Wilcox (1998) 'Lamb wave inspection of large structures using permanently attached transducers.'	Thesis: Includes analytical and experimental development of piezoelectric transducers for defect identification of large structures using Lamb waves.
Worlton (1961) 'Experimental confirmation of Lamb waves at megacycle frequencies.'	Journal: One of the first works to identify the usefulness of Lamb waves for NDE applications.
Rizzo and di Scalea (2005) 'Ultrasonic inspection of multi-wire steel strands with the aid of the wavelet transform.'	Journal: Used discrete wavelet transforms to filter (denoise) data and compress data for feature extraction; applied to multi-wire steel strands.
Sundararaman, *et al.* (2004a) 'Incipient damage identification using elastic wave propagation through a friction stir welded al-li interface for cryogenic tank applications.'	Conference: Guided wave experimental investigation using acoustic emission transducers and piezoelectric actuators.
Sundararaman, *et al.* (2004b) 'Structural health monitoring studies of a friction stir welded Al-Li plate for cryotank application.'	Conference: Presented wavelet and statistical analysis techniques for defect identification in a friction stir welded Al-Li plate.
Purekar and Pines (2001) 'Interrogation of beam and plate structures using phased array concepts.'	Conference: Presented a phased array method using a sweep sine broadband signal to identify damage in beam and plate structures.
Purekar, *et al.* (2004) 'Directional piezoelectric phased array filters for detecting damage in isotropic plates.'	Journal: A detailed numerical and experimental presentation of the phased array method for defect localization in an aluminum plate.

(*Continued*)

Table B.13   *(Continued)*

Reference	Summary
Giurgiutiu and Bao (2002) 'Embedded ultrasonic structural radar with piezoelectric wafer active sensors for the NDE of thin-wall structures.'	Conference: A detailed experimental presentation for defect identification using phased arrays consisting of piezoelectric wafers.
Yu and Giurgiutiu (2005) 'Improvement of damage detection with the embedded ultrasonics structural radar for structural health monitoring'	Conference: Presented new techniques for improving defect identification using unitized phased arrays.
Bardouillet, P. (1984) 'Application of electronic focusing and scanning systems to ultrasonic testing.'	Journal: One of the early works to use ultrasonic phased arrays for detecting defects in welds.
Ihn and Chang (2004) 'Detection and monitoring of hidden fatigue crack growth using a built-in piezoelectric sensor/actuator network: I. diagnostics.'	Journal: Used spectrograms to process guided wave signals obtained from an array of piezoelectric transducers to detect and monitor fatigue crack growth.
MacLauchlan, *et al.* (1998) 'Phased array EMATs for flaw sizing.'	Conference: Used phased array EMATs to generate and direct high frequency shear horizontal (SH) waves for defect identification of weld samples.
McNab and Campbell (1987) 'Ultrasonic phased arrays for nondestructive testing.'	Journal: Conducted a feasibility study (cost vs sample rate vs instrumentation) for using ultrasonic phased arrays for NDE.
Sundararaman and Adams (2002) 'Phased transducer arrays for structural diagnostics through beamforming.'	Conference: Developed a spatio-temporal directional filtering methodology for defect localization in isotropic structures.
Sundararaman, *et al.* (2005a) 'Biologically inspired structural diagnostics through beamforming with phased transducer arrays.'	Journal: Presented an experimental study for directional filtering using antisymmetric (Ao) mode Lamb waves in steel and woven composites.
Sundararaman, *et al.* (2005b) 'Structural damage identification in homogeneous and heterogeneous structures using beamforming.'	Journal: Presented an experimental study for directional filtering using antisymmetric (Ao) mode Lamb waves in steel and woven composites.
Tua, *et al.* (2004) 'Detection of cracks in plates using piezo-actuated Lamb waves.'	Journal: Used the Hilbert Huang transform to detect cracks in plates interrogated by piezo-actuated Lamb waves.
Li and Rose (2001) 'Implementing guided wave mode control by use of a phased transducer array.'	Journal: Use of guided waves for inspection of long pipes with a phased transducer array.
Lin (2000) 'Structural health monitoring using geophysical migration technique with built-in piezoelectric sensor/actuator arrays.'	Thesis: Presented a NDE technique based on ultrasonic sensor arrays using the ideas of geophysical migration.
Lin and Yuan (2001) 'Diagnostic Lamb waves in an integrated piezoelectric sensor/actuator plate: analytical and experimental studies.'	Journal: Modeled guided waves in an infinite isotropic plate (incorporating Mindlin plate theory) using a pair of circular actuators.
Wang (2004) 'Elastic wave propagation in composites and least-squares damage localization technique.'	Thesis: Used a least squares approach with iterative minimization for damage localization using distributed arrays.

Table B.13   (*Continued*)

Reference	Summary
Wang and Yuan (2005) 'Damage identification in a composite plate using prestack reverse-time migration technique.'	Journal: A pre-stack migration technique was used to locate damage in composite structures.
Wilcox, *et al*. (2001) 'The effect of dispersion on long-range inspection using ultrasonic guided waves.'	Journal: Studied the effects of dispersion and mode sensitivity for defect identification in order to develop design guidelines for guided wave testing.
Wilcox, *et al*. (2000) 'Lamb and SH wave transducer arrays for the inspection of large areas of thick plates.'	Conference: Presented a method of using antisymmetric Lamb and shear horizontal waves for defect identification over large areas of thick plates.
Wilcox (2003) 'A rapid signal processing technique to remove the effect of dispersion from guided wave signals.'	Journal: Used the symmetric (So) mode Lamb wave and attempted to compensate for signal dilation due to dispersion.
Wilcox (2003) 'Omni-directional guided wave transducer arrays for the rapid inspection of large areas of plate structures.'	Journal: Incorporated a dispersion compensation technique and developed a guided wave compact phased transducer technique; holes and notches.
Wilcox, *et al*. (2005) 'Omnidirectional guided wave inspection of large metallic plate structures using an EMAT array.'	Journal: Extended the work to using an EMAT array for defect identification in large metallic structures.
Rajagopalan, *et al*. (2006) 'A phase reconstruction algorithm for Lamb wave based structural health monitoring of anisotropic multilayered composite plates.'	Journal: Extended the work by Wilcox (2003b) to locate damage (medium sized through hole) using a single actuator and multiple sensors.
Chen, *et al*. (2003) 'Acoustic emission in monitoring quality of weld in friction stir welding.'	Conference: Used acoustic emission techniques for monitoring the quality of welds obtained through the friction stir welding process.
Lamarre and Moles (2000) 'Ultrasound phased array inspection technology for the evaluation of friction stir welds.'	Conference: Identified defects in a friction stir weld using ultrasonic phased arrays.
Raghavan and Cessnik (2007) 'Guided-wave based structural health monitoring: a review.'	Journal: A detailed review paper on work involving the use of guided waves for nondestructive testing.
Kundu, *et al*. (2001) 'Importance of the near Lamb mode imaging of multilayered composite plates.'	Journal: Showed that it was possible to detect internal defects in layers of mirror symmetry in the upper and lower halves of a plate.
Crawley and de Luis (1987) 'Use of piezoelectric actuators as elements of intelligent structures.'	Journal: Proposed a quasi-static induced strain actuation piezo actuator model that can be more effectively modeled to operate in a pinching mode.
Yang, J. and Chang, F. (2006) 'Detection of bolt loosening in C-C composite thermal protection panels: I. diagnostic principle.'	Journal: Used elastic waves to determine the preload in bolt connections of thermal protection panels.

Adamou, A.T.I. and Craster, R.V. (2004) 'Spectral methods for modeling guided waves in elastic media', *J. Acoust. Soc. Am.*, **116**(3), 1524–1535.

Alleyne, D.N. and Cawley, P. (1992a) 'The interaction of Lamb waves with defects', *IEEE Trans. Ultrasonic., Ferroelectric. Freq. Cont.*, **39**(3), 381–397.

Alleyne, D.N. and Cawley, P. (1992b) 'Optimization of Lamb wave inspection techniques', *NDT E Int.*, **25**, 11–22.

Banerjee, S., Banerji, P., Berning, F., and Eberle, K. (2003) 'Lamb wave propagation and scattering in layered composite plates', *Proceedings of SPIE, Smart NDE for Health Monitoring of Structural and Biological Systems, 8th Annual International Symposium on NDE for Health Monitoring & Diagnostics*, San Diego, California, paper no. 5047-02.

Bar-Cohen, Y. (2000) 'Emerging NDE technologies and challenges at the beginning of the 3rd millennium – part I', *Mat. Eval.*, **58**(1), 17–30.

Bar-Cohen, Y., Mal, A. and Chang, Z. (1998) 'Composite material defects characterization using leaky Lamb wave dispersion data', *Proceedings of SPIE, NDE Techniques for Aging Infrastructure & Manufacturing, Conference NDE of Materials and Composites II*, San Antonio, Texas, Vol. 3396, paper no. 3396-25.

Bardouillet, P. (1984) 'Application of electronic focusing and scanning systems to ultrasonic testing', *NDT Int.*, **17**(2), 81–85.

Beard, M.D. (2002) 'Guided wave inspection of embedded cylindrical structures', *PhD Dissertation*, University of London.

Chahbaz, A., Mustafa, V. and Hay, D.R. (1996) 'Corrosion detection in aircraft structures using guided Lamb waves', http://www.ndt.net/article/tektrend/tektrend.htm, **1**(11), Online Journal.

Chen, C., Kovacevic, R. and Jandgric, D. (2003) 'Acoustic emission in monitoring quality of weld in friction stir welding', *Proceedings of the Fourth International Symposium on Friction Stir Welding*, Park City, Utah, USA, 14–16 May 2003.

Crawley, E.F. and de Luis, J. (1987) 'Use of piezoelectric actuators as elements of intelligent structures', *AIAA J.*, **25**(10), 1373–1385, Oct 1987.

Doebling, S.W., Farrar, C.R., Prime, M.B. and Shevitz, D.W. (1996) 'Damage identification and health monitoring of structural and mechanical systems from changes in their vibration characteristics: a literature review', *Los Alamos National Laboratory Report* LA-13070-MS.

Fromme, P. (2001) 'Defect detection in plates using guided waves', *Doctoral Dissertation*, Swiss Federal Institute of Technology, Zurich. Eth: 14397.

Giurgiutiu, V. and Bao, J. (2004) 'Embedded-ultrasonics structural radar for in-situ structural health monitoring of thin-wall structures', *Structural Health Monitoring – an International Journal*, **3**(2), June 2004, pp. 121–140.

Giurgiutiu, V. (2003) 'Lamb wave generation with piezoelectric wafer active sensors for structural health monitoring', *Proceedings of the SPIE 5056*, pp. 111–122.

Giurgiutiu, V. and Bao, J. (2002) 'Embedded ultrasonic structural radar with piezoelectric wafer active sensors for the nde of thin-wall structures', *Proceedings of ASME International Mechanical Engineering Congress*, Nov. 17–22, New Orleans, LA, CDROM, paper no. IMECE 2002-39017, p. 1–8.

Grisso, B.L. (2004) 'Considerations of the impedance method, wave propagation, and wireless systems for structural health monitoring', MS Thesis, Virginia Polytechnic Institute and State University.

Ihn, J.-B. and Chang, F.-K. (2004) 'Detection and monitoring of hidden fatigue crack growth using a built-in piezoelectric sensor/actuator network: I. Diagnostics', *Smart Mat. Struct.*, **13**, 609–620.

Kessler, S.S. (2002) 'Piezoelectric-based in-situ damage detection of composite materials for structural health monitoring systems', *Ph.D. Dissertation*, Department of Aeronautics and Astronautics, Massachusetts Institute of Technology.

Kundu, T., Potel, C. and de Belleval, J.F. (2001) 'Importance of the near Lamb mode imaging of multilayered composite plates', *Ultrasonics*, **39**, 283–290.

Lakshmanan, K.A. and Pines, D.J. (1997) 'Modeling damage in rotorcraft flexbeams using wave mechanics', *Smart Mat. Struct.*, **6**, 383–392.

Lamarre, A. and Moles, M. (2000) 'Ultrasound Phased array inspection technology for the evaluation of friction stir welds', *Annual Conference of the British Institute of Non-Destructive Testing Proceedings*, pp. 56–61.

Lamb. H. (1917) 'On waves in an elastic plate', *Proceedings of the Royal Society*, London, Vol. 93, pp. 114–128.

Li, J. and Rose, J. L. (2001) 'Implementing guided wave mode control by use of a phased transducer array', *IEEE Transactions on Ultrasonics, Ferroelectrics, and Frequency Control*, **48**(3), 761–768.

Lin, X. and Yuan, F.G. (2001) 'Diagnostic Lamb waves in an integrated piezoelectric sensor/actuator plate: analytical and experimental studies', *Smart Materials and Structures*, **10**, 907–913.

Lin, X. (2000) 'Structural health monitoring using geophysical migration technique with built-in piezoelectric sensor/actuator arrays', *PhD Dissertation*, North Carolina State University.

Liu, W. (1997) 'Multiple wave scattering and calculated effective stiffness and wave properties in unidirectional fiber-reinforced composites', *PhD. Dissertation*, Engineering Mechanics, Virginia Polytechnic.

Lord-Rayleigh (1889) 'On the free vibrations of an infinite plate of homogeneous isotropic matter', *Proceedings of the London Mathematical Society*, Vol. 20, pp. 225–234.

Lowe, M.J.S. (1995) 'Matrix techniques for modeling ultrasonic waves in multilayered media', *IEEE Trans. Ultrasonic., Ferroelectric. Freq. Cont.*, **42**, 525–542.

Lui, G. and Qu, J. (1998) 'Guided circumferential waves in a circular annulus', *J. Appl. Mech.*, 65, 424–430.

MacLauchlan, D.T., Schlader, D.M., Clark, S.P. and Latham, W.M. (1998) 'Phased array EMATs for flaw sizing', *EPRI Phased Array Inspection Seminar 99-01*, Portland, Maine.

McNab, A. and Campbell, M.J. (1987) 'Ultrasonic phased arrays for nondestructive testing', *NDT Int.*, 6, 333–337.

Mustafa, V., Chahbaz, A., Hay, D.R., Brassard, M. and Dubois, S. (1997) 'Imaging of disbond in adhesive joints with Lamb waves', http://www.ndt.net/article/tektren2 /tektren2.htm, 2(3), Online Journal.

Pavlakovic, B. (1998) 'Leaky guided ultrasonic waves in NDT', *Doctoral Dissertation*, Imperial College, University of London.

Pavlakovic, B. and Lowe, M.J.S. (1999) 'A general purpose approach to calculating the longitudinal and flexural modes of multi-layered, embedded, transversely isotropic cylinders', in *Review of Progress in Quantitative Nondestructive Evaluation*, Thompson, D.O. and Chimenti, D.E. (eds), Vol. 18A, Plenum Press, New York, pp. 239–246

Pavlakovic, B., Lowe, M.J.S., Alleyne, D. and Cawley, P. (1997) 'Disperse: a general purpose program for creating dispersion curves', in *Review of Progress in Quantitative Nondestructive Evaluation*, Thompson D.O. and Chimenti D.E. (eds), Vol. 16A, pp. 185–192, Plenum Press, New York.

Pines, D.J. (1997) 'The use of wave propagation models for structural damage identification', *Structural Health Monitoring: Current Status and Perspectives*, International Workshop on Structural Health Monitoring, Stanford CA, Chang, F.-K. (ed), Boca Raton, CRC Press Inc., Florida, pp. 664–677.

Prosser, W.H., Jackson, K.E., Kellas, S., Smith, B.T., McKeon, J. and Friedman, A. (1995) 'Advanced, waveform based acoustic emission detection of matrix cracking in composites', *Mat. Eval.*, 53(9), 1052–1058.

Purekar, A.S. and Pines, D.J. (2005) 'Damage detection in plate structures using Lamb waves with directional filtering sensor arrays', *Proceedings of the Fifth International Workshop on Structural Health Monitoring*, Stanford, CA, pp. 1025–1032.

Purekar, A.S. and Pines, D.J. (2002) 'A phased sensor/actuator array for detecting damage in 2-D structures', *AIAA/ASME/ASCE/AHS/ASC Structures, Structural Dynamics, and Materials Conference* (No 2002-1547), pp. 1–9.

Purekar, A.S. and Pines, D.J. (2001) 'Interrogation of beam and plate structures using phased array concepts', *Proc. of the 12th International Conference on Adaptive Structures and Technologies (ICAST)*, University of Maryland, MD, pp. 275–288.

Purekar, A.S., Pines, D.J., Sundararaman, S. and Adams, D.E. (2004) 'Directional piezoelectric phased array filters for detecting damage in isotropic plates', *Smart Mat. Struct.*, 13, 838–850.

Raghavan, A. and Cesnik, C.E.S. (2005) 'Piezoelectric-actuator excited-wavefield solutions for guided-wave structural health monitoring', *Proceedings of the SPIE 5765*, pp. 1–11.

Rajagopalan, J., Balasubramanian, K. and Krishnamurthy, C.V. (2006) 'A Phase reconstruction algorithm for Lamb wave based structural health monitoring of anisotropic multilayered composite plates', *J. Acoust. Soc. Am.*, 119(2), 872–878.

Rizzo, P. and di Scalea, F.L. (2005) 'Ultrasonic inspection of multi-wire steel strands with the aid of the wavelet transform', *Smart Mat. Struct.*, 14, 685–695.

Rose, J.L. (1999) *Ultrasonic Waves in Solid Media*, Cambridge University Press, London.

Saravanos, D.A. and Heyliger, P.R. (1995) 'Coupled layerwise analysis of composite beams with embedded piezoelectric sensors and actuators', *J. Intelligent Mat. Sys. Struct.*, 6, 350–363.

Schmerr Jr., L.W. (1999) *Fundamentals of Ultrasonic Nondestructive Evaluation: A Modeling Approach*, Plenum Press, New York.

Shah, S.P., Popovics, J.S., Subramaniam, K.V. and Aldea, C. (2000) 'New directions in concrete health monitoring technology', *J. Engin. Mech.*, 126(7), 754–760.

Sohn, H., Farrar, C.R., Hemez, F.M., Shunk, D.D., Stinemates, D.W. and Nadler, B.R. (2001) 'A review of structural health monitoring literature: 1996–2001', *Los Alamos National Laboratory Report LA-13976-MS*.

Sohn, H., Wait, J.R., Park, G. and Farrar, C.R. (2004) 'Multi-scale structural health monitoring for composite structures', *Proceedings of the Second European Workshop on Structural Health Monitoring*, July 7–9, Munich, Germany, pp. 721–729.

Sundararaman, S. (2003) 'Structural diagnostics through beamforming of phased arrays: characterizing damage in steel and composite plates', *MS Thesis*, Purdue University.

Sundararaman, S., Adams, D.E. and Jata, K.V. (2004b) 'Structural health monitoring studies of a friction stir welded Al–Li Plate for cryotank application', *Materials Damage Prognosis*, edited by TMS (The Minerals, Metals and Materials Society).

Sundararaman, S., Adams, D.E. and Rigas, E. (2005a) 'Biologically inspired structural diagnostics through beamforming with phased transducer arrays', *Int. J. Engin. Sci.*, May 2005, 756–778.

Sundararaman, S., Adams, D.E. and Rigas, E.J. (2005b) 'Structural damage identification in homogeneous and heterogeneous structures using beamforming', *Struct. Health Monitor. Int. J.*, 171–190.

Sundararaman, S. and Adams, D.E. (2002) 'Phased transducer arrays for structural diagnostics through beamforming', *Proc. of the American Society for Composites (ASC) 17th Technical Conference*, W. Lafayette, IN, Sun, C.T. and Kim, H. (eds), CD-ROM, paper 177.

Sundararaman, S., Haroon, M., Adams, D.E. and Jata, K.V. (2004a) 'Incipient damage identification using elastic wave propagation through a friction stir welded Al–Li interface for cryogenic tank applications', *Proceedings of the Second European Workshop of Structural Health Monitoring*, Munich, Germany, DESTech Publications Inc., PA, USA, pp. 525–532.

Tua, P.S., Quek, S.T. and Wang, Q. (2004) 'Detection of cracks in plates using piezo-actuated Lamb waves', *Smart Mat. Struct.*, **13**, 643–660.

Tucker, B.J. (2001) 'Ultrasonic waves in wood-based composite panels', *PhD Dissertation*, Department of Civil and Environmental Engineering, Washington State University.

Viktorov, I.A. (1967) '*Rayleigh and Lamb Waves: Physical Theory and Applications*', Plenum Press, New York.

Wang, L. (2004) 'Elastic wave propagation in composites and least-squares damage localization technique', MS Thesis, North Carolina State University, Raleigh.

Wang, L. and Yuan, F.G. (2005) 'Damage identification in a composite plate using prestack reverse-time migration technique', *Struct. Health Monitor. Int. J.*, **4**(3), 195–217.

Wevers, M. (1997) 'Listening to the sound of materials: acoustic emission for the analysis of material behavior', *NDT&E Int.*, **30**(2), 99–106.

Wilcox, P., Lowe, M. and Cawley, P. (2005) 'Omnidirectional guided wave inspection of large metallic plate structures using an EMAT array', *IEEE Trans. Ultrasonic., Ferroelectric. Frequency Cont.*, **52**(4), 653–665.

Wilcox, P., Lowe, M., Cawley, P. (2000) 'Lamb and SH wave transducer arrays for the inspection of large areas of thick plates', *Review of Progress in Quantitative Nondestructive Evaluation*, Thomson, D.O. and Chimenti, D.E. (ed), CP509, Vol. 18A, pp. 1049–1056.

Wilcox, P.D. (1998) 'Lamb wave inspection of large structures using permanently attached transducers', *PhD Dissertation*, Imperial College of Science Technology and Medicine, University of London.

Wilcox, P.D. (2003) 'A rapid signal processing technique to remove the effect of dispersion from guided wave signals', *IEEE Trans. Ultrasonic. Ferroelectric. Frequency Cont.*, **50**(4), 419–427.

Wilcox, P.D. (2003) 'Omni-directional guided wave transducer arrays for the rapid inspection of large areas of plate structures', *IEEE Trans. Ultrasonic. Ferroelectric. Frequency Cont.*, **50**(4), 699–709.

Wilcox, P.D., Dalton, R.P., Lowe, M.J.S. and Cawley, P. (1999) 'Mode selection and transduction for structural monitoring using Lamb waves', *Structural Health Monitoring 2000*, 2nd International Workshop on Structural Health Monitoring, Stanford, CA, Chang, F.-K. (ed), Boca Raton, CRC Press Inc., FL, pp. 703–712.

Wilcox, P.D., Lowe, M. and Cawley, P. (2001) 'The Effect of dispersion on long-range inspection using ultrasonic guided waves', *NDT&E Int.*, **34**, 1–9.

Worlton, D.C. (1961) 'Experimental confirmation of Lamb waves at megacycle frequencies', *J. App. Phys.*, **32**, 967–971.

Yu, L. and Giurgiutiu, V. (2005) 'Improvement of damage detection with the embedded ultrasonics structural radar for structural health monitoring', *Proceedings of the Fifth International Workshop on Structural Health Monitoring*, Fu-kuo Chang (ed), pp. 1081–1090.

Yang, J. and Chang, F. (2006) 'Detection of bolt loosening in C-C composite thermal protection panels: I. Diagnostic principle', *Smart Mat. Struct.* 15, pp. 581–590.

Table B.14   References on temporal data analysis

Reference	Summary
Samuel and Pines (2005) 'A review of vibration-based techniques for helicopter transmission diagnostics.'	Journal: A detailed review paper on statistical techniques in conjunction with signal processing for helicopter transmission diagnostics.
Staszewski and Worden (2004) 'Signal processing for damage detection.'	Book Chapter: Includes a summary of data analysis methods for damage identification with illustrations of data compression and denoising.
Box, *et al.* (1994) 'Time series analysis: forecasting and control.'	Book: Detailed account of time series analysis methods including different auto-regressive and moving average models.
Castillo, *et al.* (2005) 'Extreme value and related models with applications in engineering and science.'	Book: Implementation and mathematical background for extreme value and reliability models.
Montgomery (2001) 'Design and analysis of experiments.'	Book: Illustrates methods of combining and analyzing data using experimental design and hypothesis testing.
McLachlan (1992) 'Discriminant analysis and statistical pattern recognition.'	Book: Seminal work in using temporal/ transformed temporal data for feature extraction and discrimination using pattern recognition.
Webb (2002) 'Statistical pattern recognition.'	Book: Includes basic and advanced statistical tools used for feature extraction and data/ feature discrimination using pattern recognition.
Sohn, *et al.* (2000) 'Structural health monitoring using statistical process control.'	Conference: Experimental investigation of statistical process control to identify damage during a vibration experiment.
Todd and Nichols (2002) 'Structural damage assessment using chaotic dynamic interrogation.'	Conference: Used a single factor analysis-of-variance (ANOVA) with Bonferroni confidence interval generation to as a damage sensitive feature.
Monaco, *et al.* (2000) 'Experimental and numerical activities on damage detection using magnetostrictive actuators and statistical analysis.'	Journal: Used a t-test to determine the effectiveness of damage indices obtained from changes in the frequency response functions.
Worden, *et al.* (2003) 'Extreme value statistics for damage detection in mechanical structures.'	Report: Detailed report on unsupervised learning methods based on extreme value statistical analysis using statistical process control.
George, *et al.* (2000) 'Identifying damage sensitive features using nonlinear time series and bispectral analysis.'	Conference: Multivariate analysis method that compares groups of data by a weighted linear combination known as the canonical variate analysis.
Kantz and Schreiber (1997) 'Nonlinear time series analysis.'	Book: Detailed review on nonlinear time series analysis methods.
Yu and Giurgiutiu (2005) 'Advanced signal processing for enhanced damage detection with piezoelectric wafer active sensors.'	Journal: Detailed literature review of recent works using temporal and frequency domain methods.

Box, G., Jenkins, G.M. and Reinsel, G. (1994) '*Time Series Analysis: Forecasting and Control*', 3rd edition, Prentice-Hall, New Jersey.

Castillo, E., Hadi, A.S., Balakrishnan, N., Sarabia, J.M. (2005) '*Extreme Value and Related Models with Applications in Engineering and Science*', John Wiley and Sons Inc., New Jersey.

George, D., Hunter, N., Farrar, C.R., Deen, R. (2000) 'Identifying damage sensitive features using nonlinear time series and bispectral analysis', *Proc. of the 18th International Modal Analysis Conference*, San Antonio, Texas, pp. 1–7.

Kantz, H., Schreiber, T. (1997) 'Nonlinear time series analysis', in *Cambridge Nonlinear Science Series 7*, Cambridge University Press, Cambridge, UK.

McLachlan, G.J. (1992) '*Discriminant Analysis and Statistical Pattern Recognition*', John Wiley and Sons, New York.

Monaco, E., Franco, F. and Lecce, L. (2000) 'Experimental and numerical activities on damage detection using magnetostrictive actuators and statistical analysis', *J. Intelligent Mat. Struct.*, **11**, 567–578.

Montgomery, D.C. (2001) '*Design .Anal. Exper.*', Fifth Edition, John Wiley and Sons, New York.

Samuel, P.D. and Pines, D.J. (2005) 'A review of vibration-based techniques for helicopter transmission diagnostics', *J. Sound Vib.*, **282**, 475–508.

Sohn, H., Czarnecki, J.A. and Farrar, C.R. (2000) 'Structural health monitoring using statistical process control', *J. Struct. Engin.*, Nov. 2000, 1356–1363.

Staszewski, W. and Worden, K. (2004) 'Signal processing for damage detection', in *Health Monitoring of Aerospace Structures*, Staszewski, W., Boller, C. and Tomlinson, G. (eds), John Wiley & Sons, UK, pp. 163–206.

Todd, M.D. and Nichols, J.M. (2002) 'Structural damage assessment using chaotic dynamic interrogation', *Proc. of 2002 ASME International Mechanical Engineering Conference and Exposition*, Vol. 71, pp. 613–620.

Webb, A. (2002) 'Statistical pattern recognition', 2nd edition, John Wiley and Sons, West Sussex, UK.

Worden, K., Allen, D.W., Sohn, H., Stinemates, D.W. and Farrar, C.R. (2003) 'Extreme value statistics for damage detection in mechanical structures', *Los Alamos National Laboratory Report LA-13903-MS*.

Yu, L. and Giurgiutiu, V. (2005) 'Advanced signal processing for enhanced damage detection with piezoelectric wafer active sensors', *Smart Sys. Struct.*, **1**(2), 185–215.

Table B.15   References on time-frequency data analysis

Reference	Summary
Staszewski, W.J. (1998) 'Wavelet based compression and feature selection for vibration analysis.'	Journal: Used wavelet analysis to extract features from vibration time series to detect damage.
Prosser, *et al.* (1999) 'Time-frequency analysis of the dispersion of Lamb modes.'	Journal: Lamb mode signals were processed using a pseudo Wigner Ville distribution for determining material properties (i.e., dispersion).
Cao, X. (2002) 'Adaptability and comparison of the wavelet-based with traditional equivalent linearization method and potential application for damage detection.'	Thesis: Presented background for time-frequency analysis and compared a wavelet based equivalent linearization method with traditional method.
Yuan, *et al.* (2004) 'A new damage signature for composite structural health monitoring.'	Journal: Introduced a damage signature based on wavelet analysis to determine the presence and extent of damage.
Peng, *et al.* (2005) 'A comparison study of improved hilbert-huang transform and wavelet transform: application to fault diagnosis for roller bearing.'	Journal: Compared the results obtained by processing data using the Hilbert Huang transform (HHT) and wavelet analysis.
Shinde (2004) 'A wavelet packet based sifting process and its application in structural health monitoring.'	Thesis: Extended the HHT by using wavelet packet principles; also included details and background about obtaining the HHT and wavelet transform.
Cohen (1995) 'Time-frequency analysis.'	Book: Outline and mathematical background for time-frequency methods used for signal analysis.
Auger, *et al.* (1996) 'Time frequency toolbox – for use with MATLAB: tutorial.'	Online Report: Review article and tutorial in the use of time, frequency and time-frequency analysis (including wavelet analysis) with MATLAB®.
Huang, *et al.* (1998) 'The empirical mode decomposition method and the hilbert spectrum for non-linear and non-stationary time series analysis.'	Journal: Detailed literature review of time frequency analysis and extends the Hilbert transform by implementing empirical mode decomposition.
Daubechies, I. (1992) 'Ten lectures in wavelets.'	Journal & Book: Seminal works on wavelet analysis; used quadrature mirror filters associated with the scaling function and the mother wavelet function.
Daubechies, I. (1990) 'The wavelet transform, time-frequency localization and signal analysis.'	
Donoho, D.L. (1995) 'De-noising by soft-thresholding.'	Journal: Presented a soft thresholding method for denoising data using the wavelet transform.
Jensen and la Cour-Harbo (2001) 'Ripples in mathematics: the discrete wavelet transform.'	Book: Review, background and implementation of time-frequency analysis (wavelet transforms).
Mallat (1999) 'A wavelet tour of signal processing.'	Book: Review, background and implementation of time-frequency analysis (wavelet transforms).

Auger, F., Flandrin, P., Goncalves, P. and Lemoine, O. (1996) 'Time Frequency Toolbox – For Use with MATLAB: Tutorial', Web: Matlab File Exchange.

Cao, X. (2002) 'Adaptability and comparison of the wavelet-based with traditional equivalent linearization method and potential application for damage detection', *MS Thesis* (Advisor: Mohammad N. Noori), North Carolina State University.

Cohen, L. (1995) '*Time-Frequency Analysis*', Prentice Hall, Englewood Cliffs, NJ.

Daubechies, I. (1992) 'Ten lectures in wavelets', *CBMS-NSF Regional Conference Series in Mathematics*, Society for Industrial and Applied Math (SIAM), Philadelphia, PA.

Daubechies, I. (1990) 'The wavelet transform, time-frequency localization and signal analysis', *IEEE Trans. Infor. Theo.*, **36**(5), 961–1005.

Donoho, D.L. (1995) 'De-noising by soft-thresholding', *IEEE Trans. Info. Theo.*, **41**(3), 613–627.

Huang, N.E., Shen, Z., Long, S.R., Wu, M.C., Shih, H.H., Zheng, Q., Yen, N.-C., Tung, C.C., Liu, H.H. (1998) 'The empirical mode decomposition method and the hilbert spectrum for non-linear and non-stationary time series analysis', *Proc. of the Royal Society London*, Vol. 454, pp. 903–995.

Ihn, J.-B. and Chang, F.-K. (2004) 'Detection and monitoring of hidden fatigue crack growth using a built-in piezoelectric sensor/actuator network: I. Diagnostics', *Smart Mat. Struct.*, **13**, 609–620.

Jensen, A., la Cour-Harbo, A. (2001) '*Ripples in Mathematics: The Discrete Wavelet Transform*', Springer International, New Delhi.

Mallat, S. (1999) '*A Wavelet Tour of Signal Processing*', Second Edition, Academic Press.

Peng, Z.K., Tse, P.W., Chu, F.L. (2005) 'A comparison study of improved hilbert-huang transform and wavelet transform: application to fault diagnosis for roller bearing', *Mech. Sys. Signal Proc.*, **19**, 974–988.

Prosser, W.H., Seale, M.D. and Smith, B.T. (1999) 'Time-frequency analysis of the dispersion of Lamb modes', *J. Acoustic. Soc. Am.*, **105**(5), 2669–2676.

Raghavan, A. and Cesnik, C.E.S. (2005) 'Piezoelectric-actuator excited-wavefield solutions for guided-wave structural health monitoring', *Proceedings of the SPIE 5765*, pp. 1–11.

Rizzo, P. and di Scalea, F.L. (2005) 'Ultrasonic inspection of multi-wire steel strands with the aid of the wavelet transform', *Smart Mat. Struct.*, **14**, 685–695.

Shinde, A.D. (2004) 'A wavelet packet based sifting process and its application in structural health monitoring', *MS Thesis*, Worcester Polytechnic Institute.

Staszewski, W.J. (1998) 'Wavelet based compression and feature selection for vibration analysis', *J. Sound Vib.*, **211**(5), 735–760.

Yuan, S., Wang, L. and Wang, X. (2004) 'A new damage signature for composite structural health monitoring', *Proceedings of the 2nd European Workshop on Structural Health Monitoring*, Munich, Germany, July 7–9, 2004, p. 1–8.

Hou, Z., Noori, S. and Amand, St.R. (2000) 'A wavelet-based approach for structural damage detection', *ASCE J Engin. Mech.*, **126**, 667–683.

Table B.16   References on triangulation for damage location

Reference	Summary
White, *et al.* (2005) 'Modeling and material damage identification of a sandwich plate using MDOF modal parameter estimation and the method of virtual forces.'	Conference: Developed a distributed sensor array technique for detecting and locating damage.
Sundararaman (2003) 'Structural diagnostics through beamforming of phased arrays: characterizing damage in steel and composite plates.'	Thesis: Outlined a phased array directional filtering algorithm for damage localization in steel and woven composite structures.

Sundararaman, S. (2003) 'Structural diagnostics through beamforming of phased arrays: characterizing damage in steel and composite plates', *MS Thesis*, Purdue University.

White, J., Adams, D.E., Jata, K.V. (2005) 'Modeling and material damage identification of a sandwich plate using MDOF modal parameter estimation and the method of virtual forces', *Proceedings of the International Mechanical Engineers Congress and Exposition*, Nov 5–11, 2005, Orlando, FL, paper no. 80472.

Table B.17   References on transfer path, other types of data analysis and non-contact sensing

Reference	Summary
Donskoy, D. *et al.* (2001) 'Nonlinear acoustic interaction on contact interfaces and its use for nondestructive testing.' Donskoy, D.M., *et al.* (1998) 'Vibro-acoustic modulation nondestructive evaluation technique.'	Journal: Used the modulation of a high-frequency ultrasonic wave by low frequency vibration to detect defects.
Ballad, E.M., *et al.* (2004) 'Nonlinear modulation technique for NDE with air-coupled ultrasound.'	Journal: Studied a new air-coupled nonlinear acoustic modulation method that used non-contact ultrasound excitation.
Rek, R., *et al.* (2006) 'Ultrasonic C-scan and shearography NDI techniques evaluation of impact defects identification.'	Journal: Compared the ultrasonic C-scan with laser shearography method in the impact damage identification of sandwich panels.
Edwards, R.S., *et al.* (2006) 'Dual EMAT and PEC non-contact probe: applications to defect testing.'	Journal: Applied a dual-probe combining electromagnetic acoustic transducers and a pulsed eddy current sensor to detect defects.
Cho, H., *et al.* (1996) 'Non-contact laser ultrasonics for detecting subsurface lateral defects.'	Journal: Employed non-contact and non-destructive laser ultrasonics to identify subsurface lateral defects.
Warnemuende, K., *et al.* (2004) 'Actively modulated acoustic nondestructive evaluation of concrete.'	Journal: Studied nonlinear frequency analysis methods for concrete damage detection and evaluation using actively modulated acoustic signals.
Moussatov, A., *et al.* (2002) 'Frequency up-conversion and frequency down-conversion of acoustic waves in damaged materials.'	Journal: Investigated correlation between nonlinear signatures and amount of damage.
Li, T.Y., *et al.* (2004) 'Vibrational power flow characteristics of circular plate structures with peripheral surface crack.'	Journal: Investigated the vibrational power flow of circular plates with a surface crack.
Sun, J.Q. (1995) 'Vibration and sound radiation of non-uniform beams.'	Journal: Presented an analytical method for studying vibration and acoustic radiation problems of non-uniform beams.
Lu, Y., *et al.* (2005) 'A methodology for structural health monitoring with diffuse ultrasonic waves in the presence of temperature variations.'	Journal: Applied diffuse ultrasonic waves to the problem of detecting structural damage in the presence of unmeasured temperature changes.
Wevers, M. (1997) 'Listening to the sound of materials: acoustic emission for the analysis of material behaviour.'	Journal: Used acoustic emission to do detect damage in different types of composite materials.
Gudmundson, P. (1999) 'Acoustic emission and dynamic energy release rate for steady growth of a tunneling crack in a plate in tension.'	Journal: Studied acoustic emission and dynamic steady state growth of tunneling cracks in membrane loaded isotropic Kirchhoff plates.
Toutountzakis, T., *et al.* (2003) 'Observation of acoustic emission activity during gear defect diagnosis.'	Journal: Applied acoustic emission as a non-destructive technique for damage detection in rotating machinery.
Rippert, L., *et al.* (2000) 'Optical and acoustic damage detection in laminated CFRP composite materials.'	Journal: Used an intensity-modulated fibre-optic sensor as an alternative to the piezoelectric transducers for acoustic emission monitoring.

(Continued)

Table B.17 (*Continued*)

Reference	Summary
Tong, F., *et al.* (2006) 'Impact-acoustics-based health monitoring of tile-wall bonding integrity using principal component analysis.'	Journal: Used the impact-acoustic signature in tile-wall inspection to mitigate the adverse influence of surface non-uniformity.

Donskoy, D., Sutin, A., Ekimov, A. (2001) 'Nonlinear acoustic interaction on contact interfaces and its use for nondestructive testing', *NDT&E Int.*, **34**, 231–238.

Donskov, D., Sutin, A. (1998) 'Vibro-acoustic modulation nondestructive evaluation technique' *J. Int. Mat. Sys. Struct.*, **9**(9), 765–771.

Ballad, E.M., Vezirov, S.Yu., Pfleiderer, K., Solodov, I.Yu., Busse, G. (2004) 'Nonlinear Modulation Technique for NDE with Air-Coupled Ultrasound'. *Ultrasonics*, **42**, 1031–1036.

Roman Růžek, Radek Lohonka, Josef Jiroč, (2006) 'Ultrasonic c-scan and shearography ndi techniques evaluation of impact defects identification', *NDT&E Int.*, **39**, 132–142.

Edwards, R.S., Sophian, A., Dixon, S., Tian, G.-Y., Jian, X. (2006) 'Dual EMAT and PEC non-contact probe: applications to defect testing', *NDT&E Int.*, **39**, 45–52.

Cho, H., Ogawa, S. and Takemoto, M. (1996) 'Non-contact laser ultrasonics for detecting subsurface lateral defects', *NDT&E Int.*, **29**(5), 301–306.

Warnemuende, K., Hwai-Chung, Wu. (2004) 'Actively modulated acoustic nondestructive evaluation of concrete', *Cement Concrete Res.*, **34**, 563–570.

Moussatov, A., Bernard Castagnède, Vitalyi Gusev[c1] (2002) 'Frequency up-conversion and frequency down-conversion of acoustic waves in damaged materials', *Phys. Lett. A*, 301, 281–290.

Li, T.Y., Liu, J.X., Zhang, T. (2004) 'Vibrational power flow characteristics of circular plate structures with peripheral surface crack', *J. Sound Vib.*, **276**, 1081–1091.

Sun, J. (1995) 'Vibration and sound radiation of non-uniform beams', *J. Sound Vib.*, **185**(5), 827–843.

Lu, Y., Michaels, J.E. (2005) 'A methodology for structural health monitoring with diffuse ultrasonic waves in the presence of temperature variations', *Ultrasonics*, **43**, 717–731.

Wevers, M. (1997) 'Listening to the sound of materials: acoustic emission for the analysis of material behaviour', *NDT&E Int.*, **30**(2), 99–106.

Gudmundson, P. (1999) 'Acoustic emission and dynamic energy release rate for steady growth of a tunneling crack in a plate in tension', *J. Mech. Phys. Solids*, **47**, 2057–2074.

Toutountzakis, T., David, Mba (2003) 'Observation of acoustic emission activity during gear defect diagnosis', *NDT&E Int.*, **36**, 471–477.

Rippert, L., Wevers, M., Van Huffel, S. (2000) 'Optical and acoustic damage detection in laminated CFRP composite materials', *Comp. Sci. Techno.*, **60**, 2713–2724.

Tong, F., Tso, S.K., Hung, M.Y.Y. (2006) 'Impact-acoustics-based health monitoring of tile-wall bonding integrity using principal component analysis', *J. Sound Vib.*, **294**, 329–340.

Table B.18 References on variability analysis in health monitoring

Reference	Summary
Lew, J.-S. (1995) 'Using transfer function parameter changes for damage detection of structures.'	Journal: Developed an interval modeling technique to investigate how environmental variations alter natural frequencies.
Cornwell, P.J., *et al.* (1999) 'Environmental variability of modal parameters.'	Journal: Investigated how temperature changes influence modal properties using data from the Alamosa Canyon Bridge.
Sohn, H., *et al.* (1998) 'Adaptive modeling of environmental effects in modal parameters for damage detection in civil structures.'	Conference: Applied an adaptive filter to establish a linear correlation between temperature and natural frequencies.
Peeters, B., *et al.* (2001) 'Vibration-based damage detection in civil engineering: excitation sources and temperature effects.'	Journal: Used a single-input single-output ARX model to fit baseline data and then extrapolated the influence caused by thermal variations.
Sohn, H., *et al.* (2003) 'Statistical damage classification under changing environmental and operational conditions.'	Journal: Showed that an AR-ARX model was able to detect damage in the presence of wide operational and environmental ranges.
Yan, A.-M., *et al.* (2005) 'Structural damage diagnosis under varying environmental conditions – part 1: a linear analysis.'	Journal: Uses principle component analysis to monitor systems under varying environmental conditions.
Gawronski, W. (1999) 'Simultaneous placement of actuators and sensors.'	Journal: Presents a sensor/actuator placement algorithm based on modal norming.
Shi, Y.Z., *et al.* (2000) 'Optimum sensor placement for structural damage detection.'	Journal: Uses an eigenvector sensitivity analysis to eliminate potential sensor locations.

Lew, J.-S. (1995) 'Using transfer function parameter changes for damage detection of structures', *AIAA J.*, **33**(11), 2189–2193.

Cornwell, P.J., Farrar, C.R., Doebling, S.W. and Sohn, H. (1999) 'Environmental variability of modal parameters' *Exper. Tech.*, **39**(6), 45–48.

Sohn, H., Dzwonczyk, M., Straser, E.G., Law, K.H., Kiremidjian, A.S. and Meng, T. (1998) 'Adaptive modeling of environmental effects in modal parameters for damage detection in civil structures', *Proceedings of SPIE – The International Society for Optical Engineering*, Vol. 3325(1), 127–138.

Peeters, B., Maeck, J. and De Roeck, G. (2001) 'Vibration-based damage detection in civil engineering: excitation sources and temperature effects', *Smart Mat. Struct.*, **10**(1), 518–527.

Sohn, H., Worden, K. and Farrar, C.R. (2003) 'Statistical damage classification under changing environmental and operational conditions', *J. Intel. Mat. Sys. Struct.*, **13**(9), 561–574.

Yan, A.-M., Kerschen, G., De Boe, P. and Golinval, J.-C. (2005) 'Structural damage diagnosis under varying environmental conditions - Part 1: a linear analysis, *Mech. Sys. Signal Proc.*, **19**(1), 847–864.

Gawronski, W. (1999) 'Simultaneous placement of actuators and sensors', *J. Sound Vib.*, **228**(4), 915–922.

Shi, Z.Y., Law, S.S. and Zhang, L.M. (2000) 'Optimum sensor placement for structural damage detection', *J. Engin. Mech.*, **126**(11), 1173–1179.

Index

Acoustic intensity, 149
Active sensing, 8, 10, 148, 159, 164, 179,
  333, 338, 369, 376
Actuator, 8, 146, 159–162, 289, 340, 343,
  350, 360, 375
Adhesives, 137. *See* Attachments.
Aircraft systems, 18
  Alaska Airlines, 3, 280
Aliasing, 176, 185, 220
Aluminum beam, 293, 295
Aluminum plate, 364, 371
Amplitude distortion, 142, 181
Analog to digital converter (ADC),
  171, 172, 174
AND operation, 251–253
Applications, 16, 44, 159, 198, 201
  commercial, 3, 4
  defense, 4
Area reduction, 109
Array gain factor, 200, 366, 367
Attachments, 141, 272
  adhesive, 138, 141, 143
    epoxy, 29, 137, 138, 329
    glue, 29, 137, 139
  strap, 139
  stud mount, 143
Auto-correlation, 206, 209
Automotive systems
  passenger cars, 4
  suspensions, 4
  tires, 4, 114
  trucks, 3, 4
Autonomic logistics, 3, 5

Autoregressive, 72
Averaging, 222, 234, 236, 237, 241, 242
  asynchronous, 234, 299
  cyclic, 234, 235, 240, 241
  synchronous, 234, 237, 299

Ball joint, 319. *See* Fastener.
Bearings, 1, 3, 322
Bilinear stiffness, 111
Bispectrum, 224, 225
Bit dropout, 174
Body armor, 26
Bolt, 30, 107, 108, 319, 322, 350
Broadband excitations, 93, 159
Buckling, 50, 52, 116, 211

Cables, 129, 155, 159
Calibration, 152, 162
Central limit theorem, 269
Ceramic tile, 94, 356
Charge amplifier, 157
Chasle's theorem, 309
Chirp, 226
Clearance, 11, 14, 109
Clevis, 110, 262
Coherence function, 239
Cold welding, 13, 14, 318
Combined loading, 322, 323
Compliance method, 42
Composite materials
  constitutive law, 118
  cylinder, 115, 118, 126, 283, 297, 379
  plate, 62, 64, 254, 275, 367

Condition-based maintenance, 3, 24
Conferences, 23
Continuous parameter, 31
Control charts, 270
Controllability, 166, 167
Convolution, 218, 219
Corrosion, 116, 117, 120, 311
Coulomb friction, 48–50, 69
Coupling, 40, 88, 126
Cracks, 109–111, 115, 117, 121, 354
    link up, 20
    models, 109
Creep, 94, 115, 123, 124
Curvature, 76, 116, 245, 360

Damage
    accumulation, 2, 23, 383, 393
    definition, 12, 13
    global, 12, 24
    local, 24
    models
        dynamic, 125
        static, 124, 125
Damage identification, 7, 9, 10, 24, 257, 263,
        287, 299, 371
    detection, 9, 13, 161, 222, 264, 287, 288,
        373, 374
        vibration-based, 329
        wave propagation-based, 375
    location, 10, 290, 367
    quantification, 10, 292, 293, 333
Damage growth, 7, 121, 391, 392
Damping, 31, 36, 60, 70, 76, 85, 100,
        113, 243, 336
Data analysis, 21, 24, 187, 189, 208, 298
Degrees of freedom (DOF), 31, 375
Delamination, 11, 39, 50, 108, 113
Detection. *See* Damage identification,
        detection.
Diagnostics, 9, 16, 24
Digital to analog converters (DAC),
        129, 159
Dimensionality, 263
Direct parameter models, 101
Discrete Fourier transform, 220.
        See Fourier analysis.
Discrete frequency model, 349
Discrete time models, 71, 101, 212
Dispersion
    meaning, 59, 64, 101
    relation, 53

Drop tower, 341
Durability, 17, 24, 137, 138, 164
    suspensions, 4
    tires, 4
Dye penetrant, 16, 26

Eddy current, 16, 147, 159
Eigenvalue problem, 33
Embedded sensitivity, 78, 79, 101, 258,
        274, 293, 296, 333
Embedded sensors, 350
Engine valve, 12, 13, 316
Equations of motion, 32, 33, 45, 48, 68,
        84, 309
Equilibrium, 31
Erosion, 11, 112
Evanescent wave, 60
Excitation, 31
Extrapolate, 325–327

Failure
    definition, 15
    models, 18, 122, 125
False alarms, 10
Fast Fourier transform (FFT), 219
Fastener, 48
    bolt, 30, 107, 108, 319, 322, 350
    rivet, 313, 314
Fatigue test, 391
Faults, 9, 20, 287. *See* Damage.
Feature extraction, 9, 24, 257, 258, 358
Features, 129, 257, 260, 263, 265, 280, 299
Fiber breakage, 90, 118, 124, 292
Fiber pullout, 108, 117, 118
Filters, 129
    frequency-domain, 193, 194, 298
        bandpass, 193
        finite impulse response (FIR), 193, 194
        high-pass, 156, 193
        infinite impulse response (IIR), 193, 194
        low-pass, 192, 312
    spatial, 187, 194, 196, 298
        beamformer, 197–199
    temporal, 200
Finite element model, 56, 60, 209, 351
    beams, 62
    mass matrix, 33, 56
    nodes, 55, 254, 255
    plates, 64
    rods, 55, 62, 208
    stiffness matrix, 33, 55, 110, 351

Fourier analysis, 10
  discrete Fourier transform, 218, 219
  Fourier integral transform, 91
Free body diagrams, 32
Free response, 35, 36, 100, 161
Frequency response function (FRF), 37, 44, 101, 177
Frequency spectrum, 60, 75, 90, 189
  power spectral density (PSD), 132, 226
  power spectrum, 224, 225
Functional degradation, 14, 25

Gas turbine engine, 7, 344. *See* Wire harness
  and connector.
Gaussian distribution, 96
Grating lobes, 200
Group velocity, 62, 63, 100

Harmonic response, 176
Health and usage monitoring (HUMS), 16
Health monitoring, 19, 23, 24, 111, 266,
    334, 344, 367
  definition, 1
  process, 94, 297
Higher order spectra (HOS), 224, 225, 228, 298
Humidity, 95, 136, 271, 274
Hypothesis testing, 266, 299, 315

Impact hammer, 159, 166, 275, 283, 311, 323, 333
Impedance method, 41
Impulsive load, 305, 306
Inertia, 31, 100, 110, 111, 116, 152, 280
Integrated systems health management (ISHM),
    16
Intelligent maintenance, 16
Interference, 100, 163, 196
Interpolate, 192
Inverse problems, 282, 299

Lamb waves, 64
Laser ultrasound, 149
Launch environments, 249
Leakage, 178, 179, 189, 191, 219, 234, 298
Least squares solution, 101, 281, 309, 312, 327
Loads
  cyclic, 18
  interaction effects, 4, 18, 287
  redistribution, 2
  transient, 316, 328
Loads identification, 7, 8, 24, 266, 280–282,
    299, 313, 327
  location, 380, 381

Location, 327, 363, 366, 367. *See* Damage
    identification, location.
Longitudinal waves, 60, 62, 208. *See* Wave
    propagation.
Lumped parameter, 80, 100, 187

Mass, 337, 363, 375. *See* Inertia.
MATLAB®
  'adcerrors.m', 172
  'aliasdemo.m', 174
  'attachtrans.m', 142, 184
  'averagingdemos.m', 234, 235
  'bbandpdf.m', 93
  'beamsteer.m', 200, 301
  'frfestimation.m', 239, 302
  'filterdemo.m', 193, 301
  'onedofesens.m', 79, 105
  'ninedofdata.m', 183, 184, 303, 328, 377, 394
  'ninedofenviron.m', 136, 184, 328
  'ninedoffeature.m', 263, 264
  'ninedofselect.m', 165, 168, 185
  'ninedofspatial.m', 243, 244
  'panelsimprestemp.m', 98
  'piezoaccelfrf.m', 153
  'piezosensorfrf.m', 156
  'platefemsim.m', 254, 303, 328, 377,
      378, 394
  'rod1dtranswave.m', 60, 112
  'rod1dwave.m', 104
  'rodtemporal.m', 209, 301, 302
  'thermalobserver.m', 202, 301
  'threedofesense.m', 81, 106
  'threedoffrfload.m', 74
  'threedofmodal.m', 86, 87
  'threedoftrans.m', 46
  'threedofvforce.m', 82
  'timefreqanalysis.m', 228, 302, 378
  'twodofdft.m', 220, 302
  'twodofdpe.m', 67
  'twodofdtm.m', 73, 74
  'twodofeig.m', 33, 34, 36
  'twodofembed.m', 215, 302
  'twodoffree.m', 36
  'twodoffrf.m', 38, 103
  'twodofhos.m', 225
  'twodofrf.m', 69
  'twodoftemporal.m', 206, 212, 301–303
  'twoDshapesensor.m', 144, 184
  'windowdemo.m', 189
Matrix cracking, 119
Measurement degree of freedom, 163

Measurements, 127
  channel, 127
  errors, 127, 145
  variability, 271, 272
Membranes, 62
Metallic materials, 122
Microstructural changes, 119, 120
Missile casings, 20
Modal curvatures, 245, 262
Modal deflection shapes, 33, 244
Modal matrix, 84
Modal modeling, 243
Mode of vibration, 36
Modeling, 21, 24, 29, 31, 84, 108
  data-driven, 325
  physics-based, 293, 310
Moving average, 72
Multiplexer, 179
Multi-site damage, 367

Narrowband excitation, 60
Natural frequencies, 33, 34
Neural network models, 88
Newton's laws, 32
Nodes, 55. See Finite element model.
Nodes of vibration, 164
Noise, 14, 73, 74
  60 Hz, 172, 174
Noncontact sensors, 149. See Sensing.
Nondestructive evaluation, 16
  offline, 16, 17
  online, 16, 17
Nondestructive testing, 16, 17
Nonlinearity, 48, 292
  damage detection, 9, 13, 161, 222, 264, 287,
    288, 373, 374
  modeling, 21, 24, 29, 31, 84, 108
Nonproportional damping, 111. See Proportional
    viscous damping.
Nonstationary, 133, 135, 136, 181

Observability, 167
One-dimensional waves. See Wave propagation.
OR operation, 251
Overdetermined measurements, 281
Overload, 112, 177
Oxidation, 39

Paris' law, 120–123
Passive sensing, 146

Peak pick, 85
Penetration, 112
Performance prediction, 11, 379
Phase distortion, 142, 181
Phase plane, 68, 70, 101,
    344, 347
Phase velocity, 53, 63
Phased arrays, 196, 298
Phasors, 37
Phenomenological damage
    models, 379
Pipe, 125
Plane wave, 197, 198
Plastic deformation, 112
Poincare sections, 216
Polymer beam, 166
Power spectral density (PSD), 226.
    See Fourier analysis.
Preamplifier, 157
Prediction, 210, 213, 379
Principal coordinates, 84
Probability density function, 93
Probability of detection, 287
Product life-cycle management, 16
Prognostic health management, 16
Prognostics, 16, 19
Proportional viscous damping, 55
Pseudo-inverse, 67

Quadratic stiffness, 51, 70
Quantification, 292. See Damage
    quantification.
Quantization, 174

Rayleigh waves, 64, 65
Rayleigh's criterion, 179, 181
Readiness, 4, 5, 7, 20
Regression analysis, 295
Reliability forecasting, 16
Residual strength, 382, 383
Residual stress, 116, 253
Resonant frequencies, 38, 100.
    See Natural frequencies.
Response, 35, 37
Restoring force models, 68,
    101
Return map(s), 213, 215.
    See Fastener.
Rivet, 313, 314
Rotating machines, 133

Sampling, 171, 174, 181
Sandwich panel, 115, 322, 369
Seismic loads, 246
Self loosening, 107, 329
Sensors, 1, 8, 9
    acceleration, 12
    contact, 149
    durability, 137, 164
    footprint, 138
    frequency range, 279
    models, 29
    noncontact, 149
    piezoelectric, 155, 157, 181
    robustness, 151, 181
    sensitivity, 8, 151
    stability, 139
    strain, 137, 166
Separation of variables, 100
Shannon's sampling criterion, 176, 181
Ships, 5
Shock, 8, 9, 116, 129
Signal processing, 9, 24
SIMULINK®
    'two_dof_model_nl.mdl', 49, 69
Sine sweep, 262, 296, 384
Singular value decomposition (SVD),
        168
Smart materials, 146
Solenoid, 13, 317, 318
Sonic fatigue, 94
Spectrogram, 176, 226, 228
Spectrum, 75. *See* Frequency
        spectrum.
Speed of sound, 100
Spindle, 7, 110, 272, 351–355
Stabilizer bar, 260
State awareness, 9
State inference, 201
State–space model, 167, 169
Stationary data, 130, 133
Statistical analysis, 205
Statistical model, 29
Stiffness, 32, 41
Strain gauge, 150
Structural health monitoring, 16.
        See Health monitoring.
Structures
    parasitic, 12
    unitized, 12
Suspensions, 4

Tank, 12, 13, 371
Temperature effects, 1, 116
Thermal protection systems, 329
    ceramic, 26
    metallic, 12, 305, 329
Thermocouple, 146
Thermography, 16
Thermo-mechanical loads, 21
Threshold, 151, 267, 287, 290, 364
Time delay embeddings, 213, 298
Time–frequency analysis, 226
Tires, 4, 114
Transducers, 127, 137, 146, 181, 356
    orientation, 163
    placement, 163, 166
    types, 146, 147
Transmissibility, 44, 222, 258, 273–275,
        279, 289
Transverse waves, 58, 254. *See* Wave
        propagation.
Triangulation, 243, 254, 299
Trispectrum, 224, 225
Trucks, 3, 4. *See* Automotive systems.
Two-dimensional waves, 64. *See* Wave
        propagation.

Ultraviolet radiation, 95, 271
Uncertainties, 18
Underdetermined, 282, 312, 327
Unmanned aerial vehicle, 1, 2, 126
Unsupervised, 288

Variability, 21, 27, 270, 271, 273, 274,
        279, 299, 327
Vibration, 31, 36, 51, 95, 129
Vibration modeling. *See* Modeling *and*
        Vibration.
Virtual force model, 82, 83

Wave equation, 52, 54, 64, 65, 100
Wave number, 53, 59, 100, 198
Wave propagation, 52, 57, 63, 64, 109,
        148, 351, 376
    bulk waves, 62
    guided waves, 62
    one-dimensional, 52
        longitudinal, 54, 64, 228
        transverse, 59, 254
    shear waves, 62
    two-dimensional, 62

Wavelet transform, 231
Weld, 11, 119
Wheel end, 351. *See* Spindle *and* Tires.
Windmill, 27, 91
Windows, 191, 219, 298
   boxcar window, 191, 218
   flattop (P301) window, 191
   Hanning window, 191, 222, 235, 330, 356

Wire harness and connector, 20,
   132, 344
Wireless sensing, 163
Wound composite, 288, 297

X-ray tomography, 16

Yield strength, 123